Phyllanthus Species

Scientific Evaluation and Medicinal Applications

Traditional Herbal Medicines For Modern Times

Each volume in this series provides academia, health sciences, and the herbal medicines industry with in-depth coverage of the herbal remedies for infectious diseases, certain medical conditions, or the plant medicines of a particular country.

Series Editor: Dr. Roland Hardman

Volume 1
Shengmai San, edited by Kam-Ming Ko

Volume 2
Rasayana: Ayurvedic Herbs for Rejuvenation and Longevity, by H.S. Puri

Volume 3
Sho-Saiko-To: (Xiao-Chai-Hu-Tang) Scientific Evaluation and Clinical Applications, by Yukio Ogihara and Masaki Aburada

Volume 4
Traditional Medicinal Plants and Malaria, edited by Merlin Wilcox, Gerard Bodeker, and Philippe Rasoanaivo

Volume 5
Juzen-taiho-to (Shi-Quan-Da-Bu-Tang): Scientific Evaluation and Clinical Applications, edited by Haruki Yamada and Ikuo Saiki

Volume 6
Traditional Medicines for Modern Times: Antidiabetic Plants, edited by Amala Soumyanath

Volume 7
Bupleurum Species: Scientific Evaluation and Clinical Applications, edited by Sheng-Li Pan

Volume 8
Herbal Principles in Cosmetics: Properties and Mechanisms of Action, by Bruno Burlando, Luisella Verotta, Laura Cornara, and Elisa Bottini-Massa

Volume 9
Figs: The Genus Ficus, by Ephraim Philip Lansky and Helena Maaria Paavilainen

Traditional Herbal Medicines for Modern Times

Phyllanthus Species
Scientific Evaluation and Medicinal Applications

Edited by
Ramadasan Kuttan
K. B. Harikumar

CRC Press
Taylor & Francis Group
Boca Raton London New York

CRC Press is an imprint of the
Taylor & Francis Group, an **informa** business

CRC Press
Taylor & Francis Group
6000 Broken Sound Parkway NW, Suite 300
Boca Raton, FL 33487-2742

© 2012 by Taylor and Francis Group, LLC
CRC Press is an imprint of Taylor & Francis Group, an Informa business

No claim to original U.S. Government works

Printed in the United States of America on acid-free paper
10 9 8 7 6 5 4 3 2 1

International Standard Book Number: 978-1-4398-2144-2 (Hardback)

This book contains information obtained from authentic and highly regarded sources. Reasonable efforts have been made to publish reliable data and information, but the author and publisher cannot assume responsibility for the validity of all materials or the consequences of their use. The authors and publishers have attempted to trace the copyright holders of all material reproduced in this publication and apologize to copyright holders if permission to publish in this form has not been obtained. If any copyright material has not been acknowledged please write and let us know so we may rectify in any future reprint.

Except as permitted under U.S. Copyright Law, no part of this book may be reprinted, reproduced, transmitted, or utilized in any form by any electronic, mechanical, or other means, now known or hereafter invented, including photocopying, microfilming, and recording, or in any information storage or retrieval system, without written permission from the publishers.

For permission to photocopy or use material electronically from this work, please access www.copyright.com (http://www.copyright.com/) or contact the Copyright Clearance Center, Inc. (CCC), 222 Rosewood Drive, Danvers, MA 01923, 978-750-8400. CCC is a not-for-profit organization that provides licenses and registration for a variety of users. For organizations that have been granted a photocopy license by the CCC, a separate system of payment has been arranged.

Trademark Notice: Product or corporate names may be trademarks or registered trademarks, and are used only for identification and explanation without intent to infringe.

Library of Congress Cataloging-in-Publication Data

Phyllanthus species : scientific evaluation and medicinal applications / editors: Ramadasan Kuttan and K.B. Harikumar.
      p. cm. -- (Traditional herbal medicines for modern times ; v. 10)
    ISBN 978-1-4398-2144-2 (alk. paper)
    1. Phyllanthus. 2. Phyllanthus--Therapeutic use. I. Harikumar, K. B. II. Kuttan, Ramadasan. III. Series: Traditional herbal medicines for modern times ; v. 10.

QK495.E9P49 2012
583'.79--dc22   2011008553

Visit the Taylor & Francis Web site at
http://www.taylorandfrancis.com

and the CRC Press Web site at
http://www.crcpress.com

Contents

Series Preface ix
Preface xiii
About the Editors xv
Contributors xvii

Chapter 1 Taxonomy of the Genus *Phyllanthus* 1

                Sheeja T. Tharakan

                Identification Manual for Some Species of the Genus *Phyllanthus* L. of Phyllanthaceae with Special Reference to the Indian Subcontinent 23

                A. Lalithamba

Chapter 2 Current Pharmacopoeial Status of *Phyllanthus* Species: *P. emblica*, *P. amarus*, and *P. fraternus* 37

                Raman Mohan Singh and Vivekanandan Kalaiselvan

Chapter 3 Cultivation, Economics, and Marketing of *Phyllanthus* Species 47

                B. R. Rajeswara Rao

Chapter 4 Phylogenetic Analysis of *Phyllanthus* Species 71

                Srinivasu Tadikamalla

Chapter 5 Genetic Resources of *Phyllanthus* in Southern India: Identification of Geographic and Genetic Hot Spots and Its Implication for Conservation 97

                G. Ravikanth, R. Srirama, U. Senthilkumar, K. N. Ganeshaiah, and R. Uma Shaanker

Chapter 6 Phytochemistry of the Genus *Phyllanthus* 119

                Lutfun Nahar, Satyajit D. Sarker, and Abbas Delazar

Chapter 7 Hyphenated Techniques in the Study of the Genus *Phyllanthus* 139

                Satyajit D. Sarker, Lutfun Nahar, and Abbas Delazar

Chapter 8	Anti-inflammatory Activity of Various Species of *Phyllanthus* 149	
	K. B. Harikumar and Ramadasan Kuttan	
Chapter 9	Hepatoprotective Effects of Plants in the Family Phyllanthaceae ... 157	
	V. V. Asha	
Chapter 10	Anticancer Studies of *Phyllanthus amarus* 171	
	K. B. Harikumar and Ramadasan Kuttan	
Chapter 11	Anticancer Activity of *Phyllanthus emblica* 183	
	Jeena Joseph and Ramadasan Kuttan	
Chapter 12	The *In Vivo* and *In Vitro* Proapoptotic and Antiangiogenic Effects of *Phyllanthus urinaria* 193	
	Jong-Hwei S. Pang, Sheng-Teng Huang, Rong-Chi Yang, and Hsiao-Ting Wu	
Chapter 13	*Phyllanthus* and Hepatitis B, Hepatitis C, and HIV Infections 205	
	S. P. Thyagarajan	
Chapter 14	Antiviral Activities of *Phyllanthus orbicularis*, an Endemic Cuban Species 219	
	Gloria del Barrio and Francisco Parra	
Chapter 15	Diabetes and Diabetic Complications and *Phyllanthus* species 235	
	Geereddy Bhanuprakash Reddy and Palla Suryanarayana	
Chapter 16	Chemoprotective, Genotoxic, and Antigenotoxic Effects of *Phyllanthus* Sp. 255	
	Rakesh K. Johri	
Chapter 17	Antiaging Effects of *Phyllanthus* Species 267	
	Vasudevan Mani and Shanmugapriya Thulasimani	
Chapter 18	Toxicity Studies of *Phyllanthus* Species 279	
	K. N. S. Sirajudeen	

Contents

Chapter 19 Clinical Trials Involving *Phyllanthus* Species 289

Mulyarjo Dirjomuljono and Raymond R. Tjandrawinata

Chapter 20 Immunomodulatory Activity of Brahma Rasayana, an Herbal Preparation Containing *Phyllanthus emblica* as the Main Ingredient 315

Praveen K. Vayalil, Ramadasan Kuttan, and Girija Kuttan

Chapter 21 Triphala: An Ayurvedic Drug Formulation 325

Sandhya T. Das and K. P. Mishra

Chapter 22 Kalpaamruthaa: A Successful Drug against Various Ailments 331

P. Sachdanandam and P. Shanthi

Index 349

Series Preface

Global warming and global travel are contributing factors in the spread of infectious diseases such as malaria, tuberculosis, hepatitis B, and HIV. These are not well controlled by the present drug regimes. Antibiotics also are failing because of bacterial resistance. Formerly less well-known tropical diseases are reaching new shores. A whole range of illnesses, such as cancer, for example, occurs worldwide. Advances in molecular biology, including methods of *in vitro* testing for a required medical activity, give new opportunities to draw judiciously on the use and research of traditional herbal remedies from around the world. The reexamining of the herbal medicines must be done in a multidisciplinary manner.

Since 1997, there have been 49 volumes published in the book series *Medicinal and Aromatic Plants—Industrial Profiles* (Volumes 47–49 have been on vanilla, sesame, and citrus oils, respectively). The series continues.

The same series editor is also covering *Traditional Herbal Medicines for Modern Times*. Each volume reports on the latest developments and discusses key topics relevant to interdisciplinary health sciences research by ethnobiologists, taxonomists, conservationists, agronomists, chemists, pharmacologists, clinicians, and toxicologists. The series is relevant to all these scientists and will enable them to guide business, government agencies, and commerce in the complexities of these matters. The background to the subject is outlined next.

Over many centuries, the safety and limitations of herbal medicines have been established by their empirical use by the "healers" who also took a holistic approach. The healers are aware of the infrequent adverse effects and know how to correct these when they occur. Consequently and ideally, the preclinical and clinical studies of an herbal medicine need to be carried out with the full cooperation of the traditional healer. The plant composition of the medicine, the stage of the development of the plant material, when it is to be collected from the wild or when from its cultivation, its postharvest treatment, the preparation of the medicine, the dosage and frequency, and much other essential information is required. A consideration of the intellectual property rights and appropriate models of benefit sharing may also be necessary.

Wherever the medicine is being prepared, the first requirement is a well-documented reference collection of dried plant material. Such collections are encouraged by organizations like the World Health Organization and the United Nations Industrial Development Organization. The Royal Botanic Gardens at Kew (United Kingdom) is building its collection of traditional Chinese dried plant material relevant to its purchase and use by those who sell or prescribe traditional Chinese medicine in the United Kingdom.

In any country, the control of the quality of plant raw material, of its efficacy, and of its safety in use is essential. The work requires sophisticated laboratory equipment and highly trained personnel. This kind of "control" cannot be applied to the locally produced herbal medicines in the rural areas of many countries, on which millions of people depend. Local traditional knowledge of the healers has to suffice.

Conservation and protection of plant habitats are required, and breeding for biological diversity is important. Gene systems are being studied for medicinal exploitation. There can never be too many seed conservation "banks" to conserve genetic diversity. Unfortunately, such banks are usually dominated by agricultural and horticultural crops, with little space for medicinal plants. Developments such as random amplified polymorphic DNA enable the genetic variability of a species to be checked. This can be helpful in deciding whether specimens of close genetic similarity warrant storage.

From ancient times, a great deal of information concerning diagnosis and the use of traditional herbal medicines has been documented in the scripts of China, India, and elsewhere. Today, modern formulations of these medicines exist in the form of powders, granules, capsules, and tablets. They are prepared in various institutions, such as government hospitals in China and Korea, and by companies such as the Tsumura Company of Japan, with good quality control. Similarly, products are produced by many other companies in India, the United States, and elsewhere with a varying degree of quality control. In the United States, the Dietary Supplement and Health Education Act of 1994 recognized the class of physiotherapeutic agents derived from medicinal and aromatic plants. Furthermore, under public pressure, the U.S. Congress set up an Office of Alternative Medicine. In 1994, this office assisted in the filing of several Investigational New Drug (IND) applications required for clinical trials of some Chinese herbal preparations. The significance of these applications was that each Chinese preparation involved several plants and yet was handled with a *single* IND. A demonstration of the contribution to efficacy, of *each* ingredient of *each* plant, was not required. This was a major step forward toward more sensible regulations with regard to phytomedicines.

The subject of Western herbal medicines is now being taught again to medical students in Germany and Canada. Throughout Europe, the United States, Australia, and other countries, pharmacy and health-related schools are increasingly offering training in phytotherapy. Traditional Chinese medicine clinics are now common outside China. An Ayurvedic hospital now exists in London with a BSc Honors degree course in Ayurveda available: Professor Shrikala Warrier, Resistrar/Dean, MAYUR, Ayurvedic University of Europe, 81 Wimpole Street, London, WIG 9RF, e-mail sw@unifiedherbal.com. This is a joint venture with a university in Manipal, India.

The term *integrated medicine* is now being used, which selectively combines traditional herbal medicine with "modern medicine." In Germany, there is now a hospital in which traditional Chinese medicine is integrated with Western medicine. Such comedication has become common in China, Japan, India, and North America by those educated in both systems. Benefits claimed include improved efficacy, reduction in toxicity and the period of medication, as well as a reduction in the cost of the treatment. New terms such as *adjunct therapy, supportive therapy*, and *supplementary medicine* now appear as a consequence of such comedication. Either medicine may be described as an adjunct to the other depending on the communicator's view. Great caution is necessary when traditional herbal medicines are used by doctors not trained in their use and likewise when modern medicines are used by traditional herbal doctors. Possible dangers from drug interactions need to be stressed.

In Volume 2 of this series, *Rasayana: Ayurvedic Herbs for Rejuvenation and Longevity*, by Dr. H. S. Puri, line drawings are given in 58 chapters of plants and concise medical data covering all the important Rasayanas. Probably the most commonly used herb is Amalaki, the fruit of *Phyllanthus emblica*. Since Dr. Puri's book (2003), research reports from around the world have appeared about many species of *Phyllanthus*, so justifying this review of the genus in Volume 10 of the series.

For all their hard work, I am most grateful to the editors, Professor Ramadasan Kuttan, PhD and K. B. Harikumar, PhD. Hari has been most diligent in replying to my e-mails and to those of all the chapter contributors. These also I thank for their enthusiasm and expert information. My thanks are due to the steadfast support of the staff of CRC Press: Barbara Norwitz, executive editor and Jill Jurgensen, senior project coordinator.

Roland Hardman, BPharm, BSc (Chemistry), PhD (London), FRPharmS
Head of Pharmacognosy (retired), School of Pharmacy and Pharmacology
University of Bath, United Kingdom

Preface

Astanga Hrdaya, one of the earliest textbooks in Ayurveda (the traditional system of medicine in India), written by Vagbhatta (AD 500), mentioned the following regarding medicinal plants.

> jagatyevam anoushadham
> na kinchit vidyate dravyam
> vashaannaarthayagayoh'

This means "There is nothing in this universe, which is nonmedicinal, which cannot be made use of for many purposes and by many modes." This illustrates the importance of plants and their uses.

The genus *Phyllanthus* has over 1,000 species, and many of them are reported to possess a wide array of pharmacological activities. Species are distributed throughout the world. Various species of *Phyllanthus* have been reportedly used for the treatment of a variety of aliments around the world.

The uses of *Emblica* (Indian gooseberry), which is a component of Rasayana, have been found in Siddha, Ayurvedic, Unani, Arabic, Tibetan, and Egyptian texts. *Emblica* is known by different names in various parts of the world: Amla (India), Melaka (Malaysia), Malaka (Sudan), Makam paun (Thailand), and Amlaj (Arabic). The uses of *Emblica* were described in the ancient Ayurvedic text *Charaka Samhita* (3rd century BC) as a rejuvenating drug. *Emblica* is also known for its high level of ascorbic acid content. According to Ayurveda, *Emblica* has the ability to maintain the balance in all three doshas (energies believed to govern physiological activity) that is vital for proper functioning of the body.

Phyllanthus amarus is another well-known member of this genus. The plant is known by different names, including Bhoomi amalaki, Bhui amla (India); Bhuinamla (Pakistan); gale-wind grass, hurricane weed, cane peas senna, carry me seed (West Indies); Graine en bas fievre (French Guiana); Jar amla (Fiji); Chanca piedra (Peru); Creole senna (Virgin Islands); Deye do (Haiti); Elrageig (Sudan); Mapatan (Papua-New Guinea); Shka-nin-du (Mexico); Viernes santo (Puerto Rico); Ya-tai-bai (Thailand); and Yerba de san pablo (Philippines). This plant is referred to as the stone breaker because of its ability to dissolve kidney and gallbladder stones in the body. In the Indian system, it is one of the major remedies for liver-related and gastric disorders. Other uses include, but are not limited to, fever, pain, obesity, vaginitis, malaria, and bacterial infections.

Many of these uses of the plants were derived from folklore knowledge. Based on this knowledge, scientists have developed different ways to analyze the potential of these plants in a scientific manner. These studies demonstrated the pharmacological action and various chemical entities present in each plant. They validated the folklore claims and helped in designing cost-effective and reliable sources of medicine for humans.

This book describes in detail the taxonomy, cultivation and marketing, identification of geographic and genetic hot spots, chemistry, scientific evaluation of various pharmacological properties, clinical trials, and formulations containing various species of *Phyllanthus*. This is the first book of its kind solely dedicated to the genus *Phyllanthus*. This book definitely will serve as firsthand information for those in academia, especially teachers and research scholars, and those in industry, agriculture, as well as the general public, providing up-to-date references.

We would like to thank Dr. Roland Hardman, series editor, for giving us an opportunity to edit this book and for his continuous support and valuable suggestions. We would like to thank all contributors for their valuable efforts and time. Special thanks to Kanni Das for proofreading the manuscripts. Our sincere thanks are due to Jill Jurgensen and Barbara Norwitz of CRC Press for their unfailing help.

About the Editors

Ramadasan Kuttan, PhD, has worked at the Amala Cancer Research Center, Thrissur, Kerala, India since 1984 and presently is the research director of the Center. He earned his doctoral degree from the University of Madras in 1973, receiving a Gold Medal for the outstanding thesis of the year. From 1973 to 1984, he conducted research in the United States, including stints at Roche Institute of Molecular Biology, Nutley, New Jersey, and M. D. Anderson Hospital and Cancer Center in Houston, Texas. His major areas of research include cancer drugs from plant sources and chemoprevention, chemoprotection, radioprotection, immunomodulation, and the like. He has done extensive work on the use of *Phyllanthus amarus* in cancer using animal models and cell culture.

K. B. Harikumar, PhD, received his bachelor's and master's degrees in biochemistry from Nagpur University in India. His doctoral research on cancer chemoprevention by natural products was conducted under the direction of Ramadasan Kuttan at Amala Cancer Research Center, affiliated with Mahatma Gandhi University at Kottayam in India. Presently, he is a postdoctoral associate in the Department of Biochemistry and Molecular Biology at Virginia Commonwealth University, Richmond, Virginia. He has over 40 peer-reviewed publications and several book chapters to his credit. His major research focuses are cancer chemoprevention, ubiquitination and nuclear factor kappa B signaling, and the role of bioactive lipids in cell signaling.

Contributors

V. V. Asha
Rajiv Gandhi Center for Biotechnology
Molecular Ethopharmacology
 Laboratory
Thiruvananthapuram, Kerala, India

Sandhya T. Das
John Hopkins University
School of Public Health
Baltimore, Maryland, USA

Abbas Delazar
Tabriz University of Medical Sciences
School of Pharmacy
Tabriz, Iran

Gloria del Barrio
University of Habana
Departament of Microbiology and
 Virology
Faculty of Biology
Ciudad de La Habana, Cuba

Mulyarjo Dirjomuljono
University of Airlangga/Dr. Soetomo
 General Hospital
Department of Ear, Nose, & Throat
Faculty of Medicine
Surabaya, Indonesia

K. N. Ganeshaiah
Ashoka Trust for Research in Ecology
 and Environment, Srirampura,
Jakkur Post, Bangalore, India

University of Agricultural Sciences
School of Ecology and Conservation
Bangalore, India

K. B. Harikumar
Virginia Commonwealth University
Department of Biochemistry and
 Molecular Biology
Richmond, Virginia, USA

Sheng-Teng Huang
Chang Gung Memorial Hospital
Kaohsiung Medical Center
Department of Chinese Medicine
Kaohsiung, Taiwan, Republic of China

Rakesh K. Johri
Indian Institute of Integrative Medicine
Division of Pharmacology
Jammu-Tawi, India

Jeena Joseph
University of Michigan
Research Associate
Ann Arbor, Michigan, USA

Vivekanandan Kalaiselvan
Indian Pharmacopoeia Commission
 (Ministry of Health and Family
 Welfare)
Raj Nagar, Ghaziabad, India

Girija Kuttan
Amala Cancer Research Center
Amala Nagar, Thrissur, Kerala, India

Ramadasan Kuttan
Amala Cancer Research Center
Amala Nagar, Thrissur, Kerala, India

A. Lalithamba
D.K. Government College for Women
Department of Botany
Nellore, India

Vasudevan Mani
University Teknologi MARA (UiTM),
  Campus Puncak
Faculty of Pharmacy
Brain Research Laboratory
Alam, Malaysia

K. P. Mishra
Nehru Gram Bharati University
Allahabad, India

Lutfun Nahar
University of Wolverhampton
Drug Discovery and Design Research
  Division
Department of Pharmacy
West Midland, UK

Jong-Hwei S. Pang
Chang Gung University, Tao-Yuan
Graduate Institute of Clinical Medical
  Sciences
Taiwan, Republic of China

Francisco Parra
University of Oviedo
Institute of Biotechnology of Asturias
Department of Biochemistry and
  Molecular Biology
Oviedo, Spain

B. R. Rajeswara Rao
Central Institute of Medicinal and
  Aromatic Plants (CIMAP)
Resource Center, Boduppal, Uppal Post
Hyderabad, India

G. Ravikanth
Ashoka Trust for Research in Ecology
  and Environment, Srirampura
Jakkur Post Bangalore, India

University of Agricultural Sciences
School of Ecology and Conservation
Bangalore, India

Geereddy Bhanuprakash Reddy
National Institute of Nutrition
Biochemistry Division
Hyderabad, India

P. Sachdanandam
University of Madras, Taramani
  Campus
Department of Medical Biochemistry
Chennai, India

Satyajit D. Sarker
University of Wolverhampton
School of Applied Sciences
Department of Pharmacy
West Midland, UK

U. Senthilkumar
Ashoka Trust for Research in Ecology
  and Environment, Srirampura
Jakkur Post, Bangalore, India

R. Uma Shaanker
Ashoka Trust for Research in Ecology
  and Environment, Srirampura
Jakkur Post, Bangalore, India

University of Agricultural Sciences
Department of Crop Physiology
Bangalore, India

P. Shanthi
University of Madras Taramani Campus
Department of Pathology
Chennai, India

Raman Mohan Singh
Indian Pharmacopoeia Commission
  Ministry of Health and Family
  Welfare
Raj Nagar, Ghaziabad, India

K. N. S. Sirajudeen
University Sains Malaysia
Department of Chemical Pathology
Kelantan, Malaysia

Contributors

R. Srirama
Ashoka Trust for Research in Ecology and Environment, Srirampura
Jakkur Post, Bangalore, India

Palla Suryanarayana
National Institute of Nutrition
Biochemistry Division
Hyderabad, India

Srinivasu Tadikamalla
Rashtrasant Tukadoji Maharaj Nagpur University
Department Botany
Nagpur, India

Sheeja T. Tharakan
Vimala College
Department of Botany
Thrissur, Kerala, India

Shanmugapriya Thulasimani
Vellalar College for Women, Erode
Department of Food and Nutrition
Tamil Nadu, India

S. P. Thyagarajan
Pro-Chancellor, Research
Sri Ramachandra University
Chennai, India

Raymond R. Tjandrawinata
Dexa Laboratories of Biomolecular Sciences
Dexa Medica Group
Tangerang, Indonesia

Praveen K. Vayalil
University of Alabama, Birmingham
Department of Pharmacology and Toxicology
Birmingham, Alabama, USA

Hsiao-Ting Wu
Chang Gung Memorial Hospital
Department of Chinese Medicine
Kaohsiung Medical Center
Kaohsiung, Taiwan, Republic of China

Rong-Chi Yang
Chang Gung Memorial Hospital Tao-Yuan
Chinese Herbal Pharmacy
Taiwan, Republic of China

1 Taxonomy of the Genus *Phyllanthus*

Sheeja T. Tharakan

CONTENTS

1.1 Taxonomy of the Genus *Phyllanthus* 3
    1.1.1 Introduction 3
    1.1.2 Botanical System of Classification 3
        1.1.2.1 Natural System of Classification 3
        1.1.2.2 APG II System 4
    1.1.3 General Characters of Genus *Phyllanthus* 4
    1.1.4 General Characters of *Phyllanthus* Species 4
        1.1.4.1 *Phyllanthus Amarus* Schum. & Thonn. 4
        1.1.4.2 *Phyllanthus emblica* L. (*Emblica officinalis* Gaertner) 5
        1.1.4.3 *Phyllanthus acidus* L. 5
        1.1.4.4 *Phyllanthus niruri* 6
        1.1.4.5 *Phyllanthus urinaria* L. 8
        1.1.4.6 *Phyllanthus polyphyllus* Willd. 8
        1.1.4.7 *Phyllanthus myrtifolius* Willd. 9
        1.1.4.8 *Phyllanthus kozhikodianus* 9
        1.1.4.9 *Phyllanthus reticulatus* Poir. 9
        1.1.4.10 *Phyllanthus rheedii* Wight. 9
        1.1.4.11 *Phyllanthus tenellus* Roxb. 10
        1.1.4.12 *Phyllanthus lawii* Grah. 10
        1.1.4.13 *Phyllanthus maderaspatensis* L. 10
        1.1.4.14 *Phyllanthus narayanswamii* 11
        1.1.4.15 *Phyllanthus virgatus* Forster F. (*P. simplex* Retz.) 11
        1.1.4.16 *Phyllanthus gardnerianus* Baill. 12
        1.1.4.17 *Phyllanthus macraei* Muell. 12
        1.1.4.18 *Phyllanthus rotundifolius* Klein 12
        1.1.4.19 *Phyllanthus debilis* Klein ex Willd. 13
        1.1.4.20 *Phyllanthus missionis* 13
        1.1.4.21 *Phyllanthus speciosus* Jacq. 13
        1.1.4.22 *Phyllanthus pinnatus* (Wight) Webster, J. 13
        1.1.4.23 *Phyllanthus acuminates* 14
        1.1.4.24 *Phyllanthus caroliniensis* 14
        1.1.4.25 *Phyllanthus mirabilis* 14
        1.1.4.26 *Phyllanthus caesiifolius* 14

1.1.4.27 *Phyllanthus gentryi* 14
1.1.4.28 *Phyllanthus muellerianus* (Kuntze) Exell 14
1.1.4.29 *Phyllanthus* pulcher Wall. Ex Müll. Arg. 14
1.1.4.30 *Phyllanthus fraternus* Webster 14
1.1.4.31 *Phyllanthus abnormis* 15
1.1.4.32 *Phyllanthus odontadenius* Mull. Arg. 15
1.1.4.33 *Phyllanthus muellerianus* (O Ktze) Exell 15
1.1.4.34 *Phyllanthus capillaris* Schum. & Thonn. 15
1.1.4.35 *Phyllanthus sublanatus* Schum. & Thonn. 15
1.1.4.36 *Phyllanthus beillei* Hutch. 16
1.1.4.37 *Phyllanthus indofischeri* Bennet. 16
1.1.4.38 *Phyllanthus scabrifolius* Hook. 16
1.1.4.39 *Phyllanthus gradyi* 16
1.1.4.40 *Phyllanthus longipedicellatus* 17
1.1.4.41 *Phyllanthus salesiae* 18
1.1.4.42 *Phyllanthus gongyloides* 18
1.1.4.43 *Phyllanthus indicus* Muell. 19
1.1.4.44 *Phyllanthus gunnii* Hook. F. (Synonym: *Phyllanthus gasstroemii* Muell. Arg.) 19
1.1.4.45 *Phyllanthus lacunarius* F. Muell. 19
1.1.4.46 *Phyllanthus carpentariae* Muell. Arg. (Synonyms: *Phyllanthus hebecarpus* Benth., *Phyllanthus grandisepalus* F. Muell.) 20
1.1.4.47 *Phyllanthus fuernrohrii* F. Muell. 20
1.1.4.48 *Phyllanthus hirtellus* F. Muell. Ex Muell. Arg. 20
1.1.4.49 *Phyllanthus subcrenulatus* F. Muell. 21
1.1.4.50 *Phyllanthus similis* Muell. Arg. 21
Acknowledgments 21
References 21
1.2 Identification Manual for Some Species of the Genus *Phyllanthus* L. of Phyllanthaceae with Special Reference to the Indian Subcontinent 23
    1.2.1 Introduction 23
        1.2.1.1 The Scientific Classification 23
    1.2.2 The Main Characters of the Genus 23
    1.2.3 Synoptic Key 24
        1.2.3.1 Herbs 24
        1.2.3.2 Shrubs or Trees 25
    1.2.4 Descriptions 25
        1.2.4.1 *Phyllanthus acidus* L 25
        1.2.4.2 *Phyllanthus amarus* Schumach & Thonn. 25
        1.2.4.3 *Phyllanthus debilis* Klein ex Willd. 27
        1.2.4.4 *Phyllanthus emblica* L. 28
        1.2.4.5 *Phyllanthus fraternus* G. L. Webster 28
        1.2.4.6 *Phyllanthus gardnerianus* (Wt.) Baill. 29
        1.2.4.7 *Phyllanthus indo-fischeri* Gamble. 29
        1.2.4.8 *Phyllanthus maderaspatensis* L. 30

        1.2.4.9   *Phyllanthus niruri* L. 31
        1.2.4.10 *Phyllanthus pinnatus* (Wight) G. L. Webster. 31
        1.2.4.11 *Phyllanthus polyphyllus* Willd.. 32
        1.2.4.12 *Phyllanthus reticulatus* Poir. 32
        1.2.4.13 *Phyllanthus rheedii* Wight.. 33
        1.2.4.14 *Phyllanthus rotundifolius* Klein ex Willd 33
        1.2.4.15 *Phyllanthus urinaria* L. 34
        1.2.4.16 *Phyllanthus virgatus* G. Forst. 34
Acknowledgment 35
References.. 35

1.1 TAXONOMY OF THE GENUS *PHYLLANTHUS*

1.1.1 Introduction

The genus *Phyllanthus* belongs to the family Phyllanthaceae. The number of species varies widely, from 750 to 1,200. It has a remarkable diversity, including annual and perennial herbs, shrubs, climbers, floating aquatics, and succulents. It has a wide variety of floral morphologies and a wide range of pollen types. Almost all *Phyllanthus* species express a specific type of growth called phyllanthoid branching, in which vertical stems bear deciduous, flower-bearing horizontal or oblique stems. It is distributed mainly in tropical and subtropical regions. Leaf flower is the common name for all *Phyllanthus* species.

The name *Phyllanthaceae* was first validly published by Ivan Ivanovich in 1820 in a Russian book titled *Tekhno-botanico Slovar.* A proposal to conserve this name was published in 2007 (Reveal et al., 2007). Euphorbiaceae is now defined as a much smaller family than it had been in the twentieth century (Tokuoka, 2007). Pandaceae, Phyllanthaceae, Picrodendraceae, Putranjivaceae, Peraceae, and Centroplaceae have been removed. The obsolete, older concept of Euphorbiaceae, known as Euphorbiaceae sensu lato, is sometimes still used for continuity and convenience.

1.1.2 Botanical System of Classification

1.1.2.1 Natural System of Classification

George Bentham and Sir Joseph Dalton Hooker have taken De Candolle's and Lindley's views regarding the basic categories and principles. The description of plants is original and is based on personal observations. Bentham made one of the most valuable contributions, the *Genera Plantarum,* listing 97,205 specimens. According to Bentham and Hooker, the systematic position of *Phyllanthus* is as follows.

> Class: Dicotyledonae; subclass: Monochlamydeae; series: Unisexuales; family: Euphorbiaceae; genus: *Phyllanthus*

1.1.2.2 APG II System

A modern system of plant taxonomy, the APG II system of plant classification was published in April 2003 by the Angiosperm Phylogeny Group (APG). The APG II system recognized 45 orders, 5 more than the APG system. It also recognized 457 families, 5 fewer than the APG system. Thirty-five of the APG II families were not placed in any order. The APG III system is a modern system of plant taxonomy for flowering plant classification. It was published in 2009 by the Angiosperm Phylogeny Group. In October 2009, members of the Linnaean Society proposed an accompanying forma phylogenetic classification of all land plants, compatible with the APG III system of classification (Chase & Reveal, 2009).

> Domain: Eukaryota; regnum: Plantae; clade: Angiospermae; clade: Eudicots; clade: Core eudicots; clade: rosids; clade: eurosidsI; order: Malpigiales; family: Phyllanthaceae; genus: *Phyllanthus*

1.1.3 General Characters of Genus *Phyllanthus*

Habit: Herbs or shrubs
Leaves: Small, alternate, distichous, the branchlets resembling pinnate leaves, stipules narrow
Flowers: Very small, monoecious, in axillary clusters or solitary, bracteate, disk in male flowers of small glands and in female flowers of glands or annular
Calyx: Lobes 5–6, imbricate
Petals: 0
Stamens: 3, more or less free or the filaments combined in a column, anthers oblong, didynamous, dehiscing vertically or transversely
Ovary: 3 celled, styles 3, free or connate at base
Fruit: A capsule with crustaceous or thin 2-valved cocci
Seeds: Trigonous, rounded at back

1.1.4 General Characters of *Phyllanthus* Species

The number of species in this genus varies widely, from 750 to 1,200. Various species of this genus and its characters are listed in the next sections.

1.1.4.1 *Phyllanthus Amarus* Schum. & Thonn.

> Common name: *Phyllanthus*, carry me seed

Phyllanthus Amarus Schum. & Thonn. is a herb and is seen in moist deciduous, forest plantations and in plains. It is distributed mainly in the tropics. It is a weed of gardens and cultivated land. It is seen in all districts of Kerala. It is a branching annual herb reaching 12–18 inches high (Figure 1.1a). Flowering and fruiting of this plant will be in July–October. It is often used in native medicine (Gamble, 1925).

Leaves: Membranous; usually glaucous beneath; usually broadly obtuse at apex; very variable in size, but usually under 0.5 inches long; elliptic-obovate or oblong; prominently distichous so that the branchlets resemble pinnate leaves; stipules not peltate, lanceolate and oblong, 1 mm, subsessile, scarious.

Flowers: Monoecious, male and female flowers occur solitarily or in clusters of 2 or 3 in axils of lower leaves of a branch, bracteate, subsessile, actinomorphic, hypogynous, cyclic, minute, green. Tepals: 5; ovate, valvate; a glandular disk is present within the perianth; each lobe has a distinct green midrib. Stamens: 5, exerted, monadelphous, filaments connate in a column with three bithecous anther lobe borne at its tip, anthers sessile, introse, dehiscing by slits. Carpel: 3, syncarpous, ovary superior, trilocular, each loculus with two ovules, placentation axile, ovary globular. Styles: 3, short, not dilated, bilobed, at maturity ovary becomes hexalocular.

Fruit: Capsule dry, dehiscent, more or less verrucose, and glandular.

Seeds: Regular lines of very minute tubercles joined by minute crossbars, muriculate, triquetrous.

1.1.4.2   *Phyllanthus emblica* L. (*Emblica officinalis* Gaertner)

Common name: Indian Gooseberry

Phyllanthus emblica L. (*Emblica officinalis* Gaertner) is a tree. It is seen in dry and moist deciduous and cultivated in plains. It is distributed throughout the tropics and is seen in all districts of Kerala. Flowering and fruiting of this plants will be in February–May (Gamble, 1925).

Leaves: Small, distichous, linear, obtuse, appearing like pinnate, stipules ovate, acute.

Flowers: Greenish yellow, in axillary fascicles on leaf-bearing branches, often on the naked portion below the leaves, with fimbricate bracts at base. Male flowers many on short slender pedicels and disk with glands. Female flowers few, subsessile, disk cup-like. Tepals: 6, valvate, oblong in male flowers. Stamens: 3 on a short central column in male flowers, anthers connate, dehiscence vertical. Ovary: 3, carpellate, 1.5 mm. Styles: 3, bifid, erect, broadly fimbriate.

Fruit: Drupe, fleshy, globose with 6, indehiscent, more than 1.5 cm across, obscure vertical furrows.

Seed: Pale yellow of 3 two-seeded crustaceous cocci (Figure 1.1c).

1.1.4.3   *Phyllanthus acidus* L.

Common name: Star gooseberry, Tahitian gooseberry tree

Phyllanthus acidus L. is a tree. It was considered a native of Brazil. It is cultivated in all districts of Kerala, is monoecious and semievergreen, and fruits are juicy (Gamble, 1925).

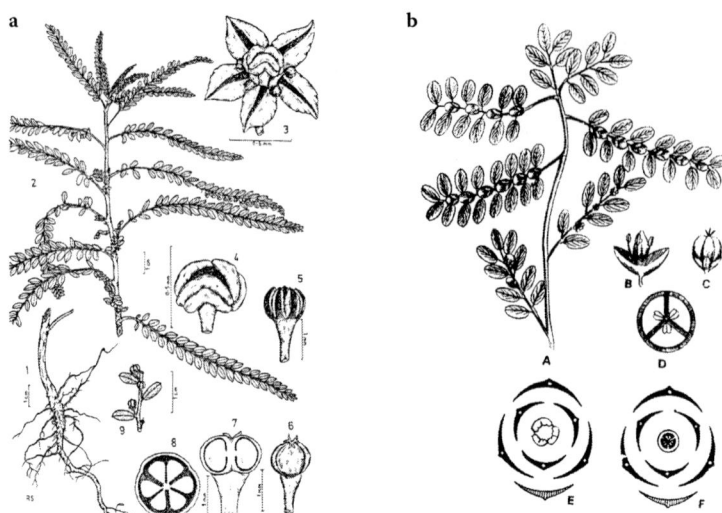

FIGURE 1.1 Floral characters of different species of *Phyllanthus*. (a) *P. amarus*: 1, roots; 2, twig; 3, flower; 4, androecium; 5, perianth; 6, pistil entire; 7, pistil L.S; 8, ovary T.S; 9, branchlet with fruits. (b) *P. niruri*: A, twig; B, male flower; C, female flower; D, ovary c.s; E, male floral diagram; F, female floral diagram. (Figure 1.1a is adapted from Mathew 1982, 1988a,b, and Figure 1.1b from Sukla and Misra, 1997.) *(continued)*

Leaves: Elliptic to obovate, base rounded, margin entire, apex acute or acuminate, and petiole to 0.4 cm, stipules toothed.

Flowers: Pedicel to 5 mm, inflorescence an axillary fascicle, disk without an inner corona in female flowers. Tepals: 4, unequal 2 + 2, ovate, 2 mm, 1 nerved, subacute. Stamens: 4, exerted, free, stamens not inserted in disk, filaments recurved, anthers oblong, dehiscence vertical. Ovary: 2 or 3 locular, subglobose. Styles: 3 or 4, reflexed, forked into subacute arms, not dilated, ovules 2 per locule.

Fruits: Drupe, indehiscent, juicy, 1 cm across, 6–8 lobed, angular, fruits notably enlarged, endocarp hard.

1.1.4.4 *Phyllanthus niruri*

Common name: Stonebreaker

Phyllanthus niruri is an annual herb. It is a widespread tropical plant commonly found in coastal areas. The root is a branched taproot, the stem is erect, cylindrical, branched, solid, and glabrous (Gamble, 1925).

Leaves: Stipulate, lanceolate and oblong, subsessile, alternate, entire, apex rounded, glaucous ventrally.

Flowers: Monoecious, male flowers occur solitary or in clusters of 2 or 3 in axils of lower leaves of a branch, bracteate, subsessile, actinomorphic, hypogynous, cyclic, minute, green. Tepals: 6, in two whorls of 3 each,

Taxonomy of the Genus *Phyllanthus* and Identification Manual

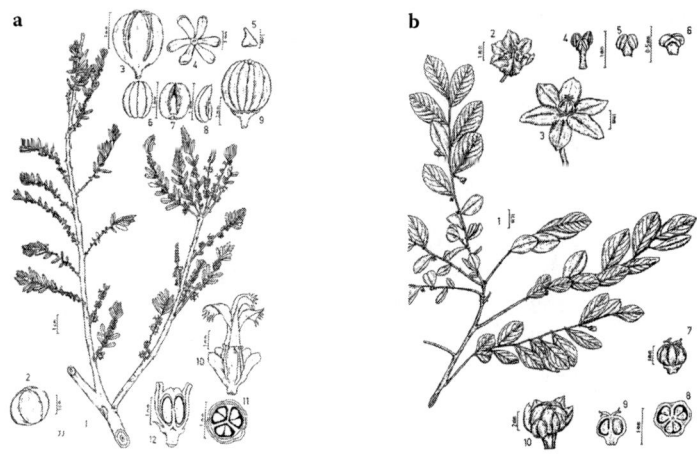

c. *P.emblica*: 1. Twig; 2. Fruit; 3. Male flower; 4. Perianth; 5.Bract; 6–8.Anthers; 9.Androecium; 10 Female flowers; 11. Ovary T.S; 12. Ovary L. S. d. *P. deblis*: 1. Twig; 2. Male flower; 3. Female flower; 4. Stamens; 5&6. Anther; 7. Pistil; 8. Ovary T. S; 9. Ovary L. S; 10. Capsule.

FIGURE 1.1 (continued) Floral characters of different species of *Phyllanthus*. (c) *P. emblica*: 1, twig; 2, fruit; 3, male flower; 4, perianth; 5, bract; 6–8, anthers; 9, androecium; 10, female flowers; 11, ovary T.S; 12, ovary L. S. (d) *P. deblis*: 1, twig; 2, male flower; 3, female flower; 4, stamens; 5 and 6, anther; 7, pistil; 8,ovary T. S; 9, ovary L. S; 10, capsule. (Adapted from Mathew 1982, 1988a,b.)

e. *P.polyphyllus*: 1. Twig; 2. Pistil; 3. Flower; 4. Ovary L. S; 5. Ovary T.S; 6. Male flower; 7. Stamens; 8. Seed.
f. *P.urinaria*: 1. Habit; 2. Leaf; 3. Male flower; 4. Stamens; 5. Female flower; 6. Pistil entire; 7. Pistil T.S; 8. Pistil L. S; 9. Capsule; 10. Seed.

FIGURE 1.1 (continued) Floral characters of different species of *Phyllanthus*. (e) *P. polyphyllus*: 1, twig; 2, pistil; 3, flower; 4, ovary L. S; 5, ovary T. S; 6, male flower; 7, stamens; 8, seed. (f) *P. urinaria*: 1, habit; 2, leaf; 3, male flower; 4, stamens; 5, female flower; 6, pistil entire; 7, pistil T.S; 8, pistil L. S; 9, capsule; 10, seed. (Adapted from Mathew 1982, 1988a,b.)

valvate, a glandular disk is present within the perianth, each lobe has a distinct green midrib. Stamens: 3, exerted, monadelphous, filaments connate in a column with three bithecous anther lobes borne at its tip, anthers sessile, introse, dehiscing by slits. Female flowers occur singly in the axils of upper leaves of a branch, rarely in clusters, bracteolate, subsessile, actinomorphic, hypogynous, small, and green. Perianth: 3 + 3, free, a glandular disk is present internal to the base of perianth lobes, persistent with a distinct green midrib. Carpels: 3, syncarpous, ovary superior, trilocular, each loculus with two ovules, placentation axile, ovary globular. Styles: 3, short, not dilated, bilobed, at maturity ovary becomes hexalocular.
Fruit: Depressed and globose capsule.
Seeds: Endospermic and trigonous.

This species is often confused with *P. amarus*. *Phyllanthus amarus* has 5 perianth lobes, while *P. niruri* has 6. *Phyllanthus amarus* has 3 stamens while *P. niruri* has 6 (Figure 1.1b).

1.1.4.5 *Phyllanthus urinaria* L.

Common name: Chamberbitter

Phyllanthus urinaria L. is an annual or perennial erect herb with more or less sensitive leaflets that are sometimes pink when young. It is seen in all districts of plains and deciduous forests. It is a native of tropical East Asia. Now, it is regarded as a weed. In Kerala, it is noted in all districts. Flowering and fruiting of this plant will be in July–October. Although of Asian origin, the weed is widely found in all tropical regions of the world. In the United States, it is found in southern states such as Florida, Georgia, Alabama, South Carolina, New Mexico, and Texas (Gamble, 1925).

Leaves: Glabrous or hispid on the margins, chartaceous, oblong, apiculate, up to 0.75 inches long, stipules subulate.
Flowers: Male flowers very minute, female flowers larger, sessile, disk annular. Tepals: 6. Stamens: 3, anthers sessile, connate, dehiscence vertical, styles spreading, not fimbriate.
Fruit: Capsules 3 valved, verrucose, dry, dehiscent, less than 0.5 cm across.
Seeds: Prominently transversely ridged and with faint crossbars (Figure 1.1f).

1.1.4.6 *Phyllanthus polyphyllus* Willd

Phyllanthus polyphyllus Willd is a shrub or small tree somewhat resembling *Emblica officinalis* in leaf but quite different in fruit. It is seen in rocky areas in semievergreen forests. It grows in Deccan; hill forests of Kurnool, Cuddapah, Chittoor, and Nellore; Kambakam Hill in Chingleput; Carnatic; Javadi Hills in South Arcot; southern hills of Tinnevelly, up to 4,000 feet; and eastern slopes of Nilgiris. It is distributed in India and Sri Lanka. In Kerala, it is seen only in Kollam district. Flowering and fruiting of this plant will be in February–June (Gamble, 1925).

Leaves: Leaf branchlets 2–6 inches long; leaves oblong, obtuse, or apiculate; 0.4–0.75 inches long; main nerves 6–8 pairs.
Flowers: Disk with glands in male flowers. Tepals: 6. Stamens: 3. Anthers: Erect, connate, the cells dehiscing vertically, the filaments united in a column, ovary 3 locular, styles spreading.
Fruit: Capsule, dry, dehiscent, less than 0.5 cm across.
Seed: Foveolate, seed pits very minute (Figure 1.1e).

1.1.4.7 *Phyllanthus myrtifolius* Willd

Common name: Mousetail plant

Phyllanthus myrtifoslius Willd is a shrub. It is a hedge plant grown in gardens. It is a native of Sri Lanka. It is seen in all districts of Kerala.

1.1.4.8 *Phyllanthus kozhikodianus*

Phyllanthus kozhikodianus is a herb. It is seen in dry and moist deciduous, semievergreen, and forest plantations. It is distributed in Western Ghats and the eastern Himalayas. It is seen in Kozhikode, Malappuram, Idukki, Thrissur, and Palakkad districts of Kerala. Flowering and fruiting of this plant will be in July–October (Manilal and Sivarajan, 1982).

1.1.4.9 *Phyllanthus reticulatus* Poir.

Phyllanthus reticulatus Poir. is a shrub 3 m high; it is seen in stream banks, lakeshores, and moist deciduous and semievergreen forests. It is seen in all districts of Kerala, savanna forest, and often on riverbanks, throughout the region from Senegal to northern and southern Nigeria; it is widespread elsewhere in tropical Africa. Flowering and fruiting of this plant will be in August–December (Mathew, 1982).

Leaves: Oblong-elliptic, obtuse or acute, thin, glabrous, at both ends, leaf margin entire, distichous, petioles long and slender, stipules long, ovate, acute, bristle pointed.
Flowers: Axillary, male flowers in fascicles, female is solitary. Tepals: 5 in male and female flowers, unequal, obovate, imbricate, obtuse, disk with glands in male flowers. Stamens: 5, connectives connate, outer free, anthers subsessile or raised by filaments, dehiscence vertical. Ovary: 5 or 12 locular, and ovules superposed, 2 per locule, styles as many as or fewer than locules. Disk glands 5.
Fruits: Berry, fleshy, indehiscent, dark blue, fruits not glochidiate.
Seeds: Trigonous, superposed, unitegmic, testa crustaceous.

1.1.4.10 *Phyllanthus rheedii* Wight.

Phyllanthus rheedii Wight. is a slender, branching, erect herb or small undershrub. It is seen in Western Ghats, most districts from south Canara to Nilgiris, Annamalais, and the hills of Tinnevelly above 5,000 feet. It is distributed in India and Sri Lanka. In

Kerala, it is seen in Kannur, Palakkad, Idukki, Kottayam, Pathanamthitta, and Kollam. Flowering and fruiting of this plant will be in November–January (Gamble, 1925).

> Leaves: glabrous, membranous, elliptic or ovate, acute up to 1.25 inches long, stipules lanceolate, decurrent.
> Flower: Male flowers minute, fascicled; female flowers solitary, on thickened pedicels. Calyx: Calyx lobes usually white margined. Stamen: Anthers free above.
> Fruit: Capsules smooth.
> Seeds: With concentric lines of minute tubercles and minute crossbars.

1.1.4.11 *Phyllanthus tenellus* Roxb.

Common name: Mascarene Island leaf-flower

Phyllanthus tenellus Roxb. is a herb and is seen in waste places for discarded materials and plantations. It is reported in Australia. In Kerala, it is seen in Malappuram, Alappuzha, and Thiruvananthapuram. Flowering and fruiting of this plant will be in June–December.

> Leaves: Elliptic to obovate, 6–25 mm long, 2–10 mm wide, margins flat, wavy, lower surface paler.
> Flowers: Solitary or 2 or 3 males and 1 or 2 females together, peduncle to 5 mm long, perianth segments narrow-ovate, about 1 mm long, margins broad, white, not enlarging under fruit. Stamens: 5, filaments free.
> Fruit: Capsule 1.5–2 mm diameter, greenish; seeds about 1 mm long, orange-brown, dorsally minutely tuberculate in longitudinal rows.

1.1.4.12 *Phyllanthus lawii* Grah.

Phyllanthus lawii Grah. was reported in Northern Circars, in Godavari and Hrishna, Carnatic hills of Salem, Western Ghats, Wynaad to Travancore, gregarious on the banks and in the beds of rocky rivers. It is a straggling shrub with long purplish branches armed with stipular tubercles bearing small thorns (Gamble, 1925).

> Leaf: Leaf branchlets 1–3 inches long or a little more, leaves elliptic-oblong, obtuse or apiculate, 0.2–0.3 inches long, main nerves obscure.
> Flowers: Axillary, males solitary or a few together, female solitary, white. Stamen: 3, anthers erect, the cells dehiscing vertically, the filaments united in a column. Ovary: styles 3, short, bifid, short lobes recurved.
> Fruit: Capsule.
> Seed: Foveolate, seed-pits very minute.

1.1.4.13 *Phyllanthus maderaspatensis* L.

Phyllanthus maderaspatensis L. is seen in northern Circars, Deccan, and Carnatic on dry lands, especially black cotton soils and near the seacoast. It is an erect or decumbent herb or small undershrub (Gamble, 1925).

Leaves: Linear, to obovate, glabrous, subcoriaceous, obovate or oblanceolate, cuneate, rounded or retuse at apex, mucronate, glaucous, up to 1.25 inches long, stipules lanceolate, peltate with white margins.

Flower: Male flowers above, minute, fascicled, disk with glands; female flowers solitary, on filiform pedicels, bracts 2, acute, disk 6 lobed. Tepals; 6, obovate, margin scarious, obtuse. Stamen: 3, anthers subsessile, filaments connate, dehiscence vertical, styles spreading, not fimbricate. Ovary: 3 lobed, styles horizontally spreading, stigma obtuse.

Fruit: Capsules dry, dehiscent, less than 0.5 cm across, smooth, the seeds with concentric lines of minute tubercles and minute crossbars.

Seeds: Triquetrous, vertically muriculate. Capsule: With longitudinal rows of minute tubercles.

1.1.4.14 Phyllanthus narayanswamii

Phyllanthus narayanswamii is seen in northern Circars, Rampa Hills of Godavari at 4,500 feet. It is a small, wiry undershrub with many branchlets from a stout rootstock (Gamble, 1925).

Leaves: Elliptic, obtuse, apiculate, the margins thickened, the nerves 4–5 prominent, joining in arches, 0.2–0.4 inches long, 0.1–0.3 inches broad, stipules peltate, subsagittate. Flower: Male flowers subsessile, female flowers pedicelled, disk of male flowers are large flat glands and female flowers saucer shaped, thin and wavy. Stamen: Anthers subglobose, the cells dehiscing transversely or on a slant, filaments free. Ovary: Style lobes recurved, flattened on the ovary.

Fruit: Capsule more or less verrucose, glandular.

Seeds: Minutely tubercled.

1.1.4.15 Phyllanthus virgatus Forster F. (*P. simplex* Retz.)

Phyllanthus virgatus Forster F. (*P. simplex* Retz.) grows in northern Circars and Carnatic from the Chilka Lake to Madras, Deccan, and North Coimbatore on hot dry soils up to 3,000 feet in a hilly area. It is a stiff, almost woody, herb with long flattened branches (Gamble, 1925).

Leaves: Linear-oblong, obtuse or acute at apex, apiculate, the margins thickened, the nerves invisible, 0.5–0.75 inches long, 0.1–0.3 inches broad, stipules peltate, subsagittate. Flower: Male flowers few, minute, subsessile; female flowers more numerous on filiform pedicels; disks of male flowers are large flat glands and female flowers saucer shaped, thin and wavy. Tepals: 6, oblong, obtuse, mucronate. Stamen: 3, free, included, anthers subglobose, the cells dehiscing transversely or on a slant, filaments free. Ovary: Style lobes recurved, flattened on the ovary.

Fruit: Capsule more or less verrucose, glandular.

Seeds: Minutely tubercled.

1.1.4.16 *Phyllanthus gardnerianus* Baill.

Phyllanthus gardnerianus Baill. is seen in Western Ghats, in all districts from south Canara to Tinnevelly, Nilgiris, Anamalai, and Pulneys. It is a slender shrub with woody rootstock and long branches and leaves smaller in size. Plants of dry hilltops were found to be dwarf with very small leaves (Gamble, 1925).

> Leaves: Upper branchlets elliptic, obtuse, about 0.3 inches long; lower stems elliptic-oblong up to 0.75 inches long, 0.4 inches broad; the nerves about 5 pairs, faint, glaucous apiculate, the margins thickened, the nerves 4–5 prominent, joining in arches, 0.2–0.4 inches long, 0.1–0.3 inches broad, stipules peltate, subsagittate.
> Flower: Male flowers subsessile, female flowers pedicelled, disks of male flowers are large flat glands and female flowers saucer shaped, thin and wavy. Stamen: Anthers subglobose, the cells dehiscing transversely or on a slant, filaments free. Ovary: Style lobes recurved, flattened on the ovary.
> Fruit: Capsule more or less verrucose, glandular.
> Seeds: Minutely tubercled.

1.1.4.17 *Phyllanthus macraei* Muell.

Phyllanthus macraei Muell. is seen in Western Ghats, Sholas of the Pulney hills at 5,000–7,000 feet, Agastiamalai Peak, and Tinnevelly. It is a shrub, apparently reaching 2–3 feet in height with long, weak, flattened branchlets and capsules prominently warted when wet, leaf margins sometimes ciliate (Gamble, 1925).

> Leaves: Elliptic or elliptic-oblong, obtuse, up to 2 inches long, 1 inch broad, the nerves about 7 pairs, glaucous beneath, glabrous or hispid, stipules peltate, subsagittate.
> Flower: Male flowers fascicled, shortly pedicelled, female flowers long pedicelled, disks of male flowers are large flat glands and female flowers cushion shaped, thick. Stamen: Anthers subglobose, the cells dehiscing transversely or on a slant, filaments free. Ovary: Style lobes erect, spreading.
> Fruit: Capsule more or less verrucose.
> Seeds: Minutely tubercled.

1.1.4.18 *Phyllanthus rotundifolius* Klein

Phyllanthus rotundifolius Klein is an annual herb, about 40 cm high, found across the West African Sahel, particularly in proximity to rivers, and extending to Northern Circars and Carnatic and sands on the seacoast. It is a prostrate or slightly ascending fleshy herb with stout rootstock and long trailing branches (Gamble, 1925).

> Leaves: Coriaceous or fleshy, orbicular or obvate, obtuse or apiculate, scarcely 0.25 inches in diameter, stipules not peltate, lanceolate.
> Flower: Male flowers subsessile, female flowers pedicelled, disks of male flowers of minute glands and female flowers cushion shaped, thin and wavy.

Stamen: Anthers subglobose, the cells dehiscing transversely or on a slant, filaments connate in a column.
Ovary: Style arms recurved with short lobes.
Fruit: Capsule more or less verrucose, glandular.
Seeds: Regular lines of very minute tubercles joined by minute crossbars.

1.1.4.19 Phyllanthus debilis Klein ex Willd

Phyllanthus debilis Klein ex Willd was reported from Northern Circars and Carnatic and west to the eastern slopes of the Ghats. It grows in shady places in the hill forests. It may be a herb or an undershrub.

> Leaves: Membraneous, usually glaucous beneath, acute, elliptic-ovate, sometimes rounded at apex, elliptic or obovate up to at most 0.75 inches long, stipules lanceolate, long acuminate.
> Flowers: Rather large, disks of male flowers of star-like glands and female flowers saucer shaped, crenulate or lobed, tepal 6, lobes with prominent scarious margins. Stamens: 3, anther free, slits transverse, filaments connate in a column, staminal column long. Ovary: style erect, shortly bifid.
> Fruit: Capsule, 3 valved.
> Seeds: Regular lines of very minute tubercles joined by minute crossbars (Figure 1.1d).

1.1.4.20 Phyllanthus missionis

Phyllanthus missionis is seen in Western Ghats, on eastern slopes and Coimbatore to Tinnevelly at low levels. It is an erect undershrub with rather distant leaves (Gamble, 1925).

> Leaves: Membraneous, usually glaucous beneath, acute, sometimes rounded at apex, elliptic or obovate up to at most 0.75 inches long, stipules ovate, acute or acuminate, small.
> Flowers: Very small, disks of male flowers with peltate glands and female flowers cushion like, broadly lobed, tepal lobes with obscure scarious margins, anther reniform, filaments connate in a column, staminal column slender, style erect, with slender lobes.
> Seeds: Regular lines of very minute tubercles joined by minute crossbars.

1.1.4.21 Phyllanthus speciosus Jacq

Phyllanthus speciosus Jacq is a shrub with flattened branches often found in gardens in the plains.

1.1.4.22 Phyllanthus pinnatus (Wight) Webster, J.

Phyllanthus pinnatus (Wight) Webster, J. is a subshrub. Branchlets are glaucous (Mathew, 1982).

> Leaves: Elliptic to (sub)orbicular, base cuneate, apex acute, peiole to 0.4 cm.

Flowers: Male flowers numerous, female flowers fewer, long stalked, tepals 6, free, ovate, entire to fimbriate, (sub)acute. Stamens: 6, free, exerted, filaments filiform, anthers oblong, parallel, dehiscence vertical. Ovary: 3 lobed, styles 3, forked, horizontally appressed in fruit, disk 6, free in male flowers, connate in female flowers.

Fruit: Capsule dry, depressed-globose, of 3, bivalved cocci.

1.1.4.23 Phyllanthus acuminates

Common name: Jamaican gooseberry tree

Phyllanthus acuminates is a herb. It is seen in Central America and South America. This plant is used by the local population as a piscicide.

1.1.4.24 Phyllanthus caroliniensis

Phyllanthus caroliniensis is a flowering plant native to the Americas, from the southeastern United States all the way to Argentina. It may have medical uses, specifically in reducing pain. Flowers are small and axillary (Catapan et al., 2000).

1.1.4.25 Phyllanthus mirabilis

Phyllanthus mirabilis is a plant species endemic to Thailand. It is the only *Phyllanthus* to be caudiciform. The leaves fold together at night.

1.1.4.26 Phyllanthus caesiifolius

Phyllanthus caesiifolius is a species endemic to Cameroon. Its natural habitat is subtropical or tropical moist lowland forests. It is threatened by habitat loss.

1.1.4.27 Phyllanthus gentryi

Phyllanthus gentryi is a species endemic to Panama. It is threatened by habitat loss.

1.1.4.28 Phyllanthus muellerianus (Kuntze) Exell

Phyllanthus muellerianus (Kuntze) Exell is a native of Africa.

1.1.4.29 Phyllanthus pulcher Wall. Ex Müll. Arg.

Common name: Tropical leaf flower

Phyllanthus pulcher Wall. Ex Müll. Arg. is a native of Malaysia. Shrubs are 0.5–1.5 m tall, monoecious, stem and branches terete; branches to 40 cm, puberulent. Leaves are distichous, 15–30 pairs along each branchlet; stipules, membranous, abaxially gray-green, adaxially green, margins slightly revolute; inflorescence a bisexual axillary fascicle, usually with several male and 1 female flower.

1.1.4.30 Phyllanthus fraternus Webster

Phyllanthus fraternus Webster. is an erect, herbaceous weed of roadsides, cultivated land, waste places of the forest and savanna, generally rare in Senegal, Ivory Coast,

Ghana, and Nigeria. It is widely distributed in Asia and in the West Indies. It is probably native to western India and Pakistan (Gamble, 1925).

> Leaves: Leaf blade oblong-elliptic, obtuse at apex, rounded or subcordate at base, stipules lanceolate, basally attached.
> Flower: Perianth lobes 5, filaments connate, anthers subglobose, dehiscing transversely.
> Seeds: Trigonous, minutely tubercled in regular concentric lines.

1.1.4.31   *Phyllanthus abnormis*

Phyllanthus abnormis is a North American plant that contains an unidentified toxin that causes liver and kidney damage, manifested by compulsive walking, tenesmus, rectal prolapse, petechiation, and death.

1.1.4.32   *Phyllanthus odontadenius* Mull. Arg.

Phyllanthus odontadenius Mull. Arg. is a subwoody herb to 1 m high, a common weed of the forested area from Guinea-Bissau to West Cameroons and Fernando Po and to Sudan and Angola. The way in which the flowers and fruits are borne on the underside of the leaves gives rise to the Ghanaian names likening the plant to a child carried pickaback. The leaves are chewed with guinea grains in Ghana to cure cough. The Ijo of the Niger Delta have a superstitious use of the plant to drive away bad spirits: A person who suffers fever every evening that is attributed to the spirit of a dead person should urinate on the plant, pick the leaves for addition to a bath with local soap, and be relieved of the spirits (Burkill, 1985).

1.1.4.33   *Phyllanthus muellerianus* (O Ktze) Exell

Phyllanthus muellerianus (O Ktze) Exell is a shrub or climber, occasionally arborescent, deciduous, in secondary forests from Guinea-Bissau, Mali to West Cameroons, Fernando Po, and widespread in other areas of tropical Africa. The stem seldom becomes large. In Bendel State of Nigeria, it is reported as a weed of rice fields, plainly by lack of timely cultivation. In Kenya, it is said to yield excellent firewood (Burkill, 1985).

1.1.4.34   *Phyllanthus capillaris* Schum. & Thonn.

Phyllanthus capillaris Schum. & Thonn. is a shrub, 1.70 m high, and is widespread from Guinea to West Cameroon, Fernando Po, and widespread elsewhere in tropical Africa. No usage is recorded for the West African region. In Kenya, a decoction of the whole plant is taken as a remedy for vomiting, and crushed roots are eaten for stomachache (Burkill, 1985).

1.1.4.35   *Phyllanthus sublanatus* Schum. & Thonn.

Phyllanthus sublanatus Schum. & Thonn. is a semiwoody herb that grows to 50 cm high and is widespread from Mali to South Nigeria. It is a weed of cultivation and is found in Sierra Leone invading cultivated swampland (Burkill, 1985).

1.1.4.36 *Phyllanthus beillei* Hutch.

Phyllanthus beillei Hutch. is a shrub that grows to 3 m tall, of foothill savanna in a few localities from Senegal to South Nigeria and the Cameroon Republic and East Africa. No use is recorded within the region. The root is reportedly used in Tanganyika as an aphrodisiac (Burkill, 1985).

1.1.4.37 *Phyllanthus indofischeri* Bennet.

Phyllanthus indofischeri Bennet. is a tree; its young branchlets are white tomentose, and it grows in peninsular India (Gamble, 1925).

> Leaves: Obtuse, rounded, or subcordate at the base, less than 50 per branchlet, oblong or elliptic.
> Flowers: Usually dioecious, disk of six small glands.

1.1.4.38 *Phyllanthus scabrifolius* Hook.

Phyllanthus scabrifolius Hook. is an annual leafy herb; its stem is erect, branched from the base and upward, and branches are angular.

> Leaves: Broadly elliptic or obovate at the apex, pale when dry, scaberulous beneath and with undulate margins, main nerves 4–5 pairs, distinct on both surfaces, petioles minute, stipules lanceolate, subulate, membranous.
> Flowers: Very shortly pedicellate, sepals oblong or obovate-oblong, with white scarious margins, female flowers long, male flowers short, filaments short, united to the middle, the apex recurved, anther cells at length confluent. Styles: 3, distinct, long, bifid, the lobes recurved. Disk of male flowers is rounded, female a low crenate cup.
> Fruit: Capsule, depressed globose, smooth or slightly granulate.
> Seeds: Long, broad, trigonous, rounded, with 7–9 parallel ribs on the back and concentric ribs on the faces.

1.1.4.39 *Phyllanthus gradyi*

The *Phyllanthus gradyi* species appears to be restricted to humid forests in northeastern Brazil. It has been found in the states of Pernambuco and Alagoas, as well as in montane forests. As a tree, it is monoecious, with stem densely branched, the branching nonphyllanthoid, and branches cylindrical, gray, densely tomentose on young parts; trichomes are rust colored (Silva and Sales, 2006).

> Leaves: Stipules 2–2.8 mm long, lanceolate, acuminate, rigid, densely tomentose on external face, glabrous on internal face, midvein evident, and margin ciliate, green to vinaceous. Petiole 1.3–2.1 mm long, cylindrical, hirsute, leaf blade 3.5–11 and 1.5–4.5 cm, firmly chartaceous, elliptic to widely elliptic, base obtuse, apex acuminate and mucronate, margin entire, adaxial surface dark green, abaxial surface light green to brown when young, hirsute at base in midvein, brochidodromous, principal and secondary veins prominent only in the abaxial surface.

Flower: Fascicles axillary, 6 to 13 staminate and 2 to 3 pistillate flowers, bracts 6 to 9 per fascicle, 1.2–2 and 0.5–0.6 mm, widely triangular, external pubescent. Staminate flowers with pedicel 0.7–1.3 cm long, filiform, glabrous, greenish, sepals 4, ovate, obtuse-rounded at the tip, yellowish, midvein evident; disk glandular, cupuliform, fleshy, margin tetragonal, stamens 2, free, facing each other, filaments thickened, anthers with enlarged connective, dehiscing horizontally; pistillate flowers with pedicel 1.2–3.1 cm long, filiform, vinaceous to whitish near receptacle; sepals 6, oblong to lanceolate, apex obtuse to rounded, midvein evident, yellow-green to light green; disk glandular, upuliform, fleshy, margin irregularly lobed; ovary 1.1–1.2 and 2.5–2.6 mm, oblong; styles 3, free, bifurcate, recurved, stigmas acute.

Fruit: Capsule 2.5–2.6 and 5–5.2 mm, spheroid, styles persistent, fruiting pedicel 2.2–2.8 mm, long, glabrous to glabrescent.

Seeds: 3–3.1 and 2.5–2.8 mm, trigonous, areolate.

1.1.4.40 *Phyllanthus longipedicellatus*

Phyllanthus longipedicellatus is known only from the type collection, from southern Bahia State in Brazil. It grows in the coastal rain forest in shaded, humid areas on clay soils covered by leaf litter. The plant is monopodial, a shrublet, 30 cm high, with stems erect, terete, papillose in the young parts with blackish trichomes; branching is phyllanthoid; cataphylls and cataphyllary stipules are 0.8–1 and 0.5–0.7 mm, triangular, not auriculate, escariose, glabrous, margins hyaline, plagiotropic branchlets 6–14 cm with 30 to 57 leaves, axis about 0.2 mm wide, slightly flattened and with dark papillae (Silva, 2009).

Leaves: Subsessile, stipules 1–1.1 mm, lanceolate, acuminate, escarious, glabrous on both surfaces, petiole, 1 mm, leaf blades 5–6 and 2.9–3.9 mm, oblong-falcate to falcate-asymmetrical, obtuse-mucronulate, oblique at base, margins obscurely serrulate, membranaceous, adaxial surface dark green, abaxial surface light grayish green, dull, glabrous, venation, brochidodromous, midvein slightly prominent abaxially, secondary veins impressed abaxially.

Flower: Staminate flowers 2 or 3 in cymules at proximal axils, pistillate flowers solitary at distal axils, staminate pedicel 9–9.2 mm, capillary, finely papillose, sepals 5, 2.1–2.2 and 1.4–1.5 mm, obovate, rounded, membranaceous, 1 nerved, disk segments 5, alternisepalous, obtriangulate and finely papillose; stamens 3, filaments free, 2.4–2.5 mm, anthers 0.2 mm, connective not enlarged, thecae not deeply emarginate, dehiscing horizontally, pistillate pedicel 10–10.2 mm, capillary, finely papillose, sepals 5, 2.9–3 and 2–2.1 mm, widely elliptic, acute, membranaceous, venation pinnate; disk patelliform, margins finely undulate, ovary 0.8–0.9 and 0.9–0.91 mm, style branches 1.2–1.3 mm, free, bipartite, tips obtuse.

Fruits and seeds are unknown.

1.1.4.41 Phyllanthus salesiae

Phyllanthus salesiae is known only from the araucaria forest in Minas Gerais State, Brazil, growing at an altitude of 1525 m at Pico do Itaguaré, with riparian vegetation in clay soils. It is a shrublet, 30 cm high, and is dioecious; stems are erect, sparsely to densely ramified, papillose, branching, nonphyllanthoid with lateral branches, persistent and distichous. Branchlets are terete to slightly flattened, papillose, with cataphylls and cataphyllary stipules lacking (Silva, 2009).

> Leaves: Stipules 2–2.1 mm, triangular, acuminate, auriculate at base, lacerate margins, escarious, glabrous on both faces, reddish, petioles 1.9–2 mm, cylindrical, greenish, papillose, leaf blades 1.9–4.3 and 0.9–1.9 cm, elliptic, acute, obtuse at base, margins minutely papillose, membranaceous, adaxial surface dark green, abaxial surface opaque, grayish green with papillae concentrated around the midvein, venation brochidodromous, midvein slightly prominent and secondary veins impressed abaxially.
> Flower: Pistillate cymules with 1 or 2 flowers, bracteoles 1 mm, triangular, acuminate, margins slightly serrulate, pedicels 1.7–1.8 mm, terete, visibly articulate; sepals 5, 1.1–1.2 mm, widely elliptic to obovate, acute to obtuse, pinnate venation with whitish margins, disk deeply 5-segmented, margins slightly undulate; ovary 0.4–0.5 and 0.5–0.6 mm, depressed ovate, smooth; styles 3, incurved, bifid, stigma subcapitate.
> Fruit: Capsule 2–2.1 and 3.1–3.2 mm, depressed globose, light brown when dry, glabrous, dehiscing loculicidally and septifragally, calyx and style persistent in fruit.
> Seeds: 1.8–1.9 and 1.1–1.2 mm, trigonous, brownish, longitudinally, finely dark punctate.

1.1.4.42 Phyllanthus gongyloides

Phyllanthus gongyloides is a subshrub; it is dioecious, glabrous, 40–50 cm high; stems are persistent, sparsely branching, with branchlets that are terete to slightly flattened, smooth (Cordeiro and Carneiro-Torres, 2004).

> Leaves: Oval to orbicular blades, subcoriaceous, rounded at apex, mucronulate, rounded at base, glabrous, 7–12 mm long, 6–11 mm broad; lateral veins 4–6 per side, arching, obscure adaxially, prominulous abaxially, petiole 1–1.5 mm long; stipules widely deltate, acute, glandulose, reddish, glabrous, 1 mm long.
> Flower: Inflorescence axillary, bracts deltate, 1 mm long, staminate cymules with 2–4 flowers, pedicel 1.5–2.5 mm long; sepals 6, widely elliptic, obtuse, 3 mm, long, disk segments 6, obconics; stamens 3, filaments, completely connate into a column, tecae distinct, divergent, dehiscing horizontally; pollen grains prolate, 4 colporate, sexine reticulate; pistillate cymules with 2–3 flowers, pedicel 3–4 mm long; sepals 6, widely elliptic, obtuse, 2 mm long; disk segments 6, obconics; ovary glabrous, globose, styles spreading, bifid, branches acute.
> Fruit: Capsule 2.5 mm long.

Seeds: Trigonous, brownish, 1.5 mm long, with longitudinally finely puncticulate striae.

1.1.4.43 Phyllanthus indicus Muell.

Phyllanthus indicus Muell. is a very branched deciduous tree that is 30–40 feet high; the bark is white, scaly, smooth, and exfoliating in plates.

Leaves: Membranous, distichous, elliptic or elliptic-lanceolate, obtuse, acute or acuminate, apiculate, glabrous, glaucous beneath, base acute, deciduous.
Flowers: Dioecious, pedicellate, pale-green, males in fascicles, in the axils and on the branches on capillary pedicels, long, female flowers larger, umbellate cluster, stout pedicel, tepal 4, rarely 5, oblong, obtuse, two out larger, in male flowers stamens 4, filaments free, anthers oblong, adnate, extrose, dehiscing longitudinally, disk annular, fleshy, large, rudimentary ovary, in female flowers stamens absent, disk small, narrow, ovary 3 celled, ovules 2 in each cell, styles very short, deeply bifid, the lobes recurved.
Fruit: Capsule, globose, reticulately rugose, cocci bivalved, two seeded.
Seeds: Trigonous, blue, surrounded by an aril.

1.1.4.44 Phyllanthus gunnii Hook. F. (Synonym: Phyllanthus gasstroemii Muell. Arg.)

Common name: Scrubby spurge

Phyllanthus gunnii Hook. F. (synonym: *Phyllanthus gasstroemii* Muell. Arg.) is an erect shrub to 2 m high; it is glabrous and grows in dry sclerophyll forest on rocky slopes and along riverbanks and frequently on sandstone in Australia (James and Harden, 2010).

Leaves: Lamina broad-ovate to circular, sometimes obovate-oblong, mostly 10–20 mm long and 8–12 mm wide, apex notched.
Flower: Male flowers in clusters of 3–7 on peduncles 1–4 mm long, female flowers solitary on peduncles lengthening to 8 mm in fruit, perianth segments ovate, 1.5–2 mm long, margins whitish, stamens 3, filaments free, slender, glands prominent, ovary glabrous, styles mostly linear, entire.
Fruit: Capsule, 4 mm diameter, often reddish brown; seeds with irregular, longitudinal ridges.

1.1.4.45 Phyllanthus lacunarius F. Muell.

Phyllanthus lacunarius F. Muell. is an annual herb with prostrate to ascending stems to 25 cm long and glaucous.

Leaves: Obovate to cuneate, 4–20 mm long, 1–7 mm wide; stipules subulate, 1–1.5 mm long, whitish (James and Harden, 1999).
Flowers: Clustered with 1 female and 2 or 3 males on peduncles to 1 mm long, lengthening to 2.5 mm in fruit, perianth segments narrow, reddish

with white margins, 1 mm long in fruit, often reflexed, stamens 3, filaments free.

Fruit: Capsule 3–4 mm diameter.

Seeds: 1.5–2 mm long, finely longitudinally striate, brown.

1.1.4.46 Phyllanthus carpentariae Muell. Arg. (Synonyms: *Phyllanthus hebecarpus* Benth., *Phyllanthus grandisepalus* F. Muell.)

Phyllanthus carpentariae Muell. Arg. (synonyms: *Phyllanthus hebecarpus* Benth., *Phyllanthus grandisepalus* F. Muell.) is a procumbent to erect shrub that grows to 1 m high; it is tomentose or villous and is reported in Australia (James and Harden, 1999).

> Leaves: Oblong to obovate, mostly 10–25 mm long, 5–10 mm wide, apex obtuse.
> Flowers: Solitary or 1 male and 1 female together on peduncles to 1.5 mm long, lengthening to 5 mm in fruit, in male flowers perianth segments narrow, 1.5 mm long, female segments broader, herbaceous, with a narrow white margin, 5 mm long in fruit, stamens 3, filaments free, glands large, styles divided to the middle.
> Fruit: Capsule 5 mm diameter.
> Seeds: 2 mm long, may or may not be smooth with fine, horizontal striations and obscure, longitudinal bands.

1.1.4.47 Phyllanthus fuernrohrii F. Muell.

Phyllanthus fuernrohrii F. Muell. is a many-stemmed subshrub that grows to 40 cm high and is hoary-tomentose.

> Leaves: Broad-obovate or obovate to oblong, 8–29 mm long, 3–10 mm wide, apex obtuse, minutely mucronate, very shortly petiolate (James and Harden, 1999).
> Flower: Male flowers 1 or 2 together on peduncles 2–3 mm long, female flowers on slender peduncles 4–9 mm long, solitary or with male flowers, perianth segments 1–1.5 mm long, pubescent with scarious margins, female segments 2 mm long, extending to 3–4 mm in fruit, stamens 3, filaments free, ovary pubescent, styles 3, divided to midway.
> Fruit: Capsule 3–5 mm wide.
> Seeds: 1.5 mm long, smooth.

1.1.4.48 Phyllanthus hirtellus F. Muell. Ex Muell. Arg.

Phyllanthus hirtellus F. Muell. Ex Muell. Arg. is common in heath and dry sclerophyll forest of Australia (James and Harden, 1999).

> Leaves: Broad-obovate to narrow-oblanceolate, 2–8 mm long, apex obtuse, mucronate, truncate or emarginate, margins may or may not be flat, recurved or revolute, midrib prominent below; may or may not be sessile.
> Flower: Male flowers 2 or 3 together, female flowers solitary, peduncles to 2 mm long, perianth segments ovate, 1.5–2 mm long, stamens 3, filaments free, ovary pubescent; styles deeply bifid.

Fruit: Capsule 4 mm diameter, pubescent.
Seeds: 2 mm long, finely reticulate.

1.1.4.49 *Phyllanthus subcrenulatus* F. Muell.

Phyllanthus subcrenulatus F. Muell. is a shrub, 2 m high, but often less than 1 m; branchlets are angular, glabrous. It grows in dry rain forest or eucalypt woodland in rocky places of Australia (James and Harden, 1999).

> Leaves: Ovate to lanceolate, mostly 8–40 mm long, 4–15 mm wide, minutely crenate.
> Flowers: Solitary or a few together, peduncles 3 mm long, lengthening to 10 mm long in fruit, in female flowers perianth segments herbaceous with white margins, to 3 mm long in fruit, male segments petaloid, shorter, stamens 3; filaments free, glands conspicuous, styles 3, slender, divided to midway.
> Fruit: Capsule globose, glabrous, 6 mm diameter.
> Seeds: 2–2.5 mm long, slightly striate longitudinally.

1.1.4.50 *Phyllanthus similis* Muell. Arg.

Phyllanthus similis Muell. Arg. is an erect subshrub that grows to 60 cm high; it is rhizomatous, glabrous, well branched with distinctly reddish stems and branches. It grows along creeks or at the edge of rain forest, mainly on the coast of Australia (James and Harden, 1999).

> Leaves: 2 ranked, lamina obovate to elliptic, mostly 10–20 mm long, 6–10 mm wide, apex obtuse to acute or mucronate.
> Flower: Male flowers in 2- or 3-flowered clusters on peduncles to 1 mm long, females usually solitary, on peduncles lengthening to 3 mm long in fruit. Perianth segments ovate, 1.5 mm long, slightly enlarging in fruit, stamens 3, filaments free, styles short, deeply bifid.
> Fruit: Capsule 3.5–4 mm diameter, pale yellow-orange; seeds orange-brown, striated.

ACKNOWLEDGMENTS

I am thankful to Dr. John Britto, Director, The Rapinat Herbarium, St. Joseph's College, Thiruchirapalli, India for allowing me to use the figures from the book *The Flora of the Tamilnadu Carnatic*.

REFERENCES

Taxonomy of the Genus *Phyllanthus*

Burkill, H. M. 1985. *The useful plants of west tropical Africa, volume 2*. Chicago: The University of Chicago Press.
Catapan, E., Otuki, M. F., Viana, A. M., et al. 2000. Pharmacological activity and chemical composition of callus culture extracts from selected species of *Phyllanthus*. *Die Pharmazie* 55: 945–946.

Chase, M. W., and Reveal, J. L. 2009. A phylogenetic classification of the land plants to accompany APG III. *Bot. J. Linn. Soc.* 122–127.

Cordeiro, I., and Carneiro-torres, D. S. 2004. A new species of *Phyllanthus* (Phyllanthaceae) from chapada diamantina, Bahia, Brazil. *Bot. J. Linn. Soc.* 146: 247–250.

Gamble, J. S. 1925. *Flora of presidency of Madras, volume 2*, 1286–1290. London: West, Newman and Adlard.

James, T. A., and Harden, G. J. 1999. Taxon concept: flora of New South Wales flora, Suppl. 1. Available at http://plantnet.rbgsyd.nsw.gov.au/cgibin/NSWfl.pl?page=nswfl&lvl= sp&name= Phyllanthus~gunnii (accessed December 27, 2010).

Manilal, K. S., and Sivarajan, V. V. 1982. *Flora of Calicut*. Dehra Dun, India: Saujanya Books.

Mathew, K. M. 1982. *The flora of the Tamilnadu Carnatic*. Madras: Ranipat Herbareum, St. Joseph's College, Tiruchirapalli.

Mathew, K. M. 1988a. *Further illustrations on the flora of the Tamilnadu Carnatic*, 578–581. Madras: Ranipat Herbareum, St. Joseph's College, Tiruchirapalli.

Mathew, K. M. 1988b. *Illustrations on the flora of the Tamilnadu Carnatic*, 648–649. Madras: Ranipat Herbareum, St. Joseph's College, Tiruchirapalli. pp. 648–649.

Reveal, J. L., Hoffmann P., Doweld, A., and Wurdack, K. J. 2007. Proposal to conserve the name Phyllanthaceae. *Taxon* 56: 266.

Shukla, P., and Misra, S. P. 1997. *An introduction to taxonomy of angiosperms*. Delhi: Vikas.

Silva, M. J., and Sales, M. F. 2006. A new species of *Phyllanthus* (Phyllanthaceae) from northeastern Brazil. *Novon* 16: 421–423.

Silva, M. J. 2009. Two new Brazilian species of *Phyllanthus* (Phyllanthaceae). *Novon* 19: 229–233.

Tokuoka, T. 2007. Molecular phylogenetic analysis of Euphorbiaceae sensu stricto based on plastid and nuclear DNA sequences and ovule and seed character evolution. *J. Plant Res.* 120: 511–522.

1.2 IDENTIFICATION MANUAL FOR SOME SPECIES OF THE GENUS *PHYLLANTHUS* L. OF PHYLLANTHACEAE WITH SPECIAL REFERENCE TO THE INDIAN SUBCONTINENT

A. Lalithamba

1.2.1 INTRODUCTION

The genus *Phyllanthus* L. is distributed in tropic and subtropical areas. The genus is the largest in the family Phyllanthaceae, with more than 750 species throughout the world. (Kathriarachchi et al., 2005; Stevens, 2001; Webster, 1994). In India, about 50 species of the genus are recorded. Some species of *Phyllanthus* L. are used as medicinal herbs in different medical systems throughout the world. A total of 24 species of the genus *Phyllanthus* L. are recorded as medicinal plants by the ENVIS Center on Medicinal Plants in India (2010). The two popular and well-known taxa with potential medicinal values are the Amla and the Bhuaamla. In this section, eight species native to India and eight species naturalized in India are described.

1.2.1.1 The Scientific Classification

As per the Bentham and Hooker system (Bentham and Hooker, 1880) (this system is widely followed in India), the following is the scientific classification:

Class: Dicotyledones; subclass: Monochlamideae; series: Unisexuales; family: Euphorbiaceae; genus: *Phyllanthus*

As per the APG (Angiospermic Phylogeny Group) system, the genus belongs to the family Phyllanthaceae:

Kingdom: Plantae; division: Angiospermae; (unranked clade): Eudicots; (unranked clade): Rosids; order: Malpighiales; family: Phyllanthaceae; tribe: Phyllanthae; genus: *Phyllanthus* L. (Stevens, 2001)

The genus *Phyllanthus* L. includes the genera earlier known as *Anisonema*, *Cicca* Linnaeus, *Diasperus* Kuntze, *Emblica* Gaertner, *Epistylium* Swartz, *Kirganelia* Jussieu, *Phyllanthus* L, and *Xylophylla* based on molecular analysis (APG II classification, Hoffmann et al., 2006; Kathriarachchi, et al., 2006).

1.2.2 THE MAIN CHARACTERS OF THE GENUS

This genus is characterized by pinnate leaf-like plagiotropic branchlets that are deciduous and flower bearing. The name *Phyllanthus* L. is derived from Greek words meaning leaf-flower, an indirect reference to the apparent bearing of flowers on the leaves. The leaves on the main axis are reduced to scale leaves, which are known as cataphylls (Webster, 1994). The growth form ranges from annual and perennial herbs to shrubs and trees. Latex is absent. The branching is specific to this genus and is known as "Phyllanthoid branching"; the branchlets resemble pinnate leaves with limited growth. Foliar leaves alternate and are distichous, simple, with narrow stipules and margins entire. Plants are dioecious, flowers monoecious (unisexual). Flowers are axillary, solitary, or in fascicles; three bracteate, small, actinomorphic and are white, light yellow, green, or pink. Perianth lobes (5 or 6) are imbricate, characterized by a white scaly margin. Disk is present in male and female flowers. Stamens (3 or 6) are free or the filaments are combined into a column; anthers are bithecate and oblong, dehiscing vertically or transversely. Ovary is three celled and superior; styles are usually bifid, free or connate at the base, and the stigma with an adaxial furrow. Pistillodes or staminodes are absent. Fruit is a septicidal capsule or drupe with three crustaceous or two-valved cocci. Seeds are large, trigonous, rounded at back; the seed coat is hard, brown, exotegman (outer seed coat) with radially elongated furrows; caruncle is absent, and cotyledons are flat (Gamble, 1921; Hutchinson, 1973; Stevens, 2001; Webster, 1957).

1.2.3 SYNOPTIC KEY

1.2.3.1 Herbs

A. Perianth lobes 5:
   1. Foliar leaves distichous, branchlets horizontal, pinnate leaf-like; capsule smooth, regular longitudinal ribs on seed coat: *amarus*
   2. Leaves not distichous, pedicels filiform, male clustered, female solitary: *rheedi*
B. Perianth lobes 6:
   1. Stamens 6:
      a. Plant prostrate, fleshy: *rotundifolius*
   2. Stamens 3:
      a. Plant erect, leaves distichous, membranous, seed segmented, dark brown tubercles on one side, vertical ridges on other side: *fraternus* (native of India)
      b. Branchlets plagiotropic, leaves distichous, minute, seed coat stellate and verucose: *niruri* (native of America)
      c. Base subwoody, suberect, foliar branches not deciduous, male flowers, subsessile, female pedicellate, capsule not glandular: *gardenerianus*
      d. Slender herb, leaves prominently nervose, flowers in androgynous fascicles, perianth purple with white margins: *virgatus*
      e. Branches erect, leaves not distichous, coriaceous: *maderaspatensis*
      f. Stipules subulose, leaf acute, capsule verucose, the seeds prominently transversely ridged and with crossbars: *urinaria*
      g. Leaves thin, pedicel thickened, flowers drooping, flowers rather large: *debilis*

1.2.3.2 Shrubs or Trees

A. Perianth lobes 4: Trees, flower pinkish green, stamens 4, free: *acidus*
B. Perianth lobes 5: Shrubs, male flowers pink, stamens 3, ovary more than 3 celled: *reticulatus*
C. Perianth lobes 6:
  1. Stamens 3:
      a. Tree, fruit a fleshy drupe; glands in male flowers minute: *emblica*
      b. Small tree, leaves large and coriaceous, glands in male flowers not conspicuous: *indofischieri*
      c. Large shrub, fruit a dry dehiscent capsule: *polyphyllus*

1.2.4 DESCRIPTIONS

1.2.4.1 *Phyllanthus acidus* L

Vernacular names: Sanskrit: Lavaliphala; Hindi: Harfarauri; Tamil: Aranelli; Telugu: Raacha usiri; Bengali: Hariphal; English: Otaheite gooseberry, star gooseberry.

Description: 5–8 m tall tree with robust branches and slender deciduous leafy branches clustered at the upper part of the woody branches. Leaves alternate, simple, distichous on 30–50 cm long branches; petiole 0.5–1 cm long, leaf blade 5–8 × 2.5–3.5 cm, obliquely ovate, base rounded, apex acute, pale beneath. Flowers about 3–4 mm diameter, pinkish, monoecious, actinomorphic, subsessile, densely clustered on 5–12 cm long axillary racemes arising from nodes along the branches; male and female flowers mixed. Male flower: Calyx lobes 4, about 2–3 mm long, petals 0; disk glands conspicuous. Stamens 4, inserted, filaments free; anthers bithecate, about 1 mm long. Female flower: Calyx 4, petals 0; disk annular; ovary superior, subglobose, 3 celled, about 1–2 mm diameter; styles 3, stigma bifid. The fruits develop densely on the branches. Fruit: Fleshy drupe; pale yellow or greenish white, waxy, about 1–2 cm diameter, globose with slightly flattened poles, with 6–8 shallow vertical ribs and 1-seeded.

Flowering and fruiting season: Flowers and fruits twice a year March–May and November–January.

Distribution: It is a native of Madagascar and cultivated in gardens and in villages throughout India for fruit. Naturalized in Philippines, Indonesia, Vietnam, Laos, Malaya, Hawaii, and some other Pacific Islands; southern Mexico and Central America, Colombia, Venezuela, Surinam, Peru, and Brazil. Fruit is edible. The species is named *acidus* because the taste of the fruit is acidic (highly sour) (Figure 2.b; also see color insert).

1.2.4.2 *Phyllanthus amarus* Schumach & Thonn.

Synonymously used as *P. niruri* auct. Non. L.

Vernacular names: English: carry me seed, black catnip, stone breaker; Sanskrit: Bhuaamlaki; Hindi: Bhuaamla; Kannada: Kirunelli, Nela nelli; Malayalam: Kilanelli, Kizhararnelli; Telugu: Nelausiri.

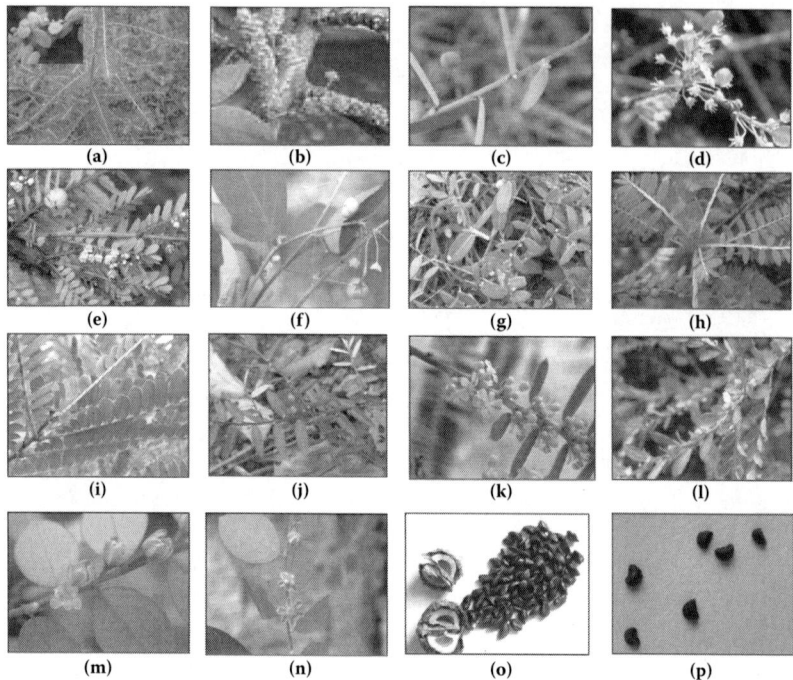

FIGURE 1.2 See color insert. Morphology of some of the common species of *Phyllanthus* (a–n): (a) *Phyllanthus amarus* (inset: a twig showing flowers); (b) *P. acidus*; (c) *P. gardnerianus*; (d) *P. pinnatus*; (e) *P. polyphyllus*; (f) *P. rheedi*; (g) *P. rotundifolius*; (h) *P. urinaria*; (i) *P. indofisheri*; (j) *P. virgatus*; (k) *P. emblica* flower; (l) *P. simplex*; (m) *P. deblis* male flower; (n) *P. deblis* female flower. (o) Fruit of *P. emblica*; (p) seeds of *P. urinaria*.

Description: Annual slender herb, 10–30 cm tall; main shoots terete, flowering branchlets horizontal. Leaves on the main axis reduced to scale leaves. Scale leaves 1–1.2 mm long, linear, subulate; stipules triangular, lanceolate, 1 mm long; foliage leaves alternate, distichous; petiole 2–3 mm long; leaf blade 5–9 × 2–4 mm, oblong, obtuse at both ends, glaucous beneath, 5–7 pairs of nerves, nerves not conspicuous on the upper surface, visible only in the lower side. Flowers axillary, 1–2 mm across; male flowers yellowish white, female flowers pale green in color; perianth with a white margin on either side in both sexes. Male flowers 1–3 together on the upper part of the branch; pedicel 1 mm long, perianth lobes 5, petaloid, triangular, apex acute, 0.5 mm long; stamens 5, alternate to perianth lobes, connate below; disk glands 5, minute; anthers 0.2–0.3 mm long, vertical; bithecate; pollen, subprolate, tectal surface fine reticulate (Perveen and Quaiser, 2005). Female flower: Solitary; on the lower part of the branch; pedicel 1 mm long extending up to 1.5–2 mm in fruit; perianth lobes 5, sepaloid, apex acute, persistent; disk 0.5 mm across, flat, 5 lobed; ovary superior, about 3–4 mm in diameter, subsessile, subglobose, 3 celled; 1–2 ovules in each cell; styles 3, about 0.3 mm long, free, spreading, stigma 2 fid. Fruit: Capsule, globose and dark brown in color when dry, 1.5–2 mm across, smooth walled; burst open and the seeds are hurled away. Seed: 1 × 0.5 mm, light brown, triquetrous; longitudinal ridges are seen on the rounded back side (Machado et al., 2006).

Flowering and fruiting season: Throughout the year but very common after monsoon.

Additional notes: The plant is bitter to taste. When the fresh juice of the plant is kept on white cotton, it turns black. This taxon is often synonymously referred to as *P. niruri* L. in Indian floras, which is native of America. It is easily distinguished from *P. niruri* L. by its equilateral leaf base and ribbed seeds (Machado et al., 2006). This taxon is used as a healing herb for jaundice by Ayurvedic (traditional system of Indian medicine) and ethnomedical practitioners throughout India and in tropical countries. It is referred to as stone breaking plant because it is also used to dissolve urinary stones in traditional medical systems in tropical countries. Unfortunately, there remains a great deal of confusion among scientists regarding plant identification, and in many cases misidentification of this taxon makes evaluation of published information difficult (Rao et al., 1999) (Figure 1.2a; also see color insert).

Distribution: Indigenous to India and widely naturalized in tropics. Common in plains, absent in hilly areas. Normally germinates after monsoon.

1.2.4.3 *Phyllanthus debilis* Klein ex Willd.

Vernacular names: Sanskrit: Bhupatri, Bhupushpi; Hindi: Bhonyaabbali, Bhuinanvalah; Tamil: Kilanelli.

Description: Annual herbs, 10–40 cm tall; branches slender. Leaves distichous; stipules lanceolate, long acuminate, about 2–3 mm long; petiole 3–4 mm long; lamina thin,1–1.5 × 0.7–1 cm, sometimes rounded at base, elliptic or obovate, acute at apex. Flowers monoecious; green, about 4–5 mm diameter, rather larger than other species, male flowers in the lower part of the branchlets, female flowers on the distal part of the branchlet, drooping, pedicel 2–2.5 mm long, thick. Male flower: Perianth lobes 6, arranged in 2 whorls of 3 each, the outer 3 larger, green with prominent scarious white margins, rounded at tip, 0.4–0.6 cm long; disk glands 3, star-like, stamens 3, filaments slightly connate below, staminal column slender; anthers erect, about 2–3 mm long, slit transverse. Female flower: Perianth lobes similar to that of male flowers; disk is saucer shaped and lobed, ovary 3 celled, smooth, about 2–3 mm diameter, globose with three white vertical stripes alternating with green stripes; styles 3, short, free, bifid at middle. Fruit: Capsule, about 2.5–3 mm diameter, globose, 3 celled. Seeds brown, obovate, and irregularly ribbed.

Flowering and fruiting season: August–December.

Distribution: Native of Indian subcontinent, but naturalized in tropical countries (Warren et al., 1999). Found in India, Bhutan, Sri Lanka, and New Guinea. Within India, it is found in northwestern India, Sikkim, Bihar,

Assam, and peninsular India on plateaus above an altitude of 600 m. This plant is a common weed in cultivated fields of hilly forest outskirts (Figure 1.2m and 1.2n; also see color insert).

1.2.4.4 Phyllanthus emblica L.

Synonym: *Emblica officinalis* Gaertn.

Vernacular names: Sanskrit: Hindi: Bengali: Amla, Amlaki; English: Emblic, Ambal, Amioki, Indian gooseberry; Telugu: Usiri; Tamil: Nelli; Malayalam: Amalakam, Nelli; Kannada: Amalaka, Chattu, Nellka, Sudhe, Aamalakee.

Description: 4–8 m tall deciduous tree, bark light gray and fissured, wood hard, red, and close grained; foliar branchlets distichous, looks like pinnate leaves, branchlets villous, about 15–40 cm long, leaves on the branchlets very many (up to 100), deciduous. Foliar leaves stipulate, stipules minute (1 mm), scarious, triangular, brown; petiole 1 mm long; lamina 1–1.5 × 0.3–0.4 cm, linear oblong, apiculate, base subcordate, subcoriaceous. Unisexual flowers arise in clusters in the axils of leaves. Before the onset of flowering, the tree almost sheds the leaves. Flowers about 3–5 mm diameter, greenish yellow, male and female mixed. Male flower: Pedicels 1 mm long; perianth lobes 6, 3–4 mm long, oblong or spatulate (spatula-like), spreading, margin white and entire, apex rounded, light yellow; disk glands 6, minute, stamens 3, connate into a column, the anthers erect, cohering by the connective, cells distinct. Female flower: Pedicel 0.5–1 mm; perianth lobes 6, 3–4 mm long, oblong, light yellow, spatulate, spreading, apex rounded, margins thin; disk cupular (cup-like); ovary ovoid, 1.5–2 mm across, 3 celled, ovules 2 in each cell, styles 3–4 mm long, thick and greenish, connate at the base and stigmas twice bifid and recurved. Fruit: Fleshy drupe, 2–3 cm across, slightly depressed, globose, pale green with 6 faint vertical lines. Seed: 4–5 × 2–3 mm, trigonous, reddish brown, exotegman hard with faint vertical furrows.

Flowering and fruiting season: March–October.

Distribution: Native of India and distributed frequently in hilly forest areas in the tropical parts of Indian subcontinent. Besides native varieties, hybrid varieties are cultivated for fruits in gardens because fruits are edible and widely used in medicinal preparations. (Gamble, 1921). Webster retained it under *Phyllanthus* L. *Phyllanthus* L. and *Emblica* Gaertn. were treated as cogeneric by Webster (1986) (JAA38:75.1975) and Airy Shaw (1980b) (KB.36.337.1981) (Suryanarayana and Rao, 2002).

1.2.4.5 Phyllanthus fraternus G. L. Webster

Vernacular names: English: Gulf leaf-flower; Sanskrit: Bhudhatri; Hindi: Jar amla/Bhu amla; Tamil: Keelanelli; Kannada: Kirunelli.

Description: A slender annual herb, 15–40 cm tall, main shoot erect, foliar branches horizontal up to 10 cm. Scale leaves 1 mm long, foliar leaves subsessile, stipules 1 mm long, linear lanceolate with one midrib, lamina 0.8–1 × 0.3–0.5 cm, elliptic-oblong or linear-oblong. Flowers axillary, unisexual; yellowish or greenish in color. Male flower: 1–3 in axillary fascicles, pedicel 0.5 mm long, flower 1.5–2 mm across; perianth lobes 6, in 2 whorls of 3 each, outer 3 apiculate, inner 3 obtuse, translucent with a green midrib; disk glands 6, flat, stamens 3, filaments connate into a column, anthers reniform, oblique, pollen grain tectal surface moderately reticulate. Female flower: Solitary, 3 mm across; pedicel 0.5–1 mm long; perianth lobes 6, about 1 mm long, slightly accrescent about 2 mm long in fruit; disk flat, minute, ovary 1.5–2 mm across, subglobose, 3 celled; styles 3, shortly bifid; stigmas recurved. Fruit: 1–1.2 × 1.5 mm capsule, 3 lobed, smooth, slightly 6 ridged, depressed. Seed: Trigonous, yellowish brown, about 1 mm long, segmented, with dark brown tubercles on one side and concentric vertical ridges on other side (Machado et al., 2006).

Distribution: This species is probably native of Pakistan or Western India, globally distributed throughout tropics except in Australia. Within India, it is found as a common winter weed on roadsides and waste places throughout the hotter parts (Rao, 1999).

Note: The species is named *fraternus* because the reproductive organs are on the lower side of the branches. *Fraternus* means to carry children on the back in local dialect in Ghana.

Phyllanthus fraternus Webster, *P. amarus, P. niruri* L., and *P. urinaria* L. are synonymously used by many pharmacists (Chatterjee and Prakashi, 2003). The pollen and seed morphology is different in *P. amarus, P. asperulatus, P. fraternus, P. urinaria,* and *P. niruri.*(Schmelzer and Gurib-Fakim, 2008, Machado et al.,2006). All five species are different.

1.2.4.6 *Phyllanthus gardnerianus* (Wt.) Baill.

Vernacular Names: Hindi: Bhiuavate; Telugu: Uchiusiri.

Description: Slender, slightly woody herb, 15–50 cm tall, branches angular, long and slender, stem reddish. Leaves alternate; stipules scaly broad or peltate, 1 mm long, brown; petiole short, leaves subsessile; lamina leathery, 1–2.5 × 0.5–0.8 cm, elliptic oblong, base slightly cordate, obtuse at apex, glacous beneath. Flowers axillary, monoecious. Male flowers in fascicles; subsessile, about 1.2–2 mm diameter, perianth lobes 6, broadly ovate, 0.5–1 mm long; disk glands 6, flat, stamens 3, filaments free, anthers about 0.5 mm diameter, subglobose. Female flower: Solitary, pedicel filiform up to 1–1.5 cm long; flowers pale green, about 2–3 mm diameter, sepals 6, ovate, 1 mm long; ovary globose, smooth, about 1–2 mm diameter, 3 celled, styles 3, bifid and recurved. Fruit: Capsule, about 2–3 mm diameter, fruiting pedicel long, capsule slightly 3 lobed, smooth. Seed: 1–2 mm long, trigonous, minutely tubercled.

Flowering and fruiting season: August–December.

Distribution: Common in waste places and moist places in laterite soils in forest foothills. Common in peninsular India (Figure 1.2c; also see color insert).

1.2.4.7 *Phyllanthus indo-fischeri* Gamble.

Synonym: *Phyllanthus indo fischeri* Bennet (Bennet, 1987).

Vernacular name: Tamil: Perunelli; Telugu: Chittiusiri.

Description: 3–4 m tall deciduous trees, plants look similar to *P. emblica* but smaller in size. Branchlets pale brown, leaves larger and thicker than those of *P. emblica*. Foliar leaves distichous; stipules brown, triangular and minute; petiole short and thick; leaf blade 1–1.8 × 0.5 cm elliptic-oblong, coriaceous, rounded or retuse at apex. Flowers also look similar to *P. emblica*, 3–4 mm across, greenish-yellow in axillary dense fascicles. Style in female flower is more slender than that of *P. emblica* and twice bifid. Drupes are smaller than those of *P. emblica*. Fruit: Drupe, about 1.5–2 cm diameter, globose, fleshy, pale green.

Flowering and fruiting season: October–February.

Distribution: Found in hilly areas in high altitudes above 350 m. Limited to peninsular India, Karnataka, Tamilnadu, and Velugonda Hills of Nellore District of Andhra Pradesh (Nayyar, 1980). Associated with *P. emblica* and *P. polyphyllus* in hilly areas (Ahmedullah and Nayar,1987). Unless closely observed, it is difficult to distinguish the two species. Fruits are not tasty as Amla and hence not preferred for eating (Figure 1.2i; also see color insert). *Phyllanthus* L. and *Emblica* Gaertn. were treated as cogeneric by Webster (1986) (JAA38:75.1975) and Airy Shaw (1980b) (KB.36.337.1981) (Suryanarayana and Rao, 2002).

1.2.4.8 *Phyllanthus maderaspatensis* L.

Vernacular names: English: Madras leaf flower; Telugu: Nelausiri; Hindi: Bazarmani, Ranavali; San: Bhuamlaki; Kanada: madaras nelli; Urdu: Kanocha (ENVIS, 2010).

Description: Annual erect herbs, 15–80 cm tall; branches all similar, erect and ascending, branchlets not horizontal and do not resemble pinnate leaves. Scale leaves minute on main axis, foliar leaves alternate, distichous nature absent, stipules about 1–2 mm long; petiole short, 1–2 mm long; leaf blade thin, subcoriaceous, glabrous, oblong-obovate, obtuse, base cuneate, rounded or retuse at apex, mucronate, glaucous, 0.7–2 × 0.3–0.8 cm, main nerve visible on upper surface. Flowers axillary, unisexual, bracteate, actinomorphic, about 2–3 mm diameter, pale yellow. Male flowers 1–3 in upper part of the branch; pedicel 1 mm long; perianth lobes 6 in 2 whorls of 3 each, about 2 mm long, suborbicular, entire; disk glands 6, small; stamens 3, about 1.5–2 mm long, filaments connate below, anthers subsessile. Female flower: Solitary, pedicel 1 mm, extending to 1.5–2 mm long in fruit, perianth similar to that of male flower in size and shape, green in color with white margin, spreading, apex rounded, accrescent in fruit up to 2 mm; disk glands 6, ovary globose, about 1.5 mm diameter, 3 celled; styles 3, free, bifid. Fruit: Capsule, about 2–3 mm diameter, epicarp smooth, globose, 3 valved, 6 seeded. Seed: 1–1.5 mm long, triquetrous, brown, with concentric lines of minute tubercles and minute crossbars.

Flowering and fruiting season: Throughout the year.

Distribution: Native of tropical Africa and globally distributed in the paleotropics. Within India, it is found throughout the drier parts in waste places as a weed and naturalized in tropical Asia. It is also used as a medicinal herb by ethnomedical practitioners (see Figure 3.1 in the color insert).

1.2.4.9 Phyllanthus niruri L.

Vernacular name: English: Chanca Piedra.

Description: An annual up to 30 cm high; stem closely sulcate, smooth; flowering branchlets up to 10 cm long, compressed or slightly winged, smooth. Leaves oblong or oblong-elliptic, rounded at both ends, 3–6 × 1.5–3 mm, membranous, glabrous on both surfaces; lateral nerves nearly invisible; petiole 1 mm; stipules lanceolate-subulate, membranous, glabrous. Flower: Minute, 1–2 mm across, axillary and unisexual. Male flower: Solitary in the lower side of the branch, sepals 6, small, 1 nerved, with membranous margins; disk glands 6, small; stamens 3, filaments connate, anthers bithecate, pollen grain 3–4 colpate (Perveen and Quaiser et al., 2005). Female flower: Solitary, arises in the upper parts of the branchlets; pedicel elongates up to 1–1.5 mm in fruit, glabrous; sepals 6, larger than that of the male; disk thin and flat, about 10 toothed; ovary subglobose, smooth; styles very short, suberect, bifid. Seed: Trigonous, stellate verucose cross bars aligned on 6 long ribs with crusts (Machado et al., 2006).

Distribution: It is a native of the Amazon region of South America and naturalized in tropics throughout the world. It is not an Indian species.

Note: The four species *P. amarus* Schumach & Thonn., *P. asperulatus* Hutch., *P. fraternus* G. L. Webster., and *P. niruri* L. are often synonymously used. These three species look very similar, are called by the same vernacular names, and hence often are misidentified by pharmacists or indigenous medical practitioners (Khatoon et al., 2006). In clinical research papers, the identification characters of the taxon worked on were never enumerated, and the names were indicated as synonymous with one another, creating confusion. The species *P. niruri* L. appears to be endemic to America (Webster, 1957) and specimens from other areas often are misidentified as *P. niruri* L. The seed morphology of the four species *P. amarus, P. asperulatus, P. fraternus,* and *P. niruri* is different (Machado et al., 2006).

1.2.4.10 Phyllanthus pinnatus (Wight) G. L. Webster.

Synonym: *P. wightianus* Muell, *Reidia floribunda* Wight.

Vernacular names: Telugu: Pachaari.

Description: 1–2 m tall shrubs; branchlets glaucous, foliar branches 6–12 cm long, pinnate leaf-like. Foliar leaves stipulate, stipules minute; pedicel 1–2 mm long; leaf blade 1–2 × 0.5–1 cm, elliptic, obovate or orbicular, obtuse, base acute. Flowers monoecious, 3–4 mm across, greenish yellow in axillary fascicles. Male flower: Pedicellate; perianth lobes 6, in 2 whorls of 3 each, about 2–3 mm diameter; disk glands 6, stamens 6, filaments 4–5 mm long, exerted, anthers 2 celled, about 1 mm diameter, bithecate. Female flower: Pedicellate; perianth lobes 6, ovary 3 celled, about 2–3 mm diameter, globose; styles 3, bifid, recurved. Fruit: Capsule, 0.8–1.5 cm across, depressed globose with papery pericarp and drooping.

Flowering and fruiting season: June–October (Figure 1.2d; also see color insert).

Distribution: Common and gregarious on latterite soils near foothills. Limited to peninsular India (Ahmedullah and Nayar, 1987).

1.2.4.11 *Phyllanthus polyphyllus* Willd

Vernacular names: Kannada: Manikanni, Krishna neli; Telugu: Kondapachaari; Malayalam: Kattunelli, Kilaranelli (ENVIS, 2010).

Description: 2–3 m tall shrub or a small tree. In vegetative phase, it resembles *P. emblica* L. except the size of the plant. Foliar branchlets 6–20 cm long, pinnate leaf-like. Foliar leaves distichous, 0.5–1.5 × 0.3–0.6 cm, oblong, obtuse or apiculate, base rounded, apex mucronate, stipules 1 mm long, brown, petiole short. Flowers monoecious, axillary, white or pale yellow. Male flowers 1–3 axillary in simple cymes, on the lower side of the leafy branches; flower white, about 5–6 mm across; pedicel 4–5 mm long; perianth lobes 6, in 2 whorls of 3 each, membranous, translucent, margin entire, apex rounded; disk glands 3, conspicuous, stamens 3, 2–3 mm long, filaments united into a column, anthers erect, 1 mm long, bithecate. Female flowers are on the upper part of the branchlets, subsessile; solitary, 4–5 mm across; perianth lobes 6, accrescent up to 3 mm in fruit; disk small, flat, ovary 3 celled, subglobose, about 2–3 mm diameter; styles 3, spreading, bifid and recurved. Fruit: Schizocarp, 4–6 mm across, light yellowish-green when young, black when dry and dehisces to three lobes. Seeds triquetrous, coarse with foveolate testa (seed pits conspicuous), pits are on longitudinal rows.

Flowering and fruiting season: May–September.

Distribution: Common on rocky boulders. Limited to peninsular India (Ahmedullah and Nayar, 1987). Found in association with *Pterocarpus santalinus* L.f., and *Anogeissus latifolia* (Roxb ex DC) Wall. in eastern ghats of Andhra Pradesh. This taxon is often mistaken as *Phyllanthus emblica* L. in the vegetative phase, which it resembles in leaf but is different in flower and fruit (also see Figure 1.2 in color insert).

1.2.4.12 *Phyllanthus reticulatus* Poir.

Vernacular names: English: black-honey shrub, black-berried featherfoil; Sanskrit: Krishna-kamboji; Hindi: Panjhuli; Bengali: Pansheuli; Tamil: Kattukilanelli, Telugu: Nallapurgudu (ENVIS, 2010).

Description: Straggling shrubs, 1–3.5 m tall; young branchlets bear leaves, branches not pinnate leaf-like. Foliar leaves alternate, not distichous, stipules 1–2 mm long, lanceolate, brown, hard and spiny when dry; petiole 2–5 mm long; leaf blade membranous or papery, 1–2.5 × 0.5 cm elliptic-oblong, rounded or acute at both ends, margin entire, lateral nerves clearly seen on both surfaces. The flowering shoots and pedicels are covered with short, velvety hairs. Flowers in axillary fascicles, monoecious, male and female mixed, pedicellate, actinomorphic, about 2–3 mm diameter. Male flower:

Pale pink, pedicel 5–8 mm, filiform; perianth lobes 5 or 6 in 2 whorls, unequal, obovate, 2–4 mm long; disk glands 5, fleshy, minute, stamens 5, in two series, inner 3 longer and connate at the base, the remaining 2 short and free. Female flower: Green, fruiting pedicel up to 1 cm long, filiform; perianth similar to that of male flower in size and shape; disk glands 5, obovate, fleshy, ovary smooth, 3–5 celled, ovules 2 per each cell; styles 3, free, bifid at apex. Fruit: Berry, 4–6 mm across, 6–10 seeded, subglobose, black or deep purple when ripe. Seed: 1–2 mm long, trigonous, and brown.

Flowering and fruiting season: Throughout the year.

Distribution: This taxon is globally distributed in the paleotropics. Within India, it is found along riverbanks and among scrubs in moist deciduous and semievergreen forests up to an altitude of 600 m.

1.2.4.13 *Phyllanthus rheedii* Wight.

Vernacular names: Not known.

Description: Annual herb, 15–30 cm tall, stems slender, terete, branchlets angular. Foliar leaves thin, stipules 1–2 mm long, lanceolate, decurrent; leaf blade about 1.2–2 × 0.8–1 cm, thin, glabrous, elliptic, apiculate or ovate. Flowers axillary; unisexual, pedicellate, solitary or in fascicles. Perianth lobes 5 in male and female flowers green with white scaly margins on either side. Male flower: Minute, about 1–1.5 mm diameter, pedicel filiform, 0.5–1 cm long; disk glandular, stamens 5, erect, the filaments united into a column, free above; anthers longish, dithecous. Female flowers solitary on thickened pendulous pedicels; disk small, flat, ovary trilocular, slightly ridged; styles 3, stigmas bifid. Fruit: Capsule, about 3–5 mm diameter, globose. Seeds with rather distant longitudinal very slender ridges and minute cross lines.

Flowering and fruiting season: July–December (Figure 1.2f; also see color insert).

Distribution: Common in forest undergrowth. Found in hilly areas at high altitudes in India.

1.2.4.14 *Phyllanthus rotundifolius* Klein ex Willd

Vernacular names: Not known.

Description: Prostrate or slightly ascending, fleshy herb with trailing branches, from a stout rootstock. Leaves coriaceous or fleshy, stipules minute, lanceolate; petiole short, leaf blade 0.8–1 × 0.4–0.8 cm orbicular or obovate, obtuse or apiculate glabrous on both surfaces. Flowers monoecious, 2–3 mm across, pale green, 2–3 males and 1 female together in each axil. Male flower: 2 mm across, perianth lobes 6, one nerved, membranous, apex rounded; disk glands 6, minute and wrinkled; stamens 6, filaments connate into a column, anthers free. Female flower: Perianth lobes 6, larger than that of male flower; disk cushion shaped, ovary trilocular, depressed-globose; styles 3 distinct, style arms recurved with short lobes. Fruit: Capsule, 3–4 mm across, globose or ovoid. Seeds trigonous with remarkable longitudinal furrows on the back.

Flower and fruiting season: July–September.

Distribution: On open sandy soils. Found in tropical Africa, Arabia, Sri Lanka, and peninsular India (Figure 1.2g; also see color insert).

1.2.4.15 *Phyllanthus urinaria* L.

Vernacular names: Sanskrit: Tamlaki; Hindi, Bengali: Hazarmani; Tamil: Shivappukinelli; Malayalam: Chirukizhukanelli, Chukanna-kizha-nelli; Marathi: Bhumyavli, Bhooyimabi; Kannada: Kempu nelanelli.

Description: Annual or perennial erect herbs up to 30 cm tall, stem woody at base, longitudinally sulcate, subterete and foliar branches ascending, 5–6 cm long, angled, wings asperulate. Leaves variable in size, distichous; oblong, stipules pinkish, broad and auriculate at base, up to 1 mm long; petiole short and compressed; leaf blade 1–1.5 × 0.4–0.6 cm glabrous, dark green above, glacous beneath, chartaceous, margins hispid, base unequal sided, apex shortly pointed, lateral nerves looped close to the margin. Flowers monoecious, axillary. Male flowers 2–4 together in leaf axils on the lower part of branchlets, minute, up to 2 mm across, pedicels 1 mm long, perianth lobes 6 in 2 whorls of 3 each, greenish white, elliptic-oblong, apex rounded; disk minute, stamens 3, filaments subsessile, united into a column, anthers erect. Female flower solitary on the axils of leaves on the abaxial side of branchlets; larger up to 3 mm across; sessile; perianth lobes 6, greenish white, persistent, slightly accrescent and reflexed in fruit; disk entire, ovary depressed globose and warty; styles 3, free, short, adjacent to the ovary, once bifid at the apex and recurved. Fruit: Capsule, reddish brown in color, 2–3 mm across, 6 seeded, pericarp verucose with 6 faint vertical lines. Seed: Trigonous, about 1–1.2 mm long, dark brown, prominently transversely ridged and with faint crossbars on the rounded back side (Figure 1.2h; also see color insert).

Flowering and fruiting season: July–December.

Distribution: Native of East Asia. Common in India in all plains, in the forest undergrowth along moist, shady localities. This taxon is also used as an alternative to *P. amarus* Schumach & Thonn. and *P. fraternus* G. L. Webster. in indigenous medicine. Because of its diuretic property it is named *urinaria* (Satyavathy et al., 1987).

1.2.4.16 *Phyllanthus virgatus* G. Forst.

Synonymously used for *P. simplex* Retz.

Vernacular names: Telugu: Uchchi usiri; Hindi: Bhiuavate; Kannada: Kaadu nelli; Malayalam: Niruri (ENVIS, 2010).

Description: Slender, branched, glabrous suffruticose herb up to 60 cm tall; branchlets angled. Leaves alternate, distichous, subsessile; stipules peltate, 1 mm long, brown, petiole short, leaf blade slightly leathery, elliptic-oblong or sublinear, prominently nervose, lateral nerves obscure, base obliquely rounded or acute at apex, apiculate, about 1–1.5 cm long and 2–4 mm wide. Flowers in androgynous axillary fascicles, male flowers few, female flowers many, often associated with 1 or 2 males; pedicellate. Male flower: Pedicel 2 mm long, perianth lobes 6, oblong, disk glandular; stamens 3 free, filaments short, anthers subglobose. Female flowers on longer pedicels, pedicel 6–9 mm long, perianth lobes 6, oblong, reflexed, purple with whitish membranous margin, persistent, disk annular, undivided; ovary globose, 3 celled, styles 3, free, bifid, short, recurved. Fruit: Capsule, depressed-globose, about 3 mm wide, smooth or slightly rugose (Benjamin, 1970; Stone, 1970). Seeds trigonous and minutely tubercled.

Flowering and fruiting season: April–May to November (Figure 1.2j; also see color insert).

Distribution: Widely distributed in the Old World tropics and presumably indigenous in Asia or Malaysia. Its introduction throughout the southern Pacific was doubtless aboriginal and inadvertent (Smith, 1981).

ACKNOWLEDGMENT

I am thankful to Dr. K. S. Murty, scientist retired, CCRAS, New Delhi, for his valuable suggestions and encouragement in preparing the chapter.

REFERENCES

Identification Manual for Some Species of the Genus *Phyllanthus* L. of Phyllanthaceae with Special Reference to the Indian Subcontinent

Ahmedullah, M., and Nayar, M. P. 1987 *Endemic plants of the Indian region,* 212. Howrah, India: Botanical Survey of India.

Airy Shaw, H. K. 1980a. The Euphorbiaceae of New Guinea. *Kew Bull. (Additional series)* 8: 185–186.

Airy Shaw, H. K. 1980b. *Notes on the Euphorbiaceae* from Indomalesia, *Australia* and the Pacific. *Kew Bull.* 35: 383–399.

Albert, S. C. 1981. *Flora Vitiensis nova: a new flora of Fiji, volume 2,* 810. Lawaii, Hawaii: Pacific Tropical Botanical Garden.

The wealth of India: a dictionary of Indian raw materials and industrial products, volume 8, 34–36. 1948–1976. New Delhi: Council of Scientific and Industrial Research, India.

Bennet, S. S. R. 1987. *Name changes in flowering plants in India and adjacent regions,* 429–430. Dehradun, India: Triseas.

Bentham, G., and Hooker, J. D. 1880. *Phyllanthus* L. In *Genera plantarum,* 272–275. London: Reeve.

Chatterjee, A., and Prakashi, S. C. 2003. *Treatise on Indian medicinal plants, volume 3,* 49–55. New Delhi: NISCAIR.

ENVIS. 2010. *Encyclopedia on Indian medicinal plants.* Available at http://envis.frlht.org.in/ (accessed November 11, 2010).

Gamble, J. S. 1921. *Flora of the presidency of Madras, 3,* 1286–1290. London: Adlard.

Hoffmann, P., Kathriarachchi, H., and Wurdack, K. J. 2006. A phylogenetic classification of Phyllanthaceae. *Kew Bull.* 61: 37–53.

Hooker, J. D. 1872. *The flora of British India,* 288–298. London: Reeve.

Hutchinson, J. 1973. *The families of flowering plants: arranged according to a new system based on their probable phylogeny,* 3rd ed. London: Oxford University Press.

Kathriarachchi, H., Hoffmann, P., Samuel, R., Wurdack, K. J., and Chase, M. W. 2005. Molecular phylogenetics of Phyllanthaceae inferred from five genes (plastid atpB, matK, 3'ndhF, rbcL, and nuclear PHYC). *Mol. Phylogenet. Evol.* 36: 112–134.

Kathriarachchi, H., Samuel, R., Hoffmann, P., et al. 2006. Phylogenetics of tribe Phyllantheae (Phyllanthaceae; Euphorbiaceae sensu lato) based on nrITS and plastid matK DNA sequence data. *Am. J. Bot.* 93: 637–655.

Khatoon, S., Rai, V., Rawat, A. K., and Mehrotra, S. 2006. Comparative pharmacognostic studies of three *Phyllanthus* species. *J. Ethnopharmacol.* 104: 79–86.

Linne, C. V., and Willdenow, C. L. 1753. Caroli a Linné. *Species Plantarum* 2: 981–982.

Machado, C. A., de Oliveira, P. L., and Mentz, L. A. 2006. SEM observations on seeds of some herbaceous *Phyllanthus* L. species (Phyllanthaceae). *Rev. Bras. Farmacogn.* [online] 16: 31–41.

Nayar, M. P., Ahmed, M., and Raju, D. C. S. 1984. Endemic and rare plants of Eastern Ghats. *Indian J. Forestry* 7: 35–42.

Nayyar, M. P. 1980. Endemic plants of peninsular India and its significance. *Bull. Bot. Surv. India* 22: 12–23.

Perveen, P., and Quaiser, M. 2005. Pollen flora of Pakistan XLVII—Euphorbiaceae. *Pak. J. Bot.* 785–796.

Rao, R. S., Sudhakar, S., and Venkanna, P. 1999. Flora of East Godavari District, Andhra Pradesh, India. Hyderbad, India: Indian National Trust for Art and Cultural Heritage, p. 632.

Rao, R. S., Sudhakar, S., and Venkanna, P. 1999. *Flora of East Godavari District, Andhra. Pradesh, India.* Hyderbad, India: Indian National Trust for Art and Cultural Heritage, p. 632.

Satyavati, G. V., Gupta, A. K., and Tandon, N. 1987. *Medicinal plants of India, volume 2,* 405. New Delhi: Indian Council of Medical Research.

Schmelzer, G. H., and Gurib-Fakim, A. 2008. Medicinal plants. Plant Resources of Tropical Africa (Program) PROTA, pp. 425–426.

Stevens, P. F. 2001 onward. Angiosperm phylogeny Web site. Version 9, June 2008 [updated since]. Available at http://www.mobot.org/MOBOT/research/APweb/.

Stone, B. C. 1970. The flora of Guam. A manual for the identification of the vascular plants of the Island. *Micronesica* 6: 1–659.

Suryanarayana, B., and Rao, A. S. 2002. *Flora of Nellore District: Eastern Veligonda hill ranges and Sriharikota Island,* 490–495. Shri Rampur, India: Gurudev Prakashan.

Warren, L. W., Derral, R. H., and Sohmer, S. H. 1999. *Manual of the flowering plants of Hawaii,* rev. ed. Bernice P. Bishop Museum special publication. Honolulu: University of Hawai'i Press/Bishop Museum Press.

Webster, G. L. 1957. A monographic study of the West Indian species of *Phyllanthus. J. Arnold Arboretum* 38: 51–80, 170–198, 295–373.

Webster, G. L. 1994. Classification of the Euphorbiacea. *Ann. Mo. Bot. Gard.* 81: 3–32.

Webster, G. L. 1986. A revision of *Phyllanthus* (Euphorbiaceae) in Eastern Melanesia. *Pacific Sci.* 40: 1–4.

2 Current Pharmacopoeial Status of *Phyllanthus* Species
P. emblica, P. amarus, and *P. fraternus*

Raman Mohan Singh and Vivekanandan Kalaiselvan

CONTENTS

2.1	Introduction	38
2.2	*Phyllanthus emblica*	38
	2.2.1 Description	38
	2.2.2 Identifications	39
	2.2.2.1 Macroscopic	39
	2.2.2.2 Microscopic	39
	2.2.2.3 By Thin-Layer Chromatography	39
	2.2.3 Assay	39
	2.2.4 Pharmacopoeial Comparison	40
2.3	*Phyllanthus amarus*	40
	2.3.1 Description	41
	2.3.2 Identifications	41
	2.3.2.1 Macroscopic	41
	2.3.2.2 Microscopic	42
	2.3.2.3 By Thin-Layer Chromatography	42
	2.3.3 Assay	42
	2.3.4 Pharmacopoeial Comparison	43
2.4	*Phyllanthus fraternus*	43
	2.4.1 Identifications	44
	2.4.1.1 Macroscopic	44
	2.4.1.2 Microscopic	44
2.5	Other Parameters	45
2.6	Conclusion	45
References		46

2.1 INTRODUCTION

Pharmacopoeial standards are prescribed to control the quality of raw herbal drugs and finished formulations and so to maintain safety and efficacy. Various pharmacopoeial bodies, such as the Indian Pharmacopoeia (2010), The Ayurvedic Pharmacopoeia of India, British Pharmacopoeia (2010), and Chinese Pharmacopoeia (2005), prescribed the standards for *Phyllanthus* species, including their description, identification test, ethanol- and water-soluble extractive values, microbial limit, and assay. The standards given in the pharmacopoeias would be applied to differentiate species-to-species identification and to detect adulterants. High-performance thin-layer chromatography (HPTLC) and high-performance liquid chromatography (HPLC) are invaluable quality assessment tools for the evaluation of herbal products and marker compounds. The Indian Pharmacopoeia and the British Pharmacopoeia applied HPTLC and HPLC techniques for identification and assay of certain *Phyllanthus* species. This chapter focuses on the different pharmacopoeial standards and comparative status of *Phyllanthus* species. Monographs of *Phyllanthus emblica* and *Phyllanthus amarus* are available in the fist edition of the *Ayurvedic Pharmacopoeia of India*, 1990 and the monograph of *Phyllanthus fraternus* is available in the Indian Herbal Pharmacopoeia.

2.2 PHYLLANTHUS EMBLICA

The drug consists of the dried fruit pericarp of *Phyllanthus emblica*.

> Family: Phyllanthaceae
> Vernacular names:
> English: Emblic myrobalan, Indian gooseberry
> Sanskrit: Aamalaki
> Hindi: Amla
> Kannada: Nelli kayi
> Marathi: Amla
> Gujarati: Ambla
> Malayalam: Nellikka
> Tamil: Nelli
> Telugu: Usirikaya
> Kashmir: Aonla

2.2.1 DESCRIPTION

The dried fruit has a highly shriveled and wrinkled external surface. The taste is sour and astringent followed by a delicately sweet tinge.

2.2.2 IDENTIFICATIONS

2.2.2.1 Macroscopic

The dried fruit shows a broad, highly shriveled and wrinkled convex surface, with the lateral surface transversely wrinkled. The external surface exhibits a few whitish specks; occasionally, some pieces show a portion of stony testa.

2.2.2.2 Microscopic

The epicarpic cells are rectangular, and their walls are highly cuticularized. The anomocytic type of stomata is found rarely. Collateral fibrovascular bundles are scattered throughout the inner mesocarp. Pitted and helical tracheids with tapering ends are seen. At places in the phloem, large cavities filled with crystal mass are present.

2.2.2.3 By Thin-Layer Chromatography

Determination by thin-layer chromatography (TLC) is made by coating the plate with silica gel GF 254.

> *Mobile phase:* A mixture of 20 volumes of toluene, 45 volumes of ethyl acetate, 20 volumes of glacial acetic acid, and 5 volumes of formic acid is used.
> *Test solution:* Reflux 2 g of coarsely powdered substance under examination with 50–75 ml methanol for 15 min; cool and filter. Reflux the residue an additional two times with 75 ml methanol; cool and filter. Combine all the filtrates and concentrate under vacuum to 50 ml.
> *Reference solution (RS):* Reflux 0.4 g of coarsely powdered *amalaki RS* with 50–75 ml methanol for 15 min; cool and filter. Reflux the residue an additional two times with 75 ml methanol; cool and filter. Combine all the filtrates and concentrate under vacuum to 10 ml.

Apply to the plate 10 µl of each solution as bands 10 mm by 2 mm. Allow the mobile phase to raise 8 cm. Dry the plate in air and examine in ultraviolet light at 254 nm and 365 nm; spray with anisaldehyde sulfuric acid reagent. Heat the plate at 100°C for 5–10 min and examine in daylight. The chromatographic profile of the test solution is similar to that of the reference solution.

2.2.3 ASSAY

Determination by liquid chromatography is as follows:

> *Test solution:* Weigh accurately about 0.5 g of coarsely powdered substance under examination; add 50 ml of water, sonicate for 3 min, and heat on a boiling water bath for 15 min; cool and dilute to 100 ml with water and filter.
> *Reference solution:* A 0.01% w/v (weight/volume) solution of *gallic acid* RS in water.

Chromatographic system:
- A stainless steel column 25 cm × 4.6 mm packed with octadecylsilane bonded to porous silica (5 µm)
- Mobile phase:
  - **A.** A solution prepared by dissolving 0.136 g of potassium dihydrogen orthophosphate in 500 ml of water; add 0.5 ml of orthophosphoric acid and dilute to 1,000 ml with water.
  - **B.** Acetonitrile
- A linear gradient program using the following conditions:
  - Flow rate 1.5 ml/min
  - Spectrophotometer set at 270 nm
  - Injection volume 20 µl

Time (in min)	Mobile Phase A (% v/v)	Mobile Phase B (% v/v)
0	100	0
18	55	45
25	20	80
30	100	0

- Inject the reference solution. The test is not valid unless the relative standard deviation for the replicate injections is not more than 2.0%. Inject the test solution and the reference solution.
- Calculate the content of gallic acid.

2.2.4 PHARMACOPOEIAL COMPARISON

Available data for the *Phyllanthus emblica* standards are compared for the Indian Pharmacopoeia, British Pharmacopoeia, and Chinese Pharmacopoeia in Table 2.1. The data show a limited difference in physical parameters, but the assay of active marker compounds is available by a liquid chromatographic (LC) method only in the Indian Pharmacopoeia and the British Pharmacopoeia; no assay method is available in the Chinese Pharmacopoeia.

2.3 PHYLLANTHUS AMARUS

The drug consists of aerial tender branches of *Phyllanthus amarus* Schum & Thon.

Family: Phyllanthaceae
Vernacular names:
  Sanskrit: Bhoomyaamalakee, Taamalakee
  Bengali: Bhuiamla, Sadahazuramani
  Gujarati: Bhonyaanvali
  Hindi: Bhuiavla, Jangli amla
  Kannada: Nelanelli, Kirunelli
  Malayalam: Kizhkkayinelli, Keezharnelli

TABLE 2.1
Standards of *Phyllanthus emblica* According to Different Pharmacopoeias

Monograph Details	Indian Pharmacopoeia 2010	British Pharmacopoeia 2010	Chinese Pharmacopoeia 2005
Monograph name	Amalaki	*Phyllanthus emblica* pericarp	*Fructus phyllanthi*
Assay	Amalaki contains not less than 1.0% w/w gallic acid calculated on dried basis (LC method)	It contains not less than 6.0% tannins (LC method)	—
Ethanol-soluble extractive	Not less than 30.0%	Not less than 15.0%	—
Foreign matter	—	Not more than 5%	—
Water-soluble extractive	Not less than 40%	Not less than 50%	Not less than 30%
Total ash	Not more than 5.0%	Not more than 7.0%	Not more than 5.0%
Acid-insoluble ash	Not more than 2.0%	—	Not more than 1.5%
Loss on drying	Not more than 12.0%	Not more than 10.0%	—
Foreign organic matter	Not more than 3.0%	—	—
Heavy metals	1.0 g complies with the limit test for heavy metals (20 ppm)	—	—
Microbial contamination	Complies with microbial contamination tests	—	—
Water	—	—	Not more than 13.0%

Marathi: Bhuiavala
Tamil: Keela nelli, Kilkkayanelli
Telugu: Nela virika, Nelavusuri

2.3.1 Description

The powdered drug of *Phyllanthus amarus* is green to greenish yellow in color, and the taste is slightly bitter.

2.3.2 Identifications

2.3.2.1 Macroscopic

Stem teret is 1–4 mm in diameter. Leaves are oblong 5 × 3 mm, short stalked, and greenish brown in color.

2.3.2.2 Microscopic

Chlorenchymetous cells are present in the inner cortex of the stem. Leaf stomata are mostly paracytic; the epidermal cell wall is markedly sinuous; rosette and prismatic crystals of calcium oxalate appear along the veins and midrib.

2.3.2.3 By Thin-Layer Chromatography

Determine by TLC, coating the plate with silica gel GF 254.

> *Mobile phase:* A mixture of 6 volumes of toluene, 2 volumes of ethyl acetate, 1 volume of formic acid, and 0.2 volume of methanol is used.
> *Test solution:* Reflux 2 g of coarsely powdered substance under examination with 50 ml methanol on a boiling water bath for 30 min; cool and filter. Reflux the residue again twice with 50 ml methanol; cool and filter. Combine all the filtrates and concentrate under vacuum to 10 ml.
> *Reference solution:* Reflux 1 g of Bhuiamla RS with 50 ml methanol on a boiling water bath for 30 min; cool and filter. Reflux the residue again twice with 50 ml methanol; cool and filter. Combine all the filtrates and concentrate under vacuum to 5 ml.

Apply 10 µl of each solution to the plate as bands 10 by 2 mm. Allow the mobile phase to raise 8 cm. Dry the plate in air and examine in ultraviolet light at 254 and 365 nm; spray with methanolic sulfuric acid (10% v/v). Heat the plate at 120°C for 5–10 min and examine in daylight. The chromatographic profile of the test solution is similar to that of the reference solution.

2.3.3 Assay

Determine by LC as follows:

> *Test solution:* Reflux 2 g of coarsely powdered substance under examination with 50 ml of methanol on a water bath for 15 min; cool and filter. Reflux the residue again with methanol until the last extract turns colorless; cool and filter. Combine all the filtrates and concentrate to 10 ml.
> *Reference solution a:* This is a 0.020% w/v solution of phyllanthin RS in methanol.
> *Reference solution b:* This is a 0.020% w/v solution of hypophyllanthin RS in methanol.
> *Chromatographic system:*
> - A stainless steel column 25 cm × 4.6 mm packed with octadecylsilane bonded to porous silica (5 µm)
> - Mobile phase: A mixture of 65 volumes of methanol and 35 volumes of water
> - Flow rate 1.5 ml/min
> - Spectrophotometer set at 230 nm
> - Injection volume 20 µl

TABLE 2.2
Comparative Standards for *Phyllanthus amarus* in the Indian Pharmacopoeia and the Indian Herbal Pharmacopoeia

Monograph Details	Indian Pharmacopoeia 2010	Indian Herbal Pharmacopoeia
Assay	Not less than 0.25% w/w total phyllanthin and hypophyllanthin calculated on dried basis (LC method)	—
Foreign organic matter	Not more than 2.0%	Not more than 2.0%
Ethanol-soluble extractive	Not less than 6.0%	—
Water-soluble extractive	Not less than 15.0%	Not less than 15.0%
Total ash	Not more than 8.0%	Not more than 8.0%
Acid-insoluble ash	Not more than 5.0%	Not more than 5.0%
Heavy metals	1.0 g complies with the limit test for heavy metals (20 ppm)	—
Loss on drying	Not more than 12.0%	—
Microbial contamination	Complies with microbial contamination tests	—
n-Hexane-soluble extractive	—	Not less than 15.0%

Inject reference solutions a and b. The test is not valid unless the relative standard deviation for the replicate injections for both the analyte peaks corresponding to phyllanthin and hypophyllanthin is not more than 2.0%. Inject the test solution, reference solution a and b. Calculate the contents of phyllanthin and hypophyllanthin.

Available data on the standards for *Phyllanthus emblica* are compared next for the Indian, British, and Chinese Pharmacopoeias.

2.3.4 PHARMACOPOEIAL COMPARISON

Available data on the standards for *Phyllanthus amarus* are compared for the Indian Pharmacopoeia and the Indian Herbal Pharmacopoeia in Table 2.2. The data show that the assay of active marker compounds is available by the LC method only in the Indian Pharmacopoeia; also, some other parameters, like heavy metals (20 ppm) and microbial contamination tests, are available in the Indian Pharmacopoeia only.

2.4 *PHYLLANTHUS FRATERNUS*

The drug consists of the root, stem, and leaf of *Phyllanthus fraternus* Webst. (Syn. *Phyllanthus miruri* Hook. F. non Linn.).

> Family: Phyllanthaceae; an annual herb, 20–60 cm high, found in central and southern India and extending to Ceylon

Vernacular names:
  Sanskrit: Mahidhatrika, Bhumyamalaki, Bahuphala
  Assam: Bhuin Amla
  Bengali: Bhumamla, Bhumi amalaki
  Gujarati: Bhoi Amali, Bhony amari, Bhonyamali
  Hindi: Bhui Amala
  Kannada: Nelanelli
  Malayalam: Kizanelli, Keezhanelli, Ajjhada
  Marati: Bhuiawali
  Orissa: Bhuin Amla
  Tamil: Kizhukai nelli, Kizanelli
  Telugu: Nela usirika

2.4.1 Identifications

2.4.1.1 Macroscopic

Root: The root is small, 2.5–11.0 cm long and nearly straight, gradually tapering, with a number of fibrous secondary and tertiary roots; external surface is light brown; fracture is short.

Stem: The stem is slender, glabrous; light brown, cylindrical, 20–75 cm long; branching is profuse toward upper region, bearing 5–10 pairs of leaves, with internode 1–3.5 cm long; odor is indistinct; taste is slightly bitter.

Leaf: The leaf is compound, and leaflets are arranged in two rows with a rachis; alternate, opposite and decussate almost sessile, stipulate, oblong, entire; up to 1.5 cm long and 0.5 cm wide; greenish-brown in color; odor is indistinct; taste is slightly bitter.

2.4.1.2 Microscopic

Root: The transverse section shows 4–6 layers of cork consisting of thin-walled, rectangular, tangentially elongated, and radially arranged cells filled with reddish-brown contents; the secondary cortex consists of 8–10 layers of thin-walled, tangentially elongated, parenchymatous cells; the secondary phloem narrow consists of sieve elements, phloem parenchyma, and is traversed by narrow phloem rays; secondary xylem is represented by a broad zone of tissue composed of vessels, tracheids, fibers, and parenchyma, all elements being thick walled and lignified with simple pits; xylem rays are uniseriate.

Stem: A transverse section shows a single-layer epidermis composed of thick-walled, flattened, tangentially elongated cells; an older stem shows four or five layers of cork composed of thin-walled, tabular, tangentially elongated, and radially arranged cells filled with reddish-brown contents; the cortex is composed of four to six layers of oval, tangentially elongated, thin-walled, parenchymatous cells and some cortical cells filled with yellowish-brown contents; the endodermis is quite distinct; the pericycle represented by a discontinuous ring composed of several tangentially elongated strands of lignified fibers with thick walls and narrow lumen; secondary phloem is narrow, composed of sieve elements, dispersed in a mass of phloem parenchyma; secondary xylem is composed of vessels, fibers, and parenchyma and is traversed by numerous uniseriate rays; vessels are mostly simple pitted, with a few show spiral thickenings; fibers are narrow elongated, with narrow or sometimes blunt ends with simple pits; the center is occupied by a pith composed of thin-walled, circular-to-oval parenchymatous cells; occasionally, cluster crystals of calcium oxalate are present in parenchymatous cells of ground tissue.

Leaf: The transverse section of leaf shows a biconvex outline; epidermis on either side is a single layer covered externally by a thick cuticle; a palisade layer is present beneath the upper epidermis, intercepted by a few parenchymatous cells in the middle; meristele is composed of small strands of xylem toward the upper surface and phloem toward the lower surface; the rest of the leaf tissue is composed of thin-walled, parenchymatous cells, some having cluster crystals of calcium oxalate; lamina shows a dorsiventral structure, with mesophyll differentiated into palisade and spongy parenchyma; epidermis on either side is composed of thin-walled, tangentially elongated cells covered externally by a thick cuticle; anisocytic-type stomata are present on both epidermises; single-layer palisade; mesophyll is composed of three to five layers of loosely arranged cells with a number of veins traversing this region; a few clusters of crystals of calcium oxalate are present in spongy parenchyma.

Powder: The drug powder is brown color; under the microscope it shows fragments of cork cells, vessels, and fibers.

2.5 OTHER PARAMETERS

Foreign matter: Not more than 2%.
Total ash: Not more than 16%.
Acid-insoluble ash: Not more than 7%.
Alcohol-soluble extractive: Not less than 3%.
Water-soluble extractive: Not less than 13%.

2.6 CONCLUSION

The current pharmacopoeial standards for *Phyllanthus* species as prescribed in different pharmacopoeias shows that the assay method of active marker compound is available in the Indian and British Pharmacopoeias only. However, the other physicochemical parameters are also prescribed in these pharmacopoeias and show minor differences in their limits.

REFERENCES

Ayurvedic Pharmacopoeia of India, Part I, Vol. 1, pp. 111–112. Department of AYUSH, Ministry of Health and Family Welfare, Government of India. Published by the Controller of Publications, New Delhi.

British Pharmacopoeia, Volume 3, pp. 3692–3693. 2010. M. G. Lee, MHRAA (Medicines and Healthcare Products Regulatory Agency), London.

Chinese Pharmacopoeia, Volume 1, p. 105. 2005. People's Medical Publishing House, Beijing.

Indian Pharmacopoeia, Vol. 3, pp. 2471–2472, 2488. 2010. Ghaziabad: Indian Pharmacopoeia Commission, Ministry of Health and Family Welfare, Government of India, Ghaziabad.

3 Cultivation, Economics, and Marketing of *Phyllanthus* Species

B. R. Rajeswara Rao

CONTENTS

3.1 Introduction 48
3.2 *Phyllanthus emblica* L. Syn. *Emblica officinalis* Gaertner............................... 49
    3.2.1 Origin and Distribution 49
    3.2.2 Botanical Classification 50
        3.2.2.1 Cronquist System 50
        3.2.2.2 APGII System (APG: Angiosperm phylogeny group).......... 50
    3.2.3 Habit... 50
    3.2.4 Genetic Diversity and Conservation 51
    3.2.5 Cultivars... 51
    3.2.6 Soil and Climate 52
    3.2.7 Propagation 52
    3.2.8 Planting.. 53
    3.2.9 Pruning 53
    3.2.10 Irrigation 54
    3.2.11 Cropping System and Weeding 54
    3.2.12 Manures and Fertilizers.. 54
    3.2.13 Pests and Diseases 54
    3.2.14 Fruit Development, Harvesting, and Yield 55
    3.2.15 Economics and Marketing.. 56
    3.2.16 Grading, Packing, and Storage 57
    3.2.17 Products from Indian Gooseberry 57
    3.2.18 Chemical Composition 57
    3.2.19 Uses... 58
    3.2.20 Nutritive Value of the Fruits 58
        3.2.20.1 China.. 58
    3.2.21 Limits for Quality Parameters 59
3.3 *Phyllanthus amarus* Schumach. and Thonn. 59
    3.3.1 Species, Origin, and Distribution 59
    3.3.2 Botanical Classification 60
        3.3.2.1 Cronquist System 60
        3.3.2.2 APGII System 60

3.3.3　Habit 61
　　　3.3.4　Genetic Diversity and Conservation 61
　　　3.3.5　Cultivars 62
　　　3.3.6　Soil and Climate 62
　　　3.3.7　Propagation 62
　　　3.3.8　Transplanting 62
　　　3.3.9　Irrigation 63
　　　3.3.10　Weed Control 63
　　　3.3.11　Fertilizers and Manures 63
　　　3.3.12　Pests and Diseases 63
　　　3.3.13　Harvesting and Yield 63
　　　3.3.14　Economics and Marketing 64
　　　3.3.15　Chemical Composition 64
　　　3.3.16　Uses 64
　　　3.3.17　Safety Issues and Adulteration 64
　3.4　Cultivation of Other *Phyllanthus* Species 65
　　　3.4.1　*Phyllanthus urinaria* L. and Related Species 65
　　　3.4.2　*Phyllanthus acidus* (L.) Skeel (Syn. *Cicca acida* (L.) Merr.,
　　　　　　　Averrhoa acida L.) 65
　　　3.4.3　*Phyllanthus indofischeri* Bennet (Syn. *Emblica fischeri* Gamble) 66
　　　3.4.4　*Phyllanthus reticulatus* Poir. (Syn. *Kirganelia reticulata* (Poir.)
　　　　　　　Baill.) 66
　　　3.4.5　*Phyllanthus piscatorum* Kunth 66
　　　3.4.6　*Phyllanthus sellowianus* Mull. Arg., *Phyllanthus stipulatus*
　　　　　　　(Raf.) Webster 66
　3.5　Conclusions 67
Acknowledgment 67
References 67

3.1　INTRODUCTION

Medicinal plants (MPs) and their products constitute a treasury of immense value to humankind. Nearly 72,000 MP have been used in diverse human cultures, and many are contemporary local and external trade commodities. The majority of MP supplies are sourced from the wild, and not more than 900 are cultivated (Schippmann et al., 2006). Loss of habitat through conversion of forestland to agriculture, forest fires, and encroachment for habitation is causing a rapid decline of native MP populations. Emphasis is placed on systematic MP cultivation for conservation and sustainable supplies (Schippmann et al., 2006). The genus *Phyllanthus* of Euphorbiaceae (Phyllanthaceae) comprises more than 800 species of aquatic plants, trees, shrubs, climbers, and annual and perennial herbs distributed in tropical and subtropical regions of both hemispheres (Webster, 1994). A number of species provide food, fodder, fruit, fuel, timber, dyes, and pharmaceutical, cosmeceutical, nutraceutical, and industrial products. Several species find extensive use in local medicine systems, and a few are cultivated in countries of origin on a small scale. Cultivation of *Phyllanthus emblica, P. amarus,* and other species is described.

3.2 PHYLLANTHUS EMBLICA L. SYN. EMBLICA OFFICINALIS GAERTNER

Phyllanthus emblica (amlaki, amla, aonla, Indian gooseberry, emblic myrobalan, and emblica) has been known in India for 3,500 years (Murthy and Joshi, 2007), with mention in ancient texts of religion and Ayurveda (medicine system) and is worshipped with the belief that it nurtures humankind. In China, it is called Yuganzi. It is popular in the East, where a number of countries utilize diverse plant parts in local medicine systems. It is grown in the Indian subcontinent, China, Taiwan, Indonesia, Malaysia, Thailand, Sri Lanka, Honduras, Costa Rica, and Reunion Island (Murthy and Joshi, 2007) in orchards, agricultural field bunds, home gardens, avenues, wastelands, and forests. Wild fruits are important nontimber forest products of trade for tribal/ethnic groups and food for wild animals. In India, *P. emblica* is cultivated in Uttar Pradesh, Gujarat, Rajasthan, Tamilnadu, Andhra Pradesh, Maharashtra, and Madhya Pradesh in over 50,000 hectares (ha) with more than 200,000 tonnes (t) fruit production, which includes wild collections. Figure 3.1 (also see color insert) shows the morphology of common *Phyllanthus* species.

3.2.1 ORIGIN AND DISTRIBUTION

Emblica is native to tropical southeastern Asia, specifically central and southern India (Firminger, 1947), and is found growing in dry, hot, deciduous, and moist forests up to 2,000 m altitude from Myanmar to Afghanistan, Sri Lanka to China, and in Cuba, Hawaii, Iran, Iraq, Malaysia, West Indies, Indonesia, Puerto Rico, Singapore, Thailand, and Trinidad and Tobago.

FIGURE 3.1 **See color insert.** Morphology of some of the common species of *Phyllanthus*. (a) *Phyllanthus emblica* tree (insert: a fruiting branch); (b) Emblica orchard; (c) *P. maderaspatensis*; (d) *P. reticulatus* (insert: a twig with fruit).

3.2.2 Botanical Classification

3.2.2.1 Cronquist System
Domain: Eukaryota
Kingdom: Plantae
Subkingdom: Viridaeplantae
Phylum: Tracheophyta
Subphylum: Euphyllophytina
Superdivision: Spermatophyta
Division: Magnoliophyta
Class: Magnoliopsida
Subclass: Rosidae
Order: Euphorbiales
Family: Euphorbiaceae
Subfamily: Phyllanthoideae
Tribe: Phyllantheae
Subtribe: Flueggeinae
Genus: *Phyllanthus*
Species: *emblica* L.
Synonyms: *Emblica officinalis* Gaertner, *Cicca emblica* Kurz., *Mirobalanus emblica* Burm., *Phyllanthus mairei* Lev.

3.2.2.2 APGII System (APG: Angiosperm phylogeny group)
Domain: Eukaryota
Regnum: Plantae
Clade: Angiospermae
Clade: Eudicots
Clade: Core eudicots
Clade: Rosids
Clade: Eurosids I
Order: Malpighiales
Family: Phyllanthaceae
Subfamily: Phyllanthoideae
Tribe: Phyllantheae
Subtribe: Flueggeinae
Genus: *Phyllanthus*
Species: *emblica* L.

3.2.3 Habit

Emblic is a small-to-medium (8–18 m tall) size tree (Hong Kong and Borneo: 23–32 m tall) having a crooked trunk with smooth, grayish-brown, thin, exfoliating bark. The tree produces two types of shoots: The short, determinate type, which are shed annually, bear flowers; the long, indeterminate shoots add annual growth. Branchlets are round, glabrous or finely pubescent, deciduous. Light green, lemon-like smelling, oblong, feathery, subsessile leaves are simple, alternate, and distichous and are closely arranged; the apex is obtuse to acute, base rounded, asymmetrical, glaucous below; the margins are entire and thickened. Stipules are oval. Emblic flowers during February to May. Warm temperature is conducive for floral bud initiation. Small, apetalous, unisexual, greenish-yellow flowers appear in clusters of 6–10 in leaf axils as cymules. Basal flowers are pedicellate male flowers with six calyx lobes and three connate stamens; apical flowers are sessile, yellow, with the female flowers having six perianth lobes and an ovary crowned by three styles deeply bifid at the apex. Male flowers open in the morning and/or the afternoon, dehiscing anthers soon after. Female flowers open in stages, requiring 72 h for completion. The stigma becomes receptive on the third day of anthesis. Flowering lasts for 3 weeks. Flowers are pollinated by wind and honey bees. Self-pollination results in low fruit set. Open and cross pollinations produce 45–82% fruit set with 10–50% retention at maturity. Flower and fruitlet shedding is common due to lack of pollination/fertilization, adverse weather, and physiological reasons. After fruit set in March–June, fruitlets pass through auxin-induced dormancy for 3½ months before resuming growth during the monsoon; they ripen during October–February (7–8 months). A dry spell during monsoon induces fruit shedding and delays fruit growth and maturity. Smooth and hard fruit, a capsule with six striations extending from base to the apex, is globose, depressed at the poles, 1–5 cm in diameter with fleshy, acidulous pulp; shining yellowish-green/greenish-yellow when ripe, the fruit encases a hard, hexagonal, stony endocarp. Fruit skin is thin and translucent and is firmly attached to the crisp, juicy flesh. The fruit weighs 15–50 g. The stone, tightly set in the center of the flesh, is six ribbed with fleshy pericarp, enclosing two hard, trigonous seeds, each in three crustaceous cocci. The seed weighs 0.4–2.0 g. Dry fruits have a shriveled and wrinkled surface with whitish specks of mucic acid crystals. The fruit tastes sour and astringent with sweet, bitter, and pungent secondary tastes. Somatic chromosome number $2n = 28, 98$, or 104 (Ammal and Bhagawan, 1957; Jansen, 2005).

3.2.4 GENETIC DIVERSITY AND CONSERVATION

Western Ghats, Eastern Ghats, Central India, and Yunnan Province in China (Li and Zhao, 2007) are rich sources of genetic diversity. Establishment of forest gene banks for *in situ* conservation of genetic diversity was suggested, but the unsustainable harvest methods of local communities are quickly eroding the genetic base. Alternate conservation strategies need to be urgently devised.

3.2.5 CULTIVARS

Several cultivars with differing maturity periods are grown in India. Banarsi, Krishna, and Balwant are early maturing (October–November); Francis, Kanchan, and Amrit are medium-early maturing (November–December); Neelam, BSR-1, and Chakaiya are late maturing (December–January). BGK-1, Faizabad, Gujarat *amla*-1, Anand-1,2,3, NA-8,9,18, Mehrun, Dongri, Banarsired, Agrabold, and Modibagh are also grown with local types. In Pakistan the Desi, Shisa, and Banarsi varieties and in China, Lanfeng, Fen'gan, Liuyuebai, Bian'gan, Quibai, and Shan'gan varieties (Wang et al., 2006) are commercially cultivated. Balwant and Kanchan are suitable for rain-fed, dry areas; Banarsi, Kanchan, BSR-1, Chakaiya, Amrit, Neelam, and Balwant are suitable for saline/sodic soils.

3.2.6 Soil and Climate

The deep root system and reduced foliage make emblic a hardy tree suitable for subtropical, tropical, dry, arid, semiarid, and humid regions. Annual rainfall of 600–800 mm produces a good yield. Young trees are sensitive to dry, hot winds and low temperatures. Mature trees tolerate freezing and high temperatures (46°C). Emblic is susceptible to frost but survives forest fires through regeneration and is suitable for growing in sandy loam to clay, poor-to-marginal soils with pH 6.5–9.5, 30 ESP: (exchangeable sodium percentage), ds/m: deciSiemens meter, and up to 10 dS/m EC (electrical conductivity). well-drained, deep, fertile sandy loams are ideal for cultivation. Waterlogged, heavy, and sandy soils are unsuitable (Tiwari et al., 2007).

3.2.7 Propagation

Seed and vegetative propagation is feasible. Seed propagation results in a heterogeneous population bearing small-size fruits. Natural regeneration in forests continues through seeds. Orchards are raised with grafted or budded plants. Ripe fruits collected in December–February are sun dried until they release seeds, which are floated in water to discard floaters and are utilized for raising rootstock. Germination (35–50%) starts within 20 days and is complete in 40 days after sowing. Seeds treated with 100–500 ppm GA_3, (gibberellic acid) 100 ppm kinetin, 1% thiourea, 0.5–1% potassium nitrate, VAM (vesicular arbiscular micorrhizal) fungi, *Azospirillum brasilens*, *Azotobacter chroococcum*, and *Trichoderma viride* exhibit enhanced germination (70–93%), seedling growth, and vigor (Aseri and Rao, 2004; Kumari et al., 2007; Tiwari et al., 2007). Seeds are sown in March–April in raised beds or polyethylene tubes/bags. Seedlings that are 6–12 months old are grafted or budded. Pesticide-treated seeds (8–10% moisture) packed in a cloth bag can be stored for 21 months under ambient conditions and for 24 months in 700-gauge polyethylene bags at 5°C (Srimathi and Sujatha, 2007).

Vegetative propagation through budding (patch, shield, ring, eye, T) (Singh et al., 2005), grafting (approach, cleft, wedge, veneer, softwood), and cuttings (softwood or hardwood) is practiced with 60–90% success. Inarching is impractical due to erect habit and availability of a limited number of shoots. Pencil thick, 8- to 10-cm long scion sticks with four to six activated buds cut from semihardwood or softwood branches are good for grafting with 95% success (Panchbhai et al., 2006). Scion sticks can be preserved for 5–7 days. Grafting is performed 2–15 cm above the collar region depending on rootstock thickness and age. Pencil-thick shoots of the previous season with four or five plump buds swollen up to 2 mm are suitable for budding (Panchbhai et al., 2006). Storing buds even for a day results in poor sprouting (33%). Spring (February–March) and rainy (July–September) seasons sustain successful grafting and budding. Grafting or budding is performed in the nursery, a room (bench grafting), or the field (*in situ*). Grafts and buds start sprouting 10–30 days after grafting and budding, reaching measurable growth in 2 months. Seedling growth can be hastened by applying biofertilizers. Old or unproductive orchards are rejuvenated with top working through patch budding or wedge grafting (Tiwari et al., 2007). *In situ* grafting on field-grown rootstock is ideal for dry and rain-fed areas. Softwood and semihardwood stem cuttings with four to six leaves are amenable for rooting with hormones (5,000 ppm IBA: (Indole butyric acid) + 5,000 ppm NAA (naphthylacetic acid)) but are rarely utilized for raising orchards. Micropropagation-mediated plant multiplication techniques have been standardized in India and China but are yet to be commercialized.

3.2.8 Planting

Rainy (July–September) or spring (February–March) season planting is adopted by digging 50- to 100-cm^3 pits that are kept open for a few weeks. Pits are filled with dug soil mixed with 20 kg FYM (farmyard manure), 1 kg neem cake, 500 g bone meal, 1 kg single superphosphate, and 100 g 10% BHC (benzene hexachloride) or carbaryl (+5–8 kg gypsum in saline/sodic soils) and watered (Tiwari et al., 2007). Budded or grafted plants are transplanted, maintaining a population density of 156–494 plants/ha, adjusting inter- and intrarow distance from 4 to 8 m (4.5 × 4.5, 6 × 6, 7 × 7, 8 × 8 m or 8 ×4 or 5 m, etc.) in square or rectangular (hedge) planting methods. Self-incompatibility and lack of pollination results in 70% flower shedding in single-cultivar orchards. Two or more cultivars are planted in suitable ratios (Tiwari et al., 2007), often including plants raised through seeds to encourage open and cross pollinations for high fruit set.

3.2.9 Pruning

Branches are brittle; therefore, to avoid breakage from heavy fruit loads, to develop a strong framework, and to facilitate easy harvesting trees are pruned to medium head size. Young trees are pruned to grow straight up to 75–100 cm without branches and are trained to a modified central leader system. Two to four opposite branches with wide crotch angles are retained during early years, and four to six branches are retained in subsequent years. During March to April, crowded branches are clipped. Pruning fruit-bearing trees induces growth and early flowering and increases fruit set, retention, fruit size, volume, and yield (Singh and Singh, 2008). Dead, infested, broken, overlapping, twisted branches and rootstock suckers are removed periodically.

3.2.10 Irrigation

Indian gooseberry is drought hardy. Mulching with pruned branches, organic material (paddy husk, maize straw, grasses), or 200-gauge black polyethylene sheet is very effective in establishing orchards in sodic soils, ravines, and rain-fed, semi-arid, and dry areas and for conserving moisture and increasing yield (Shukla et al.,

2000; Tiwari et al., 2007). Mulching over time helps in building soil organic matter, infiltration rate, and biological activity. Pitcher watering is followed in water-scarce areas. Young plants are watered immediately after planting and at 10- to 15-day intervals in summer and the nonrainy season. Established fruit-bearing plantations are irrigated once every 15–20 days during summer, fruit setting, and development stages to reduce fruit drop and increase fruit yield. Irrigation is withheld during flowering. Plants need 10 (1–2 years), 15 (3–5 years), and 20–30 (6 years and older) liters water/day. Flooding, basin, and ring methods are in practice. Drip irrigation conserves water (40–50%), regulates weed growth, and allows simultaneous fertilizer application. Alternate-day drip irrigation is superior to basin irrigation (Shukla et al., 2000) and is practiced in recently established orchards.

3.2.11 Cropping System and Weeding

Orchards are kept weed free through mechanical and manual methods for good plant growth and to control pests harboring on weeds. Wide-spaced emblic is suitable for a two- or three-tier cropping system. Flower, vegetable, spice, agricultural, medicinal, and aromatic crops (Awasthi et al., 2009; Singh, 2006; Singh et al., 2008c; Tiwari et al., 2007) are intercropped for additional revenue. A hortipastoral system with *Dichanthium annulatum*, *Chrysopogon fulvus*, *Cenchrus ciliaris* and *Stylosanthes hamata* provides fodder, controls soil and water runoff, and conserves moisture and nutrients. Emblic is cultivated as a companion crop in coconut, guava, ber, tea, and other orchards.

3.2.12 Manures and Fertilizers

To sustain high fruit yields during productive life (50–75 years), manure and fertilizer application is essential. A 1-year-old tree is supplied with 10 kg FYM, 100 g nitrogen, 50 g phosphorus, and 75 g potassium, and the rate is increased with age up to 10 years in required proportions. Ten- to 12-year-old trees need 0.5–1.5 kg nitrogen, 0.25–1.0 kg phosphorus, 0.375–0.8 kg potassium, 50.0 kg FYM, and 5.0 kg pressmud per year. Half of the fertilizer rate is applied before flowering and the rest in the rainy season or after fruit set (Singh et al., 2008a; Tiwari et al., 2007). Boron deficiency-induced fruit necrosis is prevented by three sprays of 0.6% borax. Nitrogen, potasisum, zinc, copper, boron (0.2–0.6%), 10 ppm NAA, calcium nitrate spraying improves plant growth, fruit retention, yield, and physicochemical properties (Singh et al., 2008b). Vermicompost, VAM fungi (*Glomus fasciculatum*), *Azospirillum brasilens*, and *Azotobacter chroococcum* application is beneficial. Alkali soils need the addition of gypsum.

3.2.13 Pests and Diseases

Sucking, boring insect pests, nematodes, and fungal and bacterial diseases (Table 3.1) adversely affect fruit yield and quality (Tiwari et al., 2007). Insect pests are controlled by contact, systemic insecticides 0.05% endosulfan, 0.05% monocrotophos, phorate10G 10 g/plant, 0.05% quinalphos, and 0.04% dimethoate.

TABLE 3.1
Pests and Diseases of Indian Gooseberry

Pest/Disease	Causal Organism
Bark-eating caterpillar	*Indarbela tetraonis/quadrinotata*
Shoot gall maker	*Betousa (Hypolamprus) stylophora*
Leaf-rolling/hairy caterpillar	*Garcillaria acidula, Selepa celtis, Tonica ziziphy*
Aphid	*Schoutedenia (Cerciaphis) emblica/bougainvilliae (ralumensis)*
Mealybug	*Nipaecoccus vastator/viridis, Drosicha mangiferae*
Plant bug	*Scutellaria nobilis*
Anar butterfly	*Virachola (Deudorix) isocrates*
Termite	*Odontotermes* species
Fruit midge	*Clinodiplosis* species
Stone borer	*Cuaulio* species
Aonla rust	*Ravenelia emblicae* var. *pinnular/fructiocolar*
Fruit rot/blue mold	*Penicillium oxalicum/islandicum/citinum, Aspergillus niger/terreus, Pestalotia cruenta, Nigospora sphaerica*
Leaf/fruit rust	*Phakospora phyllanthii*
Seedling root rot	*Rhizoctonia solani*
Anthracnose	*Glomerella cingulata*
Dieback of branches	*Lasiodiplodia (Botryodiplodia) theobromae*
Dry fruit rot	*Phoma emblicae/putaminum, Cladosporium tenuissimum, Alternaria alternata*
Soft rot	*Phomopsis phyllanthii, Syncephalastrum racemosum*
Brown fruit rot	*Colletotrichum* sp., *Aspergillus luchuensis, Fusarium acuminatum, Fusarium moniliforme* var. *subglutinans, Fusarium equiseti, Penicillium funiculosum*

Fungal diseases are controlled by 1% bordeax mixture, 0.3% mancozeb, 0.1% carbendazim, and 0.3% copper oxychloride. Natural enemies *Cheilomenes sexmaculata/septempunctata, Cotesia ruficrus, Charops obtusus,* and *Apanteles* species control aphids and sucking pests. Four spider species (*Neoscona, Peucetia, Argiope, Oxyopes*) feed on mealybugs. Mantids prey on different pests. Emblic trees suffer damage through plant parasite mistletoe (*Taxillus tomentosus,* family: Loranthaceae); in China, this occurs through brown spot (*Phyllosticta emblicae*), false anthracnose (*Kabatiella emblicae*), powdery mildew (*Oidium* sp.), and leaf spot (*Pestalotiopsis heterocornis*).

3.2.14 Fruit Development, Harvesting, and Yield

Emblica is a long-day tree, producing flowers in March–May in northern India and June–July and February–March in southern India. High temperatures and hot, dry winds adversely affect fruit set and add to flower and fruitlet shedding. Adequate humidity is essential for fruit growth initiation. Initial growth is rapid, and fruit development follows a double sigmoid growth pattern (Singh et al., 2008d). Grafted, budded trees start fruiting 3–5 years after planting and produce commercial yield in 8–10 years. Seed-raised trees have a 6- to 12-year gestation period. Fruit maturity is judged by changes in physicochemical properties. Maturity indices provide primary armor against deterioration of physical appeal, quality loss, and microbial decay and determine harvesting stage, shelf life, marketability, and price (Tiwari et al., 2007). During development, the fruit size and weight increase, and the color changes from green to yellow; the seed color changes from creamy white to brown or black; specific gravity, total soluble solids (TSS), total and reducing sugars, and ascorbic acid contents increase until maturity; acidity decreases as the fruits reach maturity (Singh et al., 2008d). Fruit length and diameter increase for up to 90 days; volume increases for up to 75 days, and color change appears 120 days after fruit set. Forest fruits are hand picked by climbing trees or lopping, pollarding, and coppicing the branches. In extreme cases, trees are cut. Market demand-driven competition triggers harvesting immature fruits. For tanning, unripe fruits are harvested, branches are coppiced, and the bark is quickly sun dried. Orchards are manually harvested in the morning using bamboo, a ladder, or a harvester, exercising care not to injure the fruits. Shaking trees and spreading a plastic or canvas sheet are also in practice. Delayed harvesting affects yield in the following year. Fruit yield is cultivar dependent. Potentially, a tree can yield 300 kg of fruit. Calculated potential yields with 156 (8 × 8 m) and 494 (4.5 × 4.5 m) plants are 46.8 and 148.2 t/ha, respectively. Average fruit yield is 25–50 kg/tree in rainfed, arid, semiarid areas and 50–70 kg/tree or 15–25 t/ha with good management. At provincial and national levels, average yield is 3–5 t/ha. In China, 8 t/ha yield is reported.

3.2.15 Economics and Marketing

Initial orchard establishment cost, excluding land cost, ranges from Rs 17,000 to 54,500/ha ($350–1,250) and recurring cost for the first 3 years from Rs 32,000 to 35,000/ha ($700–800). Production cost from the 4th year ranges from Rs 41,000 to 49,000/ha ($900–1,100). Returns start from the 4th year, reaching a maximum in the 15th year and remaining profitable thereafter. The estimated gross and net returns, respectively, are as follows: Rs 99,000–118,000 ($2,200–2,700) and Rs 58,000–68,700/ha ($1,300–1550). The net return is highest in the 15th year (Rs 152,600/ha or $3,400) (Gondalia and Patel, 2007). Net returns ranging from Rs 20,000 to 49,250/ha ($450–1,100) for pure emblica and more than Rs 90,000/ha ($2,000) for intercropped emblica (Singh, 2006) have been reported. It is more profitable to maintain a nursery with a maintenance cost of Rs 293,600 ($6,500) and a net profit of Rs 246,400 ($5,500) (Bhatia et al., 2007). The fruit price is low (Rs 4–5/kg) in the north and high (Rs 15–30) in southern India, where Kerala Province, with number of Ayurvedic pharmacies, is the major market. In the north, herbal pharmacies consume large quantities. A small quantity (30 t) is exported from southern India to Singapore and Malaysia. Lack of interest in the West restricts the market to producing countries, although products and extracts are offered for sale by several companies on the Internet, with no trade data provided.

3.2.16 Grading, Packing, and Storage

Fruits are graded into three sizes; large fruits (4-cm diameter) are utilized for manufacturing preserve (*murabba*), candy, pickles; medium fruits are used for making other products, and small fruits are used for pharmaceutical products. A considerable proportion of fruits is lost through spoilage due to imperfect packing and storage. Proper fruit packing and storage should arrest metabolic deterioration and extend the shelf life for marketing and processing. Harvested fruits of different cultivars can be stored at room temperature for 4–7 days in gunnysacks and baskets lined with newspapers. Shelf life can be extended to 9 days by storing fruits in perforated polyethylene bags, 12–18 days in a low-energy cool chamber (Kumar and Nath, 1993), 30 days in 200-gauge high-density polyethylene bags kept in cardboard boxes, 60 days at 5–7°C, 90 days in 15% brine solution at room temperature (Kumar et al., 1992), and 24 months in polyethylene bags after gamma-ray irradiation. For long-distance transport, polyethylene-lined wooden boxes or corrugated fiberboard boxes are best suited (Pathak et al., 1989; Singh et al., 2009). Spraying chemicals/fungicides during fruit development; dipping fruits in boiling water (blanching), diphenyl, potassium metabisulfite, calcium nitrate, growth hormones (GA_3, kinetin), and fungicides extends shelf life. Dehydration or drying fruits or fruit segments, slices, shreds, flakes, powder through diverse methods (sun, shade, solar, oven, osmo-air, osmo-vacuum, vacuum, cabinet air, fluidized bed, alternating current high electric field exposure) and treating them with chemicals, growth regulators, and wax emulsion coating were tested for retaining ascorbic acid content and extending shelf life and provided with different degrees of success. Reduction in physiological mass, ascorbic acid content, and acidity and increase in TSS, total and reducing sugars, total phenols, and carotenoids occur during storage; decrease in mass, chemical constituents; browning and spoilage occur due to disease infection.

3.2.17 Products from Indian Gooseberry

Emblica is used in 35 ways (Murthy and Joshi, 2007) as a vegetable, for quenching thirst, seasoning, and flavoring; for making food, pharmaceuticals, cosmeceuticals, nutraceuticals, and industrial products. Different cultivars lend themselves for preparing different products: Amrit and Neelam for candy, sweets, jam, chutney, *chyawanprash*, pickles, squash; Kanchan and Krishna for candy, jam, pickles; Banarsi for drying and preserves; Chakaiya for pickles, syrup, chutney; Francis for preserves; and Balwant for dehydration and pickles. Herbal tea, toffee, dehydrated fruit segments/slices, jelly, juice, beverages, powder, sauce, *supari*, *gulkand*, pomace, hair oil/dye, ink, face cream/pack, shampoo, tooth powder, and fabric dye are also made from fruits.

3.2.18 Chemical Composition

Tannins (18–35% unripe fruit, 8–21% stem bark, 12–24% twig bark, 22–28% leaves), and ascorbic acid are major fruit constituents. The sensitivity of water-soluble, volatile ascorbic acid to heat, oxygen, and light makes it liable to be lost first by oxidation to dehydroascorbic acid and then to 2,3-keto gulonic acid during cleaning, drying, storing, cooking, and processing. Under ambient conditions and at 4°C, fruits lose 45–60% and 32–45% ascorbic acid, respectively in 4 days. Complete oxidation is prevented by tannins in the fruit. The presence of ascorbic acid in the fruit was questioned (Ghosal et al., 1996), but the controversy was put to rest by proving its presence by precise analytical techniques (Raghu et al., 2007). Alkaloids, flavonoids, saponins, glycosides, terpenoids, phenolics, and polyphenols have also been isolated from plant parts. Seeds contain essential oil (β-caryophyllene, β-bourbonene, pentadecanone, heptadecanol, eugenol, nerol, thymol, borneol) and brownish-yellow fixed oil (16–22% with 44% linoleic, 28% oleic, 6–9% linolenic, 3% palmitic, 2% steric, and 1% myristic acids).

3.2.19 Uses

Fruit is acidic and cooling, refrigerant, carminative, laxative, alexiteric, antipyretic, and diuretic and an antioxidant and is used in treating diabetes, cough, chronic dysentery, diarrhea, dyspepsia, peptic ulcer, hemorrhage, anemia, jaundice; diseases of the chest, head, heart, reproductive organs; and metabolic and aging disorders. The seed is employed for treating asthma, bronchitis, biliousness, nausea, and diabetes, and burnt seed oil is used for skin afflictions. The bark is utilized for tanning (reddish-brown, less-flexible leather), dyeing fabric, and curing gonorrhea, diarrhea, jaundice, and myalgia; fresh bark juice is used for gonorrhea. Root bark is astringent and is useful for ulcerative stomatitis and mouth inflammation. Flowers are refrigerant and aperient. Leaves (4.5% ash, 48.0% carbon, 1.5–2.1% nitrogen, 2.0% calcium, 15–36 ppm zinc, 4–8 ppm copper, 45–72 ppm manganese, 318–972 ppm iron) are used as fodder for cattle, for green manure, mulch, tanning, dyeing fabric, and alleviating dysentery, dyspepsia, conjunctivitis, and inflammation. Wood (red, hard, and flexible, undergoes warping and splitting, 720–930 kg/m^3 at 15% moisture) is used for minor construction and for making furniture, implements, water-conducting pipes, water clarification; it serves as fuel and as a source of charcoal.

3.2.20 Nutritive Value of the Fruits

The fruit is the richest source of vitamin C next to Barbados cherry, containing 20 times that of two oranges and 160 times that of apple. The fruit contains higher concentrations of minerals (Table 3.2) and amino acids (5% alanine, 5% lysine, 14% proline, and 8% aspartic and 29% glutamic acids) than apple (Murthy and Joshi, 2007).

3.2.20.1 China

In China, nutritive values of the fruits are as follows: 79.8–87.0% moisture, 6.6–14.1% carbohydrates, 0.2–1.1% fats, 0.7–2.2% protein, 6.7% reducing sugar, 0.3–0.6% sugar, 1.1–2.9% glucose, 1.8–2.9% fructose, 200–1,561 mg/100 g ascorbic acid, and 13.0 g fruit weight (Chen et al., 2003).

TABLE 3.2
Physicochemical Characteristics of Indian Gooseberry Fruit

Factor	Fruit Size (cm²)	Volume (cm³)	Length (cm)	Breadth (cm)
Value	8–16	30–57	3.0–4.5	3.5–5.0

Factor	Value (%)	Factor	Value (%)	Factor	Value (mg/100 g)
Moisture	77.0–84.0 (6.0–9.0)	Pulp	86.0–95.0	Iron	0.5–1.2
Protein	0.5–0.75 (2.5–3.5)	Pectin	2.4–3.1	Nicotinic acid	0.2
Fat	0.1–0.5 (1.2–2.0)	Seed	4.0–6.0	Vitamin C	200–1,814
Carbohydrate	14.1–21.9 (53.0)	TSS	8.0–23.0	Carotene	0.01
Fiber	1.1–4.3 (17.0)	Acidity	1.4–3.2	Thiamine	0.03
Minerals	0.5–2.4 (4.0–6.0)	Phenols	2.0–5.0	Riboflavin	0.05
Calcium	0.012–0.05	Reducing sugars	3.4–11.0	Tryptophan	3.0
Phosphorus	0.03–0.2	Nonreducing sugars	1.1–3.2	Methionine	2.0
Ash	0.5–0.6	Total sugars	3.5–16.7	Lysine	17.0
Miscellaneous	0.5–0.7 (2.5–3.5)	Specific gravity	1.0–1.2	—	—

Source: Several published reports.
Note: Values in parentheses are for dried fruit.

3.2.21 LIMITS FOR QUALITY PARAMETERS

For industrial use and export, the limits for quality are as follows: not more than 3% foreign matter; not more than 5% total ash; not more than 2% acid-insoluble ash; not less than 40% alcohol-soluble extractive; and not less than 40% water-soluble extractive.

3.3 PHYLLANTHUS AMARUS SCHUMACH. AND THONN.

Phyllanthus amarus (carry me seed, *Bhumyamalaki*) and related species have been known in India for over 2,000 years for their medicinal value. *Phyllanthus amarus* is derived from the Greek words *phyllon* or *phulon* (leaf) and *anthos* (flower or flower cluster) and the Latin word *amara* (bitter). The name alludes to flowers produced in bitter leaf axils. The plant shot into prominence after its activity against hepatitis B and related hepadna viruses was scientifically proved. It is cultivated in southern India in less than 100 ha with less than 100 t biomass production.

3.3.1 SPECIES, ORIGIN, AND DISTRIBUTION

Ambiguity prevailed in the past on the true identity of *Phyllanthus amarus*, *P. debilis* Klein ex Wild. (niruri), *P. fraternus* Webster (gulf leaf flower), *P. niruri* L. (stone breaker, chanca piedra), and *P. urinaria* L. (hurricane weed, chamber bitter). Recent studies of macromorphology, micromorphology, and histochemistry and electron microscopic, chemical, and DNA profiles clearly differentiated these species. *Phyllanthus niruri* is indigenous to Amazon and other tropical regions. *Phyllanthus amarus* originated in the Caribbean as a vicarious species of *P. abnormis* and has spread around the tropics (Webster, 1957). *Phyllanthus debilis* is native to the Indian subcontinent and is naturalized in many countries. *Phyllanthus fraternus* is native to western India and Pakistan and is distributed in Asia and the West Indies. *Phyllanthus urinaria* is indigenous to India, China, and the Bahamas. *Phyllanthus amarus* and *P. urinaria* occur throughout India; *P. amarus* is more common in Eastern Ghats and *P. urinaria* in northern India. *Phyllanthus fraternus* is more abundant in northwestern India and *P. debilis* in Western and Eastern Ghats. These species grow as weeds in cultivated fields and abandoned terrains.

3.3.2 BOTANICAL CLASSIFICATION

3.3.2.1 Cronquist System
Domain: Eukaryota
Kingdom: Plantae
Subkingdom: Viridaeplantae
Phylum: Tracheophyta
Subphylum: Euphyllophytina
Superdivision: Spermatophyta
Division: Magnoliophyta
Class: Magnoliopsida
Subclass: Rosidae
Order: Euphorbiales
Family: Euphorbiaceae
Subfamily: Phyllanthoideae
Tribe: Phyllantheae
Subtribe: Phyllanthinae
Genus: *Phyllanthus*
Subgenus: *Phyllanthus*
Section: *Phyllanthus*
Subsection: *Swartziani*
Species: *amarus* Schumach. and Thonn.
Synonyms: *Phyllanthus nanus* Hook.
Phyllanthus niruri auct. non L.

3.3.2.2 APGII System
Domain: Eukaryota
Regnum: Plantae
Clade: Angiospermae
Clade: Eudicots
Clade: Core eudicots

Clade: Rosids
Clade: Eurosids I
Order: Malpighiales
Family: Phyllanthaceae
Subfamily: Phyllanthoideae
Tribe: Phyllantheae
Subtribe: Phyllanthinae
Genus: *Phyllanthus*
Subgenus: *Phyllanthus*
Section: *Phyllanthus*
Subsection: *Swartziani*
Species: *amarus* Schumach. and Thonn.

3.3.3 HABIT

Phyllanthus amarus is an erect, annual herb, 10–60 cm tall, attaining 1.0 m height at maturity. Mature plants have a slightly woody, simple or branched, smooth, terete, straminous, or brownish stem. Cataphylls are stipulate, broadly deltoid, acuminate, entire, scarious, brownish; the blade is subulate, acuminate. Branchlets are subterete, smooth, or sometimes slightly scabridulous. Leaves are 5–11 mm long, 3–6 mm broad; somewhat thickened; closely arranged; stipulate; elliptic oblong, obovate-oblong, or even obovate; entire, obtuse or rounded and apiculate at apex; obtuse or rounded, slightly inequilateral at base; bright green above; grayish beneath; margins are plane, smooth, or obscurely, minutely roughened. Flowers are monoecious, with the proximal one or two axils with unisexual cymules of one or two male flowers; all succeeding axils have bisexual cymules, each with one male and one female flower. The plant flowers throughout its lifetime. The male flower is pedicellate with five calyx lobes, five disc segments, three stamens, and divergent anther sacs, which dehisce obliquely or horizontally; pollen grains are finely reticulate. The female flower is pedicellate with five calyx lobes, five lobed disc, styles are free, and it is shallowly bifid. Pollination is through wind, small insects, and ants (Sharma et al., 2009). Capsules are oblate, 1.8–2.1 mm in diameter, smooth, and stramineous. There are six seeds, which are sharply trigonous; light brown, tan, or yellow; 0.9–1.0 mm long; 0.7–0.8 mm radially and tangentially; and with five or six ribs on the back (Bagchi et al., 1999). Seeds attain maturity 55 days after anthesis (Reddy et al., 2007). The somatic chromosome number $2n = 26$.

3.3.4 GENETIC DIVERSITY AND CONSERVATION

Plants collected from diverse Indian regions exhibited less than 70% variation and polymorphism (Jain et al., 2003; Meenakshisundaram and Maheswaran, 2008). Wide distribution of the species and ability to produce abundant seeds do not warrant conservation of the species at present.

3.3.5 Cultivars

Red, green, and intermediate stem color types were identified in Florida; the intermediate type produced higher yields (Unander et al., 1993). Local, Navyakrit (78 cm tall, 2.0 g/plant dry biomass yield, 0.52% phyllanthin), and CIM-Jeevan (65 cm tall, 11.0 g/plant dry biomass yield, 0.7–0.77% phyllanthin and 0.32–0.37% hypophyllanthin) (Gupta et al., 2003) varieties are cultivated in India. *Agrobacterium tumefaciens*-mediated transgenic plants were developed using shoot tip explants (Banerjee and Chattopadhyay, 2009). Somaclone development through tissue culture was attempted. Calli, calli-mediated shoots, and roots contained higher concentrations of alkaloids, saponins, tannins, and other compounds compared to mother plants (Marimuthu, 2007).

3.3.6 Soil and Climate

Phyllanthus amarus grows well under semitemperate to tropical conditions at up to 800 m altitude. It rarely survives under dry or very low temperatures but tolerates waterlogging. Plant growth is restricted under shade. The plant is well adapted to calcareous, well-drained, light-textured soils with acidic to alkaline pH. Soils contaminated with heavy metals are unsuitable due to toxicity (Rai et al., 2005; Rai and Mehrotra, 2008).

3.3.7 Propagation

The crop is seed propagated. Optimum temperature for germination is 20–35°C. Seed germination is less than 50% and is complete within 10 days after sowing. Germination of freshly collected seeds is slower than older seeds. Freezer-stored (–20°C) seeds retain viability for 30 months (Unander et al., 1995). Density grading by floating in petroleum ether or acetone and discarding the lighter seeds improves germination (92–94%) (Kalavathi et al., 2001). Soaking seeds in 200 ppm GA_3, 150 ppm thiourea, or 4% potassium nitrate enhances germination, root and shoot growth, and vigor (Balakumbahan et al., 2008). Seeds need light to germinate and should not be covered at sowing (Unander et al., 1995). Direct seeding results in poor stands; therefore, seeds are sown in April–May in nursery beds mixed with FYM. Seeds are mixed with dry soil, sand, or cellulose gel for uniform dispersal on the nursery bed. Adequate moisture is maintained until seedlings are ready for planting. For a hectare, 1 kg seeds is sufficient. Being self-seeding, stands can be established in cultivated fields with weed management strategies.

3.3.8 Transplanting

Seedlings that are 30–40 days old and 10–15 cm tall are transplanted with 15–25 × 10–20cm spacing (Bagchi et al., 1999; Chezhiyan et al., 2003) during the rainy season. October planting with 40 × 20 cm spacing for Brazil (Figueira et al., 2006) and December or March planting with 20–40 cm intrarow spacing for Florida (Unander et al., 1993) are suggested.

3.3.9 IRRIGATION

In southern India, the rainy season harvest needs no irrigation; subsequent harvests are flood or sprinkler irrigated once every 10–15 days. In northern India, fortnightly irrigations are necessary. Drip irrigation is advocated in Florida (Unander et al., 1993).

3.3.10 WEED CONTROL

The field is kept weed free through regular manual weeding. Plastic mulch controlled weeds in Florida (Unander et al., 1993).

3.3.11 FERTILIZERS AND MANURES

Fertilizer application increases biomass yield without affecting activity against hepatitis B and related hepadna viruses (Unander et al., 1993). The recommended rates of manure and fertilizer applications for harvesting yields with high biomass are depicted in Table 3.3.

3.3.12 PESTS AND DISEASES

Whiteflies, thrips, aphids, stem blight (*Corynespora cassiicola*), and little leaf (phytoplasma) damage the crop. Spraying 5% neem seed kernel extract reduced insect pest populations by 54.0–58.0% (Rathikannu and Sivasubramanian, 2008). Foliar spray and seedling dipping in talc-based *Bacillus subtilis* formulation controlled stem blight (Mathiyazhagan et al., 2004).

3.3.13 HARVESTING AND YIELD

Plants are harvested when they are green and herbaceous as leaf yield decreases with age due to leaf fall and stems tend to become woody. Since lignans are concentrated

TABLE 3.3
Recommended Rates of Manure and Fertilizer Applications for Carry Me Seed

Manure/Fertilizer Rate	Reference
67 kg N + 134 kg P + 134 kg K/ha	Unander et al., 1993
5 t poultry manure + 2 kg *Azospirillum* + 2 kg phosphobacteria/ha for highest biomass yield	Chezhiyan et al., 2003
15 t FYM for highest herb yield and 12 t FYM + 2.5 kg *Azospirillum* + 2.5 kg phophobacteria/ha for highest seed yield	Annamalal et al., 2004
75 kg N + 37.5 kg P + 2 kg *Azophos*/ha (fresh biomass 7.2 t/ha, dry biomass 3.3 t/ha, phyllanthin 0.68–0.70%, and hypophyllanthin 0.21–0.23%)	Balakumbahan et al., 2005
10 t/ha each of FYM + pressmud + fly ash (39.7 g herbage yield/plant)	Arumugum and Rajeswari, 2006

more in the leaves, production of higher leaf mass is desired. In India, plants are harvested in September and subsequently at 70- to 90-day intervals. Three harvests are feasible. In Florida, 6- to 7-month-old plants (planted in December and harvested in July–August) yield 40.5 kg/ha dry biomass, and two crops planted during winter and summer are possible (Unander et al., 1993). In Brazil, an 80-day-old crop yielded 16.9–20.7 g/plant (2,110–2,580 kg/ha) fresh biomass having 0.52–1.0% phyllanthin and 0.13–0.24% hypophyllanthin (Figueira et al., 2006).

3.3.14 Economics and Marketing

In India, cost of cultivation is Rs 10,000 ($200–250)/ha. Gross and net returns, respectively, with 1,500 kg/ha dry biomass yield and Rs 20/kg market price are Rs 30,000 ($600–650)/ha and Rs 20,000 ($400–450)/ha. The entire production, including wild collections, is locally utilized for herbal drug manufacturing. Yield is low in the United States and hence is not economical for large-scale cultivation. The ubiquitous nature of carry me seed and related species makes them commodities of local trade, and export is limited, although products and extracts are offered for sale by several companies on the Internet, with no trade data available.

3.3.15 Chemical Composition

Alkaloids, benzenoids, flavonoids, terpenoids, lignans, tannins, coumarins, sterols, and lipids have been isolated from different plant parts. Important constituents are lignans: bitter phyllanthin (0.5% herb, 1.56% leaf, 0.01% root, and 0.007% stem) and nonbitter hypophyllanthin.

3.3.16 Uses

Roots, shoots, and the whole plant are used in folk medicine. The herb is bitter and is reported to possess hepatoprotective, antiviral, astringent, deobstruent, stomachic, antidiabetic, diuretic, febrifugal, antipyretic, anti-inflammatory, and antiseptic properties. The plant is used in the treatment of jaundice, kidney and gall bladder stones, dyspepsia, diarrhea, dysentery, dropsy, diseases of the genitourinary system, edema, ulcers, ophthalmia, diabetes, and skin affections. Wild and cultivated plants possess identical activities (Unander et al., 1993).

3.3.17 Safety Issues and Adulteration

For industrial use and export, the limits for quality parameters are as follows: not more than 2.0% foreign organic matter; not more than 8.0% total ash; not more than 5.0% acid-insoluble ash; not less than 15.0% water-soluble extractive; not less than 3.0% n-hexane-soluble extractive. *Phyllanthus amarus* is easily confused and mixed with similar-looking species. Adulteration is detected with the help of the bitterness value or SCAR (sequence characterized amplified region) markers (Jain et al., 2008). *Phyllanthus amarus* is 70 times more bitter than *P. fraternus*, *P. simplex* Retz., and *P. urinaria* and 14 times more bitter than *P. maderaspatensis* L.

3.4 CULTIVATION OF OTHER *PHYLLANTHUS* SPECIES

Several *Phyllanthus* species are commercially important for local trade in countries of origin. Most of them are collected by the local population from natural sources or grown on a limited scale in home gardens or as companion crops in agricultural or horticultural crops. A literature survey revealed scattered attempts to discern such aspects as isolation of chemicals from plant parts, biological activities of extracts, micropropagation protocols, isolation of chemicals from callus cultures, seed germination and seedling growth, pest and disease incidence, VAM fungal association, accumulation of heavy metals, and taxonomic and nutrient studies. Cultivation protocols are lacking.

3.4.1 PHYLLANTHUS URINARIA L. AND RELATED SPECIES

Phyllanthus debilis, *P. fraternus*, *P. niruri*, *P. urinaria*, *P. maderaspatensis*, and *P. mimicus* Webster are collected for their therapeutic value similar to *P. amarus*. *Phyllanthus urinaria* is cultivated in China in warm, well-drained, sandy soils; is fertilized with sources high in nitrogen and potassium; and is susceptible to nutrient and moisture stresses; *P. debilis* is more hardy (Unander and Blumberg, 1991). Seed stratification, scarification, and treatment with growth hormones improve germination in *P. niruri*, *P. fraternus*, and *P. urinaria*. *In vitro* regeneration protocols are in place for *P. niruri* and *P. fraternus*. *Phyllanthus niruri* responds to nitrogen application (Becker et al., 2000) and organic manures (12.5 t FYM + 2.5 t vermicopost + 3% panchagavya gave 44.2 g/plant yield) (Ponni and Shakila, 2007). With fine-tuning, *P. amarus* cultivation practices can be adopted for these species.

3.4.2 PHYLLANTHUS ACIDUS (L.) SKEEL (SYN. CICCA ACIDA (L.) MERR., AVERRHOA ACIDA L.)

Commonly known as Otaheite/Tahitian/star/country/West Indian/Malay gooseberry, the tree of Madagascan origin reaches 5–9 m high and is distributed in southeastern Asian countries. Yellow, waxy, crisp, juicy, acrid, single-seeded fruits with six to eight ribs are borne in thick clusters and are edible. The tree is raised in home gardens in India, Thailand, Taiwan, Philippines, Malaysia, Brazil, Indonesia, Sri Lanka, Australia, and Venezuela (Murthy and Joshi, 2007); it is grown with mango in Trinidad and Tobago (Roberts, 2004). The tree prefers moist soil and is propagated by seeds or by budding, greenwood cuttings, and air layering. Trees bear two crops (April–May, August–September) in southern India; elsewhere, the crop is harvested in January. In India, fruit is eaten fresh or pickled. In Grenada, leaf tea finds use as a mouthwash to treat a sore throat; fruits are eaten fresh or as stew. In Vietnam, the fruit is cooked in sugar or salted, and juice is consumed as a beverage. In Puerto Rico, fruit is eaten as a dessert. Phyllanthusol A and B and triterpenoids have been isolated.

3.4.3 PHYLLANTHUS INDOFISCHERI BENNET (SYN. EMBLICA FISCHERI GAMBLE)

The tree of *Phyllanthus indofischeri* Bennet (up to 12 m tall, monoecious) is endemic to peninsular India. Leaves (2.8 × 1.3 cm) and fruits (pale or marble green, drupaceous, globose, capsule, 2.5–4.0 cm across with marked striations along the septa and containing six gray seeds) are larger than emblic. Trees grow in dry, deciduous, scrub forests at low altitudes. Tribal communities collect fruits along with emblic for trading in local markets. Fruits are pickled; they are sun dried, stored, and eaten with minor millets and are used as a substitute for emblic. Two varieties, Champakkad large and Krishna, are released for cultivation in southern India. Indiscriminate fruit collection is leading to a rapid decline in the population, necessitating conservation (Ganesan, 2003).

3.4.4 PHYLLANTHUS RETICULATUS POIR. (SYN. KIRGANELIA RETICULATA (POIR.) BAILL.)

Commonly known as seaside laurel, the potato plant *Phyllanthus reticulatus* Poir. is a monoecious, deciduous, branched shrub or small tree reaching 4–5 m high. It is widespread in the tropics, from Africa to Indochina and southeastern Asia. Flowers have a potato smell. The fruit is a fleshy berry, reddish purple or bluish black when mature, and contains six to many seeds; it is sour and edible. Leaves are used for treating sores, burns, fever, paralysis, bleeding gums, diabetes, asthma, sore throat, snakebites, diarrhea, and mental problems; the fruits are used for infantile diarrhea; stem sap is used for conjunctivitis; stem bark as a treatment for dysentery; root powder, sap, and decoction are used for ear infections, pain, spasms, headache, dysmenorrhea, gonorrhea, and abscesses. Stems are used as roof binders, twigs as chew sticks, and wood as firewood and for local construction. Red and black dyes are made from fruits, leaves, bark, and roots to dye fishing nets, cotton, and cloth. Black ink is made from ripe fruits. The plant contains tannins and triterpenoids and is occasionally grown by Indian tribal communities. Root bark, stem bark, and leaves are collected and traded in Africa (Arbonnier, 2004).

3.4.5 PHYLLANTHUS PISCATORUM KUNTH

Phyllanthus piscatorum Kunth is a shrub grown for its piscidal property by Yanomami Amerindian ethnic women groups in Venezuela. Aerial parts are used as fish poison and to treat wounds and fungal infections, and leaves are used as a substitute for tobacco. The plant holds promise as a remedy for skin infection caused by *Candida albicans* (Gertsch et al., 2004).

3.4.6 PHYLLANTHUS SELLOWIANUS MULL. ARG., PHYLLANTHUS STIPULATUS (RAF.) WEBSTER

Attempts were made to grow *Phyllanthus sellowianus* Mull. Arg. and *Phyllanthus stipulatus* in Brazil. *Phyllanthus stipulatus* responds to 4 kg/m^2 compost application; it is used to reduce the blood uric acid level and to eliminate kidney stones

(Silva et al., 1997). Leaves and stems of *P. sellowianus* possess antidiabetic, diuretic, laxative, antiseptic, and analgesic properties. The plant is propagated by stem cuttings to control bund erosion in water courses (Sutili et al., 2004).

3.5 CONCLUSIONS

In spite of a large number of economically important species in the genus *Phyllanthus*, scientific data on cultivation techniques are not available except for emblic and carry me seed. Due to their widespread nature and relative abundance, *Phyllanthus* species collected from natural sources are traded locally in the producing countries. Gradual loss of habitat due to anthropological reasons and changing weather patterns as a consequence of global warming is rapidly declining natural populations, calling for urgent conservation and cultivation strategies for sustainable supplies. Products and extracts of *Phyllanthus* species are offered for sale on the Internet by agencies of different countries. In 2008, global export of pharmaceutical herbs was valued at US$1,781.54 million, with a 42.1% growth rate between 2004 and 2008 (U.N. comtrade database (http://comtrade.un.org/db/ce/cesearch.aspx). As many herbs are exported without declaring their species identity, the contribution of *Phyllanthus* species to global trade is difficult to ascertain. The increasing global demand for medicinal herbs warrants development of cultivation practices for sustainable harvests and exports.

ACKNOWLEDGMENT

I am beholden to the director, CIMAP, Lucknow, for facilities and encouragement.

REFERENCES

Ammal, E.K.J., and Bhagawan, R.S. 1957. Physiology and vitamin C in *Emblica officinalis*. *Proc. Indian Acad. Sci.* B-46: 312–314.

Annamalai, A., Lakshmi, P.T.V., Lalithakumari, D., and Murugesan, K. 2004. Optimization of biofertilizers on growth, biomass and seed yield of *Phyllanthus amarus* (Bhumyamalaki) in sandy loam soil. *J. Med. Arom. Plant Sci.* 26: 717–720.

Arbonnier, M. 2004. *Trees, shrubs and lianas of West African dry zones*. Paris: CIRAD, Margraf.

Arumugum, S., and Rajeswari, R. 2006. Effect of FYM and industrial wastes on productivity of *Phyllanthus amarus*. *Adv. Plant Sci.* 19: 525–529.

Aseri, G.K., and Rao, A.V. 2004. Effect of bioinoculants on seedlings of Indian gooseberry (*Emblica officinalis* Gaertn.). *Indian J. Microbiol.* 44: 109–112.

Awasthi, O.P., Singh, I.S., and More, T.A. 2009. Performance of intercrops during establishment phase of aonla (*Emblica officinalis*) orchard. *Indian J. Agric. Sci.* 79: 587–591.

Bagchi, G., Chaudhuri, P.K., and Kumar, S. 1999. *Cultivation of Bhumyamalaki Phyllanthus amarus in India*. Farm Bulletin No. 10. Lucknow, India: Central Institute of Medicinal and Aromatic Plants.

Balakumbahan, R., Sadasakthi, A., Kumanan, K., and Rajamani, K. 2008. Role of growth regulators on seedling characters of *Phyllanthus amarus*. *Plant Arch.* 8: 709–710.

Balakumbahan, R., Sadasakthi, A., Kumar, S., and Sarvanan, A. 2005. Effects of inorganic and biofertilizers on biomass and alkaloids yield of keelanelli (*Phyllanthus amarus*). *J. Med. Arom. Plant Sci.* 27: 449–453.

Banerjee, A., and Chattopadhyay, S. 2009. Genetic transformation of a hepatoprotective plant, *Phyllanthus amarus*. *In Vitro Cell Mol. Biol. Plant* 45: 57–64.

Becker, L., Furtini, N.A.E., Pinto, J.E.B.P., et al. 2000. Growth and total alkaloid production of *Phyllanthus niruri* in relation to lime and nitrogen fertilizer application. *Hort. Brasiliera*. 18: 100–104.

Bhatia, S.K., Dudi, S.S., and Kumar, O.P.M. 2007. Economic analysis of raising aonla (*Emblica officinalis* Gaertn.) buddlings in nursery. *Environ. Ecol.* 25 (special issue): 951–952.

Chen, Z.Y., Liu, X.M., Wu, J.J., Li, S.F., and Zhou, Y.X. 2003. The biological characters and the fruit nutrients of yuganzi tree. *South China Fruits* 32: 71–73.

Chezhiyan, N., Saraswathy, S., and Vasumathi, R. 2003. Studies on organic manures, biofertilizers and plant density on growth, yield and alkaloid content of bhumyamalaki (*Phyllanthus amarus* Schum., and Thonn.). *South Indian Hort.* 52: 96–101.

Figueira, G.M., de Magalhaes, P.M., Rehder, V.I.G., Sartoratto, A., and Vaz, A.P.A. 2006. Chemical preliminary evaluation of selected genotype of *Phyllanthus amarus* Schumach. grown in four different counties of Sao Paulo State. *Rev. Bras. Pl. Med. Botucatu.* 8: 43–45.

Firminger, T.A. 1947. *Firminger's manual of gardening for India* (8th ed.). Calcutta, India: Thacker, Spink.

Ganesan, R. 2003. Identification, distribution and conservation of *Phyllanthus indofischeri*, another source of Indian gooseberry. *Curr. Sci.* 84: 1515–1518.

Gertsch, J.N., Roost, K.N.D., and Sticher, O. 2004. *Phyllanthus piscatorum*, ethnopharmacological studies on a women's medicinal plant of the Yanomami Amerindians. *J. Ethnopharm.* 91: 181–188.

Ghosal, S., Tripathi, V.K., and Chauhan, S. 1996. Active constituents of *Emblica officinalis*: part 1—the chemistry and antioxidative effects of two new hydrolysable tannins, emblicanin A and B. *Indian J. Chem. Sec B.* 35: 941–948.

Gondalia, V.K., and Patel, G.N. 2007. An economic evaluation of investment on aonla (*Emblica officinalis* G.) in Gujarat. *Agric. Econ. Res. Rev.* 20: 385–394.

Gupta, A.K., Khanuja, S.P.S., Gupta, M.M., et al. 2003. High herb, phyllanthin and hypophyllanthin yielding variety of Bhumyamalaki (*Phyllanthus amarus*): CIM-Jeevan. *J. Med. Arom. Plant Sci.* 25: 743–745.

Jain, N., Shasany, A.K., Singh, S., Khanuja, S.P.S., and Kumar, S. 2008. SCAR markers for correct identification of *Phyllanthus amarus*, *P. fraternus*, *P. debilis* and *P. urinaria* used in scientific investigations and dry leaf bulk herb trade. *Planta Med.* 74: 296–301.

Jain, N., Shasany, A.K., Sundaresan, V., et al. 2003. Molecular diversity in *Phyllanthus amarus* assessed through RAPD analysis. *Curr. Sci.* 85: 1454–1458.

Jansen, P.C.M. 2005. *Phyllanthus emblica* L. Record from Protabase. *PROTA (Plant Resources of Tropical Africa)*, ed. P.C.M. Jansen and D. Cardon. http://database.prota.org/PROTAhtml/Phyllanthus%20emblica_En.htm (accessed January 7, 2010).

Kalavathi, D., Karivaratharaju, T.V., and Nargis, S. 2001. Fixing up of optimum sieve for seed processing in some important medicinal plants. In *Conservation and utilization of medicinal and aromatic plants*, ed. S. Sahoo, D.B. Ramesh, Y.R. Rao, B.K. Debata, and V.K. Misra, 95–98. Bhubaneswar, India: Regional Research Laboratory.

Kumar, S., and Nath, V. 1993. Storage stability of aonla fruit—a comparative study of zero-energy cool chamber versus room temperature. *J. Food Sci. Technol. (Mysore)* 30: 202–203.

Kumar, S., Nath, V., Singh, I.S., and Nainwal, N.C. 1992. Preliminary studies on storage of aonla fruit in salt solution. *Progressive Hort.* 24: 83–86.

Kumari, R, Sindhu, S.S., Sehrawat, S.K., and Dudi, O.P. 2007. Germination studies in aonla (*Emblica officinalis* Gaertn.). *Haryana J. Hort. Sci.* 36: 9–11.

Li, Q.M., and Zhao, J.L. 2007. Genetic diversity of *Phyllanthus emblica* populations in dry-hot valleys in Yunnan. *Biodiv. Sci.* 15: 84–91.

Marimuthu, J. 2007. Somaclonal variation studies on *Phyllanthus amarus* Schum. & Thonn. *Indian J. Biotechnol.* 5: 240–245, 253.
Mathiyazhagan, S., Kavitha, K., Nakkeeran, S., et al. 2004. PGPR mediated management of stem blight of *Phyllanthus amarus* Schum. and Thonn. caused by *Corynespora cassiicola* (Berk and Curt) Wei. *Arch. Phytopathol. Plant Protect.* 37: 183–199.
Meenakshisundaram, P., and Maheswaran, M. 2008. RAPD and ISSR analyses reveal low levels of genetic diversity in *Phyllanthus amarus. Acta Hort.* 765: 179–187.
Murthy, Z.V.P., and Joshi, D. 2007. Fluidized bed drying of aonla (*Emblica officinalis*). *Drying Technol.* 25: 883–889.
Panchbhai, D.M., Athavale, R.B., Jogdande, N.D., and Dalal, S.R. 2006. Soft wood grafting—aonla propagation made easy. *Agric. Sci. Digest* 26: 71–72.
Pathak, S., Pathak, P.K., and Singh, I.S. 1989. Effect of packing containers on losses of aonla fruits during transportation. *Indian J. Hort.* 46: 468–469.
Ponni, C., and Shakila, A. 2007. Effect of certain organic manures and biostimulants on growth and yield of *Phyllanthus niruri. Asian J. Hort.* 2: 148–150.
Raghu, V., Patel, K., and Srinivasan, K. 2007. Comparison of ascorbic acid content of *Emblica offcinalis* fruits determined by different analytical methods. *J. Food Comp. Anal.* 20: 529–533.
Rai, V., Bisht, S.K., and Mehrotra, S. 2005. Effect of cadmium on growth, ultramorphology of leaf and secondary metabolites of *Phyllanthus amarus* Schum. and Thonn. *Chemosphere* 61: 1644–1650.
Rai, V., and Mehrotra, S. 2008. Chromium induced changes in ultramorphology and secondary metabolites of *Phyllanthus amarus* Schum. & Thonn.—an hepatoprotective plant. *Environ. Monit. Assess.* 147: 307–315.
Rathikannu, S., and Sivasubramanian, P. 2008. Efficacy of botanicals against sucking pests of *Phyllanthus amarus. J. Plant Protect. Environ.* 5: 95–101.
Reddy, H.S.S., Rame Gowda, D.M.V., and Vishwanath, K. 2007. Studies on physiological maturity in *Phyllanthus amarus* Schum. and Thonn. *Seed Res.* 35: 202–204.
Roberts, N.L.B. 2004. Evaluation of young fruit tree performance in hillside trials in Trinidad and Tobago. *Acta Hort.* 638: 459–464.
Schippmann, U., D. Leaman, and A.B. Cunningham. 2006. A comparison of cultivation and wild collection of medicinal and aromatic plants under sustainability aspects. In *Medicinal and aromatic plants*, ed. R.J. Rogers, L.E. Craker, and D. Lange, 75–95. The Netherlands: Springer.
Sharma, I., Sharma, N., and Kour, H. 2009. Studies on the role of ants in reproductive efficiency of three species of *Phyllanthus* L. *Curr Sci.* 96: 283–287.
Shukla, A.K., Pathak, R.K., Tiwari, R.P., and Nath, V. 2000. Influence of irrigation and mulching on plant growth and leaf nutrient status of aonla (*Emblica officinalis* G.) under sodic soil. *J. Appl. Hort. (Lucknow)* 2: 37–38.
Silva, F.D.F., da Noda, H., Clement, C.R., and Machado, F.M. 1997. Effect of organic manure on biomass production of quebra-pedra (*Phyllanthus stipulatus*, Euphorbiaceae) in Manuas, Amazonas, Brazil. *Acta Amazonica* 27: 73–80.
Singh, A., Yadav, A.L., Yadav, D.K., and Misra, S.K. 2008a. Effect of integrated nutrient management on yield and quality of aonla (*Emblica officinalis* Gaertn.) cv. NA-10. *Plant Arch.* 8: 473–474.
Singh, B.K., Sharma, S., and Niwas, R. 2005. Effect of methods and time of budding on shoot development pattern in aonla (*Emblica officinalis* Gaertn.) cv. Chakaiya. *Haryana J. Hort. Sci.* 34: 16–17.
Singh, J.K., Prasad, J., Singh, H.K., and Singh, A. 2008b. Effect of micronutrients and plant growth regulators on plant growth and fruit drop in aonla (*Emblica officinalis* Gaertn.) fruits cv. Narendra Aonla-10. *Plant Arch.* 8: 911–912.

Singh, R.K. 2006. Studies on economic feasibility and suitability of intercrops in aonla plantation under sodic soil. PhD diss., Naredra Deva University of Agriculture and Technology.

Singh, R.K., Singh, J.K., Prasad, J., and Singh, H.K. 2008c. Effect of intercrops on plant growth, yield and quality of aonla (*Emblica officinalis* Gaertn.) fruits cv. NA-6. *Plant Arch.* 8: 903–905.

Singh, S., Singh, A.K., and Joshi, H.K. 2008d. Standardization of maturity indices in aonla (*Emblica officinalis* Gaertn.) under semi-arid environment of western India. *Ann. Arid Zone.* 47: 209–212.

Singh, S., Singh, A.K., Joshi, H.K., Bagale, B.G., and Dhandar, D.G. 2009. Evaluation of packages for transportation and storability of aonla (*Emblica officinalis*) under semi-arid environment of western India. *J. Food Sci. Technol. (Mysore)* 46: 127–131.

Singh, S.K., and Singh, H.K. 2008. Pruning behavior in aonla (*Emblica officinalis* Gaertern.) cv. Narendra Aonla-7. *Environ. Ecol.* 26: 1039–1041.

Srimathi, P., and Sujatha, K. 2007. Influence of biocides on stability of amla (*Emblica officinalis*) seeds. *J. Ecobiol.* 20: 19–24.

Sutili, F.J., Durlo, M.A., and Bressan, D.A. 2004. Biotechnical capability of "sarandi-branco" (*Phyllanthus sellowianus* Mull. Arg.) and "vime" (*Salix viminalis* L.) for revegetation water course edges. *Ciencia Florestal.* 14: 13–20.

Tiwari, J.P., Mishra, D.S., Misra, K.K., and Mishra, N.K. 2007. Indian gooseberry. In *Medicinal and aromatic crops*, ed. Jitendra Singh, 112–124. Jaipur, India: Avishkar.

Unander, D.W., and Blumberg, B.S. 1991. *In vitro* activity of *Phyllanthus* (Euphorbiaceae) species against the DNA polymerase of hepatitis viruses: effects of growing environment and inter- and intra-specific differences. *Econ. Bot.* 45: 225–242.

Unander, D.W., Bryan, H.H., Lance, C.J., and McMillan, R.T., Jr. 1993. Cultivation of *Phyllanthus amarus* and evaluation of variables potentially affecting yield and the inhibition of viral DNA polymerase. *Econ. Bot.* 47: 79–88.

Unander, D.W., Bryan, H.H., Lance, C.J., and McMillan, R.T., Jr. 1995. Factors affecting germination and stand establishment of *Phyllanthus amarus* (Euphorbiaceae). *Econ. Bot.* 49: 49–55.

Wang, K.L., Yao, X.H., Ren, H.D., and Ding, M. 2006. Growth characteristics of bearing base branch and fruit branch of *Phyllanthus emblica*. *Forest Res. Beijing* 3: 326–330.

Webster, G.L. 1957. A monographic study of the West Indian species of *Phyllanthus*. *J. Arnold Arboric. Harv. Univ.* 39: 49–100.

Webster, G.L. 1994. Synopsis of the genus and suprageneric taxa of Euphorbiaceae. *Ann. Mo. Bot. Gard.* 81: 33–144.

4 Phylogenetic Analysis of *Phyllanthus* Species

Srinivasu Tadikamalla

CONTENTS

4.1 Introduction 71
4.2 Phylogenetics 72
    4.2.1 Variation at the DNA Level 72
    4.2.2 Internal Transcribed Spacer Region 73
    4.2.3 Phylogenetic Tree 74
    4.2.4 Tree Terminology 75
4.3 Materials and Methods 76
    4.3.1 Agarose Gel Electrophoresis 77
    4.3.2 Procedure 77
    4.3.3 Gel Documentation 78
    4.3.4 Polymerase Chain Reaction 78
    4.3.5 BLAST 79
        4.3.5.1 Protocol 79
    4.3.6 FASTA 79
        4.3.6.1 Protocol 79
4.4 Multiple-Sequence Alignment 80
4.5 Clustal W 80
    4.5.1 Protocol 80
4.6 Phylogenetic Analysis 81
    4.6.1 Protocol 81
4.7 Results and Discussion 81
    4.7.1 Sequence Obtained for *Phyllanthus tenellus* 81
    4.7.2 Sequence Obtained for *Phyllanthus fraternus* 82
References 94

4.1 INTRODUCTION

Phyllanthus is the largest genus in the family Euphorbiaceae. *Phyllanthus* has a remarkable diversity of growth forms, including annual and perennial herbaceous, arborescent, climbing, floating aquatic, pachycaulous, and phyllocladous. It has a wide variety of floral morphologies and chromosome number and has one of the widest varieties of pollen types of any plant genus. *Phyllanthus* has more than 700 species in at least 10 subgenera (Holm-Nielsen, 1979; Webster, 1956, 1957). The circumscription of this genus has been so confusing that *molecular phylogenetic analysis* of Phyllanthaceae (*Phyllanthoideae pro parte, Euphorbiaceae sensu lato*) using plastid RBCL DNA sequences (Wurdack et al., 2004) and evidence from plastid matK and nuclear PHYC sequences (Samuel et al., 2005) were carried out. This chapter deals with the phylogenetic analysis of Indian Phyllanthus species using ITS nr DNA sequences.

The word *species* literally means outward or visible form. It comprises groups of actually or potentially interbreeding natural populations that are reproductively isolated from other such groups. The Linnaean concept of species as relatively constant units with most of the variations occurring among them is different from Darwin's theory of evolution by gradual change, which states that the variations between species must be generated from variation within species. The historical review of the species and their varieties continued to the end of 19th century, and it was during this period that Mendel's concept of heredity made its appearance. Later, as the integration of Mendelian genetics and Darwin's evolutionary theory, often termed *neo-Darwinism*, was making its impact on biology, the significance of genetic variations within species invited wide attention. Studies on intraspecific (genetic) variations were initiated and further elaborated by many workers. During the 1920s, based on his epoch-making observations on genetic diversity of cultivated plants and their wild relatives, Vavilov described what are known today as the geographical centers of genetic diversity/origin, mostly in the tropical belt. Although these earlier workers described a vast wealth of previously unknown genetic variations, little attention was paid to the necessity to preserve these reservoirs of genetic diversity or "natural gene pools" (Mayr, 1940).

4.2 PHYLOGENETICS

In biology, *phylogenetics* is the study of evolutionary relatedness among various groups of organisms (e.g., species, populations). It is also known as *phylogenetic systematics* or *cladistics*. Phylogenetics treats a species as a group of lineage-connected individuals over time. *Taxonomy*, the classification of organisms according to similarity, has been richly informed by phylogenetics but remains methodologically and logically distinct (Edwards and Cavalli-Sforza, 1964).

Evolution is regarded as a branching process by which populations are altered over time and may speciate into separate branches, hybridize together, or terminate by extinction. The problem posed by phylogenetics is that genetic data are only available for the present, and fossil records are sporadic and less reliable. Our knowledge of how evolution operates is used to reconstruct the full tree (Cavalli-Sforza and Edwards, 1967).

4.2.1 Variation at the DNA Level

The ability to investigate DNA sequences directly became available to population biologists only during the late 1970s. Currently, three major DNA-based techniques have been widely used for analyzing the genetic diversity in natural populations. These include (1) restriction fragment length polymorphism (Botstein et al., 1980);

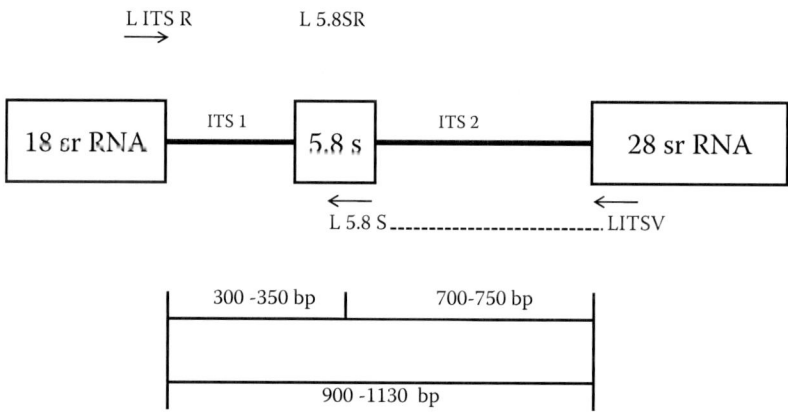

FIGURE 4.1 Internal transcribed spacer region.

(2) polymerase chain reaction (PCR; Mullis and Faloona, 1987) and its derivatives, termed random amplified polymorphic DNA (RAPD) (Williams et al., 1990) and arbitrarily primed PCR (fingerprinting of RNA) (AP-PCR; Welsh and McClelland, 1990); and (3) a hybrid of both these techniques called amplification fragment length polymorphism (Vos et al., 1995).

4.2.2 INTERNAL TRANSCRIBED SPACER REGION

The internal transcribed spacer (ITS) has been used in numerous systematic studies at the generic and specific levels of a wide array of plant taxa (Baldwin et al., 1995). The two internal spacers, ITS1 and ITS2, are located between genes encoding the 18S, 5.8S, and 28S nuclear ribosomal RNA (nrRNA) subunits (Baldwin, 1993). ITS1 and ITS2, in addition to the 5.8S nrRNA, are referred to as the ITS region (Baldwin et al., 1995). Individually, ITS1 and ITS2 are around 300 bp in length, and the 5.8S subunit is almost invariant in length within angiosperms (163–164 bp), making the entire ITS region approximately 700 bp (Figure 4.1).

Even though ITS1 and ITS2 are part of the ribosomal transcriptional unit, these sequences are not incorporated into the mature ribosome. The two ITS sequences, however, do appear to function in the maturation of nrRNAs; the specific deletions or point mutations in ITS1 can inhibit production of mature large- and small-subunit rRNAs (ribosomal RNAs), and deletions or point mutations in ITS2 prevent or reduce processing of large-subunit rRNAs. Given the short length and the highly conserved nature of the flanking ribosomal subunit genes, the ITS region is easily amplified from small amounts of genomic DNA by PCR (Baldwin, 1993).

The ITS refers to a segment of nonfunctional RNA located between structural rRNAs on a common precursor transcript. Read from 5' to 3', this polycistronic rRNA precursor transcript contains the 5' external transcribed sequence (5' ETS), 18S rRNA, ITS1, 5.8S rRNA, ITS2, 28S rRNA, and finally the 3' ETS. During rRNA maturation, ETS and ITS pieces are excised and, as nonfunctional maturation by-products, are rapidly degraded. Genes encoding rRNA and spacers occur in tandem repeats that are thousands of copies long, each separated by regions of nontranscribed DNA termed intergenic spacers (IGSs) or nontranscribed spacers (NTSs). Sequence comparison of the ITS region is widely used in taxonomy and molecular phylogeny because (due to the high copy number of rRNA genes) it is easy to amplify even from small quantities of DNA and has a high degree of variation even between closely related species. This can be explained by the relatively low evolutionary pressure acting on such nonfunctional sequences. It has proved to be useful for checking relationships among species and various genera in Asteraceae (Baldwin, 1993).

Eukaryotic ribosomal RNA genes (known as ribosomal DNA or rDNA) are found as parts of repeat units that are arranged in tandem arrays, located at the chromosomal sites known as nucleolar organizing regions (NORs). Each repeat unit consists of a transcribed region (having genes for 18S, 5.8S, and 28S rRNAs and the external transcribed spacers, i.e., ETS1 and ETS2) and an NTS region. In the transcribed region, ITSs are found on either side of the 5.8S rRNA gene and are described as ITS1 and ITS2.

The length and sequences of ITS regions of rDNA repeats are believed to be fast evolving and therefore may vary. Universal PCR primers designed from highly conserved regions flanking the ITS and its relatively small size (600–700 bp) enable easy amplification of the ITS region due to high copy number, up to 30,000 per cell (Dubouzet and Shinoda, 1999), of rDNA repeats. This makes the ITS region an interesting subject for evolutionary or phylogenetic investigations (Baldwin et al., 1995; Hershkovitz and Zimmer, 1996; Hershkovitz et al., 1999) as well as biogeographic investigations (Baldwin, 1993; Suh et al., 1993; Hsiao et al., 1994; Dubouzet and Shinoda, 1999).

4.2.3 Phylogenetic Tree

A phylogenetic tree (also known as an evolutionary tree), is a tree showing the evolutionary relationships among various biological species or other entities that are believed to have a common ancestor. In a phylogenetic tree, each node with descendants represents the most recent common ancestor of the descendants, and the edge lengths in some trees correspond to time estimate.

A rooted phylogenetic tree is a directed tree (data structure) with a unique node corresponding to the (usually imputed) most recent common ancestor of all the entities at the leaves of the tree. The most common method for rooting trees is the use of an uncontroversial out-group—close enough to allow inference from sequence or trait data but far enough to be a clear out-group. Unrooted trees illustrate the relatedness of the leaf nodes without making assumptions about common ancestry. While unrooted trees can always be generated from rooted ones by simply omitting the root, a root cannot be inferred from an unrooted tree without some means of identifying ancestry; this is normally done by including an out-group in the input data or introducing additional assumptions about the relative rates of evolution on each branch, such as an application of the molecular clock hypothesis (Maher, 2002) (Figure 4.2).

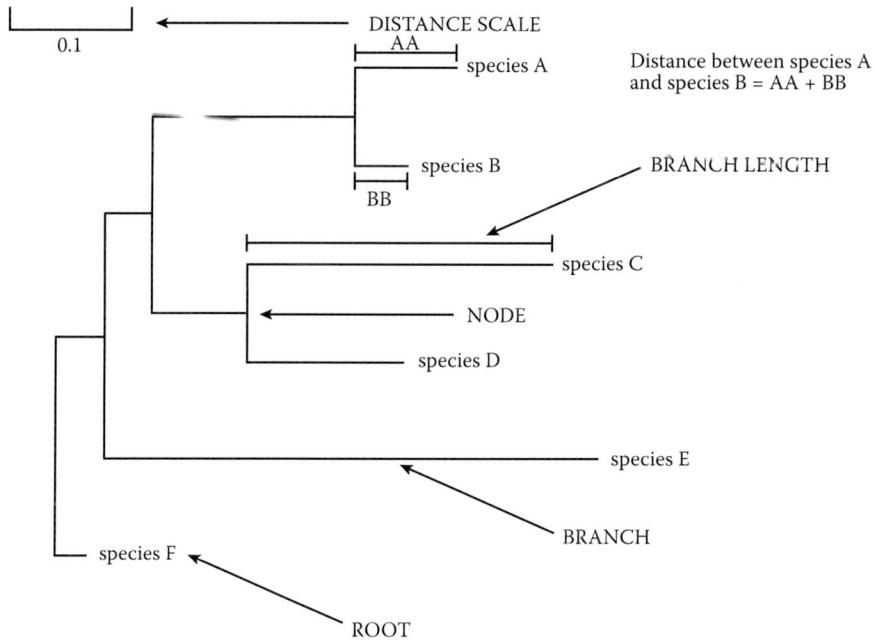

FIGURE 4.2 Explanation of various tree terminologies.

4.2.4 Tree Terminology

Node: A node represents a taxonomic unit. This can be a taxon (an existing species) or an ancestor (unknown species; represents the ancestor of two or more species).
Branch: A branch is defined as the relationship between the taxa in terms of descent and ancestry.
Topology: This is the branching pattern.
Branch length: Branch length often represents the number of changes that have occurred in that branch.
Root: The root is the common ancestor of all taxa.
Distance scale: This scale represents the number of differences between sequences (e.g., 0.1 means 10% differences between two sequences).

Phylogenetic trees can be drawn in different ways. There are trees with unscaled branches and with scaled branches, as shown in Figure 4.3.

Unscaled branches: In this case, the length is not proportional to the number of changes that occurred. Sometimes, the number of changes is indicated on the branches with numbers. The nodes represent the divergence event on a time scale.

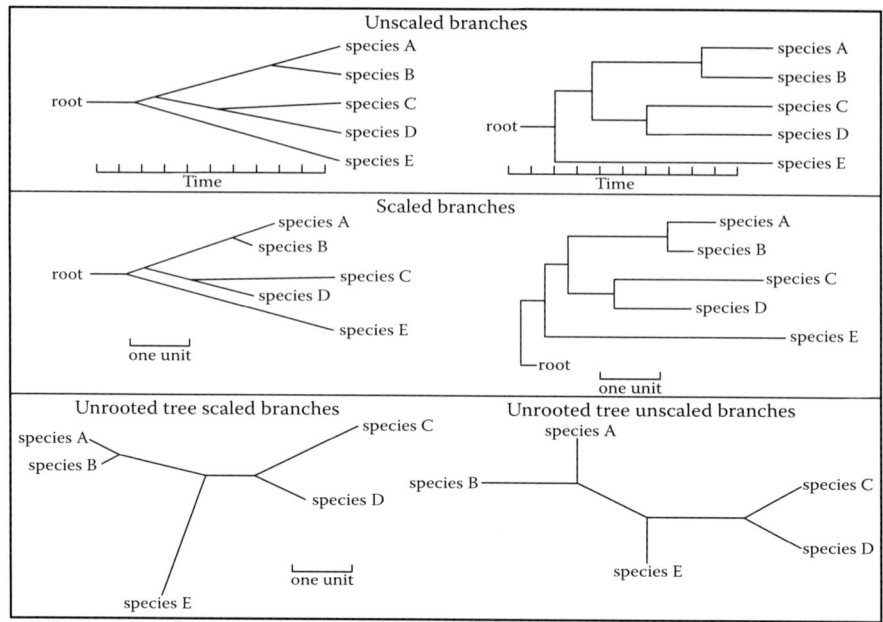

FIGURE 4.3 Different types of scaled and unscaled branches.

Scaled branches: Here, the length of the branch is proportional to the number of changes that occurred. The distance between two species is the sum of the length of all branches connecting them.

It is also possible to draw these trees with or without a root. For rooted trees, the root is the common ancestor. For each species, there is a unique path that leads from the root to that species. The direction of each path corresponds to evolutionary time. An unrooted tree specifies the relationships among species and does not define the evolutionary path (Baldwin, 1993).

4.3 MATERIALS AND METHODS

The plant materials were collected from Veer Mata Jijabai Bhonsale Udyan, Byculla, in Mumbai (India). The samples (leaves) were washed, and total cellular DNAs were isolated by a Hi-media DNA purification spin kit protocol.

Briefly, add 400 µl of lysis buffer and 20 µl of RNase A (ribonuclease A) stock solution to 100 mg of homogenated sample and vortex vigorously. Incubate the mixture for 10 min at 65°C and mix the contents two or three times by inverting the tube. Add 130 µl of precipitation buffer to the lysate, mix, and incubate for 5 min on ice. Centrifuge the lysate for 8 min at 10,000 rpm. Load lysate in a HiShredder placed in a 2-ml collection tube, and centrifuge for 5 min at 10,000 rpm, and transfer the fraction to a new 2-ml collection tube without disturbing the cell debris pellet. Add 1.5 volumes of binding buffer to the cleared lysate and mix by pipette. The proportion of lysate and binding buffer should be 450 and 675 µl. The volume can be reduced accordingly if less lysate is obtained. Add 650 µl of the mixture of lysate and binding buffer to the HiElute Spin column sitting in a 2-ml collection tube. Centrifuge for 1 min at 8,000 rpm. Discard the flow-through and repeat it with the remaining sample. Discard the flow-through liquid and the 2-ml collection tube. Place the HiElute Spin column in a new 2-ml collection tube, add 500 µl of wash buffer, and centrifuge for 1 min at 8,000 rpm. Add another 500 µl of the wash buffer to the HiElute Spin column and centrifuge for 5 min at 10,000 rpm to dry the membrane.

For DNA elution, pipette 100 µl of the elution buffer directly without spilling to the sides. Incubate for 1 min at room temperature. Centrifuge at 10,000 rpm for 1 min to elute the DNA. Repeat the step with another 100 µl of elution buffer for high DNA yield. Incubate for 5 min at room temperature to increase the elution efficiency, then centrifuge. The elute contains pure genomic DNA. It can be stored at 2–8°C for short-term or −20°C for long-term storage. The elution buffer will help stabilize the DNA at these temperatures. The quality and quantity of DNA samples were checked on agarose gel using lambda DNA as a marker.

4.3.1 Agarose Gel Electrophoresis

Agarose gel electrophoresis is a method used to separate DNA or RNA molecules by size. This is achieved by moving negatively charged nucleic acid molecules through an agarose matrix with an electric field (electrophoresis). Shorter molecules move faster and migrate further than longer ones.

Increasing the agarose concentration of a gel reduces the migration speed and enables separation of smaller DNA molecules. The higher the voltage, the faster the DNA migrates. But, voltage is limited by the fact that it heats and ultimately causes the gel to melt. High voltages also decrease the resolution (above about 5 to 8 V/cm).

The most common dye used for agarose gel electrophoresis is ethidium bromide (EtBr). It fluoresces under ultraviolet (UV) light when intercalated into DNA (or RNA). By running DNA through an EtBr-treated gel and visualizing it with UV light, distinct bands of DNA become visible.

Loading buffers are added with the DNA to visualize it (dye) and sediment it in the gel well (sucrose or glycerol). In our study, we used Orange G, a negatively charged indicator to keep track of the migration of the DNA through the gel.

4.3.2 Procedure

Our procedure is to seal the edges of a clean, dry, glass plate with autoclave tape to form a mold. Set the mold on a horizontal section of the bench (check with a level).

Prepare sufficient electrophoresis buffer to fill the electrophoresis tank and to prepare the gel. Add the correct amount of powdered agarose to a measured quantity of electrophoresis buffer in an Erlenmeyer flask or a glass bottle with a loose-fitting cap. The buffer should not occupy more than 50% of the volume of the flask or bottle.

It is important to use the same batch of electrophoresis buffer in both the electrophoresis tank and the gel. Any change in the ionic strength or pH creates fronts in the gel that can greatly affect the mobility of DNA fragments.

Heat the slurry in a microwave oven until the agarose dissolves. Cool the solution to 60°C and add EtBr (from a stock solution of 10 mg/ml in water) to a final concentration of 0.5 µg/ml and mix thoroughly.

Position the comb 0.5–1.0 mm above the plate so that the complete well is formed when the agarose is added. If the comb is closer to the glass plate, there is a risk that the base of the well may tear when the comb is withdrawn, allowing the sample to leak between the gel and the glass plate. Pour the warm agarose solution into the mold.

After the gel is completely set (30–45 min at room temperature), carefully remove the tape and the comb and mount the gel in the electrophoresis tank. Add enough electrophoresis buffer to cover the gel to a depth of about 1 mm. Mix the samples of DNA with the desired gel loading buffer and slowly load the mixture into the slots of the submerged gel using a micropipette. (Gel loading buffer serves three purposes: It increases the density of the sample, ensuring that the DNA drops evenly into the well; it adds color to the sample, thereby simplifying the loading process; and it contains dyes that, in an electric field, moved toward the anode at predictable rates.) Close the lid of the gel tank and attach the electrical leads so that the DNA will migrate toward the anode (red lead). Turn off the electric current, remove the leads and lid from the gel tank, and examine the gel by UV light.

4.3.3 Gel Documentation

Photos of gels are invaluable as records of size marker locations, DNA digestion comparison of DNA concentration, and migration distances. A convenient apparatus includes a Polaroid Land camera station and a UV transilluminator. A filter on the camera is required. The specifications for the filter unit will depend on the film and suggested guidelines for the transilluminator. EtBr-stained DNA fluoresces under UV irradiation and therefore is readily photographed. Caution should be exercised during this process to minimize the time of exposure because DNA can be damaged when stained with EtBr and exposed to UV irradiation.

4.3.4 Polymerase Chain Reaction

The PCR (Helena Biosciences, UK) usually consists of a series of 20 to 35 repeated temperature changes called *cycles*; each cycle typically consists of two or three discrete temperature steps. Most commonly, PCR is carried out with cycles that have three temperature steps. The cycling is often preceded by a single temperature step (called *hold*) at a high temperature (>90°C) and followed by one hold at the end for final product extension or brief storage. The temperatures used and the length of time they are applied in each cycle depend on a variety of parameters. These include the enzyme used for DNA synthesis, the concentration of divalent ions and dNTPs (deoxynucleotide triphosphates) in the reaction, and the melting temperature T_m of the primers.

The ITS1-5.8S-ITS2 rDNA region was amplified using the following universal primers: forward primer (GGAAGGAGAAGTCGTAACAAGG) and reverse primer (TCCTCCGCTTATTGATATGC). Amplifications were carried out in a 50-µl reaction mixture containing 33.7 μl sterile water, 5 μl of 10x PCR buffer, 3 μl of 25 mM MgCl$_2$, 4 μl of 10 mM dNTPs, 1 μl of each primer (10 pM), 0.3 μl (3 U/μl) of Taq polymerase, and 2 μl of template DNA. A PerkinElmer DNA thermal cycler was used with the following PCR profile: an initial denaturation for 5 min at 94°C; 35 thermal cycles (30 s to 1 min at 94°C, 30 s at 55°C, and 1 min at 72°C); and a final 5 min extension at 72°C. The amplified DNA was purified using a PCR purification kit following the manufacturer's instructions.

4.3.5 BLAST

In bioinformatics, the Basic Local Alignment Search Tool (BLAST) is an algorithm for comparing primary biological sequence information, such as the nucleotides of a DNA sequence or amino acid sequence of different proteins. It is one of the most widely used bioinformatics programs because it addresses a fundamental problem, and the algorithm emphasizes speed over sensitivity. This emphasis on speed is vital to make the algorithm practical on the huge genome database currently available, although subsequent algorithms can be even faster. The BLAST algorithm and the computer program implemented were developed by Altschul et al. (1997) at the U.S. National Center for Biotechnology Information (NCBI).

4.3.5.1 Protocol
1. Go to the BLAST home page (http://blast.ncbi.nlm.nih.gov/Blast.cgi).
2. Choose nucleotide BLAST program to run from the submenu.
3. Paste the query sequence in the FASTA format/accession no./GI in the dialogue box choose database type as others (nr etc.).
4. Select program to optimize for highly similarly sequences (megablast) and Click BLAST button/icon.
5. The page displays the result.

4.3.6 FASTA

FASTA is a DNA and protein sequence alignment package first developed (as FASTA P) by Lipman and William in 1985 and Pearson in 1990 the article Rapid and sensitive protein similarity searches. The original FASTA P program was designed for protein sequence similarity searching. As described in 1988, FASTA added the ability to do DNA: DNA searches. There are several programs in this package that allow the alignment of protein sequences and DNA sequences. The current FASTA package contains programs for protein: protein, DNA: DNA, protein: translated DNA (with frameshift), and ordered or unordered peptide searches. The most recent version of the FASTA package includes special translated search algorithms that correctly handle frameshift errors when comparing nucleotide-to-protein sequence data.

4.3.6.1 Protocol
1. Go to the FASTA home page (http://www.ebi.ac.uk/Tools/sss/fasta/).
2. Choose the FASTA-nucleotide tool.

3. Paste the given query sequence in the dialogue box given, keeping the rest of the parameters at the default setting.
4. Click RUN FASTA.
5. The page displayed shows a FASTA summary table with the following self-explanatory headings: Align, DBID, source, length, score, identities, positives, and E-value. Each entry is hyperlinked to its detail.

4.4 MULTIPLE-SEQUENCE ALIGNMENT

A comparison of many sequences of proteins or nucleic acid at a time is called multiple-sequence alignment (MSA). It tries to align all the sequences such that there is maximum alignment of identities or similarities possible within the sequence analyzed. The MSA determines the level of homology (relatedness) between members of the series of globally related sequences. It is important for finding similar domains in a set of sequences and for doing phylogenic analysis. There are several approaches for conducting sequence alignment. The aim of MSA is to generate a concise, information-rich summary of sequence data to aid decision making on the relatedness of sequence to a gene family. It is also useful for distinguishing between proteins that perform the same function in different species and those that perform different but related functions within one organism among sequences of homologues.

4.5 CLUSTAL W

CLUSTAL W is a program commonly used for MSA. Paula Hogeway first described CLUSTAL W in the early 1980s. Later, it was completely developed by Larkin et al. (2007).

4.5.1 Protocol

1. Paste a set of DNA sequences in FASTA format, keeping other options default and submit.
2. The page displayed shows Clustal W2 results without color. Click on show color icon, the same results are displayed in green (conserved), red colors (non-conserved nucleotides) and '-' as indels. MSA shows alignment scores on the right side in a separate column. It uses the alignment score using the neighbor-joining method.
   a. Sequence alignment: The sequences of ITS1–5.8S-ITS2 regions are aligned with the corresponding sequences already available in the database.
   b. Sequence submission: Sequences of DNA are submitted directly to GenBank through Bankit (a World Wide Web sequence submission server available at the NCBI home page, http://www.ncbi.nlm.nih.gov). The sequences are available online at this NCBI home page and can be located by accession numbers or GI numbers.

4.6 PHYLOGENETIC ANALYSIS

The field of phylogeny has the goals of working out the evolutionary relationships among species, population by analysis of family of related nucleic acid. Studies of gene evolution involve comparison of homologue sequences that have common origins but may or may not have common activity. *Homology* specifically means descent from a common ancestor, while *similarity* is the measurement of resemblance or difference, independent of the source of resemblance.

An evolutionary tree is a two-dimensional (2-D) graph showing evolutionary relationships among organisms or in certain genes from separate organisms. It represents an estimated pedigree of the inherited relationships among organisms, molecules, or both.

A *dendogram* is a broad term for the diagrammatic representation of a phylogenetic tree. Many sequence alignment methods, such as CLUSTAL W, produce both sequence alignments and phylogenic trees. Phylogenetic trees can be calculated by DISTANCE-MATRIX methods (e.g., neighbor joining); those that calculate genetic distances from multiple sequence alignments are the simplest to implement. Distance is calculated for all pairwise combinations of operational taxonomic units, and then the distances are assembled into a tree.

4.6.1 Protocol

1. Select query sequence.
2. Use CLUSTAL W to align all the sequences.
3. Obtain a dendogram of the sequences by the selection option.
4. Analyze the dendogram to determine evolutionary relationships.

4.7 RESULTS AND DISCUSSION

Phyllanthus tenellus and *P. fraternus* are two species belongs to family Phyllanthaceae. The ancestral relationship and evolutionary history of these two species were studied by sequencing the ITS of the ribosomal DNA ITS region, which is highly variable and lies between conserved regions ITS1 (which lies between 18S and 5.8S) and ITS2 (which lies between 5.8S and 28S).

4.7.1 Sequence Obtained for *Phyllanthus tenellus*

>II35_T_G1
TCGTAGGTGACCTGCGGAAGGA
TCATTGTCGAACCTGCACAGCAG
TACGACCCGCGAACAAGTTTATA
CACTGCGGAAGGTGTCTCGTGCA
CCCGATGCAAGGTCCCGTGGGGT
GCTACGCTCCTCGCGGTGGCCA
CGTAAAAAACCCCCGGCGCGGAA
AGCGCCAAGGAAAATAAACATA
CAAGCGAGAACCCTCTAATCACC

CTGGAGACTCCGGTGCGTGTTTG
GTAGGTGTTTCTCCTTTAGTAA
CAAAAACGACTCTCGGCAACGG
ATATCTCGGCTCTCGCATCGATG
AAGAACGTAGCGAAATGCGATAC
TTGGTGTGAATTGCAGAATCCCGT
GAACCATCGAGTTTTTGAACGCAA
GTTGCGCCCGAAGCCTTTTGGTC
GAGGGCACGTCTGCCTGGGTGTCA
CGCAACGTCGCTCCCTCACTCCC
TCACTCGAGGCATGTGAATTCGG
GGCGGAAAATGGCCTCCCGTGAA
CTTCGTGATCGCGGTTGGCCCAAA
CATGAGACCAATTCGGCCAATGC
CGTGGCATTCGGTGGTTGAAAATA
CCTTACTATTGCCTCGTTCATTTGT
CCGAACAAACAAGGATCTCGACGA
CCCTCTATGTATCCGACGCGACC
CCAGGTCAGGCGGGATTACCCG
CTGAGTTTAAGCTAATTAAAAGGGGGAAGGAAAGTTTT

GenBank accession numbers:
  Phyllanthus tenellus ITS1: EU580530
  Phyllanthus tenellus 5.8S: EU580531
  Phyllanthus tenellus ITS2: EU580532

4.7.2 Sequence Obtained for *Phyllanthus fraternus*

>II35_F_G1
CTCGGTGGTGACCTGCGGAGGATCA
TTGTCGAAACCTGCTCTGCAGTATGACC
CGCGAACAAGTTTATACACTGCCGAAGG
TGCCTTGTGCTCCTGACGCGAGGCCCCGT
TGGGTGCTACGCTCCCTGCGGTGGCCACG
TAACAAACCCCGGCGCGGAAAGCGCCAAG
GAAAATGAACATACAAGCGAGAACCCAAC
AGGCTCCCCGGAAACGGTGCGTGCTTTGCTG
AGTTTCTCCTTACGTAACCAAAACGACTCTC
GGCAACGGATATCTCGGCTCTCGCATCGA
TGAAGAACGTAGCGAAATGCGATACTTGG
TGTGAATTGCAGAATCCCGTGAACCATCG
AGTTTTTGAACGCAAGTTGCGCCCAACGC
CTTCGGGTCGAGGGCACGTCTGCCTGGGT
GTCACGCAACGTCGCTCCCTCACTCCCGC
GTGGAACGTGAATTTCGAGCGGAATATGGC
CTCCCGTGAACTCTTCGATCGCGGTTGGCC

Phylogenetic Analysis of *Phyllanthus* Species

TAAACACGAGACCATGTCGGCCAATGCC
GTGGCATTCGGTGGTTGAAATACCCTCAAAACGCCTCGTT
CGTTTGGCCGCGCTAAAAAGGTTTTCAA
CGACCCTCTACTATCCGACGCGACCCCAGGT
CAGGCGGGATTACCCGCTGAGTTTAAGCATAAATT
AAAGCGGCGGAAGGAAAGTTTTTGTTTGTGT
CTTGCTTCTGTGCGCTGGTGAGGGGTGTGGTGGGTGT
TGGGGGGGGGGTGGGGGGGGGTAGTGTGGGA
AATTGATATTGTAAGTTTGATCGGTGGGGGTTTT

GenBank accession numbers:
  Phyllanthus fraternus ITS1: EU580527
  Phyllanthus fraternus 5.8S: EU580528
  Phyllanthus fraternus ITS2: EU580529

Fourteen species sequences were downloaded from the NCBI GenBank to align with ITS sequences of *Phyllanthus tenellus* and *Phyllanthus fraternus* (Figure 4.4). The following are the species and their GenBank accession numbers:

Phyllanthus debilis: AY936686.1
Phyllanthus niruri: AY765286.1
Phyllanthus acidus: AY725468.1
Phyllanthus madagascariensis: AY936706.1
Phyllanthus nummulariifolius: AY936714.1
Phyllanthus rheedii: AY936729.1
Phyllanthus graveolens: AY936696.1
Phyllanthus polyphyllus: AY936725.1
Phyllanthus oxyphyllus: AY936719.1
Phyllanthus clarkei: AY765288.1
Phyllanthus reticulates: AY835843.2
Phyllanthus discolor: AY936688.1
Phyllanthus emblica: AY830087.1
Phyllanthus acuminatus: AY936667.1

MSA is a sequence alignment of three or more biological sequences, generally protein, DNA, or RNA. In general, the query sequences in the input set are assumed to have an evolutionary relationship by which they share a lineage and are descended from a common ancestor. From the resulting MSA, sequence homology can be inferred, and phylogenetic analysis can be conducted to assess the shared evolutionary origins of the sequences. Visual depictions of the alignment as in Figure 4.4 illustrate mutation events such as point mutations (single amino acid or nucleotide changes), which appear as differing characters in a single alignment column, and insertion or deletion mutations (or indels), which appear as gaps in one or more of the sequences in the alignment. MSA is often used to assess sequence conservation of protein domains, tertiary and secondary structures, and even individual amino acids or nucleotides. Thus, MSA of the set of the query sequences was performed, and phylogenetic and evolutionary analyses were carried out.

```
P.debilis              ------------------------------------------------------------
P.discolor             ------------------------------------------------------------
P.graveolens           ------------------------------------------------------------
P.niruri               ------------------------------------------------------------
P.madagascariensis     ------------------------------------------------------------
P.acidus               ------------------------------------------------------------
P.polyphyllus          ------------------------------------------------------------
P.acuminatus           ------------------------------------------------------------
P.oxyphyllus           ------------------------------------------------------------
P.sikkimensis          ------------------------------------------------------------
P.rheedii              ------------------------------------------------------------
P.harmandii            ------------------------------------------------------------
P.collinsiae           ------------------------------------------------------------
P.angkorensis          ------------------------------------------------------------
P.gracilipes           ------------------------------------------------------------
P.lingulatus           ------------------------------------------------------------
P.microcarpus          ------------------------------------------------------------
P.clarkei              ------------------------------------------------------------
P.welwitschianus       -------------------------------------ATTCAGAGAGTGTTCGGATCGAG  23
P.reticulatus          ------------------------------------------------------------
P.columnaris           ------------------------------------------------------------
P.nummulariifolius     CATCTACACCCGCCCGTCGCTCCTACCGATTGAATGGTCCGGTGAAGTGTTCGGATCGAG  60
P.orientalis           ------------------------------------------------------------
P.emblica              ------------------------------------------------------------
P.acutissimus          ------------------------------------------------------------
P.pulcher              ------------------------------------------------------------
P.myrtifolius          ------------------------------------------------------------
P.virgatus             ------------------------------------------------------------
P.roseus               ------------------------------------------------------------


P.debilis              ------------------------------------------------------------
P.discolor             ------------------------------------------------------------
P.graveolens           ------TGGGCGGTTCGCCGCCGACGACGTCGCGAGAAGTCCACTGAACC-TTATCATTT  53
P.niruri               ------------------------------------------------------------
P.madagascariensis     ------------------------------------------------------------
P.acidus               ------TGGGCGGTTCGCCGCCGACGACGTCGCGAGAAGTCCACTGAACC-TTATCATTT  53
P.polyphyllus          ------TGGGCGGTTCGCCGCCGACGACGTCGCGAGAAGTCCACTGAACC-TTATCATTT  53
P.acuminatus           ------TGGGCGGTTCGCCACCGACGACGTCGCGAGAAGTCCACTGAACC-TTATCATTT  53
P.oxyphyllus           ------TGGGCGGTTCGCCGCCGACGACGTCGCGAGAAGTCCACTGAACC-TTATCATTT  53
P.sikkimensis          ------TGGGCGGTTCGCCGCCGACGACGTCGCGAGAAGTCCACTGAACC-TTATCATTT  53
P.rheedii              -----GTGGGCGGT-CGCCGCCGACGACGTCGCGAGAAGTCCACTGAACC-TTATCATTT  53
P.harmandii            ------------------------------------------------------------
P.collinsiae           ------------------------------------------------------------
P.angkorensis          ------------------------------------------------------------
P.gracilipes           ------------------------------------------------------------
P.lingulatus           ------------------------------------------------------------
P.microcarpus          ------------------------------------------------------------
P.clarkei              ------TGGGCGGTTCGCCGCCGACGACGTCGCGAGAAGTCCACTGAACC-TTATCATTT  53
P.welwitschianus       GCGACGTGGGCGGTTCGCCGCCGACGACGTCGCGAGAAGTCCACTGAACC-TTATCATTT  82
P.reticulatus          ------TGGGCGGTTCGCCGCCGACGACGTCGCGAGAAGTCCACTGAACC-TTATCATTT  53
P.columnaris           ------------------------------------------------------------
P.nummulariifolius     GCGACGTGGGCGGTTCGCCGCCGACGACGTCGCGAGAAGTCCACTGAACCCTTATCATTT 120
P.orientalis           ------------------------------------------------------------
P.emblica              ------------------------------------------------------------
P.acutissimus          ------------------------------------------------------------
P.pulcher              ------------------------------------------------------------
P.myrtifolius          ------------------------------------------------------------
P.virgatus             ------------------------------------------------------------
P.roseus               ------------------------------------------------------------
```

FIGURE 4.4 Clustal 2.0.12 multiple-sequence alignment of the species belonging to genus *Phyllanthus*. (*continued*)

```
P.debilis           ------------------------------TCGTAGGTGA-CCTGCGGAAGGATCATTGTCG  31
P.discolor          ---------------------------------------GCGGAAGGATCATTGTCG       18
P.graveolens        AAAGGAAGGAGAAGTCTTAACAAGGTTTCCGTAGGTGAACCTGCGGAAGGATCATTGTCG   113
P.niruri            ------------------**       *--CTCGGTGGTGA-CCTGCGGA-GGATCATTGTCG 31
P.madagascariensis  ------------------------------------------------------------
P.acidus            AGAGGAAGGAGAAGTCGTAACAAGGTTTCCGTAGGTGAACCTGCGGAAGGATCATTGTCG   113
P.polyphyllus       AGAGGAAGGAGAAGTCGTAACAAGGTTTCCGTAGGTGAACCTGCGGAAGGATCATTGTCG   113
P.acuminatus        AGAGGAAGGAGAAGTCGTAACAAGGTTTCCGTAGGTGAACCTGCGGAAGGATCATTGTCG   113
P.oxyphyllus        AGAGGAAGGAGAAGTCGTAACAAGGTTTCCGTAGGTGAACCTGCGGAAGGATCATTGTCG   113
P.sikkimensis       AGAGGAAGGAGAAGTCGTAACAAGGTTTCCGTAGGTGAACCTGCGGAAGGATCATTGTCG   113
P.rheedii           AGAGGAAGGAGAAGTCGTAACAAGGTTTCCGTAGGTGAACCTGCGGAAGGATCATTGTCG   113
P.harmandii         ---------------------------TCCGTAGGTGAACCTGCGGAAGGATCATTGTCA   33
P.collinsiae        ---------------------------TCCGTAGGTGAACCTGCGGAAGGATCATTGTCA   33
P.angkorensis       ---------------------------TCCGTAGGTGAACCTGCGGAAGGATCATTGTCA   33
P.gracilipes        ---------------------------TCCGTAGGTGAACCTGCGGAAGGATCATTGTCA   33
P.lingulatus        ---------------------------TCCGTAGGTGAACCTGCGGAAGGATCATTGTCA   33
P.microcarpus       ---------------------------TCCGTAGGTGAACCTGCGGAAGGATCATTGTCA   33
P.clarkei           AGAGGAAGGAGAAGTCGTAACAAGGTTTCCGTAGGTGAACCTGCGGAAGGATCATTGTCA   113
P.welwitschianus    AGAGGAAGGAGAAGTCGTAACAAGGTTTCCGTAGGTGAACCTGCGGAAGGATCATTGTCA   142
P.reticulatus       AGAGGAAGGAGAAGTCGTAACAAGGTTTCCGTAGGTGAACCTGCGGAAGGATCATTGTCA   113
P.columnaris        ---------------------------TCCGTAGGTGAACCTGCGGAAGGATCATTGTCA   33
P.nummulariifolius  AGAGGAAGGAGAAGTCGTAACAAGGTTTCCGTAGGTGAACCTGCGGAAGGATCATTGTCG   180
P.orientalis        ---------------------------TCCGTAGGTGAACCTGCGGAAGGATCATTGTCG   33
P.emblica           --------------TCGTAACAAGGTTTCCGTAGGTGAACCTGCGGAAGGATCATTGTCG   46
P.acutissimus       ---------------------------TCCGTAGGTGAACCTGCGGAAGGATCATTGTCG   33
P.pulcher           ---------------------------TCCGTAGGTGAACCTGCGGAAGGATCATTGTCG   33
P.myrtifolius       ---------------------------TCCGTAGGTGAACCTGCGGAAGGATCATTGTCG   33
P.virgatus          ---------------------------TCCGTAGGTGAACCTGCGGAAGGATCATTGTCG   33
P.roseus            ---------------------------TCCGTAGGTGAACCTGCGGAAGGATCATTGTCG   33


P.debilis           AA-C-CTGCACAGCAGTACGACCCGCGAACAAGTTT-ATACACTGCGGA-AGGTGTCTCG    87
P.discolor          AAAC-CTGCACAGCAGTACGACCCGCGAACAAGTTT-ATACACTGCGGA-AGGTGTCTCG    75
P.graveolens        AAAC-CTGCACAGCAGTACGACCCGCGAACAAGTTT-ATACACTGCGGA-AGGTGTCTCG   170
P.niruri            AAAC-CTGCTCTGCAGTATGACCCGCGAACAAGTTT-ATACACTGCCGA-AGGTGCCTCG    88
P.madagascariensis  -----------------------------------------GCCGA-AGGTGCCTTG      15
P.acidus            AAAC-CTGCTCTGCAGTATGACCCGCGAACAAGTTT-ATACACTGCCGA-AGGTGCCTTG   170
P.polyphyllus       TAAC-CTGCAATGCAGCATGACTTGCGAACAAGTTT-AACCACCGCCGA-AGGTGCCTTG   170
P.acuminatus        AAAC-CTGCACTGCAGTACGACCCGCGAACATGTAT-AACCACTGCGGA-AGGTGCCATG   170
P.oxyphyllus        AAAC-CTGCTCTGCAGTATGACCAGCGAACAAGTTT-ATCCATAGCGGA-AGGTGCCATG   170
P.sikkimensis       AAAC-CTGCTCAGCAGTATGACCCGCGAACAAGTTT-ATCCACAGCGGA-AGGTGTCATG   170
P.rheedii           AAAC-CTGCCATGCAGTATGACCCGCGAACAAGTTT-ATCCACTGCTGA-AGGTGCCTCG   170
P.harmandii         AAACCTTGTACTG--GTATGACCCGCGAACAAGTTT-AGTCACTGCGGA-TGGTGCCTCG    89
P.collinsiae        AAACCTTGTACTG--GTATGACCCGCGAACAAGTTT-AGTCACTGCGGA-TGGTGCCTCG    89
P.angkorensis       AAACCTTGTACTG--GTATGACCCGCGAACAAGTTT-AGTCACTGCGGA-TGGTGCCTCG    89
P.gracilipes        AAACCTTGTACTG--GTATGACCCGCGAACAAGTTT-AGTCACTGCGGA-TGGTGCCTCG    89
P.lingulatus        AAAC-TTGTAGTG--GTATGGCCCGCGAACAAGTTT-ACTCACTGCGGA-TGGTGCCTCG    88
P.microcarpus       AATCCTTGTACTG--GTATGACCCGCGAACAAGTTT-ACTCACTGCGGA-TGGTGCCTCG    89
P.clarkei           AAACCTTGTACTG--GTATGACCCGCGAACAAGTTT-AATCACTGCGGA-TGGTGCCTCG   169
P.welwitschianus    AAACCTTGTACTG--GTATGACCCGCGAACAAGTTT-AGTCACTGCGGA-TGGTGCCTTG   198
P.reticulatus       AAACCTTGTACTG--GTATGACCCGCGAACAAGTTT-AGTCACTGCGGA-TGGTGCCTCG   169
P.columnaris        AAACCTTGTACTG--GTATGACTCGCGAACAAGTTT-AATCACTGCGGA-TGGTGTCTTG    89
P.nummulariifolius  AAAC-CTGCACTGCAGTACGACCCGTGAACGTGTTT-ATCCACTGCGGA-AGGTGCCTTG   237
P.orientalis        AAAC-CTGGATGGGAGCATGACCCGCGAACAAGTTT-ATCCACGGCCGA--AGTGCCTCG    89
P.emblica           AAAC-CTGCCACGCAGTACGACCCGCGAACAAGTTT-ATCCACCGCGGA-GGGTGCCCCG   103
P.acutissimus       AAACCTGCAACCGCAGAACGACCCGCGAACAAGTTTTACTCACTGCGGA-GGGTGCCCTG    92
P.pulcher           AAACCTGCAACTGCAGAATGACCCGCGAACAAGTTT-ATTCACTGCGGAAGGGTGCCTTG    92
P.myrtifolius       AAAC-CTGCCCAGCAGAACGACCCGCGAACATGTTT-ATCGACCGCGGA-AGGTGCCTTG    90
P.virgatus          AAAC CTGCCCAGCAGAACGACCCGCGAACATGTTT-ATCAATGGCGGA-AGGGGCCTTG    90
P.roseus            AACC-TCGCTCAGCAGAATGACCCGCGAACCTGTTC-ATCAACCGCGAA-AGGTGCCCTG    90
                                                        **  *   *  *  *
```

FIGURE 4.4 Clustal 2.0.12 multiple-sequence alignment of the species belonging to genus *Phyllanthus*. (*continued*)

```
P.debilis            TGCACCCGATGC-AAGGTCCCGTGGGGTGCTACGCTCCTCGCGGTGGCCACGTAAAAAAC 146
P.discolor           TGCACCCGATGC-AAGGTCCCGTGGGGTGCTACGCTCCTCGCGGTGGCCACGTAAAAAAC 134
P.graveolens         TGCACCCGATGC-AAGGTCCCGTGGGGTGCTACGCTCCTCGCGGTGGCCACGTAAAAAAC 229
P.niruri             TGCTCCTGACGC-GAGGCCCCGTTGGGTGCTACGCTCCCTGCGGTGGCCACGTAACAAAC 147
P.madagascariensis   TGCTCCTGACGC-GAGGCCCCGTTGGGTGCTACGCTCCCTGCGGTGGCCACGTAACAAAC 74
P.acidus             TGCTCCTGACGC-GAGGCCCCGTTGGGTGCTACGCTCCCTGCGGTGGCCACGAAACAAAC 229
P.polyphyllus        TGCTCCTGACGC-GAGGACCCGTTAGATGCTATGCTTCTAACGGTGGCCACGTAA-CAAA 228
P.acuminatus         CGTACCTGACGC-AAGGCCCCGTTTTTTGCTTTGCTCTTTACGGGGGCCTTGTAAACAAA 229
P.oxyphyllus         TGCTCCTGATGC-AAGGCCCCGTTAGGTGCTATGCTCCTTATCGAGGCCACGAAAACAAA 229
P.sikkimensis        TGCACCTGATGCCAAGGCCCCGTTAGGTGCTATGCTCCTTATCGAGGCCACGAAAACAAA 230
P.rheedii            TGCTCCTGATGC-GAGGCCCCGTGGGGGACAATGCTCCTTGCGGTGGCCACAAAA-CAAA 228
P.harmandii          TGGACCTGAAGC-AAGGCCACGTAGGGTGCTATGCTCCCTGCGGAGGCCACGTAATCCAA 148
P.collinsiae         TGGACCTGAAGC-AAGGCCACGTAGGGTGCTATGCTCCCTGCGGAGGCCACGTAATCCAA 148
P.angkorensis        TGGACCTGAAGC-AAGGCCACGTAGGGTGCTATGCTCCCTGCGGAGGCCACGTAATCCAA 148
P.gracilipes         TGGACCTGAAGC-AAGGCCACGTAGGGTGCTATGCTCCCTGCGGAGGCCACGTAATCCAA 148
P.lingulatus         TACACCTGAAGC-AAGGCCACGTGGGGTGCTATGCTCCCTGCAGAGGCCACGTAATCCAA 147
P.microcarpus        TGCACCTGAAGC-AAGGCCACGTGGGGTGCTATGCTCCCTGCAGAGGCCACGTAATCCAA 148
P.clarkei            TGCACCTGAAGC-AAGGCCACGTGGGGAGCTATGCTCCTTTCGGAGGCCACGTAATCCAA 228
P.welwitschianus     TGCACTTGAAGC-CAGGCCACGCCGGGTGCTATGCTCCTTGCGGAGGCCACATAAACCAA 257
P.reticulatus        TGCACCTGAAGC-AAGGCCACGTAGGGTGTTATTCTCCTTGCGGAGGCCACGTAATCCAA 228
P.columnaris         TGCACCTGAAGC-AAGGCCACGTAGGGTGCTATACTCCTTGCGGAGGCCACGTAATCCAA 148
P.nummulariifolius   TGCCCCTGACGC-AAGGCCCCGTGGGGTGCTATGCTCCTTGCGGAGGCCACGTAAACAAA 296
P.orientalis         TGCGCCCGAAGC-AAGGCCTCGTAGGGTGTTATGC-CCTTGCGTTGGCCACG-AAACAAA 146
P.emblica            TGCTCCCGACGC-GAGGCCCCGTGGGGTGCGACGCTCCCTGCGGAGGCCGCCGAAACAAA 162
P.acutissimus        TGCTCCTGACGC-AAGGCCCCGTGCGGTGCAATGCTCCATGCGGTGGCCACGAAAACAAA 151
P.pulcher            TGCTCCTGACGC-AAGGCCCCGTGTGGTGCTATGCTCCACTCGGTGGCCACAAGAACAAA 151
P.myrtifolius        TGCTCCTGACGC-AAGGCCCTGTGGGGTGGGATACACCTCACGGGTGCTTAGAAAACAAA 149
P.virgatus           TGCTCCTGCCGC-AAGGCTCTGCGAGGTGGGAAACACCACGCAGTTGCCTTGAAAACCAA 149
P.roseus             TGCCCCTGACGC-ACGGCCTCGTGGGGTGCCATGCTCCTTACGGGTGCCCAAAAAACAAA 149
                                  *    *   **    **      *         *         **     *   *

P.debilis            CCCCGGCGCGGAAAGCGCCAAGGAAAATAAACATACAAGCGAGAACCCTCTAATCACCCT 206
P.discolor           CCCCGGCGCGGAAAGCGCCAAGGAAAATAAACATACAAGCGAGAACCCTCTAATCACCCT 194
P.graveolens         CCCCGGCGCGGAAAGCGCCAAGGAAAATAAACATACAAGCGAGAACCCTCTAATCACCCT 289
P.niruri             CCC-GGCGCGGAAAGCGCCAAGGAAAATGAACATACAAGCGAGAACCCAACAGGCTCCCC 206
P.madagascariensis   CCC-GGCGCGGAAAGCGCCAAGGAAAATGAACATACAAGCGAGAACCCAACAGGCTCCCC 133
P.acidus             CCC-GGCGCGGAAAGCGCCAAGGAAAATGAACATACAAGCGAGAACCCAACAGGCTCCCC 288
P.polyphyllus        CCCCGGCGCGGAAAGCGCCAAGGAACATTAACAAAAGAGCGAGAACCTTGCAATC-TCCC 287
P.acuminatus         CCCCGGCGCGGAAAGCGCCAAGGAAAATAAACGAAAAAGCGAGACCCGCGGGCT-GCCC 288
P.oxyphyllus         CCCCGGCGCGGAAAGCGCCAAGGAAAATAAACGGACAAGCGAGATTGCCACGATCAGCTC 289
P.sikkimensis        CCCCGGCGCGGAAAGCGCCAAGGAAAACAAACGAACAAGCGAGATTGCCACGATCAGCCC 290
P.rheedii            CCCCGGCGCGGAATGCGCCAAGGAAAATGAACATACAAGAGAGAACCCTGCAATCTCCTC 288
P.harmandii          CCCCGGCGCGGAATGCGCCAAGGAAAACGAATATAAAAGAGAGAACTCTACATTCACCTC 208
P.collinsiae         CCCCGGCGCGGAATGCGCCAAGGAAAACGAATATAAAAGAGAGAACTCTACATTCACCTC 208
P.angkorensis        CCCCGGCGCGGAATGCGCCAAGGAAAACGAATATAAAAGAGAGAACTCTACATTCACCTC 208
P.gracilipes         CCCCGGCGCGGAATGCGCCAAGGAAAACGAATATAAAAGAGAGAACTCTACATTCACCTC 208
P.lingulatus         CCCTGGCGCGGATTGCGCCAAGGAAAACGAATATAAAAGAGAGAACTCTACATTCACCTC 207
P.microcarpus        CCCTGGCGCGGATTGCGCCAAGGAAAACGAATATAAAAGAGAGAACTCTACATTCACCTC 208
P.clarkei            CCCCGGCGCGGAATGCGCCAAGGAAAACGAATCTAAAAGGTAGAACTCTACATTCACCTC 288
P.welwitschianus     CCCCGGCGCGGAATGCGCCAAGGAAAACGAATCTAAAAGAGAGAACTCTACGTTCGCCTC 317
P.reticulatus        CCCCGGCGCGGAATGCGCCAAGGAAAACAAATCTAAAAGAGAGAACTCTACATTCACCTC 288
P.columnaris         CCCCGGCGCGGAATGCGCCAAGGAAAACGAATCTAAAAGAGAGAACTCTACATTCACCTC 208
P.nummulariifolius   CCCCGGCGCGGAAAGCGCCAAGGAAAATGAACAGACAAGCGAGAATCCCACAATCACCCC 356
P.orientalis         CCCCGGCGCGGAAAGCGCCAAGGAACACGAACGTATAAGCGAGAACCCGTCGAACACCCC 206
P.emblica            CCCCGGCGCGGAAAGCGCCAAGGAAAATGAACGGACGAGAGAGAGGCCCCCCATTCACCCC 222
P.acutissimus        CCCCGGCGCGGAAAGCGCCAAGGAAAAT-ATTGAGGAAGCGAGACCGACCCAAT-ACTTG 209
P.pulcher            CCCCGGCGCGGAAAGCGCCAAGGAAAAT-AACGAGGAAGAGAGATCGACCCAATTACTTG 210
P.myrtifolius        CCCCGGCGCGGAAAGCGCCAAGGAAAACTAATGAAAAGCGAGAACTCCCTCATTCACCCC 209
P.virgatus           CCCCGGCACGGAAAGTGCCAAGGAATACTATCGAAAAAGTGAATACTCCCTCATCGACCC 209
P.roseus             CCCCGGCGCGGAAAGCGCCAAGGAAAACAAACAAGAGAGTGAGACGACGCTCATCACCCT 209
                     ***  ***   ****   *  ********   *     *       **    *

P.debilis            GGAGAC--TCCGGTGCGTGTTTGGTAGGTGTTTCTCCTTTAGTAACAAAAACGACTCTCG 264
```

FIGURE 4.4 Clustal 2.0.12 multiple-sequence alignment of the species belonging to genus *Phyllanthus*. (*continued*)

Phylogenetic Analysis of *Phyllanthus* Species

```
P.discolor          GGAGAC--TCCGGTGCGTGTTTGGTAGGTGTTTCTCCTTTAGTAACAAAAACGACTCTCG 252
P.graveolens        GGAGACGGTCCGGTGCGTGTTTGGTAGGTGTTTCTCCTTTAGTAACAAAAACGACTCTCG 349
P.niruri            GGAAAC-----GGTGCGTGCTTTGCTGA-GTTTCTCCTTACGTAACCAAAACGACTCTCG 260
P.madagascariensis  GGAAAC-----GGTGCGTGCTTTGCTGA-GTTTCTCGTTACGTAACCAAAACGACTCTCG 187
P.acidus            GGAAAC-----GGTGCGTGCTTTGCTGA-GTTTCTCCTTACGTAACCAAAACGACTCTCG 342
P.polyphyllus       CGGAAC-----GGTGCGTGCTTTGCAG-CGTCTCTCATTTCTTAATCAAAACGACTCTCG 341
P.acuminatus        GGAAAC-----GGTGCGTGCTTTGTAGGTGTTTCTC-CTTCATAACAAAAACGACTCTCG 342
P.oxyphyllus        GGAAAC-----GGTGTGTGTTTCGTAGGCATTTCTCCTTTCACTACTAAAACGACTCTCG 344
P.sikkimensis       GGAAAC-----GGTGCGTGTTTTGTAGGCATTACTCCTTTCACTACTAGAACGACTCTCG 345
P.rheedii           GGAAAC-----GATGCATGATTGTTAGGTGTGTCTCCTTTCGTAACCAAAACGACTCTCG 343
P.harmandii         GGAAAC-----GATGTGTGCATCGTAGTTGCTTCTCCTTTCATAACCAAAACGACTCTCG 263
P.collinsiae        GGAAAC-----GATGTGTGCATCGTAGTTGCTTCTCCTTTCATAACCAAAACGACTCTCG 263
P.angkorensis       GGAAAC-----GATGTGTGCATCGTAGTTGCTTCTCCTTTCATAACCAAAACGACTCTCG 263
P.gracilipes        GGAAAC-----GATGTGTGCATCGTAGTTGCTTCTCCTTTCATAACCAAAACGACTCTCG 263
P.lingulatus        GGAAAC-----GATGTGTGCATCGTAGTTGCTTCTCCTTTCATAACCAAAACGACTCTCG 262
P.microcarpus       GGAAAC-----GATGTGTGCATCGTAGTTGCTTCTCTTTTCATAACCAAAACGACTCTCG 263
P.clarkei           GGAAAC-----GATGTGTGAATGGTAGTTGCTTCTTCTTTCATAACCAAAACGACTCTCG 343
P.welwitschianus    GGAAAC-----GATGTGTGCATG-TAGTTGAATCTCCTTTCATAACCAAAACGACTCTCG 371
P.reticulatus       GGAAAC-----GATGTGTGAATGGTAGTTGCTTCTCCTTTCATAACCAAAACGACTCTCG 343
P.columnaris        GGAAAC-----GATGTGTGCTTGGTAGTTGCTTCTCCTTTCATAACCAAAACGACTCTCG 263
P.nummariifolius    GGAAAC-----GGTGCGTGCTTGGTAGGTGTTTCTCCTTTCATAACCAAAACGACTCTCG 411
P.orientalis        GGAAAC-----GGTGCGTGCTCGATAGGCGTTGCTCCTTTT-TAAATGGAACGACTCTCG 260
P.emblica           GGAAAC-----GGTGCGTGCCTGGGATGCGT-TCTCCTTTCATAACCAAAACGACTCTCG 276
P.acutissimus       GGAAAC---------------------GTTCTCCTT----AA--AAAACGACTCTCG 239
P.pulcher           GGAAAT---------------------GTTCTCCCT----AA--AAAACGACTCTCG 240
P.myrtifolius       AGAAAT-----GGTGAGTGTGTGGGAGGTGTTTCTCCTTGCAATACTAAAACGACTCTCG 264
P.virgatus          AGGAAT-----GGTGCGCGCATGAGGGGTGTTTATACTTGGAAAACTAAAACGACTCTCG 264
P.roseus            GGAGAC-----AGCGTGTGCATGGGAGGACTCTCTCCTTTCAATACTAAAACGACTCTCG 264
                     *   *                                 *    *    * ***********

P.debilis           GCAACGGATATCTCGGCTCTCGCATCGATGAAGAACGTAGCGAAATGCGATACTTGGTGT 324
P.discolor          GCAACGGATATCTCGGCTCTCGCATCGATGAAGAACGTAGCGAAATGCGATACTTGGTGT 312
P.graveolens        GCAACGGATATCTCGGCTCTCGCATCGATGAAGAACGTAGCGAAATGCGATACTTGGTGT 409
P.niruri            GCAACGGATATCTCGGCTCTCGCATCGATGAAGAACGTAGCGAAATGCGATACTTGGTGT 320
P.madagascariensis  GCAACGGATATCTCGGCTCTCGCATCGATGAAGAACGTAGCGAAATGCGATACTTGGTGT 247
P.acidus            GCAACGGATATCTCGGCTCTCGCATCGATGAAGAACGTAGCGAAATGCGATACTTGGTGT 402
P.polyphyllus       GCAACGGATATCTCGGCTCTCGCATCGATGAAGAACGTAGCGAAATGCGATACTTGGTGT 401
P.acuminatus        GCAACGGATATCTCGGCTCTCGCATCGATGAAGAACGTAGCGAAATGCGATACTTGGTGT 402
P.oxyphyllus        GCAACGGATATCTCGGCTCTCGCATCGATGAAGAACGTAGCGAAATGCGATACTTGGTGT 404
P.sikkimensis       ACAACGGATATCTCGGCTCTCGCATCGATGAAGAACGTAGCGAAATGCGATACTTGGTGT 405
P.rheedii           GCAACGGATATCTCGGCTCTCGCATCGATGAAGAACGTAGCGAAATGCGATACTTGGTGT 403
P.harmandii         GCAACGGATATCTCGGCTCTCGCATCGATGAAGAACGTAGCGAAATGCGATACTTGGTGT 323
P.collinsiae        GCAACGGATATCTCGGCTCTCGCATCGATGAAGAACGTAGCGAAATGCGATACTTGGTGT 323
P.angkorensis       GCAACGGATATCTCGGCTCTCGCATCGATGAAGAACGTAGCGAAATGCGATACTTGGTGT 323
P.gracilipes        GCAACGGATATCTCGGCTCTCGCATCGATGAAGAACGTAGCGAAATGCGATACTTGGTGT 323
P.lingulatus        GTAACGGATATCTCGGCTCTCGCATCGAAGAAGAACGTAGCAAAATGCGATACTTGGTGT 322
P.microcarpus       GTAACGGATATCTCGGCTCTCGCATCGATGAAGAACGTAGCGAAATGCGATACTTGGTGT 323
P.clarkei           GCAACGGATATCTCGGCTCTCGCATCGATGAAGAACGTAGCGAAATGCGATACTTGGTGT 403
P.welwitschianus    GCAACGGATATCTCGGCTCTCGCATCGATGAAGAACGTAGCGAAATGCGATACTTGGTGT 431
P.reticulatus       GCAACGGATATCTCGGCTCTCGCATCGATGAAGAACGTAGCGAAATGCGATACTTGGTGT 403
P.columnaris        GCAACGGATATCTCGGCTCTCGCATCGATGAAGAACGTAGCGAAATGCGATACTTGGTGT 323
P.nummariifolius    GCAACGGATATCTCGGCTCTCGCATCGATGAAGAACGTAGCGAAATGCGATACTTGGTGT 471
P.orientalis        GCAACGGATATCTCGGCTCTCGCATCGAAGAAGAACGTAGCGAAATGCGATACTTGGTGT 320
P.emblica           GCAACGGATATCTCGGCTCTCGCATCGATGAAGAACGTAGCGAAATGCGATACTTGGTGT 336
P.acutissimus       GCAACGGATATCTCGGCTCTCGCATCGATGAAGAACGTAGCGAAATGCGATACTTGGTGT 299
P.pulcher           GCAACGGATATCTCGGCTCTCGCATCGATGAAGAACGTAGCGAAATGCGATACTTGGTGT 300
P.myrtifolius       GCAACGGATATCTCGGCTCTCGCATCGATGAAGAACGTAGCGAAATGCGATACTTGGTGT 324
P.virgatus          GCAACGGATATCTCGGCTCTCGCATCGATGAAGAACGTAGCGAAATGCGATACTTGGTGT 324
P.roseus            GCAACGGATATCTCGGCTCTCGCATCGATGAAGAACGTAGCGAAATGCGATACTTGGTGT 324
                    *********************** ** **** ************ ***************

P.debilis           GAATTGCAGAATCCCGTGAACCATCGAGTTTTTGAACGCAAGTTGCGCCCGAAGCCTTTT 384
P.discolor          GAATTGCAGAATCCCGTGAACCATCGAGTTTTTGAACGCAAGTTGCGCCCGAAGCCTTTT 372
```

FIGURE 4.4 Clustal 2.0.12 multiple-sequence alignment of the species belonging to genus *Phyllanthus*. (*continued*)

```
P.graveolens         GAATTGCAGAATCCCGTGAACCATCGAGTTTTTGAACGCAAGTTGCGCCCGAAGCCTTTT 469
P.niruri             GAATTGCAGAATCCCGTGAACCATCGAGTTTTTGAACGCAAGTTGCGCCCAACGCCTTCG 380
P.madagascariensis   GAATTGCAGAATCCCGTGAACCATCGAGTTTTTGAACGCAAGTTGCGCCCAACGCCTTCG 307
P.acidus             GAATTGCAGAATCCCGTGAACCATCGAGTTTTTGAACGCAAGTTGCGCCCAACGCCTTCG 462
P.polyphyllus        GAATTGCAGAATCCCGTGAACCATCGAGTTTTTGAACGCAAGTTGCGCCCAAAGCCTTCG 461
P.acuminatus         GAATTGCAGAATCCCGTGAACCATCGAGTCTTTGAACGCAAGTTGCGCCCAAGGCCTTCG 462
P.oxyphyllus         GAATTGCAGAATCCCGTGAACCATCGAGTCTTTGAACGCAAGTTGCGCCCAAAGCCTTCG 464
P.sikkimensis        GAATTGCAGAATCCCGTGAACCATCGAGTTTTTGAACGCAAGTTGCGCCCAAAGCCTTCG 465
P.rheedii            GAATTGCAGAATCCCGTGAACCATCGAGTCTTTGAACGCAAGTTGCGCCCAAAGCCTTCG 463
P.harmandii          GAATTGCAGAATCCCGTGAACCATCGAGTCTTTGAACGCAAGTTGCGCCCAAAGCCTTCG 383
P.collinsiae         GAATTGCAGAATCCCGTGAACCATCGAGTCTTTGAACGCAAGTTGCGCCCAAAGCCTTCG 383
P.angkorensis        GAATTGCAGAATCCCGTGAACCATCGAGTCTTTGAACGCAAGTTGCGCCCAAAGCCTTCG 383
P.gracilipes         GAATTGCAGAATCCCGTGAACCATCGAGTCTTTGAACGCAAGTTGCGCCCAAAGCCTTCG 383
P.lingulatus         GAATTGCAGAATCCTGTGAACCATCGAGTCTTTGAACGCAAGTTGCGCCCAAAGCCTTCG 382
P.microcarpus        GAATTGCAGAATCCTGTGAACCATCGAGTCTTTGAACGCAAGTTGCGCCCAAAGCCTTCG 383
P.clarkei            GAATTGCAGAATCCCGTGAACCATCGAGTCTTTGAACGCAAGTTGCGCCCAAAGCCTTCG 463
P.welwitschianus     GAATTGCAGAATCCCGTGAACCATCGAGTCTTTGAACGCAAGTTGCGCCCAAAGCCTTCG 491
P.reticulatus        GAATTGCAGAATCCCGTGAACCATCGAGTCTTTGAACGCAAGTTGCGCCCAAAGCCTTCG 463
P.columnaris         GAATTGCAGAATCCCGTGAACCATCGAGTCTTTGAACGCAAGTTGCGCCCAAAGCCTTCG 383
P.nummulariifolius   GAATTGCAGAATCCCGTGAACCATCGAGTCTTTGAACGCAAGTTGCGCCCAAAGCCTTCG 531
P.orientalis         GAATTGCAGAATCCCGTGAACCATCGAGTCTTTGAACGCAGGTTGCGCCCAAAGCCTTCG 380
P.emblica            GAATTGCAGAATCCCGTGAACCATCGAGTTTTTGAACGCAAGTTGCGCCCAAAGCCTTCG 396
P.acutissimus        GAATTGCAGAATCCCGTGAACCATCGAGTTTTTGAACGCAAGTTGCGCCCAAAGCCTTCG 359
P.pulcher            GAATTGCAGAATCCCGTGAACCATCGAGTTTTTGAACGCAAGTTGCGCCCGAAGCCTTCG 360
P.myrtifolius        GAATTGCAGAATCCCGTGAACCATCGAGTCTTTGAACGCAAGTTGCGCCCAAAGTCCTCG 384
P.virgatus           GAATTGCAGAATCCCGTGAACCATCGAGTCTTTGAACGCAAGTTGCGCCCAAAGCCTTCG 384
P.roseus             GAATTGCAGAATCCCGTGAACCATCGAGTCTTTGAACGCAAGTTGCGCCCAAAGCCTTCG 384
                     **************  *************** ********** ********* *  * *

P.debilis            GGTCGAGGGCACGTCTGCCTGGGTGTCACGCAACGTCGCTCCCTCACTCC-CTCACTCGA 443
P.discolor           GGTCGAGGGCACGTCTGCCTGGGTGTCACGCAACGTCGCTCCCTCACTCC-CTCACTCGA 431
P.graveolens         GGTCGAGGGCACGTCTGCCTGGGTGTCACGCAACGTCGCTCCCTCACTCC-CTCACTCGA 528
P.niruri             GGTCGAGGGCACGTCTGCCTGGGTGTCACGCAACGTCGCTCCCTCACTCC-CGCGT---- 435
P.madagascariensis   GGTCGAGGGCACGTCTGCCTGGGTGTCACGCAACGTCGCTCCCTCACTCC-TGCGT---- 362
P.acidus             GGTCGAGGGCACGTCTGCCTGGGTGTCACGCAACGTCGCTCCCTCACTCC-CGCGT---- 517
P.polyphyllus        GGTGGAGGGCACGTCTGCCTGGGTGTCACGCAACGTCGCTCCCACGCCATGCTTTGCATT 521
P.acuminatus         GGTTGAGGGCACGTCTGCCTGGGTGTCACGCAACGTCGCCCCATCACCC--CCTTCAATT 520
P.oxyphyllus         GGTCGAGGGCACGTCTGCCTGGGTGTCACGCAACGTCGCTCCCTCACTCC-CCTCATTCA 523
P.sikkimensis        GGTCGAGGGCACGTCTGCCTGGGTGTCACGCAACGTCGCTCCTTCACTCC-CCTAATCCA 524
P.rheedii            GGTTGAGGGCACGTCTGCCTGGGTGTCACGCAACGTCGCTCCCACACAC--CCTATTAAG 521
P.harmandii          GGTCGAGGGCACGTCTGCCTGGGTGTCACGCAACGTCGCTCCCTCACTTCCCACATGTAG 443
P.collinsiae         GGTCGAGGGCACGTCTGCCTGGGTGTCACGCAACGTCGCTCCCTCACTTCCCACATGTAG 443
P.angkorensis        GGTCGAGGGCACGTCTGCCTGGGTGTCACGCAACGTCGCTCCCTCACTTCCCACATGTAG 443
P.gracilipes         GGTCGAGGGCACGTCTGCCTGGGTGTCACGCAACGTCGCTCCCTCACTTCCCACATGTAG 443
P.lingulatus         GGTCGAGGGCACGTCTGCCTGGGTGTCACGCAACGTCGCTCCCTCACTTCCCACATGTAG 442
P.microcarpus        GGTCGAGGGCACGTCTGCCTGGGTGTCACGCAACGTCGCTCCCTCACTT---ACATGTAG 440
P.clarkei            GGTCGAGGGCACGTCTGCCTGGGTGTCACGCAACGTCGCTCCCTCACTTCCCTCATGTAG 523
P.welwitschianus     GGTCGAGGGCACGTCTGCCTGGGTGTCACGCAACGTCGCTCCCTCACTTCCCTCTTGTGG 551
P.reticulatus        GGTCGAGGGCACGTCTGCCTGGGTGTCACGCAACGTCGCTCCCTCACTTCCCTCATGTAG 523
P.columnaris         GGTCGAGGGCACGTCTGCCTGGGTGTCACGCAACGTCGCTCCCTCACTTCCTTCAAGTAG 443
P.nummulariifolius   GGTCGAGGGCACGTCTGCCTGGGTGTCACGCAACGTCGCTCCCTCACTCCCCTCATATGG 591
P.orientalis         GGTCGAGGGCACGTCTGCCTGGGTGTCACGCAACGTCGCTCCCTCACTCC--TGTTTGG 438
P.emblica            GGTCGAGGGCACGTCTGCCTGGGTGTCACGCAACGTCGCTCCCCAATCCCCTGCTCTGG 456
P.acutissimus        GGCTGAGGGCACGTCTGCCTGGGTGTCACGCAACGTCGCTCCCTCCAACC-CTCTGTTGG 418
P.pulcher            GGCCGAGGGCACGTCTGCCTGGGTGTCACGCAACGTCGCTCCCTCCAACC-CTTTGCTGG 419
P.myrtifolius        GGTCGAGGGCACGTCTGCCTGGGTGTCACGCAACGTCGCTCCTCATTCTTCC-CACAGGTTG 443
P.virgatus           GGCTAAGGGCACGTCTGCCTGGGTGTCACGCAACGTCGCTCCACCTTTCT-TATCAATTG 443
P.roseus             GGCTGAGGGCACGTCTGCCTGGGTGTCACGCAACGTCGCTCTATCCATCC-CATTTGTGG 443
                     **  *****************************  *

P.debilis            GGCATGT-GAATT--CGGGGCGGAAAATGGCCTCCCGTGAACTTCGTGATCGCGGTTGGC 500
P.discolor           GGCATGT-GAATT--CGGGGCGGAAAATGGCCTCCCGTGAACTTCGTGATCGCGGTTGGC 488
P.graveolens         GGCATGT-GAATT--CGGGGCGGAAAATGGCCTCCCGTGAACTTCGTGATTGCGGTTGGC 585
```

FIGURE 4.4 Clustal 2.0.12 multiple-sequence alignment of the species belonging to genus *Phyllanthus*. (*continued*)

Phylogenetic Analysis of *Phyllanthus* Species

```
P.niruri              GGAACGT-GAATT--TCGAGCGGAATATGGCCTCCCGTGAACTCTTCGATCGCGGTTGGC 492
P.madagascariensis    GGATCGT-GAGTT--TCGAGCGGAATATGGCCTCCCGTGAACTCTTCGATCGCGGTTGGC 419
P.acidus              GGATCGT-GAATT--TCGAGCGGAATATGGCCTCCCGTGAACTCTTCGATCGCGGTTGGC 574
P.polyphyllus         GGAACGT-GATTT--TGGGGCGGAATATGGTCTCCCGTGGGCTTTAGAAT-GCGGTTGGC 577
P.acuminatus          AGGAATC-GTGTT--TGTGGCGGAAATGGCGTGGCGTGATCTTTACGGTCGCGGTTGGC 577
P.oxyphyllus          GGATTGT-GAACC--AGGAGCGGAAAATGGCTTCCCGTGAACTTTGTGATTGCGGTTGGC 580
P.sikkimensis         GGGTTGT-GAACT--TGGTGCGGAAAATGGCTTCCCGTGAACTTTGTGATTGCGGTTGGC 581
P.rheedii             GGATGCT-GAATT--TGGGGCGGAATATGGCCTCCCGTGAGATTATAAATTGCGGTTGGC 578
P.harmandii           GGCTCGT-GAATA--TGGGGCGGAAAATGGCTTCCCATGAACCTCAAGATTGTGGTTGGC 500
P.collinsiae          GGCTCGT-GAATA--TGGGGCGGAAAATGGCTTCCCATGAACCTCAAGATTGTGGTTGGC 500
P.angkorensis         GGCTCGT-GAATA--TGGGGCGGAAAATGGCTTCCCATGAACCTCAAGATTGTGGTTGGC 500
P.gracilipes          GGCTCGT-GAATA--TGGGGCGGAAAATGGCTTCCCATGAACCTCAAGATTGTGGTTGGC 500
P.lingulatus          GGCTCGTAGAATA--TGGGGCGGAATATGGTTTCCCATGAACCTCAAGATTGTGGTTGGC 500
P.microcarpus         GGCTCGT-GAATA--TGGGGCGGAAAATGGCTTCCCATGAACCTCAAGATTGTGGTTGGC 497
P.clarkei             GGCTCGT-GAATT--TGGGGCGGAAAATGGCTTCCCATGAACCTCAAGATTGTGGTTGGC 580
P.welwitschianus      GGCTCGT-GAATT--GGGAGCGGAAAATGGTTTCCCATAAACTTCGAGATTGTGGTTGGC 608
P.reticulatus         GGCTCGT-GAATT--TGGGGCGGAAAATGGCTTCCCATGAACCTCAAGATTGTGGTTGGC 580
P.columnaris          GGCTCGT-GAATT--TGGGGCGGAAAATGGCTTCCCATGAACCTAACGATTGTGGTTGGC 500
P.nummulariifolius    GGCTCGT-GAATT--CGGGGCGGAAAATGGCCTCCCGTGAACTTCGTGATTGCGGTTGGC 648
P.orientalis          GGCTCGT-GAGTT--TGGTGCGGAAAATGGCCTCCCGTTAGCCCTGAGTTAGCGGTTGGC 495
P.emblica             GGCTCTT-GAGTG--CGGAGCGGAGAATGGCCTCCCGTGACCTTGTCGTCTGCGGTTGGC 513
P.acutissimus         GGTCTTTCGAAGG--TTGAGCGGAGAATGGCCTCCCGTGAGCTTATG-TTCGCGGTTGGC 475
P.pulcher             GGTCTTTCGAAGG--TTGAGCGGAAAATGGCCTCCCGTGAGCCTATGGTTTGCGGTTGGC 477
P.myrtifolius         GGATCATGAA-TGAGTTGAGCGGAAAGTGGCCTCCCGTGAGCTAGTTGTTCGTGGTTGGC 502
P.virgatus            GGATAAAAAATGAGTTGAGCGGAAAGTGGCCTCCCGTGAGCTTATTGTTCGTGGTTGGC 503
P.roseus              GTCTTG---------TGAGCGGAAAATGGCCTCCCATGAGCAGAGAACCCGTGGTTGGC 493
                                       *****   *** ****  *              * *******

P.debilis             CCAAACAT-GAGACCAATTCGGCCAATGCCGTGGCATTCGGTGGTT-GAAAATACCT-TA 557
P.discolor            CCAAACAT-GAGACCAATTCGGCCAATGCCGTGGCATTCGGTGGTT-GAAAATACCT-TA 545
P.graveolens          CCAAACAT-GAGACCAATTCGGCCAATGCCGTGGCATTCGGTGGTT-GAAAATACCT-TA 642
P.niruri              CTAAACAC-GAGACCATGTCGGCCAATGCCGTGGCATTCGGTGGTT-GAAA-TACCC-TC 548
P.madagascariensis    CTAAACAC-GAGACCATGTCGGCCAATGCCGTGGCATTCGGTGGTT-GAAA-TACCC-TC 475
P.acidus              CTAAACAC-GAGACCATGTCGGCCAATGCCGTGGCATTCGGTGGTT-GAAA-TACCC-TC 630
P.polyphyllus         CTAAACAT-GATACCATTTCGGCCAATGTCGTGGCATTCGGTGGTT-GA-AATACCC-TC 633
P.acuminatus          CCAAAAGTTGAGACCAACTCGGCCAATTGTCGTGGCATTCGGTGGAT-GACAATACCC-TC 635
P.oxyphyllus          CCAAAAAT-GAGACCTAGTCGGCCAATGTCGTGGCAATCGGTGGTA-GGAAATGCCCCTC 638
P.sikkimensis         CCAAAAAT-GAGACCAAGTCGGCCAGTGTCGTGGCAATCGGTGGTA-GAAAATGCCCCTC 639
P.rheedii             CTAAATGT-GAGACCTTGTCGGCCAGTGCCGTGGCATTCGGTGGTT-GA-AATACCC-TC 634
P.harmandii           CCAAACAT-GAGACCAAGTCGGTCAGTGCCGTGGCATTCGGTGGTTTGAAAATACCCATT 559
P.collinsiae          CCAAACAT-GAGACCAAGTCGGTCAGTGCCGTGGCATTCGGTGGTT-GAAAATACCCATT 558
P.angkorensis         CCAAACAT-GAGACCAAGTCGGTCAGTGCCGTGGCATTCGGTGGTT-GAAAATACCCATT 558
P.gracilipes          CCAAACAT-GAGACCAAGTCGGTCAGTGCCGTGGCATTCGGTGGTT-GAGCATACCCATT 558
P.lingulatus          CCACACATAGACACCAAGTCGGACAGTGCCGTGGTATTCGGTGGTT-GAAAATACCCATT 559
P.microcarpus         CCAAACAT-GAGACCAAGTCGGTCAGTGCCGTGGCATTCGGTGGTT-GAAAATACCCATT 555
P.clarkei             CCAAACAT-GAGACCAAGTCGGTCAATGCCGTGGCATTCGGTGGTT-GAAAATACCC-TA 637
P.welwitschianus      CCAAACAT-GAGACCAGGTCGGTCAGTGCCGTGGCATTCGGTGGTT-GAAAATACCC-TA 665
P.reticulatus         CCAAACAT-GAGACCAAGTCGGTCAGTGCCGTGGCATTCGGTGGTT-GAAAATACCC-TA 637
P.columnaris          CCAAACAT-GAGACCAAGTCGGTCAGTGCCGTGGCATTCGGTGGTT-GAAAATACCC-TA 557
P.nummulariifolius    CCAAACAT-GAGACCAAGTCGGCCAGTGCCGTGGCATTCGGTGGTT-GAAAATACCC-TC 705
P.orientalis          CCAAACAT-GAGACCAAGTCGGCCAGTGTCGTGGCATACGGTGGTT--GAATTACCT-TC 551
P.emblica             CCAAAAAT-GAGTCCAAGTCGGCGGGTGCCGCGGCGTTCGGTGGTT-GAAAATACCC-TT 570
P.acutissimus         CCAAAAAATGAGACAAAGTCGGTGAACGACGCGAACATTCGGTGGTTTAAAAATTACCCTT 535
P.pulcher             CCAAAAAATGAGACCAAGTCGGTGAACGATGCGACATTCGGTGGTTTAAAATTGCCCTT 537
P.myrtifolius         TAAAACAT-GAGTCCAAGTCGGCGAGTGCCGTGGCATTCGGTGGTTGTAACATCACT-AG 560
P.virgatus            TAAAACAC-GAGTCCAAGTCGGCAAGTGCCATGGCATTCGGTGGTTGTAACATCACT-TG 561
P.roseus              CTAAACAT-GAGTCCAAGTCGGCGAGTGCCGTGGCATTTGGTGGTTGAAA-ATCCT-TT 550
                       *  *  **  *   ****      *      *         ****     *   *

P.debilis             CTATTCCCTCGTTCATTTGTCCG-AACAAACAAGG-ATCTCGACGACCC-TCTATGTATC 614
P.discolor            CTATTGCCTCGTTCATTTGTCCG-AACAAACAAGG-ATCTCGACGACCC-TCTATGTATC 602
P.graveolens          CTATTGCCTCGTTCATTTGTCCG-AACCAACAAGG-ATCTCGACGACCC-TCTATGTATC 699
P.niruri              AAAACGCCTCGTTCGTTTGGCCG-CGCTAAAAAGG-TTTTCAACGACCC-TCTAC-TATC 604
```

FIGURE 4.4 Clustal 2.0.12 multiple-sequence alignment of the species belonging to genus *Phyllanthus*. (continued)

```
P.madagascariensis  AAAACGCCTCGTTCATTTGGCCG-CGCTAAAAAGG-TTTTCAACGACCC-TCTAC-TATC 531
P.acidus            AAAACGCCTCGTTCATTTGGCCG-CGCTAAAAAGG-TTTTCAACGACCC-TCTAC-TATC 686
P.polyphyllus       ACAAAGCCACGCCCATTTGGTCG-ATTTGATAAGG-TTTTCAATGATCT-TCTATT-ATT 689
P.acuminatus        AAAACGCCACGTTCATTTTGCCG-GGCCAATAACG-ATCTCAATGACCC-TTAGTTTATC 692
P.oxyphyllus        AAAATGCCTCGT-CATTTGGTCG-AACTGAAAATG-TTTTCAACGACCC-TCAACGTATT 694
P.sikkimensis       AAAACGCCTCGT-CATTTGGTCC-AACTGAAAATG-TTCTCAACGACCC-TCAACGTATC 695
P.rheedii           AAAACGCCTCGTTCATT-GCCCG-AAGCAATAAGG-TTCTCAATAACCC-TTAATGTATT 690
P.harmandii         AAAACGCCTCGTTCATTTGGCCGGAACCAACAAGGGATCTCAACGACCC-TCTATGTATT 618
P.collinsiae        AAAACGCCTCGTTCATTTGGCCG-AACCAACAAGG-ATCTCAACGACCC-TCTATGTATT 615
P.angkorensis       AAAACGCCTCGTTCATTTGGCCG-AACCAACAAGG-ATCTCAACGACCC-TCTATGTATT 615
P.gracilipes        AAAACGCCTCGTTCATTTGGCCG-AACCAACAAGG-ATCTCAACGACCC-TCTATGTATT 615
P.lingulatus        AAAACGCCTCGTTTATTTGGCCG-AACCAACGAGG-ATCTCAACGACCC-TCTATGTATT 616
P.microcarpus       AAAACGCCTCGTTCATTTG-CCA-AACCAACAAGG-ATCTCAACGACCC-TCTATGTATT 611
P.clarkei           AAAACGCCTCGTTCATTTGGCCG-AACCAACAAGG-ATCTCAACGACCC-TCTATGTATC 615
P.welwitschianus    AAAACGCCTCGTTCATTTGGCCG-AACCAGCACGG-ATCTCAACGACCC-TCTATGTATC 722
P.reticulatus       AAAACGCCTCGTTCATT-GGCCG-AACCAACAAGG-ATCTCAACGACCC-TCTATGTATC 693
P.columnaris        AAAACGCCTCGTTCATTTGGCCG-AACAAACAAGG-ATCTCAACGACCC-TCTATGTATT 614
P.nummulariifolius  AAAACGCCTCGTTCATTTGGCCG-AACGAACAAGG-ATCTCAACGACCC-TCTATGTATC 762
P.orientalis        AGAATGCCGCGTTCATTTGTCCG-AACAAAGTATGGTTCTCGACGACCC-TCTATGTATT 609
P.emblica           ACGATGCCGCTCTCATTCGGCCG-ATCCAGCATGG-ATCTCAACGACCC-TCTTTGTATC 627
P.acutissimus       ACAGTGTCGTCGTCGTAGTGCCA-GTCGAACAAGG-ATCTCAACGACCC-TCACAACTTT 592
P.pulcher           ACAGTGTCGTCGTCGCACTGCCG-ATCGAACAAGG-ATCTCAACGACCC-TCATAACTCT 594
P.myrtifolius       ATATTGCCGCTTTCATATGGCCAAA-CCAACAAGG-ATCTCTAGGACCC-TTCATCTATT 617
P.virgatus          ATATTGCCGTAATTACCTGGCCAAAACCAATAAGG-ATCTCAAGGACCCCTTTATATAAC 620
P.roseus            GGACGGATGCTCACATTTGGCCGGG-CCAACAAGG-ATTTCAAGGACCT-TTCTTACATC 607
                           *            *          *  ** *    * *   *

P.debilis           CGACGCGACCCCAGGTCAGGCGGGATTACCCGCTGAGTTTAAGCTAATTAAAAGGGGGAA 674
P.discolor          CGACGCGACCCCAGGTCAGGCGG------------------------------------ 625
P.graveolens        CGACGCGACCCCAGGTCAGGCGGGATTACCCGCTGAGTTTAAGCATATCAATAAGCGGAG 759
P.niruri            CGACGCGACCCCAGGTCAGGCGGGATTACCCGCTGAGTTTAAGCATAAATTAAAGCGGCG 664
P.madagascariensis  CGACGCGACCCCAGGTCAGGCGG------------------------------------ 554
P.acidus            CGACGCGACCCCAGGTCAGGCGGGATTACCCGCTGAGTTTAAGCATATCAATAAGCGGAG 746
P.polyphyllus       TGACGCGACCCCAGGTCAGGCGGGATTACCCGCTGAGTTTAAGCATATCAATAAGCGGAG 749
P.acuminatus        CGACGCGACCCCAGGTCAGGCGGGATTACCCGCTGAGTTTAAGCATATCAATAAGCGGAG 752
P.oxyphyllus        CGACGCGACCCCAGGTCAGGCGGGATTACCCGCTGAGTTTAAGCATATCAATAAGCGGAG 754
P.sikkimensis       CGACGCGACCCCAGGTCAGGCGGGATTACCCGCTGAGTTTAAGCATATCAATAAGCGGAG 755
P.rheedii           CGACGCGACCCCAGGTCAGGCAGGATTACCCGCTGAGTTTAAGCATATCAATAAGCGGAG 750
P.harmandii         CGACGCGACCC------------------------------------------------ 629
P.collinsiae        CGACGCGACCC------------------------------------------------ 626
P.angkorensis       CGACGCGACCC------------------------------------------------ 626
P.gracilipes        CGACGCGACCC------------------------------------------------ 626
P.lingulatus        CGACGCGACCC------------------------------------------------ 627
P.microcarpus       CGACGCGACCC------------------------------------------------ 622
P.clarkei           TGACGCGACCCCAGGTCAGGCGGGATTACCCGCTGAGTTTAAGCATATCAATAAGCGGAG 754
P.welwitschianus    CGACGCGACCCCAGGTCAG-CGGGATTACCCGCTGAGTTTAAGCATATCAATAAGCGGAG 781
P.reticulatus       CGACGCGACCCCAGGTCAGGCGGGATTACCCGCTGAGTTTAAGCATATCAATAAGCGNAG 753
P.columnaris        CGACGCGACCC------------------------------------------------ 625
P.nummulariifolius  CGACGCGACCCCAGGTCAGGCGGGATTACCCGCTGAGTTTAAGCATATCAATAAGCGAGA 822
P.orientalis        CGACGCGACCC------------------------------------------------ 620
P.emblica           CGACGCGACCCCAGGTCAGGCGGGATTACCCGCTGAGTTTAAGCATATCAATAAGCGGAG 687
P.acutissimus       TGACGCGACCC------------------------------------------------ 603
P.pulcher           TGACGCGACCC------------------------------------------------ 605
P.myrtifolius       CGACGCGACCC------------------------------------------------ 628
P.virgatus          TGACGCGACC-------------------------------------------------- 630
P.roseus            TGACGCGACCC------------------------------------------------ 618
                       *********

P.debilis           GGAAAGTTTT-------------------------------------------- 684
P.discolor          ------------------------------------------------------
P.graveolens        GAAAAGAAACTTACAAGGATTCC------------------------------- 782
P.niruri            GAAGGAAAGTTTTTGTTTGTGTCTTGCTTCTGTGCGCTGGTGAGGGGTGTGGTGGGTGTT 724
P.madagascariensis  ------------------------------------------------------
```

FIGURE 4.4 Clustal 2.0.12 multiple-sequence alignment of the species belonging to genus *Phyllanthus*. *(continued)*

```
P.acidus              GAAAAGAAACTTACAAGGATTCCCGCGGAAGGATCATTGTCGAAACCTGCTCTGCAGTAT  806
P.polyphyllus         GAAAAGAAACTTACAAGGATTCCC-----------------------------------  773
P.acuminatus          GAAAAGAAACTTACAAGGTTCCC------------------------------------  775
P.oxyphyllus          GAAAAGAAACTTACAAGGATTCCC-----------------------------------  778
P.sikkimensis         GAAAAGAAACTTACAAGGATTCCC-----------------------------------  779
P.rheedii             GAAAAGAAACTTACAAGGATTCCC-----------------------------------  774
P.harmandii           -----------------------------------------------------------
P.collinsiae          -----------------------------------------------------------
P.angkorensis         -----------------------------------------------------------
P.gracilipes          -----------------------------------------------------------
P.lingulatus          -----------------------------------------------------------
P.microcarpus         -----------------------------------------------------------
P.clarkei             GGAAAAGAACTTACAAGGATTCCC-----------------------------------  778
P.welwitschianus      G-AAAAGAACTTACAAGGATTCCCTTAGTAACGGCGAGCGAAC----------------  823
P.reticulatus         GGAAAGAGACTTACAAGGATTCCT-----------------------------------  777
P.columnaris          -----------------------------------------------------------
P.nummulariifolius    AAACGAACTCA------------------------------------------------  833
P.orientalis          -----------------------------------------------------------
P.emblica             GAAAAGAAACTTAC---------------------------------------------  701
P.acutissimus         -----------------------------------------------------------
P.pulcher             -----------------------------------------------------------
P.myrtifolius         -----------------------------------------------------------
P.virgatus            -----------------------------------------------------------
P.roseus              -----------------------------------------------------------


P.debilis             -----------------------------------------------------------
P.discolor            -----------------------------------------------------------
P.graveolens          -----------------------------------------------------------
P.niruri              GGGGGGGGGGTGGGGGGGGGTAGTGTGGGAAATTGATATTGTAAGTTTGATCGGTGGGG  784
P.madagascariensis    -----------------------------------------------------------
P.acidus              GACCCGCGAACAAGTTTATACACT-----------------------------------  830
P.polyphyllus         -----------------------------------------------------------
P.acuminatus          -----------------------------------------------------------
P.oxyphyllus          -----------------------------------------------------------
P.sikkimensis         -----------------------------------------------------------
P.rheedii             -----------------------------------------------------------
P.harmandii           -----------------------------------------------------------
P.collinsiae          -----------------------------------------------------------
P.angkorensis         -----------------------------------------------------------
P.gracilipes          -----------------------------------------------------------
P.lingulatus          -----------------------------------------------------------
P.microcarpus         -----------------------------------------------------------
P.clarkei             -----------------------------------------------------------
P.welwitschianus      -----------------------------------------------------------
P.reticulatus         -----------------------------------------------------------
P.columnaris          -----------------------------------------------------------
P.nummulariifolius    -----------------------------------------------------------
P.orientalis          -----------------------------------------------------------
P.emblica             -----------------------------------------------------------
P.acutissimus         -----------------------------------------------------------
P.pulcher             -----------------------------------------------------------
P.myrtifolius         -----------------------------------------------------------
P.virgatus            -----------------------------------------------------------
P.roseus              -----------------------------------------------------------


P.debilis             ----
P.discolor            ----
P.graveolens          ----
P.niruri              TTTT  788
P.madagascariensis    ----
P.acidus              ----
```

FIGURE 4.4 Clustal 2.0.12 multiple-sequence alignment of the species belonging to genus *Phyllanthus*. (*continued*)

```
P.polyphyllus          ----
P.acuminatus           ----
P.oxyphyllus           ----
P.sikkimensis          ----
P.rheedii              ----
P.harmandii            ----
P.collinsiae           ----
P.angkorensis          ----
P.gracilipes           ----
P.lingulatus           ----
P.microcarpus          ----
P.clarkei              ----
P.welwitschianus       ----
P.reticulatus          ----
P.columnaris           ----
P.nummulariifolius     ----
P.orientalis           ----
P.emblica              ----
P.acutissimus          ----
P.pulcher              ----
P.myrtifolius          ----
P.virgatus             ----
P.roseus               ----
```

FIGURE 4.4 (continued) Clustal 2.0.12 multiple-sequence alignment of the species belonging to genus *Phyllanthus*.

As we assume that all species evolve from a common ancestor, the phylogram shows *Phyllanthus fraternus* is closely related to *Phyllanthus niruri*, and they are sister taxa. *Phyllanthus tenellus* is closely related to *Phyllanthus clarkei,* and they form sister taxa. But, these two species (*Phyllanthus tenellus* and *Phyllanthus fraternus*) are distantly related to each other, so they do not support a monophyletic lineage. As the branch length indicates the amount of evolution, *Phyllanthus tenellus* is more evolved than *Phyllanthus clarkei,* and *Phyllanthus fraternus* is more evolved than *Phyllanthus niruri* (Figure 4.5).

A *cladogram* is a tree-like diagram showing evolutionary relationships. Any two branch tips sharing the same immediate node are most closely related. All taxa that

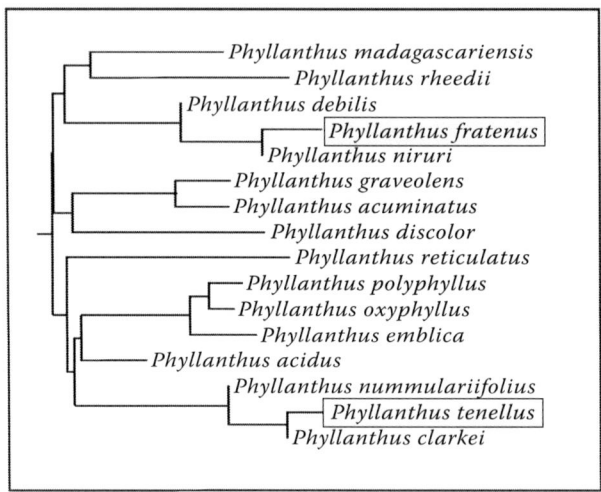

FIGURE 4.5 Phylogram.

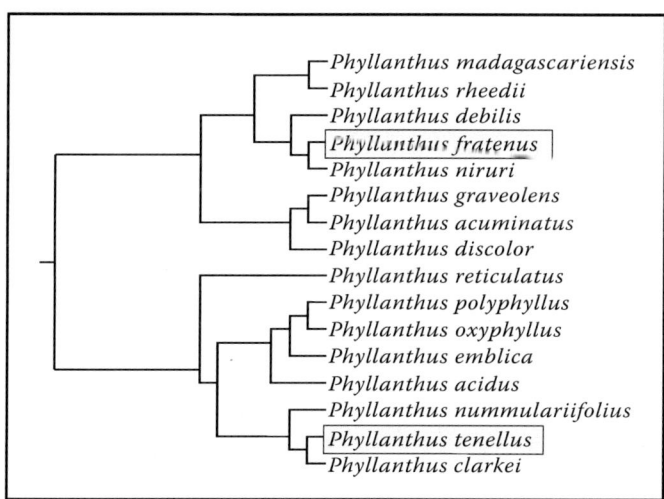

FIGURE 4.6 Cladogram.

can be traced directly to one node (that is, they are "upstream of a node") are said to be members of a monophyletic group. Each branch on a cladogram is referred to as a *clade* and can have two or more arms. Taxa sharing arms branching from the same clade are referred to as *sister groups* or *sister taxa*. Thus, *Phyllanthus clarkei* and *Phyllanthus tenellus* can be referred to as sister taxa, while *Phyllanthus fraternus* and *Phyllanthus niruri* can be termed sister taxa. But, *Phyllanthus fraternus* and *Phyllanthus tenellus* can be termed distinctly related but not *sister taxa* from both of phylogeny trees (Figure 4.6).

The sequence divergence of the ITS1 in wild barley due to substitutions ranged from 0.0 to 2.71%, and that due to substitutions plus indels ranged from 0.0 to 5.42%, which was lower than those observed in wheat (0.0 to 3.12% for substitutions and 0.44 to 7.0% for substitutions plus indels). The sequence divergence of the ITS2 in wild barley due to substitutions ranged from 0.0 to 2.28%, and that due to substitutions plus indels ranged from 0.0 to 5.0%. In wheat, substitutions ranged from 0.45 to 2.26%, and for substitutions plus indels, the range was from 0.45 to 4.07%. Thus, in both wild barley and wheat, sequence divergence was greater in the ITS1 than in the ITS2 region.

The higher level of divergence in ITS1 observed during the present study was in conformity with earlier reports in a variety of plant species (Kollipara et al., 1997; Baldwin, 1993; Moller and Cronk, 1997). The deletions within ITS1 and ITS2 were believed to interfere with rRNA processing. For instance, *in vivo* mutational studies in yeast (*Saccharomyces cerevisiae*) indicated that deletions of certain regions within ITS1 inhibited production of mature small- and large-subunit rRNAs (Musters et al., 1990; Nues et al., 1994), whereas certain deletions and point mutations in ITS2 prevented or reduced processing of large-subunit rRNA (Sande et al., 1992).

Thus, it was concluded that the data on length and sequence of an ITS may be a useful parameter for the assessment of genetic diversity at the intraspecific level in species like barley and wheat, although the level of diversity detected using ITS data at interspecific level was much higher in different groups of plants (Sharma et al., 2000).

In general, it is assumed that all the species of *Phyllanthus* evolved from a common ancestor. In the present study of *Phyllanthus* phylogenetics, the cladogram obtained on a molecular basis (based on their ITS region) justified that these species bear some common morphological features and shows that these two species are closely related to each other; at the same time, some distinguished morphological variations indicate that they are distinctly related and do not support the monophyletic lineage proposal.

REFERENCES

Altschul, S.F., Madden, T.L., Schaffer, A.A., et al. 1997. Gapped BLAST and PSI.BLAST: a new generation of protein database search programs. *Nucleic Acids Res.* 25: 3389–3402.

Baldwin, B.G. 1993. Molecular phylogenetic of Calcydenia (Compositae) based on ITS sequences of nuclear ribosomal DNA: chromosomal and morphological evolution reexamined. *Am. J. Bot.* 80: 222–238.

Baldwin, B.G., Sanderson, M.J., Porter, J.M., et al. 1995. The ITS region of nuclear ribosomal DNA: a valuable source of evidence on angiosperm phylogeny. *Ann. Mo. Bot. Gard.* 82: 247–277.

Botstein, D., White, R.L., Skolnick, M., et al. 1980. Construction of a genetic linkage map in man using restriction fragment length polymorphisms. *Am. J. Hum. Genet.* 32: 314–331.

Cavalli-Sforza, L.L., and Edwards, A.W.F. 1967. Phylogenetic analysis: models and estimation procedures. *Evolution* 21: 550–570.

Dubouzet, J.G., and Shinoda, K. 1999. Relationships among old and new world *Alliums* according to ITS DNA sequence analysis. *Theor. Appl. Genet.* 98: 422–433.

Edwards, A.W.F., and Cavalli-Sforza, L.L. 1964. Phenetic and phylogenetic classification: reconstruction of evolutionary trees. *Systematics Assoc.* 6: 67–76.

Govaerts, R., Frodin, D.G., Radcliffe-Smith, A. 2000. *World checklist and bibliography of Euphorbiaceae (and pandaceae)*. Kew: Royal Botanic Gardens.

Hershkovitz, M.A., and Zimmer, E.A .1996. Conservation patterns in angiosperm rDNA ITS$_2$ sequences. *Nucleic Acid Res.* 24: 2857–2867.

Hershkovitz, M.A., Zimmer, E.A., and Hahn, W.J. 1999. Ribosomal DNA sequences and angiosperm systematics. In *Molecular systematics and plant evolution,* ed. P.M. Hollingsworth, R.M. Bateman, and R.J. Gornall, 268–326. London: Taylor and Francis.

Holm-Nielsen, L.B. 1979. Comments on the distribution and evolution of the genus *Phyllanthus* (Euphorbiaceae). In *Tropical botany*, ed. K. Larsen and L.B. Holm-Nielsen, 277–290. New York: Academic Press.

Hsiao, C., Chatterton, N.J., Asay, K.H., et al. 1994. Phylogenetic relationships of 10 grass species: an assessment of phylogenetic utility of the internal transcribed spacer region in nuclear ribosomal DNA in monocots. *Genome* 37: 112–120.

Kollipara, K.P., Singh, R.J., and Hymowitz, T. 1997. Phylogenetic and genomic relationship in the genus *Glycine* Wild based on sequences from the ITS region rDNA. *Genome* 40: 57–68.

Larkin, M.A., Blackshields, G., Brown, N.P., et al. 2007. ClustalW2 and ClustalX version 2. *Bioinformatics* 23: 2947–2948.

Maher, B.A. 2002. Uprooting the tree of life. *The Scientist* 16: 18.

Mayr, E. 1940. Speciation phenomena in birds. *Am. Nat.* 74: 249–278.

Moller, M., and Cronk, Q.C.B. 1997. Origin and relationships of *Santipulia* (Gesneriaceae) based on ribosomal DNA internal transcribed spacer (ITS) sequences. *Am. J. Bot.* 84: 956–965.
Mullis, K.B., and Faloona, F.A. 1987. Specific synthesis of DNA *in vitro* via a polymerase-catalyzed chain reaction. *Methods Enzymol.* 155: 335–350.
Musters, W., Boon, K., Sande Van Der, C.A.F.M., et al. 1990. Function analysis of transcribed spacers of yeast ribosomal DNA. *EMBO J.* 9: 3989–3996.
Nues, R.W., Van Rientjes, J.M.J., Sande Van Der, C.A.F.M., et al. 1994. Separate structural elements within internal transcribed spacer of *Sacchromyces cerviasiae* precursor ribosomal RNA direct the formation of 17S and 26S rRNA. *Nucl. Acids Res.* 22: 912–919.
Samuel, R.H., Kathriarchchi, P., Hoffman, P.M. et al. 2005. Molecular phylogenetics of Phyllanthaceae: evidence from plastid matK and nuclear PHYC sequences. *Am. J. Bot.* 92: 132–141.
Sande Van Der, C.A.F.M., Kwa, M., Van Nues, R.W., et al. 1992. Functional analysis of internal transcribed spacer 2 of *Saccharomyces cerevisae* ribosomal DNA. *J. Mol. Biol.* 223: 899–910.
Sharma, S., Rustgi, S., Balyan, H.S., et al. 2000. Internal transcribed spacer (ITS) sequences of ribosomal DNA of wild barley and their comparison with ITS sequences in common wheat. *Barley Gen. Newsl.* 32.
Suh, Y., Thien, L.B., Reeve, H.E., et al. 1993. Molecular evolution and phylogenetic implications of ribosomal DNA in Winteraceae. *Am. J. Bot.* 80: 1042–1055.
Vos, P., Hogers, R.B., Hornes, M., et al. 1995. AFLP: a new technique for DNA fingerprinting. *Nucleic Acids Res.* 23: 4407–4414.
Webster, G.L. 1956. A monographic study of the West Indian species of *Phyllanthus*. *J. Arnold Arbor.* 37: 91–122, 340–359.
Webster, G.L. 1957. A monographic study of the West Indian species of *Phyllanthus*. *J. Arnold Arbor.* 38: 51–80.
Welsh, J., and McClelland, M. 1990. Fingerprinting genomes using PCR with arbitrary primers. *Nucleic Acids Res.* 18: 7213–7218.
Williams, J.G.K., Kubelik, A.R., Livak, K.J., et al. 1990. DNA polymorphisms amplified by arbitrary primers are useful as genetic markers. *Nucleic Acid Res.* 18: 6531–6535.
Wurdack, K.J., Hoffman, P., Samuel, R. et al. 2004. Molecular phylogenetic analysis of Phyllanthaceae (Phyllanthoideae propaste, euphorbiaceae sensu lato) using plastid RBCL DNA sequences. *Am. J. Botany.* 91: 1882–1900.

5 Genetic Resources of *Phyllanthus* in Southern India
Identification of Geographic and Genetic Hot Spots and Its Implication for Conservation

G. Ravikanth, R. Srirama, U. Senthilkumar,
K. N. Ganeshaiah, and R. Uma Shaanker

CONTENTS

5.1 Introduction	98
5.2 Distribution of *Phyllanthus* Species	104
5.2.1 Identification of Geographic Hot Spots of *Phyllanthus* in South India: Contours of Species Richness	104
5.3 Identification of Genetic Hot Spots of Economically Important *Phyllanthus* Species	107
5.3.1 *Phyllanthus emblica*	107
5.3.2 *Phyllanthus amarus*	108
5.4 Impact of Harvesting on the Genetic Variability of *P. emblica*	108
5.5 *Phyllanthus*: Taxonomic Incongruities, Species Adulteration, and DNA Bar Coding	110
5.6 Implications for Utilization and Conservation	111
Acknowledgments	112
References	113

5.1 INTRODUCTION

The genus *Phyllanthus* (family: Phyllanthaceae) is one of the most important groups of plants traded as a raw herbal drug in India (Ved and Goraya, 2008). Plants of this genus have been used in traditional medicine for a variety of uses, including as an antipyretic, laxative, tonic, antibacterial, antioxidative, immunomodulatory, antiviral, antiatherosclerotic, and antineoplasic (Unander et al., 1991, 1995; Calixto et al., 1998). In India, *Phyllanthus* is used as a common folk remedy for the treatment of jaundice and hepatitis. The genus is also used as a general tonic and to treat weakness in infants (Unander et al., 1991). A number of taxa are cultivated for their fleshy edible fruits and for preparation of herbal drugs. Among the *Phyllanthus* species in India, *P. amarus, P. debilis, P. fraternus, P. urinaria, P. kozhikodianus, P. maderaspatensis, P. emblica,* and *P. indofischeri* are widely used as herbal medicines, and some of these species are also cultivated in southern India (Table 5.1).

Phyllanthus amarus, a predominant species occurring in southern India, has been shown to suppress the growth and replication of hepatitis B virus (Venkateswaran et al., 1987; Thyagarajan et al., 1988; Yeh et al., 1993; Jayaram and Thyagarajan, 1996; Lee et al., 1996; Paranjape, 2001). A few species, such as *P. amarus, P. fraternus,* and *P. debilis,* have been reported to be extensively used for curing jaundice; *P. urinaria* has been recommended for curing urinary tract diseases (Table 5.1; Jain et al., 2008). Phyllanthin and hypophyllanthin, both present in *P. amarus,* have been shown to protect hepatocytes against carbon tetrachloride (CCl_4) and galactosamine-induced cytotoxicity in rats (Syamasundar et al., 1985).

Phyllanthus emblica is another medicinally important species widely distributed across the Indian subcontinent. It is commonly called the Indian gooseberry. Traditionally, it has been used to treat digestive disorders, constipation, fever, cough, and asthma and to stimulate hair growth. Extracts of *P. emblica* have been shown to possess several pharmacological actions, such as analgesic, anti-inflammatory, antioxidant, and chemoprotective, (Calixto et al., 1998; Vormisto et al., 1997; Khopde et al., 2001). The fruits contain diterpenes; triterpenes; lupeol; flavonoids such as kaempherol-3-O-fl-D-glucoside and quercetin-3-O-fl-D-glucoside; polyphenols such as emblicanin A and B; punigluconin and pedunculagin, and other molecules (Calixto et al., 1998; Bhattacharya et al., 1999; Summanen, 1999; Ghosal et al., 1996). *Phyllanthus emblica* fruits are used for preparations of pickles, jams, and juices. The fruits are also used by the cosmetic, hair dye and shampoo industries (Ganesan and Shetty, 2004).

The annual volume of *Phyllanthus* trade in India is estimated to be about 2,000–5,000 metric tonnes of herbaceous material and about 16,000–18,000 metric tonnes of fruits (Ved and Goraya, 2008). Several species are also exported in powder form for the extraction of a number of phytochemicals or for use in the preparation of traditional formulations in the treatment of liver disorders (Kamble et al., 2008). Because of its multifarious use and demand, *Phyllanthus* species form an important nontimber forest product resource. Most of the material for trade is sourced from the wild by forest-dwelling communities, and only a small percentage is obtained from cultivation (Ved and Goraya, 2008). Because of the often-indiscriminate harvesting,

TABLE 5.1
Phyllanthus L. Species in India and Their Pharmacological Activities

Sl No.	*Phyllanthus* Species	Habit	Status	Bioactivity	References
1	*P. acidus* (L.) Skeels	Tree	Cultivated	Immunomodulatory effect against gastrointestinal disorders	Kundu et al., 2009
2	*P. airyshawii* Brunel & Roux	Herb	Rare	No reports	
3	*P. ajmerianus* L.B. Chaudhary & R.R. Rao	Herb	Endemic	No reports	
4	*P. amarus* Schumach.	Herb	Common	Antioxidant, anticancer, nephroprotective activity, hepatoprotective activity, antibacterial activity, antiinflammatory, antiallodynic, antitumor	Abhyankar et al., 2010; Adeneye and Benebo, 2008; Chirdchupunseree and Pramyothin, 2010; Eldeen et al., 2010; Faremi et al., 2008; Kassuya et al., 2006; Kiemer et al., 2003; Kumar and Kuttan, 2005; Naaz et al., 2007; Narendranathan et al., 1997; Notka et al., 2003; Notka et al., 2004; Rajeshkumar and Kuttan, 2000; Rajeshkumar et al., 2002; Raphael and Kuttan, 2003
5	*P. anamalayanus* (Gamble) G.L. Webster	Shrub	Endemic	No reports	
6	*P. andamanicus* N.P. Balakr. & N.G. Nair	Shrub	Endemic	No reports	
7	*P. arbuscula* (Sw.) J.F. Gmelin	Shrub	Cultivated	No reports	
8	*P. baeobotryoides* Wall.	Shrub	Rare	No reports	
9	*P. baillonianus* Muell.Arg.	Shrub	Endemic	No reports	
10	*P. beddomei* Gamble	Shrub	Endemic	No reports	
11	*P. brevipes* Hook.f.	Shrub	Endemic	No reports	

(continued)

TABLE 5.1 (continued)
Phyllanthus L. Species in India and Their Pharmacological Activities

Sl No.	*Phyllanthus* Species	Habit	Status	Bioactivity	References
12	*P. chandrabosei* Govaets & Radcl.-Sm.	Shrub	Endemic	No reports	
13	*P. clarkei* Hook.f.	Shrub	Very rare	No reports	
14	*P. columnaris* Muell. Arg.	Shrub	Rare	No reports	
15	*P. debilis* Klein ex Willd.	Herb	Common in plains	No reports	
16	*P. emblica* L.	Tree	Common	Antioxidant, antisecretory, antiulcer, and cytoprotective properties; antitumor, free radical scavenging activity, hypolipidaemic activity, antitussive, antipyretic, analgesic, hepatoprotective activity	Al-Rehaily et al., 2002; Bandyopadhyay et al., 2000; Jose et al., 2001; Liu et al., 2008; Luo et al., 2009; Mathur et al., 1996; Nosál'ová et al., 2003; Perianayagam et al., 2004; Pramyothin et al., 2006; Reddy et al., 2009; Sai Ram et al., 2002; Sharma et al., 2009; Sultana et al., 2008
17	*P. fimbriatus* (Wight) Muell. Arg.	Shrub	Endemic	No reports	
18	*P. fraternus* G.L. Webster	Herb	Uncommon	Hepatoprotective activity	Gopi and Setty, 2010; Padma and Setty, 1999; Sailaja and Setty, 2006; Sebastian and Setty, 1999
19	*P. gageanus* (Gamble) M. Mohanan	Shrub	Endemic	No reports	
20	*P. glaucus* Wall.	Shrub	Common in northeast	No reports	
21	*P. gomphocarpus* Hook.f.	Shrub	Rare	No reports	
22	*P. griffithii* Muell. Arg.	Shrub	Endemic	No reports	
23	*P. heyneanus* Muell. Arg.	Shrub	Endemic	No reports	
24	*P. indofischeri* Bennet	Tree	Endemic	No reports	

(continued)

TABLE 5.1 (continued)
Phyllanthus L. Species in India and Their Pharmacological Activities

Sl No.	*Phyllanthus* Species	Habit	Status	Bioactivity	References
25	*P. juniperinoides* Muell. Arg.	Shrub	Endemic to peninsular India	No reports	
26	*P. leschenaultii* Muell. Arg.	Shrub	Very rare	No reports	
27	*P. macraei* Muell. Arg.	Shrub	Endemic	No reports	
28	*P. macrocalyx* Muell. Arg	Shrub	Endemic	No reports	
29	*P. maderaspatensis* L.	Herb	Common	Hepatoprotective activity	Asha et al., 2004; Asha et al., 2007
30	*P. megacarpus* (Gamble) Kumari & Chandrab.	Shrub	Endemic	No reports	
31	*P. myrtifolius* (Wight) Muell. Arg.	Shrub	Cultivated	Antibacterial, antioxidant, anti-HIV activity	Eldeen et al., 2010
32	*P. narayanaswamii* Gamble	Herb	Endemic	No reports	
33	*P. parvifolius* Buch.-Ham.ex D.Don	Shrub	Rare	No reports	
34	*P. pendulus* Roxb.	Shrub	Endemic	No reports	
35	*P. pinnatus* (Wight) G.L Webster	Shrub	Uncommon	No reports	
36	*P. polyphyllus* Willd.	Shrub	Common in peninsular India	Anti-inflammatory activity	Rao et al., 2006
37	*P. praetervisus* Muell. Arg.	Shrub	Rare	No reports	
38	*P. pseudoparvifolius* R.L. Mitra & Sanjappa	Shrub	Rare	No reports	
39	*P. pulcher* Wall.	Shrub	Cultivated	Antitumor, antioxidant, anti-HIV activity	Stanslas et al., 2008; Eldeen et al., 2010
40	*P. reticulatus* Poir.	Shrub	Very common	Antioxidant, antidiabetic	Eldeen et al., 2010; Kumar et al., 2008
41	*P. rheedei* Wight	Herb	Rare in peninsular India	Hepatoprotective activity	Suresh and Asha, 2008
42	*P. roeperianus* Wall.	Shrub	Rare	No reports	

(continued)

TABLE 5.1 (continued)
Phyllanthus L. Species in India and Their Pharmacological Activities

Sl No.	Phyllanthus Species	Habit	Status	Bioactivity	References
43	*P. rotundifolius* Klein ex Willd.	Herb	Common in coastal areas	No reports	
44	*P. sanjappae* Chakrab. & M. Gangop.	Shrub	Endemic	No reports	
45	*P. scabrifolius* Hook.f.	Herb	Endemic	No reports	
46	*P. sikkimensis* Muell. Arg.	Shrub	Very rare	No reports	
47	*P. simplex* Retz	Herb	Common	Antidiabetic, antioxidant activity	Shabeer et al., 2009
48	*P. singampattianus* (Sebastine & A.N Henry) Kumari & Chandrab.	Shrub	Endemic	No reports	
49	*P. talbotii* Sedgw.	Shrub	Endemic	No reports	
50	*P. tenellus* Roxb.	Herb	Naturalized	Immunomodulatory effect against microbial activity	Ignácio et al., 2001
51	*P. tetrandrus* Roxb.	Shrub	Rare	No reports	
52	*P. urinaria* L.	Herb	Common	Antibacterial, antioxidant, anti-HIV activity, hepatoprotective activity, anti-inflammatory, anticancer, antitumor, antiangiogenic, chemopreventive agent for peptic ulcer, anti-HSV	Eldeen et al., 2010; Chudapongse et al., 2010; Fang et al., 2008; Hau et al., 2009; Huang et al., 2003; Huang et al., 2004; Huang et al., 2006; Lai et al., 2008; Lin et al., 2008; Yang et al., 2005; Yang et al., 2007
53	*P. wightianus* Muell. Arg.	Shrub	Endemic	No reports	

many of the species face the risk of local or regional extinction of their populations (Uma Shaanker et al., 2002, Ravikanth et al., 2009).

In this chapter, we briefly review the status of *Phyllanthus* resources in southern India with the overall aim of understanding the spatial distribution of the species as well as its genetic diversity. This information is critical in designing strategies for the long-term utilization and conservation of *Phyllanthus* genetic resources.

TABLE 5.2
Phyllanthus Species Traded in India and Molecular Regions that Have Been Sequenced Along with Their Accession Numbers

Sl No.	*Phyllanthus* Species (Trade Name)	Molecular Regions	GenBank Accession No.	References
1	*P. amarus* (Bhumiamla)	ITS	EU623557.1	Pruesapan et al., 2008
		psbA-trnH	GU598561–65, GU598577	Srirama et al., 2010
		ndhF	EU643742	Pruesapan et al., 2008
		phyC	FJ235474.1	Kawakita and Kato, 2009
		atpB	FJ235356	Kawakita and Kato, 2009
		trnL	FJ235310.1	Kawakita and Kato, 2009
		rbcL	EU861193.1	Unpublished
		trnK, matK	AY765265.1	Lee et al., 2006
2	*P. debilis* (Bhumiamla)	*psbA-trnH*	AY936686	Srirama et al., 2010
		ndhF	AY936591	Kawakita and Kato, 2009
		phyC	FJ235265	Kawakita and Kato, 2009
		atpB	FJ235311	Kawakita and Kato, 2009
		matK	FJ235357	Kawakita and Kato, 2009
		trnK	FJ235475	Kathriarachchi et al., 2006
		ITS	GU598567–68	Kathriarachchi et al., 2006
3	*P. fraternus* (Bhumiamla)	ITS	GU598566, 69	Srirama et al., 2010
		psbA-trnH	EU876847	Unpublished
4	*P. maderaspatensis* (Kanocha)	ITS	GU598536–38	Srirama et al., 2010
		trnK	AY936609	Kathriarachchi et al., 2006
		psbA-trnH	AY936707	Kathriarachchi et al., 2006
5	*P. reticulatus* (Buinowla)	*psbA-trnH*	AY765290	Kawakita and Kato, 2009
		ndhF	AY936629	Srirama et al., 2010
		phyC	FJ235270	Kawakita and Kato, 2009
		atpB	FJ235316	Kawakita and Kato, 2009
		matK	FJ235362	Kathriarachchi et al., 2006
			FJ235480	Kawakita and Kato, 2009
		ITS	GU598539–40	Lee et al., 2006
6	*P. urinaria* (Lal-Bhuin-Anvalah)	*psbA-trnH*	AY765305	Lee et al., 2006
		rbcL	AY936736	Kathriarachchi et al., 2006
		matK, trnK	AY936637	Kathriarachchi et al., 2006
		ITS	AY765268	Lee et al., 2006
		atpB	GU598573–74	Srirama et al., 2010
7	*P. virgatus* (Niruri)	*ndhF*	FJ235485	Unpublished
		phyC	FJ235367	Kathriarachchi et al., 2006
		atpB	FJ235321	Kathriarachchi et al., 2006
		matK	FJ235275	Kawakita and Kato, 2009
		trnK, matK	AY936639	Kawakita and Kato, 2009

(continued)

TABLE 5.2 (continued)
Phyllanthus Species Traded in India and Molecular Regions that Have Been Sequenced Along with Their Accession Numbers

Sl No.	Phyllanthus Species (Trade Name)	Molecular Regions	GenBank Accession No.	Reference
		ITS	AY936738	Kawakita and Kato, 2009
		rbcL	GU441778	Kawakita and Kato, 2009
8	P. emblica (Amla)	psbA-trnH	EU643743	Srirama et al., 2010
		ndhF	FJ847837	Kawakita and Kato, 2009
		phyC	GU441788	Kawakita and Kato, 2009
		atpB	AY936689	Kawakita and Kato, 2009
		trnK, matK	AY936594	Kathriarachchi et al., 2006
		ITS	FJ235297	Kathriarachchi et al., 2006
		rbcL	FJ235343	Unpublished
		trnL	FJ235461	Unpublished
		matK	GU598547	Pruesapan et al., 2008
9	P. indofischeri (Amla/*Ittu nelli*)	psbA-trnH	GU598558–60	Srirama et al., 2010
		trnL	GU930706	Unpublished

5.2 DISTRIBUTION OF *PHYLLANTHUS* SPECIES

Phyllanthus is one of the most species-rich genera of the family Phyllanthaceae, comprising over 800 species worldwide (Calixto et al., 1998; Govaerts et al., 2000; Wurdack et al., 2004). The genus is subdivided into 11 subgenera: *Isocladus, Kirganelia, Cicca, Emblica, Conani, Gomphidium, Phyllanthodendron, Xylophylla, Botryanthus, Ericoccus,* and *Phyllanthus.* Plants of this genus are characterized by diverse growth forms, including shrubs, trees, and annual or biennial herbs and are distributed throughout the tropical and subtropical regions of both hemispheres. A few species of *Phyllanthus* are notorious weeds and occur in four continents (America, Africa, Asia, and Australia). India has 53 species of *Phyllanthus,* of which 23 species are endemic (Balakrishnan and Chakrabarty, 2007). These are distributed throughout the Indian subcontinent, with higher densities in the southern region. As many as 17 species are endemic to peninsular India, 2 species to the Andaman and Nicobar Islands, and others restricted to central and northeastern India (Balakrishnan and Chakrabarty, 2007).

5.2.1 Identification of Geographic Hot Spots of *Phyllanthus* in South India: Contours of Species Richness

One of the key strategies to the sustainable management of any natural resource is to obtain spatially explicit information on its distribution. Spatially explicit distribution maps of the species could facilitate the management and utilization of these resources, tracking the dynamics of the resources over time; aid in preparing germplasm collections; and help in assigning conservation value and priorities. Using geographic information systems, Ganeshaiah and Uma Shaanker (2003) developed a spatially explicit distribution map of all herbaceous species of *Phyllanthus* in India (Figure 5.1). Secondary data of occurrence of the species were obtained from various sources, like floras, herbaria, books, and other published sources, and then digitized. Species richness of *Phyllanthus* was summated on grids (1° latitude × 1° longitude), and contours of species richness were plotted (Figure 5.1).

Among the 53 species of *Phyllanthus,* 37 are shrubs, 13 are herbs, and 3 are trees (Balakrishnan and Chakrabarty, 2007). The 13 herbaceous species of *Phyllanthus* are primarily concentrated in the states of Tamil Nadu, Kerala, Karnataka, Maharashtra, and Andhra Pradesh. A few species of *Phyllanthus,* such as *P. amarus,* are distributed throughout the country. Two herbs (i.e., *P. rheedii* and *P. kozhikodianus*) are endemic to peninsular India and Sri Lanka. Similarly, *P. scabrifolius* is endemic to Maharashtra and northern Karnataka. *Phyllanthus ajmerianus* is endemic to

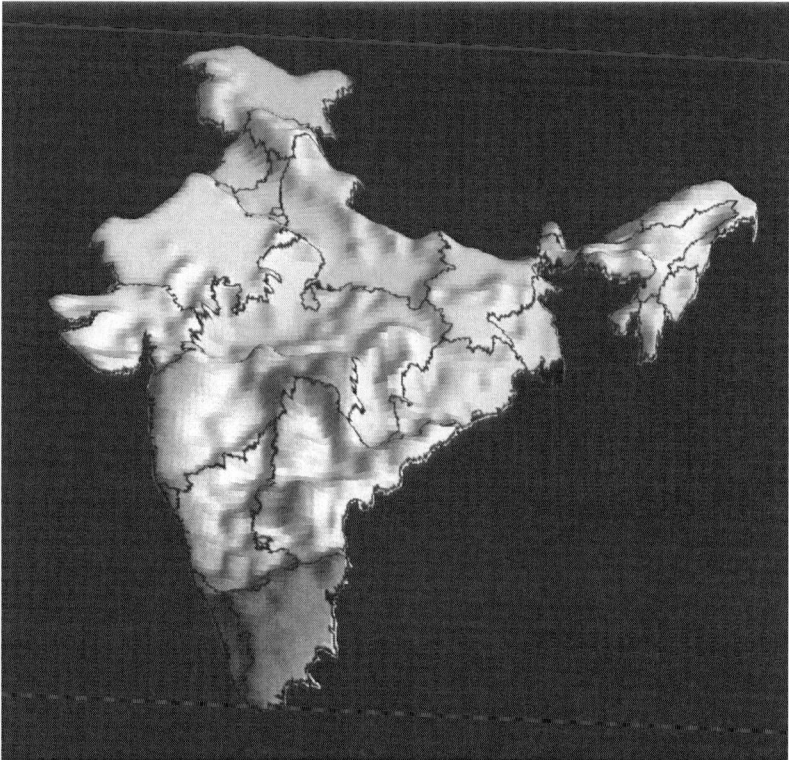

FIGURE 5.1 **See color insert.** Hypsographic view of *Phyllanthus* species richness in India. The data on the distribution of the species were obtained from diverse sources (monographs, etc.), and the latitude and longitude were assigned for each record and mapped. The density of the species in each grid of the size 10 km × 10 km was computed and the contours for the density obtained. Based on the contour data, the three-dimensional view was constructed using suitable GIS software.

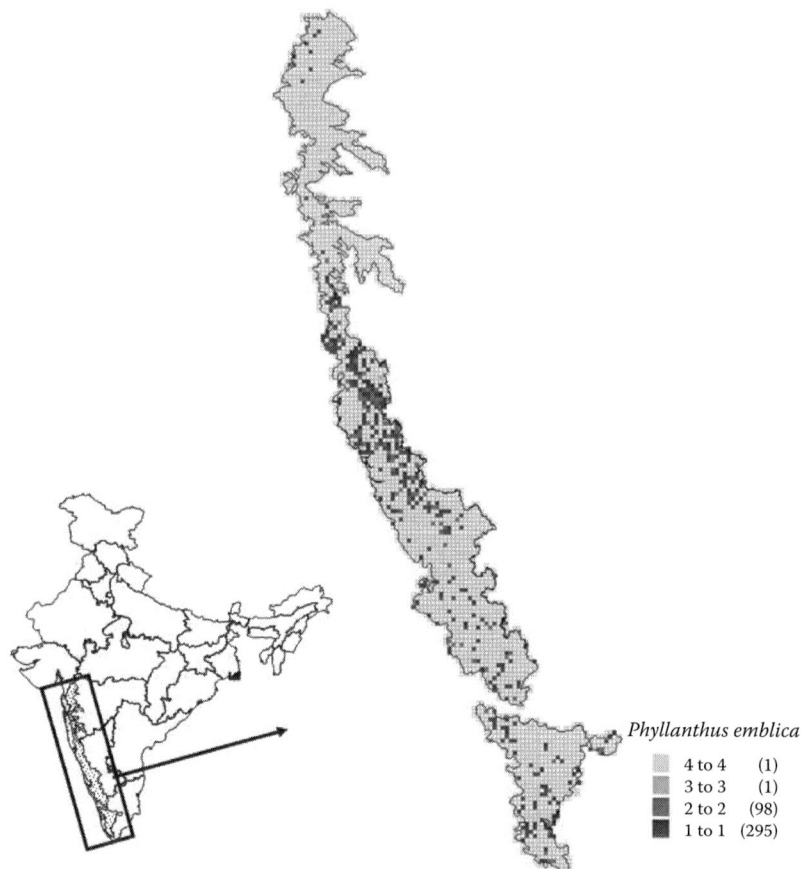

FIGURE 5.2 Resource map of *Phyllanthus emblica*. The map was developed using data collected from primary sources. Each pixel is the size of 6.25 km × 6.25 km. The lighter shades indicate higher densities. The numbers in brackets indicate the number of records at each pixel.

Rajasthan. Among the shrubs, about 7 species are endemic to India. *Phyllanthus indofischeri* is the only tree species endemic to the Deccan Plateau (Ganesan, 2003; Balakrishnan and Chakrabarty, 2007). The distribution of *P. indofischeri* overlaps with *P. emblica*; while *P. emblica* is distributed throughout the Deccan peninsular region and some parts of central India (Figure 5.2), *P. indofischeri* is restricted to southern India (Ganesan and Shetty, 2004). These contour maps depicting the relative richness of the species on the Indian landscape are extremely useful both in guiding conservation strategies and in planning the sustained utilization of the resources (Ravikanth et al., 2001, 2002).

Besides the species richness maps described, Ganeshaiah and U. Shaanker (2007) also developed species-specific maps, especially for those in trade. For example, specific maps have been developed for *P. emblica*, one of the most important species in trade (Figure 5.2). These maps, with grids resolution of 6.25 × 6.25 km, provide precise spatial information on the resource stock of the species. Further, when overlaid with other parameters, such as demand levels and production data (supply), the species specific maps can be used to develop a user-friendly resource management system that can advise on a variety of issues, including the optimal levels of harvesting, rotation schedules of harvesting over the distributional range of the species, and finally providing a dynamic inventory of the resource (Uma Shaanker et al., 2004).

5.3 IDENTIFICATION OF GENETIC HOT SPOTS OF ECONOMICALLY IMPORTANT *PHYLLANTHUS* SPECIES

One of the major limitations in planning the effective utilization and conservation of genetic resources of medicinal plant species is the lack of critical information on the spatial distribution of genetic variability of the species. Spatially explicit analysis of the genetic variability of the species could aid in (1) identification of genetic hot spots of the species, (2) appropriately designing germplasm collections, and (3) deciding on what and where to conserve. Unfortunately for many medicinally important species, there is a severe dearth of information with respect to the spatial distribution of variability. Here, we briefly review attempts that have been carried out to identify the genetic hot spots for the two most important medicinal plants in the genus *Phyllanthus*.

5.3.1 *Phyllanthus emblica*

Phyllanthus emblica L. constitutes one of the important medicinal plants in the *Phyllanthus* genus. *Phyllanthus emblica* is a medium-size tree, and all its parts are used for various medicinal applications. It is widely distributed in the deciduous forests in southern India. The tree is one of the most important nontimber forest product species and is a source of livelihood for scores of forest-dwelling communities in India (Ganesan and Shetty, 2004). In recent years, because of the increase in the demand for herbal products, there has been intense extraction of the fruits. Destructive harvesting practices could have a detrimental effect on the populations of *P. emblica* (Sinha and Bawa, 2002). This consequently can reduce the regeneration of the species, leading to a loss of genetic variability.

Uma Shaanker and Ganeshaiah (1997) assessed the genetic diversity of *P. emblica* populations in southern India spread across three states, namely, Karnataka, Tamil Nadu, and Kerala. These populations are geographically isolated and represent diverse biogeographic strata from the three states. Based on isozyme analysis of six enzyme systems, the genetic variability of seven populations was assessed. Populations in southern Kerala had the highest allelic diversity as well as allele richness compared to the other populations (Uma Shaanker and Ganeshaiah, 1997). The relative abundance of most of the alleles was also found to be high in the Kerala populations. Based on these genetic parameters, Uma Shaanker and Ganeshaiah (1997) argued that the populations in Kerala could represent a potential hot spot of genetic variability of *P. emblica*.

5.3.2 PHYLLANTHUS AMARUS

Phyllanthus amarus is traditionally used in the treatment of bile and urinary conditions, hepatitis, flu, cold, jaundice, liver cancer, tuberculosis, diabetes, hypertension, pains, and other maladies (Amaechina and Omogbai, 2007). *Phyllanthus amarus* is reported to contain lignans such as phyllanthin and hypophyllanthin, alkaloids, and flavonoids such as quercetin (Santos et al., 1995). *Phyllanthus amarus* is traded as a raw herbal drug and exported for various medicinal formulations for the treatment of liver disorders (Kamble et al., 2008). However, most of the raw trade of *P. amarus* involves widespread collection of the individuals from the wild.

Jain et al. (2003) assessed the molecular diversity of *P. amarus* across India using RAPD (random amplified polymorphic DNA) markers. The genetic variability was assessed across 33 locations covering the states of Tamil Nadu, Karnataka, Maharashtra, Gujarat, Assam, West Bengal, Tripura, Uttar Pradesh, Punjab, and Haryana. Intrapopulation variation was larger in accessions from southern India compared to other parts of the country (Jain et al., 2003).

In another study, Murthy (2004) assessed the population genetic variability of *P. amarus* in three southern states of India (Karnataka, Tamil Nadu, and Kerala) and a union territory (Pondicherry). The population genetic parameters were assessed using 10 intersequence simple repeat (ISSR) primers. Kerala populations had the highest percentage of polymorphic loci, while the Karnataka populations had the least. The highest diversity was also found to be in the populations collected from Kerala. There was a clear genetic differentiation of populations based on their sites of origin.

In summary, studies of both *P. emblica* and *P. amarus* suggested that Kerala populations in southern India are genetically most diverse. It is likely that species radiated and diverged from this region into the rest of the country.

5.4 IMPACT OF HARVESTING ON THE GENETIC VARIABILITY OF *P. EMBLICA*

Phyllanthus species are among the most highly traded medicinal plant species in the country. Fruits of the species (such as *P. emblica* and *P. indofischeri*) and whole plants (in case of herbaceous species such as *P. amarus, P. debilis,* etc.) are extracted mostly from their natural populations. In recent years, because of resurgence of the herbal market globally, there has been an upsurge in the extraction of several medicinal plant species, including *Phyllanthus*, from their natural populations (Ved and Goraya, 2008; Ravikanth et al., 2009). In most cases, harvesting of the medicinal plants goes on unabated with few regulations on the extent and the nature of harvest. Consequently, such high extraction pressures could have a severe impact on the demographic and genetic profile of the populations. For example, a seemingly low-impact use, such as harvesting of *Phyllanthus emblica* fruits, may have a high long-term effect on populations, either because of the effect on seedling recruitment or because fruit collection involves pruning of branches or sometimes even tree felling (Figure 5.3; Padmini et al., 2001; Uma Shaanker et al., 1996).

Among the possible impacts of harvesting, that on the genetic variability and structure of populations has been the least studied (Uma Shaanker et al., 2001;

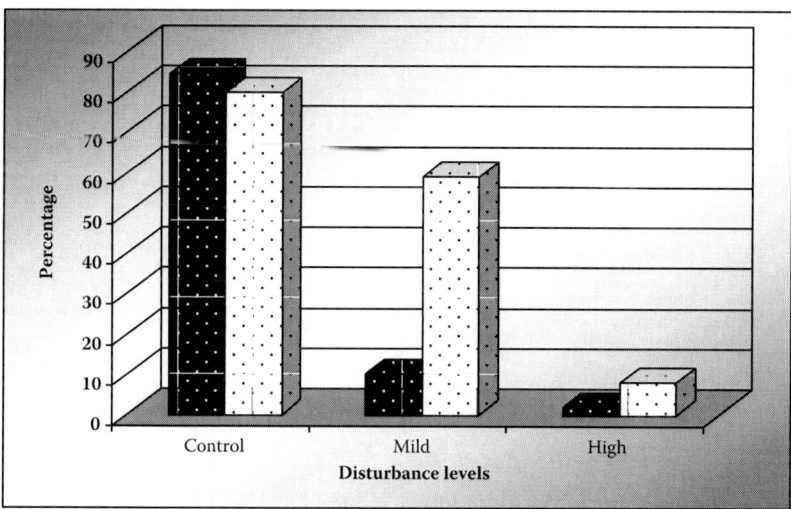

FIGURE 5.3 Percentage seedlings and saplings (<10 cm dbh (diameter at breast height) at the various disturbance levels at Biligiri Rangaswamy Temple Wildlife Sanctuary (dark columns) and Mudumalai Wildlife Sanctuary (light columns). (Adapted from Padmini, S., Nageswara Rao, M., Ganeshaiah, K.N., and Uma Shaanker, R. 2001. *J. Tropical Forest Sci.* 13: 297–310.)

Young et al., 2000). Yet, as is well realized, the impact at the level of the gene is most important. Genetic impoverishment of populations due to indiscriminate and unregulated harvesting can lead to a spiraling of effects, including low regeneration, small population sizes, inbreeding and loss of genetic diversity through genetic drift, and finally the local extinction of populations. Surprisingly, except for a few studies investigating the effect of logging on the genetic diversity of trees, hardly any studies have addressed the impact of other anthropogenic pressures on the genetic diversity of trees (Burchert et al., 1997; Wickneswari and Boyle, 2000; Young et al., 2000).

Padmini et al. (2001) and Uma Shaanker et al. (2001) examined the impact of harvesting of fruits of *P. emblica* at two forest sites in southern India along a gradient of harvesting pressures. Using isozyme markers, they examined if populations experiencing greater harvesting pressures actually differed in their genetic diversity and structure compared to populations that were less harvested or not harvested at all. Their results suggested that adult genetic diversity (as measured by percentage heterozygosity) was not affected by harvesting pressure. However, the overall genetic structure of the populations was significantly altered due to harvesting pressure; there was a significant genetic differentiation between the harvested and nonharvested populations. Alleles were differentially sensitive to harvesting pressures; frequency of a few alleles was adversely affected in the harvested population (Figure 5.4). While this might suggest that there could be a selective harvesting of plants and their fruits (e.g., of plants known to produce greater levels of ascorbic acid), Padmini et al. (2001) did not observe any evidence of selection at harvesting by local collectors. Besides these changes at the genetic level, Padmini et al. (2001)

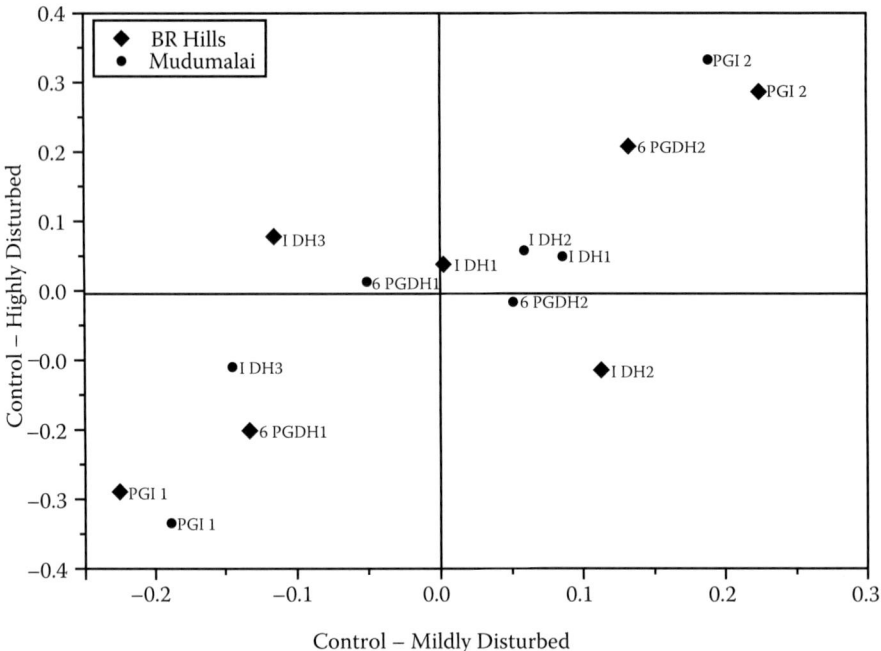

FIGURE 5.4 Relation between differences in allele frequencies between control and mildly disturbed as well as control and highly disturbed populations of *P. emblica* from Biligiri Rangaswamy Temple (B.R. Hills) Wildlife Sanctuary and Mudumalai Wildlife Sanctuary (Mudumalai). PGI, phosphogluco isomerase; IDH, isocitrate dehydrogenase; PGDH, 6-phosoglouconate dehydrogenase. (Adapted from Padmini, S., Nageswara Rao, M., Ganeshaiah, K.N., and Uma Shaanker, R. 2001. *J. Tropical Forest Sci.* 13: 297–310.)

reported a significant reduction in the net regeneration of stems in the harvested compared to the unharvested populations.

In summary, these results form one of the few quantitative examinations of the impacts of harvesting *Phyllanthus* fruits on their regeneration and genetic diversity. Both these impacts—a reduced regeneration and a change in the genetic structure of the population—can lead to a cascade of events that might endanger the local populations. The challenge lies in devising ways and methods that mitigate the harvesting-induced loss. For example, it should be possible to model the change in demographic features and genetic diversity as a function of harvesting intensities and advise on the optimal levels of harvesting. Further, the study could offer possibilities of evolving management protocols for the genetic enrichment of harvested populations to maintain the long-term productivity and adaptability of the species.

5.5 *PHYLLANTHUS*: TAXONOMIC INCONGRUITIES, SPECIES ADULTERATION, AND DNA BAR CODING

Phyllanthus species trade in India is mired by two problems: (1) taxonomic confusion among closely related species (Elvin-Lewis et al., 2002; Ganeshaiah et al., 1998) and (2) the fact that many species in trade share a common vernacular name (Srirama et al., 2010). Consequently, it is not uncommon to find substantial species admixtures in trade samples (Dymock, 1883; Dymock et al., 1893; Kirtikar and Basu, 1975; Nadkarni, 1954; van Rhede, 1690; Ganeshaiah et al., 1998; Srirama et al., 2010; Khatoon et al., 2006). For example, although *P. amarus* is a predominant species in trade, it is often found mixed with several other *Phyllanthus* species, including *P. fraternus* and *P. maderaspatensis* (Khatoon et al., 2006; Ved and Goraya, 2008).

Species admixtures may have significant implications on the quality and efficacy of the eventual phytomedicine made from these mixtures (Song et al., 2009). Khatoon et al. (2006) found significant differences in the metabolite profile between *P. fraternus* and *P. maderaspatensis* both occurring as admixtures in *P. amarus* trade samples. *Phyllanthus amarus* was the only species found to contain phyllanthin and hypophyllanthin, the two major compounds believed to be responsible for the hepatoprotective activity (Calixto et al., 1998). The admixtures of different species could lead to diluting the efficacy of the herbal drug. Using both morphotaxonomical keys and molecular techniques, Srirama et al. (2010) addressed species admixtures in southern India. Analyzing *Phyllanthus* raw herbal samples from 25 different shops from three southern Indian states (i.e., Kerala, Tamil Nadu, and Karnataka), they showed the presence of six different species of *Phyllanthus* (*P. amarus, P. debilis, P. fraternus, P. urinaria, P. maderaspatensis,* and *P. kozhikodianus*). Of the shops, 76% had *P. amarus* as the predominant species. The species identities were confirmed using species-specific DNA bar codes developed using the chloroplast *psbA-trnH*. These bar codes could be used as a diagnostic key to authenticate *Phyllanthus* species in trade as well as to identify species admixtures (Table 5.2).

5.6 IMPLICATIONS FOR UTILIZATION AND CONSERVATION

The genus *Phyllanthus* forms one of the most important nontimber forest products. In fact, a large number of forest-dwelling and forest fringe communities depend on *P. emblica* and *P. indofischeri*. A number of studies have reported that unsustainable and destructive harvesting would have an adverse impact on the regeneration of *Phyllanthus* species (Murali et al., 1996; Uma Shaanker et al., 1996; Sinha and Bawa, 2002; Ganesan, 2003). Studies by Padmini et al. (2001) have shown that the genetic resources are also adversely affected due to overharvesting. Thus, unless adequate measures are taken to protect the existing populations, the genetic resources of these species will be irreversibly lost. In recent years, there have been efforts to comprehensively address the sustainable utilization and conservation of *Phyllanthus* genetic resources. Domestication of the *Phyllanthus* species and maintaining in situ gardens would help in the long-term conservation of the economically important species. In particular, Uma Shaanker and Ganeshaiah (1997) proposed a new model, namely, the forest gene bank model, to address the long-term conservation of *Phyllanthus* species. In this model, a set of source (gene or allele donor) populations is identified from which genetic resources (through either pollen or seed) are donated to recipient "sink" populations (Figure 5.5). Sites that can serve as a "source" and those that serve as a "sink" to receive the gene pool have to be identified. The choice of the source and sink populations is guided by several criteria, including the extent

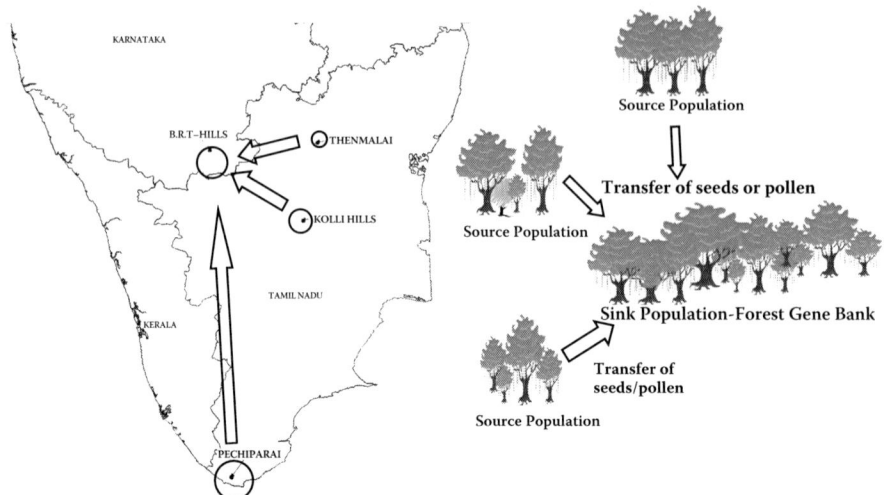

FIGURE 5.5   See color insert. Genetic diversity map of *Phyllanthus emblica*. The relative size of the circle indicates the levels of gene diversity. The arrows indicate the possible flow of genes either through seeds or pollen to the sink population from the source population (see text for further explanation). (Adapted from Uma Shaanker, R. and Ganeshaiah, K.N. 1997. *Curr. Sci.* 73: 163–168.)

and amount of genetic variation existing in the populations, the occurrence of rare alleles, threats to the population, levels of protection, extent and area of distribution, etc. The sink population should ideally be located in the protected area, should be comprised of a higher population size, and should have a broad genetic base (Uma Shaanker et al., 2002).

This model combines the virtues of both in situ and ex situ conservation techniques as this allows for the genetic diversity to "evolve" as it would in any other natural habitat of the species (Uma Shaanker et al., 2002). The forest gene bank model allows infusion of genetic material from source to sink population. This infusion encourages gene flow among the small and fragmented populations, leading to the overall genetic enrichment of the species. Such genetically enriched populations could serve as repositories of the entire spectrum of genetic variability of the species.

In the case of *Phyllanthus emblica*, Uma Shaanker and Ganeshaiah (1997) proposed that southern Indian populations (in Kerala) can be regarded as the sink into which genetic resources from other parts of southern India can be translocated. Alternatively, the populations in the state of Karnataka, which are located in a wildlife sanctuary (B.R. Hills) can serve as a sink into which population and genes/alleles from other parts of southern India can be translocated (Figure 5.5). This would ensure the conservation and maintenance of the global gene pool of *P. emblica*.

ACKNOWLEDGMENTS

This work was supported by grants from the International Plant Genetic Resources Institute (IPGRI), Center for International Forestry Research (CIFOR), Department of Forest, Ecology, and the Environment and Karnataka Forest Department, Government of Karnataka, Foundation for Revitalization of Local Health Traditions (FRLHT), and the Department of Biotechnology, Government of India.

REFERENCES

Abhyankar, G., Suprasanna, P., Pandey, B.N., et al., 2010. Hairy root extract of *Phyllanthus amarus* induces apoptotic cell death in human breast cancer cells. *Innovat. Food Sci. Emerg. Tech.* 11: 526–532.

Adeneye, A.A., and Benebo, A.S. 2008. Protective effect of the aqueous leaf and seed extract of *Phyllanthus amarus* on gentamicin and acetaminophen-induced nephrotoxic rats. *J. Ethnopharmacol.* 118: 318–323.

Al-Rehaily, A.J., Al-Howiriny, T.S., Al-Sohaibani, M.O., and Rafatullah, S. 2002. Gastroprotective effects of "Amla" *Emblica officinalis* on *in vivo* test models in rats. *Phytomedicine* 9: 515–522.

Amaechina, F.C., and Omogbai, E.K. 2007. Hypotensive effect of aqueous extract of the leaves of *Phyllanthus amarus* Schum and Thonn (*Euphorbiaceae*). *Acta Pol. Pharm. Drug Res.* 64: 547–552.

Asha, V.V., Akhila, S., Wills, P.J., and Subramoniam, A. 2004. Further studies on the antihepatotoxic activity of *Phyllanthus maderaspatensis* Linn. *J. Ethnopharmacol.* 92: 67–70.

Asha, V.V., Sheeba, M.S., Suresh, V., and Wills, P.J. 2007. Hepatoprotection of *Phyllanthus maderaspatensis* against experimentally induced liver injury in rats. *Fitoterapia* 78. 134–141.

Balakrishnan N.P., and Chakrabarty, T. 2007. *The family Euphorbiaceae in India—a synopsis of its profile, taxonomy and bibliography*. New Delhi: Eastern Book.

Bandyopadhyay, S.K., Pakrashi, S.C., and Pakrashi, A. 2000. The role of antioxidant activity of *Phyllanthus emblica* fruits on prevention from indomethacin induced gastric ulcer. *J. Ethnopharmacol.* 70: 171–176.

Bhattacharya, A.A., Chatterjee, A., Ghosal, S., and Bhattacharya, S.K. 1999. Antioxidant activity of active tannoid principles of *Emblica officinalis* (amla). *Indian J. Exp. Biol.* 37: 676–680.

Calixto, J.B., Santos, A.R.S., Filho, V.C., and Yunes, R.A. 1998. A review of the plants of the genus *Phyllanthus*: their chemistry, pharmacology, and therapeutic potential. *Med. Res. Rev.* 18: 225–258.

Chirdchupunseree, H., and Pramyothin, P. 2010. Protective activity of phyllanthin in ethanol-treated primary culture of rat hepatocytes. *J. Ethnopharmacol.* 128: 172–176.

Chudapongse, N., Kamkhunthod, M., and Poompachee, K. 2010. Effects of *Phyllanthus urinaria* extract on HepG2 cell viability and oxidative phosphorylation by isolated rat liver mitochondria. *J. Ethnopharmacol.* 130: 315–319.

Dymock, W., 1883. *The vegetable materia medica of western India*. Bombay: Education Society's Press.

Dymock, W., Warden, C.J.H., and Hooper, D. 1893. *Pharmacographia indica: a history of the principal drugs of the vegetable origin*, 261–265. London: Kegsm Paul, Trench, Trubner.

Eldeen, I.M.S., Seow, E.M., Abdullah, R., and Sulaiman, S.F. 2010. *In vitro* antibacterial, antioxidant, total phenolic contents and anti-HIV-1 reverse transcriptase activities of extracts of seven *Phyllanthus* sp. *South Afr. J. Bot.* doi: 10.1016/j.sajb.2010.05.009.

Elvin-Lewis, M., Navarro, M., Colichon, A., and Lewis, W.H. 2002. Therapeutic evaluation of hepatitis remedies: the usefulness of ethnomedical focusing techniques. In *7th International Congress of Ethnobiology,* ed. J.R. Stepp, F.S. Wyndham, and R.K. Zarger, 270–281. Athens: University of Georgia Press.

Fang, S.H., Rao, Y.K., and Tzeng, Y.M. 2008. Anti-oxidant and inflammatory mediator's growth inhibitory effects of compounds isolated from *Phyllanthus urinaria*, *J. Ethnopharmacol.* 116: 333–340.

Faremi, T.Y., Suru, S.M., Fafunso, M.A., and Obioha, U.E. 2008. Hepatoprotective potentials of *Phyllanthus amarus* against ethanol-induced oxidative stress in rats. *Food Chem. Toxicol.* 46: 2658–2664.

Ganesan, R. 2003. Identification, distribution and conservation of *Phyllanthus indofischeri*, another source of Indian gooseberry. *Curr. Sci.* 84: 1515–1518.

Ganesan, R., and Shetty, S.R. 2004. Regeneration of amla, an important non-timber forest product from southern India. *Conserv. Soc.* 2: 365–375.

Ganeshaiah, K.N., Ganesan, R., Uma Shaanker, R., and Meera, C. 1998. A taxonomic hurdle or hurdled by taxonomists. *Amurth*, August: 3–8.

Ganeshaiah, K.N., and Uma Shaanker, R. 2003. *A decade of diversity*. Bangalore, India: University of Agricultural Sciences.

Ganeshaiah, K.N., and Uma Shaanker, R. 2007. Bioresources database: Documenting life. *Biotech. News* 2: 10–13.

Ghosal, S., Tripathi, V.K., and Chauhan, S. 1996. Active constituents of *Emblica officinalis*: Part I. The chemistry and antioxidant activity of two new hydrolysable tannins, emblicanin A and B. *Indian J. Chem.* 35: 941–948.

Gopi, S., and Setty, O.H. 2010. Protective effect of *Phyllanthus fraternus* against bromobenzene induced mitochondrial dysfunction in rat liver mitochondria. *Food Chem. Toxicol.* 48: 2170–2175.

Govaerts, R.D.G., Frodin, A., and Radcliffe-Smith. 2000. *World checklist and bibliography of Euphorbiaceae*. Kew, UK: Royal Botanic Gardens.

Hau, D.K.P., Gambari, R., and Wong, R.S.M. 2009. *Phyllanthus urinaria* extract attenuates acetaminophen induced hepatotoxicity: involvement of cytochrome P450 CYP2E1. *Phytomedicine* 16: 751–760.

Huang, S.T., Yang, R.C., Chen, M.Y., and Pang, J.H.S. 2004. *Phyllanthus urinaria* induces the Fas receptor/ligand expression and ceramide-mediated apoptosis in HL-60 cells. *Life Sci.* 75: 339–351.

Huang, S.T., Yang, R.C., Lee, P.N., et al., 2006. Anti-tumor and anti-angiogenic effects of *Phyllanthus urinaria* in mice bearing Lewis lung carcinoma. *Int. Immunopharmacol.* 6: 870–879.

Huang, S.T., Yang, R.C., Yang, LJ., et al. 2003. *Phyllanthus urinaria* triggers the apoptosis and Bcl-2 down-regulation in Lewis lung carcinoma cells. *Life Sci.* 72: 1705–1716.

Ignácio, S.R.N., Ferreira, J.L.P., Almeida M.B., and Kubelka, C. F. 2001. Nitric oxide production by murine peritoneal macrophages *in vitro* and *in vivo* treated with *Phyllanthus tenellus* extracts. *J. Ethnopharmacol.* 74: 181–187.

Jain, N., Shasany, A.K., Singh, S., et al. 2008. SCAR markers for correct identification of *Phyllanthus amarus*, *P. fraternus*, *P. debilis* and *P. urinaria* used in scientific investigations and dry leaf bulk herb trade. *Planta Med.* 74: 296–301.

Jain, N., Shasany, A.K., Sundaresan, V., et al. 2003. Molecular diversity in *Phyllanthus amarus* assessed through RAPD analysis. *Curr. Sci.* 85: 1454–1458.

Jayaram, S., and Thyagarajan, S.P. 1996. Inhibition of HBs Ag secretion from Alexander cell line by *Phyllanthus amarus*. *Indian J. Pathol. Microbiol.* 39: 211–215.

Jose, J.K., Kuttan, G., and Kuttan, R. 2001. Antitumour activity of *Emblica officinalis*. *J. Ethnopharmacol.* 75: 65–69.

Kamble, M.B., Dumbre, R.K., and Rangari, V.D. 2008. Hepatoprotective studies of herbal formulations. *Int. J. Green Pharm.* 2: 147–151.

Kassuya, C.A.L., Silvestre, A., Menezes-de-Lima, O., Jr., et al. 2006. Anti-inflammatory and antiallodynic actions of the lignan niranthin isolated from *Phyllanthus amarus*: evidence for interaction with platelet activating factor receptor. *Eur. J. Pharmacol.* 546: 182–188.

Kathriarachchi, H., Samuel, R., Hoffmann, P., et al. 2006. Phylogenetics of tribe Phyllantheae (Phyllanthaceae; Euphorbiaceae sensu lato) based on *nrITS* and plastid *matK* DNA sequence data. *Am. J. Bot.* 93: 637–655.

Kawakita, A., and Kato, M. 2009. Repeated independent evolution of obligate pollination mutualism in the Phyllantheae-Epicephala association. *Proc. Biol. Sci.* 276: 417–426.

Khatoon, S., Rai, V., Rawat, A.K.S., and Mehrotra, S. 2006. Comparative pharmacognostic studies of three *Phyllanthus* species. *J. Ethnopharmacol.* 104: 79–86.

Khopde, S.M., Indira Priyadarsini, K., Mohan, H.,et al. 2001. Characterizing the antioxidant activity of amla (*Phyllanthus emblica*) extract. *Curr. Sci.* 81: 185–190.

Kiemer, A.K., Hartung, T., Huber, C., and Vollmar, A.M. 2003. *Phyllanthus amarus* has anti-inflammatory potential by inhibition of iNOS, COX-2, and cytokines via the NF-[kappa] B pathway. *J. Hepatol.* 38: 289–297.

Kirtikar, K.R., and Basu, B.D. 1975. *Indian medicinal plants*. New Connaught Place, Dehrahun: Bishen Singh Mahendra Pal Singh.

Kumar, S., Kumar, D., Deshmukh, R.R., et al. 2008. Antidiabetic potential of *Phyllanthus reticulatus* in alloxan-induced diabetic mice. *Fitoterapia* 79: 21–23.

Kumar, K.B.H., and Kuttan, R. 2005. Chemoprotective activity of an extract of *Phyllanthus amarus* against cyclophosphamide induced toxicity in mice. *Phytomedicine* 12: 494–500.

Kundu, S., Nandy, D., and Bishayi, B. 2009. Immunomodulatory effects of *Phyllanthus acidus* and *Parkia javanica* whole plant extracts on murine splenic macrophages. *Int. J. Infectious Dis.* 13: 58.

Lai, C.H., Fang, S.H., and Rao, Y.K. 2008. Inhibition of *Helicobacter pylori*-induced inflammation in human gastric epithelial AGS cells by *Phyllanthus urinaria* extracts. *J. Ethnopharmacol.* 118: 522–526.

Lee, C.D., Ott, M., Thyagarajan, S.P., et al. 1996. *Phyllanthus amarus* down-regulates hepatitis B virus mRNA transcription and replication. *Eur. J. Clin. Invest.* 26: 1069–1076.

Lee, S.K., Li, P.T., Lau, D.T., et al. 2006. Phylogeny of medicinal *Phyllanthus* species in China based on nuclear *ITS* and chloroplast *atpB-rbcL* sequences and multiplex PCR detection assay analysis. *Planta Medica* 72: 721–726.

Lin, S.Y., Wang, C.C., and Lu, Y.L. 2008. Antioxidant, anti-semicarbazide-sensitive amine oxidase, and anti-hypertensive activities of geraniin isolated from *Phyllanthus urinaria*. *Food Chem. Toxicol.* 46: 2485–2492.

Liu, X., Zhao, M., Wang, J., et al. 2008. Antioxidant activity of methanolic extract of emblica fruit (*Phyllanthus emblica* L.) from six regions in China. *J. Food Composition Anal.* 21: 219–228.

Luo, W., Zhao, M., Yang, B., et al. 2009. Identification of bioactive compounds in *Phyllanthus emblica* L. fruit and their free radical scavenging activities. *Food Chem.* 114: 499–504.

Mathur, R., Sharma, A., Dixit, V. P., and Varma, M. 1996. Hypolipidaemic effect of fruit juice of *Emblica officinalis* in cholesterol-fed rabbits. *J. Ethnopharmacol.* 50: 61–68.

Murali, K.S., Uma Shaankar, R., Ganeshaiah, K.N., and Bawa, K.S. 1996. Extraction of non-timber forest products in the forests of Biligari Rangan Hills, India, 2: Impact of NTFP extraction on regeneration, population structure and species composition. *Econ. Bot.* 50: 252–269.

Murthy, P.M.K. 2004. Intra and inter specific genetic variation in *Phyllanthus amarus-fraternus* (PAF) complex. Master of science thesis. Department of Genetics and Plant Breeding, University of Agricultural Sciences, GKVK, Bangalore, India.

Naaz, F., Javed, S., and Abdin, M.Z. 2007. Hepatoprotective effect of ethanolic extract of *Phyllanthus amarus* Schum. et Thonn. on aflatoxin B1-induced liver damage in mice. *J. Ethnopharmacol.* 113: 503–509.

Nadkarni, A.K. 1954. *Dr. K.M. Nadkarni's Indian materia medica*. Bombay: Popular Book Depot.

Narendranathan, M., Mini, P.C., and Remla, A. 1997. A trial of *Phyllanthus amarus* in acute viral hepatitis. *J. Clin. Epidemiol.* 50: S8.

Nosál'ová, G., Mokr´y, J., and Hassan, K.M. 2003. Antitussive activity of the fruit extract of *Emblica officinalis* Gaertn. (Euphorbiaceae). *Phytomedicine* 10: 583–589.

Notka, F., Meier, G.R., and Wagner, R. 2003. Inhibition of wild-type human immunodeficiency virus and reverse transcriptase inhibitor-resistant variants by *Phyllanthus amarus*. *Antiviral Res.* 58: 175–186.

Notka, F., Meier, G., and Wagner, R. 2004. Concerted inhibitory activities of *Phyllanthus amarus* on HIV replication *in vitro* and *ex vivo*. *Antiviral Res.* 64: 93–102.

Padma, P., and Setty, O.H. 1999. Protective effect of *Phyllanthus fraternus* against carbon tetrachloride-induced mitochondrial dysfunction. *Life Sci.* 64: 2411–2417.

Padmini, S., Nageswara Rao, M., Ganeshaiah, K.N., and Uma Shaanker, R. 2001. Genetic diversity of *Phyllanthus emblica* in tropical forests of south India: impact of anthropogenic pressures. *J. Trop. Forest Sci.* 13: 297–310.

Paranjape, P. 2001. *Indian medicinal plants: forgotten healers*, 48–49. Delhi: Chaukhamba Sanskrit Pratisthan.

Perianayagam, J.B., Sharma, S.K., Joseph, A., and Christina, A.J.M. 2004. Evaluation of antipyretic and analgesic activity of *Emblica officinalis* Gaertn. *J. Ethnopharmacol.* 95: 83–85.

Pramyothin, P., Samosorn, P., Poungshompoo, S., and Chaichantipyuth, C. 2006. The protective effects of *Phyllanthus emblica* Linn. extract on ethanol induced rat hepatic injury. *J. Ethnopharmacol.* 107: 361–364.

Pruesapan, K., Ian, R., Telford, H., et al. 2008. Delimitation of *Sauropus* (Phyllanthaceae) based on plastid matK and nuclear ribosomal *ITS* DNA sequence data. *Ann. Bot.* 102: 1007–1018.

Rajeshkumar, N. V., Joy, K. L., Kuttan, G., et al. 2002. Antitumour and anticarcinogenic activity of *Phyllanthus amarus* extract. *J. Ethnopharmacol.* 81: 17–22.

Rajeshkumar, N. V., and Kuttan, R. 2000. *Phyllanthus amarus* extract administration increases the life span of rats with hepatocellular carcinoma. *J. Ethnopharmacol.* 73: 215–219.

Rao, Y.K., Fang, S.H., and Tzeng, Y.M. 2006. Anti-inflammatory activities of constituents isolated from *Phyllanthus polyphyllus*. *J. Ethnopharmacol.* 103: 181–186.

Raphael, K.R., and Kuttan, R. 2003. Inhibition of experimental gastric lesion and inflammation by *Phyllanthus amarus* extract. *J. Ethnopharmacol.* 87: 193–197.

Ravikanth, G., Rao, M.N., Ganeshaiah, K.N., and Uma Shaanker, R. 2009. Genetic diversity of NTFP species: issues and implications. In *Non-timber forest products conservation, management and policies,* ed. R. Uma Shaanker, A.J. Hiremath, G.C. Joseph, and N, Rai, 53–64. Bangalore: Ashoka Trust for Research in Ecology and Environment, Bangalore and Forestry Research Support Program for Asia and the Pacific, Food and Agriculture Organisation, Bangkok.

Ravikanth, G., Uma Shaanker, R., and Ganeshaiah, K.N. 2002. Identification of hot spots of species richness and genetic variability in rattans—an approach using GIS and molecular tools. *Plant Gen. Resources Newsl.* 132: 17–21.

Reddy, V.D., Pannuru, P., Maturu, P., and Nallanchakravarthula, V. 2009. Modulatory role of *Emblica officinalis* against alcohol induced biochemical and biophysical changes in rat erythrocyte membranes. *Food Chem. Toxicol.* 47: 1958–1963.

Sailaja, R., and Setty, O.H. 2006. Protective effect of *Phyllanthus fraternus* against allyl alcohol-induced oxidative stress in liver mitochondria. *J. Ethnopharmacol.* 105: 201–209.

Sai Ram, M., Neetu, D., Yogesh, B. et al. 2002. Cyto-protective and immunomodulating properties of amla (*Emblica officinalis*) on lymphocytes: an in-vitro study. *J. Ethnopharmacol.* 81: 5–10.

Santos, A.R., Filiho, V.C., Yunes, R.A., and Calixto, J.B. 1995. Analysis of the mechanisms underlying antinociceptive effect of the extract of plants from the genus *Phyllanthus*. *Gen. Pharmacol.* 26: 1499–1506.

Sebastian, T., and Setty, O. H. 1999. Protective effect of *P. fraternus* against ethanol-induced mitochondrial dysfunction. *Alcohol* 17: 29–34.

Shabeer, J., Srivastava, R.S., and Singh, S.K. 2009. Antidiabetic and antioxidant effect of various fractions of *Phyllanthus simplex* in alloxan diabetic rats. *J. Ethnopharmacol.* 124: 34–38.

Sharma, A., Sharma, M.K., and Kumar, M. 2009. Modulatory role of *Emblica officinalis* fruit extract against arsenic induced oxidative stress in Swiss albino mice. *Chem. Biol. Interact.* 180: 20–30.

Sinha, A., and Bawa, K.S. 2002. Harvesting techniques, hemiparasites and fruit production in two non-timber forest tree species in south India. *For. Ecol. Manage.* 168: 289–300.

Song, J., Yao, H., Li, Y., et al. 2009. Authentication of the family polygonaceae in Chinese pharmacopoeia by DNA barcoding technique. *J. Ethnopharmacol.* 124: 434–439.

Srirama, R., Senthilkumar, U., Sreejayan, N., et al. 2010. Assessing species admixtures in raw drug trade of *Phyllanthus*, a hepato-protective plant using molecular tools. *J. Ethnopharmacol.* 130: 208–215.

Stanslas, J., Bagalkotkar, G., Tang, S.C., et al. 2008. New antitumour agents from *Phyllanthus pulcher*, a tropical medicinal plant. *Eur. J. Cancer* 6(Suppl): 58–59.

Sultana, S., Ahmed, S., and Jahangir, T. 2008. *Emblica officinalis* and hepatocarcinogenesis: a chemopreventive study in Wistar rats. *J. Ethnopharmacol:* 118: 1–6.

Summanen, J.O. 1999. A chemical and ethnopharmalogical study on *Phyllanthus emblica* (Euphorbiaceae). Master's thesis. Department of Pharmacy, Division of Pharmacognosy, University of Helsinki.

Suresh, V., and Asha, V.V. 2008. Preventive effect of ethanol extract of *Phyllanthus rheedii* Wight. on D-galactosamine induced hepatic damage in Wistar rats. *J. Ethnopharmacol.* 116: 447–453.

Syamasundar, K.V., Singh, B., and Thakor, R.S. 1985. Antihepatotoxic principle of *Phyllanthus niruri* herbs. *J. Ethnopharmacol.* 14: 41–44.

Thyagarajan, S.P., Subramanian, S., and Thirunalasundari, T., et al. 1988. Preliminary study: the effect of *Phyllanthus amarus* on chronic carriers of hepatitis B virus. *Lancet* 11: 764–766.

Uma Shaanker, R., and Ganeshaiah, K.N. 1997. Mapping genetic diversity of *Phyllanthus emblica*: forest gene banks as a new approach for *in situ* conservation of genetic resources. *Curr. Sci.* 73: 163–168.

Uma Shaanker, R, Ganeshaiah. K.N., Nageswara Rao, M., and Aravind, N.A. 2004. Ecological consequences of forest use: from genes to ecosystem—a case study in the Biligiri Rangaswamy Temple Wildlife Sanctuary, south India. *Conserv. Soc.* 2: 347–363.

Uma Shaanker, R., Ganeshaiah, K.N., and Rao, M.N. 2001. Genetic diversity of medicinal plant species in deciduous forest of south India: impact of harvesting and other anthropogenic pressures. *J. Plant Biol.* 28: 91–97.

Uma Shaanker, R., Ganeshaiah, K.N., Rao, N.M., and Ravikanth, G. 2002. Forest gene banks: a new approach to conserving forest tree genetic resources. In *Managing plant genetic resources,* ed. J.M.M. Engels, A.H.D. Brown, A.H.D. and M.T. Jackson, 229–235. Wallingford, UK: CABI.

Uma Shaankar, R., Murali, K.S., Uma Shaanker, R., Ganeshaiah, K.N., and Bawa, K.S. 1996. Extraction of non-timber forest products in the forests of Biligiri Rangan Hills, India. 3. Productivity, extraction and prospects of sustainable harvest of Amla, *Phyllanthus emblica* (Euphorbiaceae). *Econ. Bot.* 50: 270-279.

Unander, D.W., Webster, G.L., and Blumberg, B.S. 1991. Uses and bioassays in *Phyllanthus* (Euphorbiaceae): a compilation. II. The subgenus *Phyllanthus. J. Ethnopharmacol.* 34: 97–133.

Unander, D.W., Webster, G.L., and Blumberg, B.S. 1995. Usage and bioassays in *Phyllanthus* (Euphorbiaceae): IV. Clustering of antiviral uses and other effect. *J. Ethnopharmacol.* 45: 1–18.

van Rheede, A. 1690. *Horti Muluhorici Purs De&m de Herhis er Diversis Illurum Specienus*, 10, 29–31. Amsterdam: Someren.

Ved, D.K., and Goraya, G.S. 2008. *Demand and supply of medicinal plants in India*. Bangalore, India: FRLHT.

Venkateswaran, P.S., Millman, I., and Blumberg, B.S. 1987. Effects of an extract from *Phyllanthus niruri* on hepatitis B and woodchuck hepatitis viruses: *in vitro* and *in vivo* studies. *Proc. Natl. Acad. Sci. U S A* 84: 274–278.

Vormisto, I.A., Summanen, J., and Kankaanranta, H. 1997. Anti-inflammatory activity of extracts from leaves of *Phyllanthus emblica*. *Plant Med.* 63: 518–524.

Walaiphachara, N. 1994. Effect of *Phyllanthus amarus* on carbon tetrachloride (CCl_4)-induced hepatotoxicity in rats. Master's thesis. Department of Pathobiology, Faculty of Science, Mahidol University.

Wurdack, K.J., Hoffmann, P., Samuel, R., Bruijn, A.Y. De, et al. 2004. Molecular phylogenetic analysis of Phyllanthaceae (Phyllanthoideae pro parte, Euphorbiaceae sensu lato) using plastid *rbcL* DNA sequences. *Am. J. Bot.* 91: 1882–1900.

Yang, C.M., Cheng, H.Y., and Lin, TC. 2005. Acetone, ethanol and methanol extracts of *Phyllanthus urinaria* inhibit HSV-2 infection *in vitro*. *Antiviral Res.* 67: 24–30.

Yang, C.M., Cheng, H.Y., and Lin, T.C. 2007. The *in vitro* activity of geraniin and 1,3,4,6-tetra-O-galloyl-[beta]-d-glucose isolated from *Phyllanthus urinaria* against herpes simplex virus type 1 and type 2 infection. *J. Ethnopharmacol.* 110: 555–558.

Yeh, S.F., Hong, C.Y., Huang, Y.L., et al. 1993. Effect of an extract from *Phyllanthus amarus* on hepatitis B surface antigen gene expression in human hepatoma cells. *Antiviral Res.* 20: 185–192.

Young, A., Boshier, D., and Boyle, T. (Eds.). *Forest conservation genetics: Principles and practice*. CSIRO Publishing, Collingwood, Australia.

6 Phytochemistry of the Genus *Phyllanthus*

Lutfun Nahar, Satyajit D. Sarker, and Abbas Delazar

CONTENTS

6.1 Introduction 120
6.2 Phytochemicals Isolated from The Genus *Phyllanthus* 120
    6.2.1 Alkaloids 120
    6.2.2 Coumarins and Cinnamic Acid Derivatives 120
    6.2.3 Flavonoids 122
        6.2.3.1 Flavonols 122
        6.2.3.2 Flavones 124
        6.2.3.3 Flavanones 124
    6.2.4 Lignans 124
        6.2.4.1 Dibenzylbutanes 124
        6.2.4.2 Dibenzylbutyrolactones 124
        6.2.4.3 Epoxy and Diepoxylignans 125
        6.2.4.4 Aryltetralins 125
        6.2.4.5 Arylnapthalenes 126
        6.2.4.6 Neolignans 126
    6.2.5 Simple Lactones 126
    6.2.6 Simple Phenolics and Benzene Derivatives 127
    6.2.7 Steroidal Compounds 128
    6.2.8 Tannins 128
    6.2.9 Terpenoids 128
        6.2.9.1 Monoterpenes 130
        6.2.9.2 Sesquiterpenes 130
        6.2.9.3 Diterpenes 130
        6.2.9.4 Triterpenes 130
    6.2.10 Miscellaneous Compounds 132
6.3 Distribution and Chemotaxonomic Significance 132
6.4 Conclusion 132
References 134

6.1 INTRODUCTION

The genus *Phyllanthus* L. belongs to the family Phyllanthaceaea, but often it is placed within the Euphorbiaceae (Nozeran et al., 1984; Calixto et al., 1998; GRIN Database, 2011). It is the largest genus of this family and comprises well over 800 species of annuals, perennials, shrubs, climbers, and floating aquatics, mainly from the tropical and the subtropical countries, and distributed within 11 subgenera: *Botryanthus, Cicca, Conani, Emblica, Ericocus, Gomphidium, Isocladus, Kirganelia, Phyllanthodendron, Phyllanthus,* and *Xylophylla.*

Several species of this genus have long been used in traditional medicine, particularly in the Ayurvedic medicine system, to treat, notably, kidney problems, urinary bladder disturbances, diabetes, pain, jaundice, gonorrhea, chronic dysentery, skin ulcers, and hepatitis B (Calixto et al., 1998; Dr. Duke's Phytochemical and Ethnobotanical Databases, 2011). Because of the tremendous therapeutic potentials of the *Phyllanthus* species, numerous phytochemical and bioactivity studies have been carried out on these plants, resulting in the isolation and identification of various compounds, ranging from alkaloids, coumarins, flavonoids, lignans, and terpenes (ISI Web of Knowledge, 2011). The aims of this chapter are to look into the phytochemistry of the genus *Phyllanthus*, to document the compounds isolated from various species, to evaluate their distribution pattern within the genus, and to discuss their possible chemotaxonomic significance.

6.2 PHYTOCHEMICALS ISOLATED FROM THE GENUS *PHYLLANTHUS*

The species of the genus *Phyllanthus* produce a variety of compounds, which are discussed next.

6.2.1 ALKALOIDS

Among the phytochemically investigated *Phyllanthus* species, at least 12 species were found to contain about 22 different alkaloids (**1–22**), predominantly of the securinine/norsecurinine type (Figure 6.1; Table 6.1), which have a unique tricyclic skeleton with an α,β-unsaturated-γ-lactone ring. Securinine (**5**) is arguably the first alkaloid found in any *Phyllanthus* species and was isolated from *P. discoides* about five decades ago (Hassarajani et al., 1990; Quevauviller and Blanpin, 1959). Phyllanthine **4** and **5** were the most common alkaloids and were isolated from three different *Phyllanthus* species. *Phyllanthus amarus* (Houghton et al., 1996), *P. discoides* (Quevauviller et al., 1967; Foussard-Blanpin et al., 1967; Parello, 1968), and *P. discoideus* (Mensah et al., 1988; Calixto et al., 1998) produced the highest number of alkaloids.

6.2.2 COUMARINS AND CINNAMIC ACID DERIVATIVES

The occurrence of coumarins and simple cinnamic acid derivatives within the genus *Phyllanthus* is rather rare. To date, only the following have been reported: two

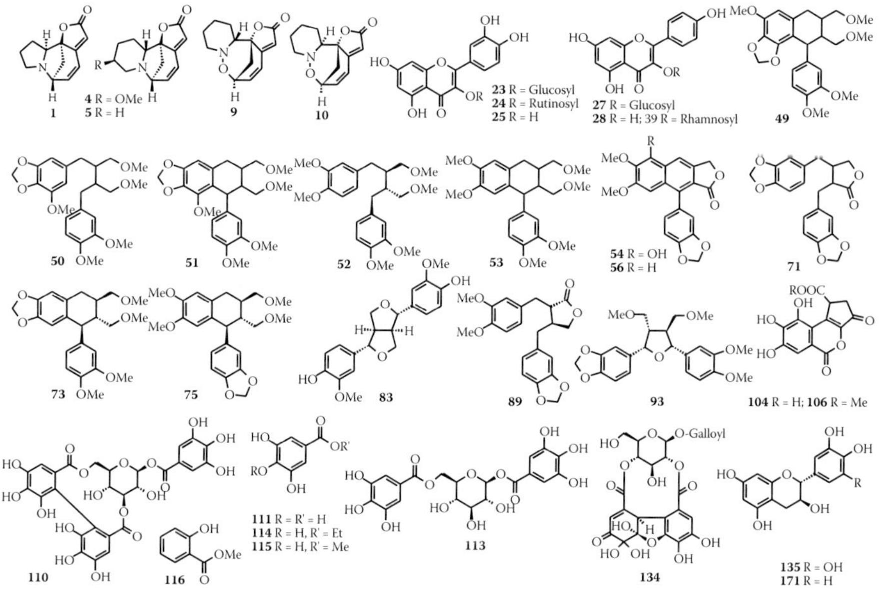

Phytochemistry of the Genus *Phyllanthus*

FIGURE 6.1A Structures of compounds that have been found in two or more species of the genus *Phyllanthus*.

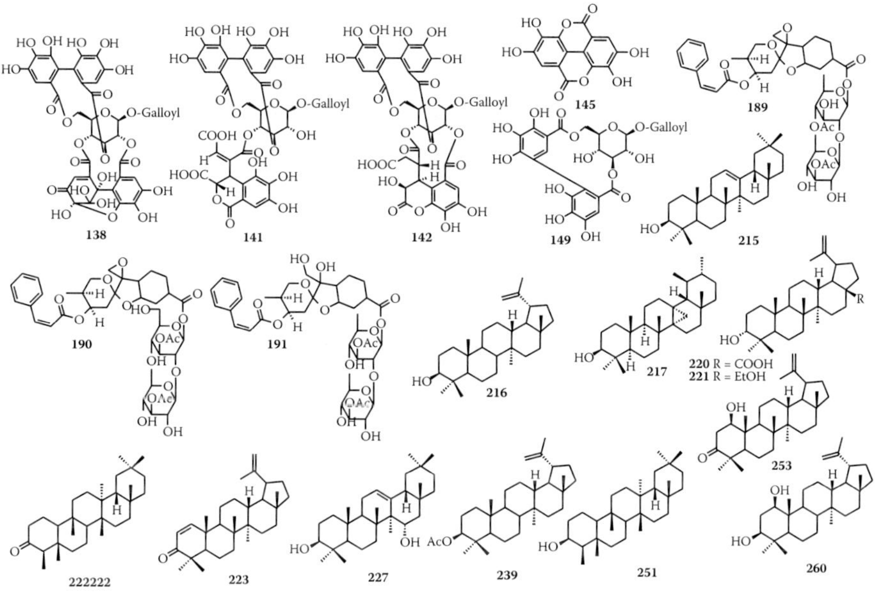

FIGURE 6.1A (continued)

TABLE 6.1
Alkaloids from the Genus *Phyllanthus*

Species (Plant Parts): Isolated Alkaloids	References
P. amarus (leaves): *epi*-bubbialine (**1**), isobubbialine (**2**), norsecurinine (**3**), phyllanthine (**4**), securinine (**5**)	Houghton et al., 1996
P. burchellii (leaves): alkaloids	Mdlolo et al., 2008
P. discoides (leaves): **4, 5**, phyllabine (**6**), phyllochrysine (**7**)	Quevauvillar et al., 1965; Quevauvillar and Blanpin, 1959; Foussard-Blanpin et al, 1967; Parello, 1968; Quevauvillar et al., 1967
P. discoideus (leaves): **3–6**, phyllocristine (**8**), phyllanthidine (**9**), phyllantidine (**10**), 14,15-dihydro-allo-securinine-15-β-ol (**11**), allo-securinine (**12**), v*iro*-allo-securinine (**13**), dihydrosecurinine (**14**)	Mensah et al., 1988; Calixto et al., 1998
P. emblica (leaves and fruits): **4, 10**, zeatine (**15**)	Khanna and Bansal, 1975; Ram and Rao, 1976
P. niruri (leaves): *ent*-norsecurinine (**16**), 4-methoxy-norsecurinine (**17**), nirurine (**18**)	Joshi et al., 1986; Mulchandani and Hassarajani, 1984; Petchnaree et al., 1986
P. niruroides (leaves): **5**, niruroidine (**19**)	Babady-Bila et al., 1996
P. parvulus var. *garipensis* (aerial parts): alkaloids	Mdlolo et al., 2008
P. sellowianus (leaves): phyllanthimide (**20**)	Tempesta et al., 1988
P. simplex (whole plant): **4**, Simplexine (**21**)	Negi and Fakir, 1988
P. urinaria (whole plant): **4, 9**	Calixto et al., 1998
P. virgatus (leaves): indole-3-carboxylic acid (**22**)	Calixto et al., 1998

simple coumarins (scopoletin and isofraxidine) from *P. sellowianus* (Calixto et al., 1998) and five cinnamic acid derivatives: cinnamic acid from *P. emblica* (Luo et al., 2009), *p*-coumaric acid from *P. reticulates* (Calixto et al., 1998), caffeic acid and chlorogenic acid from *P. sellowianus* (Calixto et al., 1998), and phyllurine from *P. urinaria* (Ueda et al., 1999).

6.2.3 Flavonoids

Flavonoids are one of the main groups of secondary metabolites produced by the *Phyllanthus* species. More than 25 different flavonoids, including flavonols, flavones, and flavanones and their glycosides, have been reported from at least 13 species (Figure 6.1; Table 6.2). In addition, a few flavan derivatives, for example, catechin (**178**), epicatechin (**179**), gallocatechin (**142**), epigallocatechin (**181**), epicatechin gallatae (**180**), and epigallocatechin gallate (**182**), which are the components of various tannins, have also been isolated from several *Phyllanthus* species (Calixto et al., 1998) and are discussed under the category of tannins (Section 6.2.8).

6.2.3.1 Flavonols

Flavonols (Table 6.2), isolated from various species of the *Phyllanthus*, mainly possess quercetin (**25**) and kaempferol (**28**) skeletons. Glycosylation, particularly at C-3,

TABLE 6.2
Flavonoids from the Genus *Phyllanthus*

Species (Plant Parts): Isolated Flavonoids	References
P. amarus (aerial parts): quercetin 3-*O*-β-D-glucopyranoside (**23**), rutin (**24**)	Foo, 1993; Londhe et al., 2008
P. burchellii (aerial parts): unidentified flavonoids	Mdlolo et al., 2008
P. caroliniensis (whole plant): quercetin (**25**)	Cechinel et al., 1996
P. debilis (aerial parts): unidentified flavonoids	Kumaran and Karunakaran, 2007
P. emblica (fruits and leaves): **23–25**, apigenin 7-*O*-(6′-butyryl-β-D-glucopyranoside (**26**), astragalin (**27**), kaempferol (**28**), kaempferol 3-*O*-α-L-(6′-ethyl)-rhamnopyranoside (**29**), kaempferol 3-*O*-α-L-(6′-methyl)-rhamnopyranoside (**30**), luteolin 4′-*O*-neohesperiodoside (**31**)	Calixto et al., 1998; Rehman et al., 2007; El-Desouky et al., 2008; Liu et al., 2008
P. maderaspatensis (aerial parts): unidentified flavonoids	Kumaran and Karunakaran, 2007
P. niruri (whole plant, roots, and leaves): **24, 25, 27**, (S)-eriodictyol 7-*O*-(6′-*O*-*trans*-*p*-coumaroyl)-β-D-glucopyranoside (**32**), (S)-eriodictyol 7-*O*-(6′-*O*-galloyl)-β-D-glucopyranoside (**33**), 2-(4-hydroxyphenyl)-8-(3-methyl-but-2-enyl)-chroman-4-one (**34**), 8-(3-methyl-but-2-enyl)-2-phenyl-chroman-4-one (**35**), kaempferol 4-*O*-α-L-rhamnopyranoside (**36**), niruriflavone (**37**), nirurin (**38**), quercitrin (**39**), quercetin 3-*O*-β-D-glucopyranosyl(1®4)-α-rhamnopyranoside (**40**), quercetin 3-*O*-β-D-glucopyranosyl-(2®1)-*O*-β-D-xylopyranoside (**41**)	Chauhan et al., 1977; Gupta and Ahmed, 1984; Calixto et al., 1998; Zhang et al., 2002; Subeki et al., 2005; Than et al., 2006; Shakil et al., 2008
P. orbiculatus (leaves): **24, 25, 27, 39**	Calixto et al., 1998
P. parvulus Sond var. *garipensis* (aerial parts): unidentified flavonoids	Mdlolo et al., 2008
P. reticulates (leaves): **23, 24, 27, 39**, tricin (**42**)	Lam et al., 2007
P. sellowianus (leaves and stems): **24, 25**, 4,4-di-*O*-methyl-cupressuflavone (**43**), 7-hydroxyflavanone (**44**)	Hnatyszyn et al., 1987; Calixto et al, 1998
P. urinaria (whole plant): **23–25, 27, 28, 39**, rhamnocitrin (**45**)	Calixto et al., 1998; Fang et al., 2008
P. virgatus (aerial parts): galangin-8-sulfonate (**46**), galangin 3-*O*-β-D-glucoside-8-sulfonate (**47**), kaempferol-8-sulfonate (**48**)	Huang et al., 1998a

is quite common in these flavonols. β-D-Glucose and α–L-rhamnose are the most common sugars found in the flavonol glycosides of this genus. Kaempferol (**28**), quercetin (**25**), and rutin (**24**) are present in most flavonoid-containing *Phyllanthus* species. Three flavonol sulfonates—galangin-8-sulphonate (**46**), galangin 3-*O*-β-D-glucoside-8-sulfonate (**47**), and kaempferol-8-sulfonate (**48**)—have been isolated only from the aerial parts of *P. virgatus* (Huang et al., 1998a). Rhamnocitrin (**45**) appears to be the only methoxylated flavonol reported from this genus (Calixto et al., 1998; Fang et al., 2008).

6.2.3.2 Flavones

There are five flavones, including a biflavone (**43**), found in the *Phyllanthus* (Table 6.2). Apigenin 7-*O*-(6′-butyryl-β-D-glucopyranoside (**26**) and luteolin 4′-*O*-neohesperiodoside (**31**) are found in *P. emblica* (Calixto et al., 1998; Rehman et al., 2007; El-Desouky et al., 2008; Liu et al., 2008); the sulfonic acid moiety containing flavone, niruriflavone (**37**), in *P. niruri* (Than et al., 2006); tricin (**42**) in *P. reticulatus* (Lam et al., 2007); and the only biflavone, 4,4-di-*O*-methyl-cupressuflavone (**43**), in *P. sellowianus* (Hnatyszyn et al., 1987; Calixto et al., 1998).

6.2.3.3 Flavanones

Six different flavanones (Table 6.2) have been reported only from two species of the *Phyllanthus*: five (**32–35** and **38**) from *P. niruri* (Calixto et al., 1998; Zhang et al., 2002; Subeki et al., 2005; Than et al., 2006; Shakil et al., 2008) and one (**44**) from *P. sellowianus* (Hnatyszyn et al., 1987; Calixto et al., 1998). There are three flavanones with prenylation at C-8 (**34, 35,** and **38**). It is interesting to note that prenylated flavonoids have only been found in *P. niruri* but in no other species of the genus *Phyllanthus* to date.

6.2.4 Lignans

About 50 different lignans (Figure 6.1; Table 6.3) belonging to dibenzylbutane, dibenzylbutyrolactone, epoxy and diepoxy, aryltetralin, arylnapthalene, and neolignan classes have been reported from at least 11 species of the *Phyllanthus* to date. *Phyllanthus myrtifolius*, *P. niruri*, and *P. urinaria* are the major lignan-producing species.

6.2.4.1 Dibenzylbutanes

Dibenzylbutane-type lignans form one of the major classes of lignans isolated from the genus *Phyllanthus* (Table 6.3). Niranthin (**50**), isolated from *P. amarus, P. niruri, P. urinaria,* and *P. virgatus,* and phyllanthin (**52**), reported from *P. amarus, P. discoideus, P. niruri,* and *P. urinaria,* are the most common dibenzylbutane-type lignans within this genus (Calixto et al., 1998; Wei et al., 2002; Kassuya et al., 2003; El-Fahmi et al., 2006; Murugaiyah and Chan, 2007). The most notable feature of the dibenzylbutane-type lignans found in the *Phyllanthus* is the extensive methoxylations, particularly the methoxylation at C-1 and C-4 of the butane chain. Also, seven of the isolated dibenzylbutane-type lignans (**50, 68–70, 77, 79,** and **88**) also contain a methylenedioxy ring system.

6.2.4.2 Dibenzylbutyrolactones

Only three dibenzylbutyrolactone-type lignans—hinokinin (**71**) from *P. niruri* (Calixto et al., 1998) and *P. virgatus* (Huang et al., 1998a) and dextrobursehernin (**89**) and heliobuphthalmin lactone (**90**) from *P. urinaria* (Calixto et al., 1998; Chang et al., 2003; Fang et al., 2008)—have ever been found in this genus (Table 6.3). Dextrobursehernin (**89**) was also reported from *P. virgatus* (Calixto et al., 1998; Huang et al., 1998).

TABLE 6.3
Lignans from the Genus *Phyllanthus*

Species (Plant Parts): Isolated Lignans	References
P. amarus (aerial parts): hypophyllanthin (**49**), niranthin (**50**), nirtetralin (**51**), phyllanthin (**52**), phyltetralin (**53**)	Kassuya et al., 2003
P. anisobulos (aerial parts): justicidin B (**54**), phyllanthostatin A (**55**)	Bachmann et al., 1993; Calixto et al., 1998
P. discoideus (aerial parts): **52**	Calixto et al., 1998
P. myrtifolius (whole plant): **54**, justicidin A (**56**), phyllamyricin A (**57**), phyllamyricin B (**58**), phyllamyricin C (**59**), phyllamyricin D (**60**), phyllamyricin E (**61**), phyllamyricin F (**62**), phyllamyricoside A (**63**), phyllamyricoside B (**64**), phyllamyricoside C (**65**), retrojuscidin B (**66**)	Lee et al., 1996; Calixto et al., 1998
P. niruri (whole plant and suspension culture): **49–53**, cubebin methyl ether (**67**), 2,3-desmethoxy-*seco*-isolintetralin (**68**), 2,3-desmethoxy-*seco*-isolintetralin diacetate (**69**), demethylenedioxyniranthin (**70**), hinokinin (**71**), hydroxyniranthin (**72**), isolintetralin (**73**), linnanthin (**74**), lintetralin (**75**), neonirtetralin (**76**), Nirphyllin (**77**), phyllnirurin (**78**), *seco*-4-hydroxylintetralin (**79**), *seco*-isolariciresinol trimethyl ether (**80**)	Row et al., 1966; Satyanarayana and Venkateswarlu, 1991; Huang et al., 1992; Calixto et al., 1998; Wei et al., 2002; El-Fahmi et al., 2006; Murugaiyah and Chan, 2007
P. polyphyllus (whole plant): **54, 56**, diphyllin (**81**)	Rao et al., 2006
P. oxyphyllus (roots): *seco*-isolariciresinol (**82**), pinoresinol (**83**)	Sutthivaiyakit et al., 2003
P. reticulates (stem): **83**	Calixto et al., 1998
P. taxodiifolius (aerial parts): cleistanthin A (**84**), cleistanthin A methyl ether (**85**), cleistanthoside A (**86**), taxodiifoloside (**87**)	Tuchinda et al., 2006
P. urinaria (whole plant): **49–53, 73, 75**, 5-demethoxyniranthin (**88**), dextrobursehernin (**89**), heliobuphthalmin lactone (**90**), urunaligran (**91**), urinatetralin (**92**), virgatusin (**93**)	Calixto et al., 1998; Chang et al., 2003; Fang et al., 2008
P. virgatus (aerial parts): **49–51, 53, 71, 73, 89, 93**, 2-(3,4-methylenedioxybenzyl)-4-(3,4-methylenedioxyphenyl)-3-butyne-1,2-diol (virgatyne) (**94**)	Calixto et al., 1998; Huang et al., 1998a

6.2.4.3 Epoxy and Diepoxylignans

Only three epoxy lignans—cubebin methyl ether (**67**), urunaligran (**91**), and virgatusin (**93**)—and a diepoxy lignan, pinoresinol (**83**), have been reported from this genus (Table 6.3). While lignans **67** and **91** and **93** were isolated, respectively, from *P. niruri* and *P. urinaria* (Calixto et al., 1998), **83** was found in *P. oxyphyllus* and *P. reticulates* (Sutthivaiyakit et al., 2003).

6.2.4.4 Aryltetralins

Seven lignans (**49, 51, 53, 73, 75, 76**, and **92**) belonging to the aryltetralin class were isolated from the genus *Phyllanthus* (Table 6.3). Like dibenzylbutane-type lignans (Section 6.2.4.2), extensive methoxylations are also present in these aryltetralin lignans. Also, five of the six isolated aryltetralin lignans possess at least one methylenedioxy ring. Urinatetralin (**92**), isolated from *P. urinaria* (Calixto et al., 1998; Chang et al., 2003; Fang et al., 2008), has two methylenedioxy rings. *Phyllanthus amarus*, *P. niruri*, and *P. urinaria* appear to be the only *Phyllanthus* species that produce aryltetralin-type lignans.

6.2.4.5 Arylnapthalenes

Arylnapthalin-type lignans constitute the largest class of lignans produced by the genus *Phyllanthus*, and there are at least 17 different lignans of this class reported from seven *Phyllanthus* species (Table 6.3). Like dibenzylbutane and aryltetralin types (Sections 6.2.4.1 and 6.2.4.4), the extensive presence of methoxyl functionalities in arylnapthalene-type lignans is quite conspicuous in the *Phyllanthus*, and except for only phyllamyricoside C (**65**), isolated from *P. myrtifolius* (Calixto et al., 1998), in all other arylnapthalin-type lignans there is at least one methylenedioxy ring. Seven of the reported arylnapthalin-type lignans (**55, 63–65, 84–86**) are glycosides for which either glucose or xylose/xylose-derivative is the glycone part.

6.2.4.6 Neolignans

Phyllnirurin (**78**) from *P. niruri* and virgatyne (**94**) from *P. virgatus* are the only neolignans reported from this genus to date (Table 6.3).

6.2.5 SIMPLE LACTONES

Lactones are cyclic esters, and five- and six-member lactones are most common in the plant kingdom. The genus *Phyllanthus* is no exception. It produces a variety of five- and six-member lactones (Table 6.4); some of them are simple as they do not have much additional functionality, and some are much more complicated as, most often, they are parts of larger molecules, like tannins. Aquilegiolide (**95**) and menisdaurilide (**96**) from the aerial parts of *P. anisobulos* and ascorbic acid (**97**) from the fruits of *P. emblica* are the examples of the simplest lactones found in this genus

TABLE 6.4
Simple Lactones from the Genus *Phyllanthus*

Species (Plant Parts): Isolated Simple Lactones	References
P. anisobulos (aerial parts): aquilegiolide (**95**), menisdaurilide (**96**)	Calixto et al., 1998
P. emblica (fruits): ascorbic acid (**97**), mucic acid 1,4-lactone 2-*O*-gallate (**98**), mucic acid 1,4-lactone-6-methyl ester 2-*O*-gallate (**99**), mucic acid 1,4-lactone 3-*O*-gallate (**100**), mucic acid 1,4-lactone 5-*O*-gallate (**101**), mucic acid 1,4-lactone-6-methyl ester 5-*O*-gallate (**102**), mucic acid 1,4-lactone 3,5-di-*O*-gallate (**103**)	Calixto et al., 1998
P. flexuosus (leaves): brevifolin carboxylic acid (**104**)	Calixto et al., 1998
P. niruri (leaves): ethyl brevifolincarboxylate (**105**), methyl brevifolincarboxylate (**106**)	Izuka et al., 2006; Bagalkotkar et al., 2006
P. simplex (aerial parts): **104**, brevifolin (**107**)	Niu et al., 2006
P. urinaria (leaves): **104, 106**, phyllantutinolactone (**108**), trimethyl-3,4-dehydrochebulate (**109**)	Ueda et al., 1995; Fang et al., 2008

(Calixto et al., 1998). The mucic acid group of lactones (**98–103**) contain galloyl functionalities attached to the parent five-member lactone skeletons. Brevifolin (**107**) and its derivatives (**104–106**) contain a six-member lactone ring and are found in *P. flexuosus, P. niruri, P. simplex,* and *P. urinaria* (Calixto et al., 1998; Bagalkotkar et al., 2006; Izuka et al., 2006; Niu et al., 2006; Fang et al., 2008).

6.2.6 SIMPLE PHENOLICS AND BENZENE DERIVATIVES

In general, phenolics are the major class of compounds found in the *Phyllanthus*, which includes flavonoids, lignans, tannins, and so on, and have been almost invariably present in all species of this genus investigated to date. It is in fact difficult to draw any distinction between simple phenolics and tannins. However, for the sake of simplicity in discussion, small- to medium-size phenolic molecules and benzene derivatives are listed in Table 6.5. Most of these are gallic acid (**111**) and its derivatives, such as corillagin (**110**), 4-*O*-galloylquinic acid (**112**), 1,6-digalloylglucopyranose (**113**), ethyl gallate (**114**), methyl gallate (**115**), 3-ethoxy-4,5-dihydroxybenzoic acid (**119**), L-malic acid 2-*O*-gallate (**1207**), mucic acid 2-*O*-gallate (**121**), mucic acid dimethyl ester 2-*O*-gallate (**122**), mucic acid 1-methyl ester 2-*O*-gallate (**123**), mucic acid 6-methyl ester 2-*O*-gallate (**124**), and 4-*O*-methylgallic acid (**128**). Xanthoxyline (**129**) is the only acetophenone reported from this genus.

TABLE 6.5
Simple Phenolics and Benzene Derivatives from the Genus *Phyllanthus*

Species (Plant Parts): Isolated Simple Phenolics and Benzene Derivatives	References
P. amarus (aerial parts): corillagin (**110**), gallic acid (**111**), 4-*O*-galloylquinic acid (**112**), 1,6-digalloylglucopyranose (**113**)	Foo and Wong, 1992; Foo, 1993, 1995
P. carolinensis (leaves): ethyl gallate (**114**), methyl gallate (**115**)	Calixto et al., 1998
P. corcovadensis (leaves): salicylic acid methyl ester (**116**)	Calixto et al., 1998
P. emblica (aerial parts, fruits, and roots): **111**, **115**, 2-carboxymethylphenol-1-*O*-β-D-glucopyranoside (**117**), 2,6-dimethoxy-4-(2-hydroxyethyl)-phenol-1-*O*-β-D-glucopyranoside (**118**), 3-ethoxy-4,5-dihydroxybenzoic acid (**119**), L-malic acid 2-*O*-gallate (**120**), mucic acid 2-*O*-gallate (**121**), mucic acid dimethyl ester 2-*O*-gallate (**122**), mucic acid 1-methyl ester 2-*O*-gallate (**123**), mucic acid 6-methyl ester 2-*O*-gallate (**124**)	Calixto et al., 1998; Zhang et al., 2001b; El-Desouky et al., 2008
P. flexuosus (leaves): **110**, bergenin (**125**)	Calixto et al., 1998
P. niruri (whole plant): **110**, **111**, phyllester (**126**)	Ueno et al., 1988; Calixto et al., 1998
P. polyphyllus (whole plant): 4-*O*-methylgallic acid (**127**)	Rao et al., 2006
P. reticulates (leaves): **111**, pyrogallic acid (**128**)	Calixto et al., 1998
P. sellowianus (stems and leaves): **114**, xanthoxyline (**129**)	Calixto et al., 1998
P. urinaria (leaves): **111**, **115**	Calixto et al., 1998; Fang et al., 2008

6.2.7 Steroidal Compounds

Stigmasterol, β-sitosterol, and daucosterol are the three main steroids found in a number of *Phyllanthus* species (Calixto et al., 1998). In addition to these three steroids, eight other steroids (amarosterol A and B, campesterol, 5,6-dihydro-stigmasta-22-en-3β-ol, estradiol, isopropyl-cholesterol, stigmasterol acetate, and stigmas-4-en-3-one) have been reported from at least 14 different *Phyllanthus* species: *P. amarus* (Ahmad and Alam, 2003); *P. carolinensis* (Calixto et al., 1998); *P. corcovadensis* (Santos et al., 1995; Calixto et al., 1998); *P. emblica* (Khanna and Bansal, 1975; Luo et al., 2009); *P. flexuosus* (Calixto et al., 1998); *P. maderaspatensis* (Jain et al., 2005); *P. muellerianus* (Calixto et al., 1998); *P. niruri* (Subeki et al., 2005); *P. oxyphyllus* (Sutthivaiyakit et al., 2003); *P. reticulates* (Hui et al., 1976); *P. sellowianus* (Calixto et al., 1998); *P. singampattiana* (Ramesh et al., 2004); *P. urinaria* (Fang et al., 2008); and *P. watsonii* (Calixto et al., 1998). However, amarosterol A and amarosterol B isolated from *P. amarus* are not exactly true steroidal compounds (Ahmad and Alam, 2003). Estradiol and isopropyl-24-cholesterol were only reported from *P. niruri*.

6.2.8 Tannins

Tannins (Figure 6.1; Table 6.6) are the largest group of compounds found in the genus *Phyllanthus* and are often attributed to various bioactivities of these plants. While there are several well-known tannins, such as catechin (**171**), epicatechin (**172**), ellagic acid (**145**), gallic acid (**111**), gallocatechin (**135**), and epigallocatechin (**174**), most of the other tannins produced by this genus are structurally unique. *Phyllanthus emblica* produces the largest variety of tannins, including **111**, chebulagic acid (**142**), chebulinic acid (**143**), corillagin (**110**), 1,6-digalloylglucopyranose (**113**), 3,6-digalloylglucopyranose (**144**), 3-ethoxy-4,5-dihydroxy-benzoic acid (**119**), ellagic acid (**145**), emblicanin A (**146**), emblicanin B (**147**), 1-*O*-galloylglucose (**148**), geraniin (**138**), isocorilagin (**149**), leucodelphinidin (**150**), 2-(2-methylbutyryl) phloroglucinol-1-*O*-(6′-*O*-β-D-apiofuranosyl)-β-D-glucopyranoside (**151**), pedunculagin (**152**), 1,2,3,4,6-penta-*O*-galloylglucose (**153**), phyllanemblinin A–F (**154–159**), punigluconin (**160**), putranjivain A (**161**), putranjivain B (**162**), epicatechin-(4β®8)-gallocatechin (**163**), and epigallocatechin (2β®7,4β→8)-gallocatechin or prodelphinidin A1 (**164**) (Calixto et al., 1998; Zhang et al., 2000a, 2001b, 2002, 2003; El-Desouky et al., 2008; Liu et al., 2008).

6.2.9 Terpenoids

The genus *Phyllanthus* produces predominantly sesquiterpenes (Table 6.7) and triterpenes (Figure 6.1; Table 6.8). However, there are a few diterpenes also reported from this genus (Table 6.9). Several well-known monoterpenes (e.g., limonene) are the constituents of the essential oils obtained from the *Phyllanthus* (Calixto et al., 1998; Li et al., 2008; Liu et al., 2009).

TABLE 6.6
Tannins from the Genus *Phyllanthus*

Species (Plant Parts): Isolated Tannins	References
P. amarus (aerial parts): **110, 111, 113**, amariin (**130**), amariinic acid (**131**), amarulone (**132**), elaeocarpusin (**133**), furosin (**134**), gallocatechin (**135**), 1-galloyl-2,3-dehydrohexahydroxydiphenoyl-glucose (**136**), 1-galloyl-2,4-(acetonyl-dehydrohexahydroxydiphenoyl)-3,6-hexahydroxydiphenoyl-glucopyranoside (**137**), geraniin (**138**), geraniinic acid (**139**), phyllanthusin D (**140**), repandusinic acid B (**141**)	Foo and Wong, 1992; Foo, 1993, 1995; Londhe et al., 2008
P. carolinensis (aerial parts): **138**	Cechinel at al., 1996; Calixto et al., 1998
P. emblica (fruits, leaves, and roots): **110, 111, 113, 116, 138**, chebulagic acid (**142**), chebulinic acid (**143**), 3,6-digalloylglucopyranose (**144**), ellagic acid (**145**), emblicanin A (**146**), emblicanin B (**147**), 1-*O*-galloylglucose (**148**), isocorilagin (**149**), leucodelphinidin (**150**), 2-(2-methylbutyryl)phloroglucinol-1-*O*-(6'-*O*-β-D-apiofuranosyl)-β-D-glucopyranoside (**151**), pedunculagin (**152**), 1,2,3,4,6-penta-*O*-galloylglucose (**153**), phyllanemblinin A–F (**154–159**), punigluconin (**160**), putranjivain A (**161**), putranjivain B (**162**), epicatechin-(4β®8)-gallocatechin (**163**), epigallocatechin (2β®7,4β®8)-gallocatechin or prodelphinidin A1 (**164**)	Calixto et al., 1998; Zhang et al., 2000b, 2001, 2002, 2003; El-Desouky et al., 2008; Liu et al., 2008
P. flexuosus (leaves): **110, 138, 141, 142, 145**, geraniinic acid (**165**), phyllanthusiin A-E (**166–170**)	Calixto et al., 1998
P. niruri (aerial parts and hairy roots): **110, 111, 135, 138, 145, 149**, catechin (**171**), epicatechin (**172**), epicatechin 3-*O*-gallate (**173**), epigallocatechin (**174**), epigallocatechin 3-*O*-gallate (**175**), 1-*O*-galloyl-6-*O*-luteoyl-α-D-glucose (**176**), 2,3,5,6-tetrahydroxybenzyl acetate (**177**), 2,4,5-trihydroxy-3-(4,6,7-trihydroxy-3-oxo-1,3-dihydroisobenzofuran-5-yl)-benzoic acid methyl ester or phyllangin (**178**)	Ueno et al., 1988; Ishimaru et al., 1992; Calixto et al., 1998; Wei et al., 2004; Subeki et al., 2005; Than et al., 2006
P. reticulates (leaves and stems): **111, 134, 138, 145**, 3,3',4-tri-*O*-methylellagic acid (**179**)	Miguel et al., 1996; Calixto et al., 1998
P. tenellus (leaves): pinocembrin-7-*O*-[4',6'-(S)-hexahydroxydiphenoyl]-β-D-glucose (**180**), pinocembrin-7-*O*-[3'-*O*-galloyl-4',6'-(S)-hexahydroxydiphenoyl]-β-D-glucose (**181**)	Huang et al., 1998b
P. urinaria (aerial parts): **110, 111, 138, 145**, hippomanin A (**182**), phyllanthusiin G (**183**), 1,3,4,6-tetra-*O*-galloyl-β-D-glucose (**184**), trimethyl ester 3,4-dehydrochebulic acid (**185**)	Zhang et al., 2000a, 2004; Yang et al., 2007a, 2007b; Fang et al., 2008
P. virgatus (aerial parts): virganin (**186**)	Huang et al., 1998a

TABLE 6.7
Sesquiterpenes from the Genus *Phyllanthus*

Species (Plant Part): Isolated Sesquiterpenes	References
P. acidus (roots): phyllanthusol A (**187**) and B (**188**)	Durham et al., 2002
P. acuminatus (roots): phyllanthostatin 1–6 (**189–194**)	Calixto et al., 1998
P. emblica (roots): phyllaemblic acid (**195**), phyllaemblic acid methyl ester (**196**), phyllaemblic acid B (**197**) and C (**198**), phyllaemblicin A–D (**199–202**)	Zhang et al., 2000c, 2001b
P. engleri (aerial parts): englerin A (**203**) and B (**204**)	Ratnayake et al., 2009
P. oxyphyllus (roots): guaiane derivative (**205**)	Sutthivaiyakit et al., 2003
P. verminatus (aerial parts): **189–191**, phyllanthoside (**206**)	Calixto et al., 1998

6.2.9.1 Monoterpenes

The volatile components and *in vitro* antimicrobial activities of *P. emblica* essential oils, obtained by hydrodistillation and supercritical fluid extraction, have been studied (Liu et al., 2009), and the presence of a number of monoterpenes has been detected. Similarly, limonene and other monoterpenes were found to be present in *P. arenarius* and *P. urinaria* (Li et al., 2008). Phytol, β-citronellol, and *trans*-geraniol were the most abundant components in the essential oils of the leaves of *P. salviaefolius* (Villarreal et al., 2008). The presence of monoterpenes has also been reported from *P. acidus* (Quijano et al., 2007).

6.2.9.2 Sesquiterpenes

About 20 different sesquiterpenes (**187–206**) have been reported from at least six species of the *Phyllanthus* (Table 6.7). Phyllanthostatin 1–3 (**189–191**) were found in *P. acuminatus* and *P. verminatus* (Calixto et al., 1998), but all other sesquiterpenes appear not to occur in more than one species. Sesquiterpenes **203–205** possess a guaiane skeleton (Sutthivaiyakit et al., 2003; Ratnayake et al., 2009), and most of the other sesquiterpenes have a highly oxygenated norbisabolane structure (**195, 196, 199** and **200**). *Phyllanthus emblica* produces the maximum number of sesquiterpenes from its roots, including phyllaemblic acid (**195**), phyllaemblic acid methyl ester (**196**), phyllaemblic acid B (**197**) and C (**198**), and phyllaemblicin A–D (**199–202**) (Zhang et al., 2000, 2001a).

6.2.9.3 Diterpenes

Phyllanthus emblica, P. flexuosus, and *P. oxyphyllus* are the only species known to produce eight different diterpenes (**207–214**) within this genus (Table 6.9).

6.2.9.4 Triterpenes

Over 50 different triterpenes (Figure 6.1; Table 6.8), predominantly of betulane, friedelane, lupine, and oleane types, have been reported from at least 15 *Phyllanthus* species. The vast majority of these triterpenes possess pentacyclic structures, but tetracyclic and acyclic triterpenes are also found in this genus. *Phyllanthus reticulatus* is known to produce the highest number of triterpenes and contains

TABLE 6.8
Triterpenes from the Genus *Phyllanthus*

Species (Plant Part): Isolated Triterpenes	References
P. acidus (aerial parts): β-amyrin (**215**), lupeol (**216**), phyllanthol (**217**)	Calixto et al., 1998
P. amarus (leaves): **216**, oleanolic acid (**218**), ursolic acid (**219**)	Ali et al., 2006
P. discoideus (aerial parts): betulinic acid (**220**)	Calixto et al., 1998
P. emblica (aerial parts and roots): **216**	Calixto et al., 1998
P. engleri (aerial parts): **217**	Calixto et al., 199
P. flexuosus (aerial parts): **215**, **216**, betulin (**221**), friedelin (**222**), glochidone (**223**), lup-20(29)-ene-3β, -diol (**224**), ocotillol (**225**), olean-12-ene-3β,5α,-triol (**226**), olean-12-ene-3β,15α-diol (**227**), olean-12-ene-3β,-diol (**228**), oleana-11,13(18)-dien-3β-ol (**229**), oleana-9(11),12-dien-3β-ol (**230**), oleana-11,13(18)-diene-3β,-diol (**231**), trichadenic acid B (**232**)	Calixto et al., 1998
P. muellerianus (aerial parts): friedel-1-ene-22β-ol (**233**), friedelin-1β,22β-diol (**234**)	Calixto et al., 1998
P. myrtifolius (aerial parts): phyllenolide A–C (**235–237**)	Lee et al., 2002
P. niruri (aerial parts): **216**, 3,7,11,15,19,23-hexamethyl 2Z, 6Z, 10Z, 14E, 18E, 22E-tetracoshenen-1-ol (**238**), lupeol acetate (**239**), phyllanthenol (**240**), phyllanthenone (**241**), phyllantheol (**242**)	Calixto et al., 1998
P. polyanthus (stem bark and leaves): **217**, δ-amyrin acetate (**243**), (20*S*)-3β-acetoxy-methylenedammaran-20-ol (**244**), (20*S*)-3α-acetoxy-methylenedammaran-20-ol (**245**), lupenone (**246**), phyllanthone (**247**)	Ndlebe et al., 2008
P. oxyphyllus (roots): **216**, 29-*nor*-3,4-*seco*-friedelan-4(23),20(30)-dien-3-oic acid (**248**), polpunoic acid (**249**)	Sutthivaiyakit et al., 2003
P. reticulates (aerial parts and roots): **220–223**, **227**, 20-*O*-acetyl-lup-1-en-3-one (**250**), *epi*-friedelanol (**251**), friedelanol (**252**), glochidonol (**253**), 21α-hydroxyfriedelan-3-one (**254**), 21α-hydroxyfriedel-4(23)-en-3-one (**255**), 21α-hydroxyfriedelin (**256**), kokoonol (**257**), *cis*-(4-hydroxycinnamoyl)-olean-12-ene (**258**), *trans*-(4-hydroxycinnamoyl)-olean-12-ene (**259**), lup-20(29)-ene-1β, 3β-diol (**260**), sorghumol (**261**), sorghumol acetate (**262**), taraxerol (**263**), taraxerol acetate (**264**), taraxerone (**265**)	Calixto et al., 1998; Jain and Nagpal, 2002
P. sellowianus (aerial parts): **217**	Calixto et al., 1998
P. urinaria (aerial parts): **215**, **239**	Calixto et al., 1998
P. watsonii (aerial parts): **216**, **222**, **223**, **251**, **253**, **260**, 26-*nor*-D:A-friedoolean-14-en-3-one (**266**), 26-*nor*-D:A-friedoolean-14-en-3β-ol (**267**)	Calixto et al., 1998

TABLE 6.9
Diterpenes from the Genus *Phyllanthus*

Species (Plant Part): Isolated Diterpenes	References
P. emblica (fruits): gibberellin A-1 (**207**), A-3 (**208**), A-4 (**209**), A-7 (**210**), A-9 (**211**)	Ram and Rao, 1978
P. flexuosus (aerial parts): *ent*-3β-hydroxykaur-16-ene (**212**)	Calixto et al., 1998
P. oxyphyllus (roots): cleistantho (**213**), spruceanol (**214**)	Sutthivaiyakit et al., 2003

betulin (**221**), betulinic acid (**220**), *epi*-friedelanol (**251**), friedelin (**222**), friedelanol (**252**), glochidone (**223**), glochidonol (**253**), 21α-hydroxyfriedelan-3-one (**254**), 21α-hydroxyfriedel-4(23)-en-3-one (**255**), 21α-hydroxyfriedelin (**256**), kokoonol (**257**), olean-12-ene-3β,15α-diol (**228**), *cis*-(4-hydroxycinnamoyl)-olean-12-ene (**258**), *trans*-(4-hydroxycinnamoyl)-olean-12-ene (**259**), lup-20(29)-ene-3β, 24-diol (**224**), lup-20(29)-ene-1β, 3β-diol (**260**), sorghumol (**261**), sorghumol acetate (**262**), taraxerol (**263**), taraxerol acetate (**264**), and taraxerone (**265**) (Calixto et al., 1998; Jain and Nagpal, 2002).

6.2.10 Miscellaneous Compounds

In addition to the compounds described in Sections 6.2. to 6.7, carbohydrates and sugars, lipids and fatty acids, and bioactive proteins, there are also a few other compounds reported from this genus: 5-hydroxymethylfurfural from *P. emblica* (Luo et al., 2009); two alkamides, *E,E*-2,4-octadienamide and *E,Z*-2,4-decadienamide, from *P. fraternus* (Sittie et al., 1998); and butenolides from *P. klotzschianus* (Kuster et al., 1997).

6.3 DISTRIBUTION AND CHEMOTAXONOMIC SIGNIFICANCE

Information on phytochemical investigations on about 35 *Phyllanthus* species of about 800 species is available to date (Table 6.10). While some of the species (e.g., *P. amarus, P. emblica, P. niuri*) have been studied extensively, investigations of a number of species (e.g., *P. acuminatus, P. arenarius, P. debilis*) are rather incomplete or partial. On the basis of the available phytochemical information, the distribution of various classes of phytochemicals within different *Phyllanthus* species is summarized in Table 6.10. While *P. emblica* produces all major classes of compounds discussed in this chapter, *P. amarus, P. niruri, P. reticulates,* and *P. urinaria* have yielded most of the major classes. Triterpenes are widely distributed within this genus and have been found in at least 15 species. The distribution of alkaloids and phenolics is also quite widespread. Many of the compounds isolated from this genus only occur in one species and therefore have limited chemotaxonomic value. However, there are a few compounds, especially alkaloids, lignans, and tannins, that occur in more than one *Phyllanthus* species, might have some chemotaxonomic significance, at least at the family level.

6.4 CONCLUSION

The available phytochemical data on the genus *Phyllanthus*, with the exception of a few species, are rather incomprehensive, and many *Phyllanthus* species demand further detailed studies because of their traditional medicinal uses. Also, less than 90% of the *Phyllanthus* species have ever been phytochemically investigated. The structural diversity found among the *Phyllanthus* compounds is unique, and many compounds isolated from this genus so far are highly bioactive. Therefore, the genus *Phyllanthus* will continue to be one of the major plant sources of new chemical entities and possibly new drug molecules.

TABLE 6.10
Distribution of Major Classes of Phytochemicals within the Genus *Phyllanthus*

Phyllanthus Species	A	B	C	D	E	F1	F2	G	H1	H2	H3	H4
P. acidus	−	−	−	−	−	−	−	−	+	+	−	+
P. acuminatus	−	−	−	−	−	−	−	−	−	+	−	−
P. amarus	+	−	+	+	−	+	+	+	−	−	−	+
P. anisobulos	−	−	−	+	+	−	−	−	−	−	−	−
P. arenarius	−	−	−	−	−	−	−	−	+	−	−	−
P. burchellii	+	−	+	−	−	−	−	−	−	−	−	−
P. caroliniensis	−	−	+	−	−	+	+	+	−	−	−	−
P. corcovadensis	−	−	−	−	−	+	−	+	+	−	−	−
P. debilis	−	−	+	−	−	−	−	−	−	−	−	−
P. discoides	+	−	−	−	−	−	−	−	−	−	−	−
P. discoideus	+	−	−	+	−	−	−	−	−	−	−	+
P. emblica	+	+	+	−	+	+	+	+	+	+	+	+
P. engleri	−	−	−	−	−	−	−	−	−	+	−	+
P. flexuosus	−	−	−	−	+	+	+	+	−	−	+	+
P. maderaspatensis	−	−	+	−	−	−	−	+	−	−	−	−
P. muellerianus	−	−	−	−	−	−	−	−	−	−	−	+
P. myrtifolius	−	−	−	+	−	−	−	−	−	−	−	+
P. niruri	+	−	+	+	+	+	+	+	−	−	−	+
P. niruroides	+	−	−	−	−	−	−	−	−	−	−	−
P. parvulus var. *garipensis*	+	−	+	−	−	−	−	−	−	−	−	−
P. polyanthus	−	−	−	−	−	−	−	−	−	−	−	+
P. polyphyllus	−	−	−	+	−	+	−	−	−	−	−	−
P. orbiculatus	−	−	+	−	−	−	−	−	−	−	−	−
P. oxyphyllus	−	−	−	+	−	−	−	+	−	−	+	+
P. reticulatus	−	+	+	+	−	+	+	+	−	−	−	+
P. salviaefolius	−	−	−	−	−	−	−	−	−	−	−	−
P. sellowianus	+	+	+	−	−	+	−	+	−	−	−	+
P. singampattiana	−	+	−	−	−	−	−	+	−	−	−	−
P. simplex	+	−	−	−	+	−	−	−	−	−	−	−
P. taxodiifolius	−	−	−	+	−	−	−	−	−	−	−	−
P. tenellus	−	−	−	−	−	+	−	−	−	−	−	−
P. urinaria	+	+	+	+	+	+	+	+	+	−	−	+
P. verminatus	−	−	−	−	−	−	−	−	−	+	−	−
P. virgatus	+	−	+	+	−	−	+	−	−	−	−	−
P. watsonii	−	−	−	−	−	−	−	+	−	−	−	+

Note: A, alkaloids; B, coumarins and cinnamic acid derivatives; C, flavonoids; D, lignans; E, simple lactones; F, phenolics; F1, simple phenolics; F2, tannins; G, steroids; H, terpenoids; H1, monoterpenes; H2, sesquiterpenes; H3, diterpenes; H4, triterpenes; +, presence; −, absence.

REFERENCES

Ahmad, B., and Alam, T. 2003. Components from whole plant of *Phyllanthus amarus* Linn. *Indian J. Chem.* Section *B* 42, 1786–1790.
Ali, H., Houghton, P.J., and Soumyanath, A. 2006. α-Amylase inhibitory activity of some Malaysian plants used to treat diabetes; with particular reference to *Phyllanthus amarus*. *J. Ethnopharmacol.* 107: 449–455.
Babady-Bila, B., Gedris, T.E., and Herz, W. 1996. Niruroidine, a norsecurinine-type alkaloid from *Phyllanthus niruroides*. *Phytochemistry* 41: 141–1443.
Bachmann, T.L., Kurt, F.G., and Torssell, B.G. 1993. Lignans and lactones from *Phyllanthus anisolobus*. *Phytochemistry* 33: 189–191.
Bagalkotkar, G., Sagineedu, S.R., Saad, M.S., and Stanslas, J. 2006. Phytochemicals from *Phyllanthus niruri* Linn. and their pharmacological properties: a review. *J. Pharm. Pharmacol.* 58: 1559–1570.
Calixto, J.B., Santos, A.R.S., Cechinel, V., and Yunes, R.A. 1998. A review of the plants of the genus *Phyllanthus*: their chemistry, pharmacology, and therapeutic potential. *Med. Res. Rev.* 18: 225–258.
Cechinel, V., Santos, A.R.S., DeCampos, R.O.P., et al. 1996. Chemical and pharmacological studies of *Phyllanthus caroliniensis* in mice. *J. Pharm. Pharmacol.* 48: 1231–1236.
Chang, C.C., Lien, Y.C., Liu, K.C.S.C., and Lee, S.S. 2003. Lignans from *Phyllanthus urinaria*. *Phytochemistry* 63: 825–833.
Chauhan, J.S., Sultan, M., and Srivastava, S.K. 1977. Two new glycoflavones from roots of *Phyllanthus niruri*. *Planta Med.* 32: 217–222.
Duke's Phytochemical and Ethnobotanical Databases. (2011). Available at http://www.ars-grin.gov/cgi-bin/duke/ethnobot.pl.
Durham, D.G., Reid, R.G., Wangboonskul, J., and Daodee, S. 2002. Extraction of phyllanthusols A and B from *Phyllanthus acidus* and analysis by capillary electrophoresis. *Phytochem. Anal.* 13: 358–362.
El-Desouky, S.K., Ryu, S.Y., and Kim, Y.K. 2008. A new cytotoxic acylated apigenin glucoside from *Phyllanthus emblica* L. *Nat. Prod. Res.* 22: 91–95.
El-Fahmi, E., Batterman, S., Koulman, A., et al. 2006. Lignans from cell suspension cultures of *Phyllanthus niruri*, an Indonesian medicinal plant. *J. Nat. Prod.* 69: 55–58.
Fang, S.H., Rao, Y.K., and Tzeng, Y.M. 2008. Anti-oxidant and inflammatory mediator's growth inhibitory effects of compounds isolated from *Phyllanthus urinaria*. *J. Ethnopharmacol.* 116: 333–340.
Foo, L.Y. 1993. Amariin, a di-dehydrohexahydroxydiphenoyl hydrolysable tannin from *Phyllanthus amarus*. *Phytochemistry* 33: 487–491.
Foo, L.Y. 1995. Amariinic acid and related ellagitannins from *Phyllanthus amarus*. *Phytochemistry* 39: 217–224.
Foo, L.Y., and Wong, H. 1992. Phyllanthusiin-D, an unusual hydrolyzable tannin from *Phyllanthus amarus*. *Phytochemistry* 31: 711–713.
Foussard-Blanpin, O., Quevauviller, A., and Bourrinet, P. 1967. On phyllochrysine, an alkaloid from *Phyllanthus discoides*, Euphorbiaceae. *Therapie* 22: 303–307.
GRIN Taxonomy Database. (2011). USDA, ARS, National Genetic Resources Program. Germplasm Resources Information Network (GRIN) [Online Database]. National Germplasm Resources Laboratory, Beltsville, Maryland. Available at http://www.ars-grin.gov/cgi-bin/npgs/html/taxdump.pl?phyllanthus (accessed March 20, 2010).
Gupta, D.R., and Ahmed, B. 1984. Nirurin—a new prenylated flavanone glycoside from *Phyllanthus nirurii*. *J. Nat. Prod.* 47: 958–963.
Hassarajani, S.A., and Mulchandani, N.B. 1990. Securinine type of alkaloids from *Phyllanthus niruri*. *Indian J. Chem. Section B*. 29: 801–803.

Hnatyszyn, O., Ferraro, G., and Coussio, J.D. 1987. A biflavonoid from *Phyllanthus sellowianus*. *J. Nat. Prod.* 50, 1156–1157.
Houghton, P.J., Woldemariam, T.Z., O'Shea, S., and Thyagarajan, S.P. 1996. Two securinega-type alkaloids from *Phyllanthus amarus*. *Phytochemistry* 43: 715–717.
Huang, Y.L., Chen, C.C., Hsu, F.L., and Chen, C.F. 1998a. Tannins, flavonol sulfonates, and a norlignan from *Phyllanthus virgatus*. *J. Nat. Prod.* 61: 1194–1197.
Huang, Y.L., Chen, C.C., Hsu, F.L., and Chen, C.F. 1998b. Two tannins from *Phyllanthus tenellus*. *J. Nat. Prod.* 61: 523–524.
Huang, Y.L., Chen, C.C., and Ou, J.C. 1992. Isolintetralin—a new lignan from *Phyllanthus niruri*. *Planta Med.* 58: 473–474.
Hui, W.H., Li, M.M., and Wong, K.M. 1976. New compound, 21a-hydroxyfriedel-4(23)-en-3-one and other triterpenoids from *Phyllanthus reticulatus*. *Phytochemistry* 15: 797–798.
Ishimaru, K., Yoshimatsu, K., Yamakawa, T., et al. 1992. Phenolic constituents in tissue cultures of *Phyllanthus niruri*. *Phytochemistry* 31: 2015–2018.
ISI Web of Knowledge. (2011). Thomson Reuters, UK. Available at http://apps.isiknowledge.com/WOS_GeneralSearch_input.do?highlighted_tab=WOS&product=WOS&last_prod=WOS&SID=P26hC5KmD6k98AN@IF@&search_mode=GeneralSearch.
Izuka, T., Moriyama, H., and Nagai, M. 2006. Vasorelaxant effects of methyl brevifolincarboxylate from the leaves of *Phyllanthus niruri*. *Biol. Pharm. Bull.* 29: 177–179.
Jain, R., Chitale, G., and Jain, S.C. 2005. Chemical constituents of *Phyllanthus maderaspatensis* Linn. *J. Indian Chem. Soc.* 82: 752–753.
Jain, R., and Nagpal, S. 2002. Chemical constituents of the roots of *Kirganelia reticulate*. *J. Indian Chem. Soc.* 79: 776–777.
Joshi, B.S., Gawad, D.H., Pelletier, S.W., Kartha, G., and Bhandary, K. 1986. Isolation and structure (X-ray-analysis) of ent-norsecurinine, an alkaloid from *Phyllanthus niruri*. *J. Nat. Prod.* 49: 614–620.
Kassuya, C.A.L., Silvestre, A.A., Rehder, V.L.G., and Calixto, J.B. 2003. Anti-allodynic and anti-oedematogenic properties of the extract and lignans from *Phyllanthus amarus* in models of persistent inflammatory and neuropathic pain. *Eur. J. Pharmacol.* 478: 145–153.
Khanna, P., and Bansal, R. 1975. Phyllantidine and phyllantine from *Emblica officinalis* Gaertn leaves, fruits and *in vitro* tissue cultures. *Indian J. Exp. Biol.* 13: 82–83.
Kumaran, A., and Karunakaran, R.J. 2007. *In vitro* antioxidant activities of methanol extracts of five *Phyllanthus* species from India. LWT *Food Sci. Tech.* 40: 344–352.
Kuster, R.M., Mors, W.B., and Wagner, H. 1997. Cyclohexenyl butenolides from *Phyllanthus klotzschianus*. *Biochem. Syst. Ecol.* 25: 675.
Lam, S.H., Wang, C.Y., Chen, C.K., and Lee, S.S. 2007. Chemical investigation of *Phyllanthus reticulatus* by HPLC-SPE-NMR and conventional methods. *Phytochem. Anal.* 18: 251–255.
Lee, S.S., Lin, M.T., Liu, C.L., et al. 1996. Six lignans from *Phyllanthus myrtifolius*. *J. Nat. Prod.* 59: 1061–1065.
Lee, S.S., Kishore, P.H., and Chen, C.H. 2002. Three novel triterpenoid dienolides from *Phyllanthus myrtifolius*. *Helv. Chim. Acta* 85: 2403–2408.
Li, X.R., Wei, W.X., and Lin, C.W. 2008. Comparative analysis of essential oil compositions from *Phyllanthus niriru*, *P. urinaria*, and *P. arenarius*. *Chem. Nat. Compounds* 44: 257–260.
Liu, X., Cui, C., Zhao, M., et al. 2008. Identification of phenolics in the fruit of emblica (*Phyllanthus emblica* L.) and their antioxidant activities. *Food Chem.* 109: 909–915.
Liu, X.L., Zhao, M.M., Luo, W., et al. 2009. Identification of volatile components in *Phyllanthus emblica* L. and their antimicrobial activity. *J. Med. Food.* 12: 423–428.
Londhe, J.S., Devasagayam, T.P.A., Foo, L.Y., and Ghaskadbi, S.S. 2008. Antioxidant activity of some polyphenol constituents of the medicinal plant *Phyllanthus amarus* Linn. *Redox Rep.* 13: 199–207.

Luo, W., Zhao, M., Yang, B., et al. 2009. Identification of bioactive compounds in *Phyllanthus emblica* L. fruit and their free radical scavenging activities. *Food Chem.* 114: 499–504.

Mdlolo, C.M., Shandhu, J.S., and Oyedeji, O.A. 2008. Phytochemical constituents and antimicrobial studies of two South African *Phyllanthus species*. *Afr. J. Biotech.* 7: 639–643.

Mensah, J.L., Gleye, J., Moulis, C., and Fouraste, I. 1988 Alkaloids from the leaves of *Phyllanthus discoideus*. *J. Nat. Prod.* 51: 1113–1115.

Mulchandani, N.B., and Hassarajani, S.A. 1984. 4-Methoxy-norsecurinine, a new alkaloid from *Phyllanthus niruri*. *Planta Med.* 50: 104–105.

Miguel, O.G., Calixto, J.B., Santos, A.R.S., et al. 1996. Chemical and preliminary analgesic evaluation of geraniin and furosin isolated from *Phyllanthus sellowianus*. *Planta Med.* 62: 146–149.

Murugaiyah, V., and Chan, K.L. 2007. Determination of four lignans in *Phyllanthus niruri* L. by a simple high-performance liquid chromatography method with fluorescence detection. *J. Chromatogr. A* 1154: 1998–204.

Ndlebe, V.J., Crouch, N.R., and Mulholland, D.A. 2008. Triterpenoids from the African tree *Phyllanthus polyanthus*. *Phytochem. Lett.* 1: 11–17.

Negi, R.S., and Fakir, T.M. 1988. Simplexine (14-hydroxy-4-methoxy-13,14-dihydronorsecurinine): an alkaloid from *Phyllanthus simplex*. *Phytochemistry* 27: 3027–3028.

Niu, X.F., He, L.C., Fan, T., and Li, Y. 2006. Protecting effect of brevifolin and 8,9-single-epoxy brevifolin of *Phyllanthus simplex* on rat liver injury. *Zhongguo Zhong Yao Za Zhi* 31: 1529–1532.

Nozeran, R., Rossignolbancilhon, L., and Mangenot, G. 1984. Studies on the genus *Phyllanthus* (Euphorbiaceae)—latest developments and perspectives. *Bot. Helv.* 94: 199–233.

Parello, J. 1968. Structure of phyllanthine, a minor alkaloid from *Phyllanthus discoides* Meull. Arg. (Euphorbiacees). *Bull. Soc, Chim. Fr.* 3: 1117–1129.

Petchnaree, P., Bunyapraphatsara, N., Cordell, G.A., et al. 1986. X-ray crystal and molecular-structure of nirurine, a novel alkaloid related to the securinega alkaloid skeleton, from *Phyllanthus niruri* (Euphorbiaceae). *J. Chem. Soc. Perkin Trans.* 1. 9: 1551–1556.

Quevauviller, A., and Blanpin, O. 1959. Pharmacodynamic study of phyllochrysin, an alkaloid from *Phyllanthus discoides* (Euphorbiaceae). *Therapie* 14: 619–624.

Quevauviller, A., Foussard-Blanpin, O., and Bourrinet, P. 1967. Pharmacodynamics of securinine, present in *Phyllanthus discoides*—Euphorbiaceae. *Therapie* 22: 297–302.

Quevauviller, A., Foussard-Blanpin, O., and Coignard, D. 1965. An alkaloid of *Phyllanthus discoides* (Euphorbiaceae) phyllalbine, a central and peripheral sympathomimetic. *Therapie* 20: 1033–1041.

Quijano, C.E., Linares, D., and Pino, J.A. 2007. Changes in volatile compounds of fermented cereza agria [*Phylianthus acidus* (L.) Skeels] fruit. *Flavour Fragr. J.* 22: 392–394.

Ram, S., and Rao, T.R. 1976. Naturally occurring cytokinins in aonla (*Emblica officinalis* Gaertn) fruit. *New Phytol.* 76: 441–448.

Ram, S., and Rao, T.R. 1978. Studies on the naturally occurring gibberellins in aonla (*Emblica officinalis* Gaertn) fruit. *New Phytol.* 81: 513–518.

Ramesh, N., Viswanathan, M.B., Selvi, V.T., and Lakshmanaperumalsamy, P. 2004. Antimicrobial and phytochemical studies on the leaves of *Phyllanthus singampattiana* (Sebastine & AN Henry) Kumari & Chandrabose from India. *Med. Chem. Res.* 13: 348–360.

Rao, Y.K., Fang, S.H., and Tzeng, Y.M. 2006. Anti-inflammatory activities of constituents isolated from *Phyllanthus polyphyllus*. *J. Ethnopharmacol.* 103: 181–186.

Ratnayake, R., Covell, D., Ransom, T.T., et al. 2009. Englerin A, a selective inhibitor of renal cancer cell growth, from *Phyllanthus engleri*. *Organ. Lett.* 11: 57–60.

Rehman, H., Yasin, K.A., Choudhary, M.A., et al. 2007. Studies on the chemical constituents of *Phyllanthus emblica*. *Nat. Prod. Res.* 21: 775–781.

Row, L.R., Srinivasulu, C., Smith, M., and Subba-Rao, G.S.R. 1966. Crystalline constituents of Euphorbiaceae—V: new lignans from *Phyllanthus niruri* Linn—the constitution of phyllanthin. *Tetrahedron* 22: 2899–2908.

Santos, A.R.S., Niero, R., Filho, A.C., et al. 1995. Antinociceptive properties of steroids isolated from *Phyllanthus corcovadensis* in mice. *Planta Med.* 61: 329–332.

Satyanarayana, P., and Venkateswarlu, S. 1991. Isolation, structure and synthesis of new diarylbutane lignans from *Phyllanthus niruri*: synthesis of 5′-desmethoxy niranthin and an antitumour extractive. *Tetrahedron* 47: 8931–8940.

Shakil, N.A., Kumar, P.J., Pandey, R.K., and Saxena, D.B. 2008. Nematicidal prenylated flavanones from *Phyllanthus niruri*. *Phytochemistry* 69: 759–764.

Sittie, A.A., Lemmich, E., Olsen, C.E., et al. 1998. Alkamides from *Phyllanthus fraternus*. *Planta Med.* 64: 1992–193.

Subeki, M.H., Takahashi, K., Yamasaki, M., et al. 2005. Anti-babesial and anti-plasmodial compounds from *Phyllanthus niruri*. *J. Nat. Prod.* 68: 537–539.

Sutthivaiyakit, S., Nakorn, N.N., Kraus, W., and Sutthivaiyakit, P. 2003. A novel 29-nor-3,4-seco-friedelane triterpene and a new guaiane sesquiterpene from the roots of *Phyllanthus oxyphyllus*. *Tetrahedron* 53: 9991–9995.

Tempesta, M.S., Corley, D.G., Beutler, J.A., et al. 1988. Phyllanthimide, a new alkaloid from *Phyllanthus sellowianus*. *J. Nat. Prod.* 51: 617–618.

Than, N.N., Fotso, S., Poeggeler, B., Hardeland, R., and Laatsch, H. 2006. Niruriflavone, a new antioxidant flavone sulfonic acid from *Phyllanthus niruri*. *Z. Naturforsch.* 61b: 57–60.

Tuchinda, P., Kumkao, A., Pohmakotr, M., et al. 2006. Cytotoxic arylnaphthalide lignan glycosides from the aerial parts of *Phyllanthus taxodiifolius*. *Planta Med.* 72: 60–62.

Ueda, M., Asano, M., Sawai, Y., and Yamamura, S. 1999. Leaf-movement factors of nyctinastic plant, *Phyllanthus urinaria* L.; the universal mechanism for the regulation of nyctinastic leaf-movement. *Tetrahedron* 55: 5781–5792.

Ueda, M., Shigemori-Suzuki, T., and Yamamura, S. 1995. Phyllanthurinolactone, a leaf-closing factor of nyctinastic plant, *Phyllanthus urinaria* L. *Tetrahedron Lett.* 36: 6267–6270.

Ueno, H., Horie, S., Nishi, Y., et al. 1988. Chemical and pharmaceutical studies on medicinal-plants in Paraguay—geraniin, an angiotensin-converting enzyme-inhibitor from paraparai mi, *Phyllanthus niruri*. *J. Nat. Prod.* 51: 357–359.

Villarreal, S., Rojas, L.B., and Usubillaga, A. 2008. Volatile constituents from the leaves of *Phyllanthus salviaefolius* H. B. K. from the Venezuelan Andes. *Nat. Prod. Commun.* 3: 275–277.

Wei, W.X., Gong, X.G., Ishrud, O., and Pan, Y.J. 2002. New lignan isolated from *Phyllanthus niruri* Linn. structure elucidation by NMR spectroscopy. *Bull. Korean Chem. Soc.* 23: 896–898.

Wei, W.X., Pan, Y.J., Zhang, H., et al. 2004. Two new compounds from *Phyllanthus niruri*. *Chem. Nat. Compounds* 40: 460–464.

Yang, C.M., Cheng, H.Y., Lin, T.C., et al. 2007a. Hippomanin A from acetone extract of *Phyllanthus urinaria* inhibited HSV-2 but not HSV-1 infection *in vitro*. *Phytother. Res.* 21: 1182–1186.

Yang, C.M., Cheng, H.Y., Lin, T.C., et al. 2007b. The *in vitro* activity of geraniin and 1,3,4,6-tetra-O-galloyl-β-D-glucose isolated from *Phyllanthus urinaria* against herpes simplex virus type 1 and type 2 infection. *J. Ethnopharmacol.* 110: 555–558.

Zhang, L.Z., Guo, Y.J., Tu, G.Z., et al. 2000a. Studies on chemical constituents of *Phyllanthus urinaria* L. *Zhongguo Zhong Yao Za Zhi* 25: 615–617.

Zhang, L.Z., Guo, Y.J., Tu, G.Z., et al. 2004. Isolation and identification of a novel ellagitannin from *Phyllanthus urinaria* L. *Yao Xue Xue Bao* 39: 119–122.

Zhang, Y.J., Abe, T., Tanaka, T., et al. 2002. Two new acylated flavanone glycosides from the leaves and branches of *Phyllanthus emblica*. *Chem. Pharm. Bull.* 50: 841–843.

Zhang, Y.J., Tanaka, T., Iwamoto, Y., et al. 2000b. Novel norsesquiterpenoids from the roots of *Phyllanthus emblica*. *J. Nat. Prod.* 63: 1507–1510.

Zhang, Y.J., Tanaka, T., Iwamoto, Y., et al. 2000c. Phyllaemblic acid, a novel highly oxygenated norbisabolane from the roots of *Phyllanthus emblica*. *Tetrahedron Lett.* 41: 1781–1784.

Zhang, Y.J., Tanaka, T., Iwamoto, Y., et al. 2001a. Novel sesquiterpenoids from the roots of *Phyllanthus emblica*. *J. Nat. Prod.* 64: 870–873.

Zhang, Y.J., Tanaka, T., Yang, C.R., and Kouno, I. 2001b. New phenolic constituents from the fruit juice of *Phyllanthus emblica*. *Chem. Pharm. Bull.* 49: 537–540.

Zhang, L.Z., Zhao, W.H., Guo, Y.J., et al. 2003. Studies on chemical constituents in fruits of Tibetan medicine *Phyllanthus emblica*. *Zhongguo Zhong Yao Za Zhi* 28: 940–943.

7 Hyphenated Techniques in the Study of the Genus *Phyllanthus*

Satyajit D. Sarker, Lutfun Nahar, and Abbas Delazar

CONTENTS

7.1 Introduction 139
7.2 Hyphenated Techniques 140
7.3 Hyphenated Techniques and The Genus *Phyllanthus* 140
    7.3.1 Hyphenated High-Performance Liquid Chromatography 140
    7.3.2 Gas Chromatography and Mass Spectrometry 145
    7.3.3 Hyphenated Capillary Electrophoresis 145
7.4 Conclusion 145
References 146

7.1 INTRODUCTION

The genus *Phyllanthus* L. is the largest genus of the family Phyllanthaceae (alt. Euphorbiaceae) (Nozeran et al., 1984; Calixto et al., 1998; GRIN Database, 2011). It comprises more than 800 species of annuals, perennials, shrubs, climbers, and floating aquatics, mainly from the tropical and the subtropical countries. The species of this genus are classified under 11 subgenera: *Botryanthus, Cicca, Conani, Emblica, Ericocus, Gomphidium, Isocladus, Kirganelia, Phyllanthodendron, Phyllanthus,* and *Xylophylla*.

    Phyllanthus species have been used in traditional medicine (e.g., the Ayurvedic medicine system) for more than 3,000 years. These plants are particularly useful in the treatment of kidney problems, urinary bladder disturbances, diabetes, pain, jaundice, gonorrhea, chronic dysentery, skin ulcers, and hepatitis B (Calixto et al., 1998; Duke's Phytochemical and Ethnobotanical Databases, 2011). Owing to the reported medicinal properties of the *Phyllanthus* species, several phytochemical and bioactivity studies have been performed; these yielded various compounds, ranging from alkaloids, coumarins, flavonoids, lignans, tannins, and terpenes (ISI Web of Knowledge, 2011). Many of the isolated compounds have been shown to possess exciting biological activities. A number of hyphenated analytical techniques have been applied in the study of *Phyllanthus* species. The aims and objectives of this chapter are to provide an overview of the hyphenated techniques used for the isolation and identification of phytochemicals from the genus *Phyllanthus*.

7.2 HYPHENATED TECHNIQUES

Hyphenated techniques refer to those techniques for which a separation technique is coupled with at least one online detection technique (Sarker et al., 2005). A liquid chromatographic (LC) method—most often high-performance liquid chromatography (HPLC), gas chromatography (GC), or capillary electrophoresis (CE)—is linked to spectroscopic detection techniques, such as Fourier transform infrared (FT-IR), UV-Vis (ultraviolet-visible) photodiode array absorbance (PDA) or fluorescence emission, mass spectroscopy (MS), and nuclear magnetic resonance (NMR) spectroscopy, to provide modern hyphenated techniques like CE-MS, GC-MS, LC-MS, and LC-NMR. Since the introduction of these techniques, they have been routinely and successfully applied for both quantitative and qualitative analysis of unknown compounds in complex natural product extracts or fractions.

7.3 HYPHENATED TECHNIQUES AND THE GENUS *PHYLLANTHUS*

Crude natural product extracts, which represent extremely complex mixtures of numerous compounds, can be analyzed successfully using appropriate hyphenated techniques (Sarker et al., 2005). Hyphenated techniques have proven useful in the analysis of crude plant extracts, and some of these techniques have also been used to phytochemically investigate the genus *Phyllanthus*. Particularly, over the past several decades, the use of various hyphenated techniques in the analysis of the *Phyllanthus* species has increased (ISI Web of Knowledge, 2011).

7.3.1 Hyphenated High-Performance Liquid Chromatography

High-performance liquid chromatography or simply LC has been successfully hyphenated to a number of detection techniques, such as UV-Vis spectroscopy, PDA, fluorescence detection, mass spectrometry, and NMR spectroscopy. These techniques have been used for the isolation and characterization of secondary metabolites from higher plants, including the plants from the genus *Phyllanthus*.

The UV-Vis spectroscopic detector is a universal detector for any LC system. The PDA detector is an advanced form of UV-Vis detector that can be coupled to HPLC to provide the hyphenated technique HPLC-PDA, also known as LC-PDA (Sarker et al., 2005). For routine isolation of compounds, especially polar to highly polar compounds (e.g., phenolics like flavonoid and their glycosides, lignans, and tannins) from the *Phyllanthus* species, HPLC coupled with UV-Vis detection has been the technique of choice (Ueda et al., 2000; De Souza et al., 2002; Yadav et al., 2008; Krithika et al., 2009). For example, the ethyl acetate fraction of the methanol extract of the fruits of *P. emblica,* after Sephadex LH-20 chromatographic cleanup, was analyzed by reverse-phase HPLC coupled with UV-Vis detection, yielding six antioxidant phenolics compounds: geraniin (**1**), quercetin 3-β-D-glucopyranoside (**2**), kaempferol 3-β-D-glucopyranoside (**3**), isocorilagin (**4**), quercetin (**5**), and kaempferol (**6**) (Liu et al., 2008) (Figure 7.1). In another similar study, Kumaran and Karunakaran (2006) used a simple HPLC method coupled with UV-Vis detection for identification of simple phenolics, such as gallic acid (**12**) in the extract of *P. emblica*. Krithika et al. (2009) used a Shimadzu reverse-phase HPLC-UV-Vis system (Phenomenex C18 250 × 4.6 mm column, 5-μm particle size; column oven at 40°C; mobile phase of isocratic 60% acetonitrile in water, 0.5 ml/min flow rate, detection at 230 nm) for confirming the identity of phyllanthin (**7**), one of the bioactive lignans, in the extract of *P. amarus*. The HPLC and HPLC-UV-Vis technique has also been used to standardize *Phyllanthus niruri* extract using a marker compound, such as corilagin

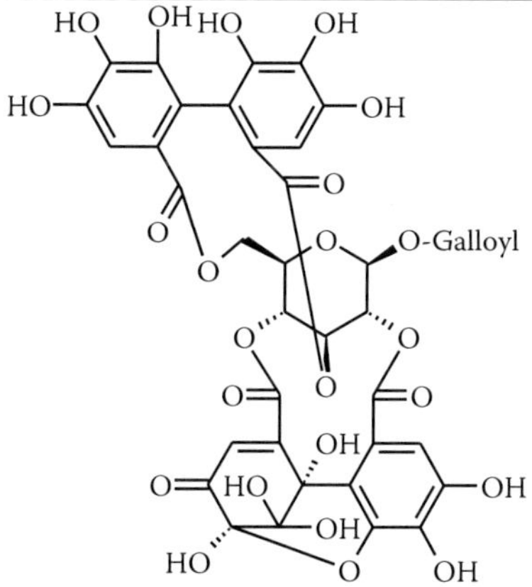

Geraniin (1)

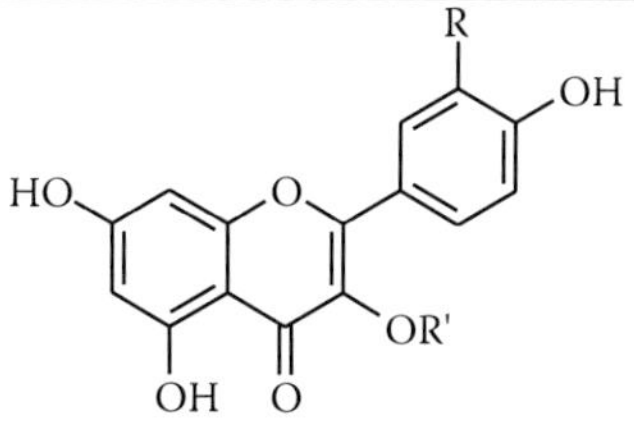

Quercetin 3-β-D-glucopyranoside (2) R = **OH; R'** = β-D-Glucopyranosyl
Kaempferol 3-β-D-glucopyranoside (3) R = **H; R'** = β-D-Glucopyranosyl,
Quercetin (5) **R = OH; R'** = H
Quercitrin (19) R = **OH; R'** = α-L-Rhamnopyranosyl
Rutin (16) **R = OH; R'** = Rutinosyl
Kaempferol (6) **R = R'** = H
Kaempferol 3-O-rutinoside (18) R = H; **R'** = Rutinosyl

FIGURE 7.1 Examples of some compounds isolated from the *Phyllanthus* species by hyphenated techniques. (*continued*)

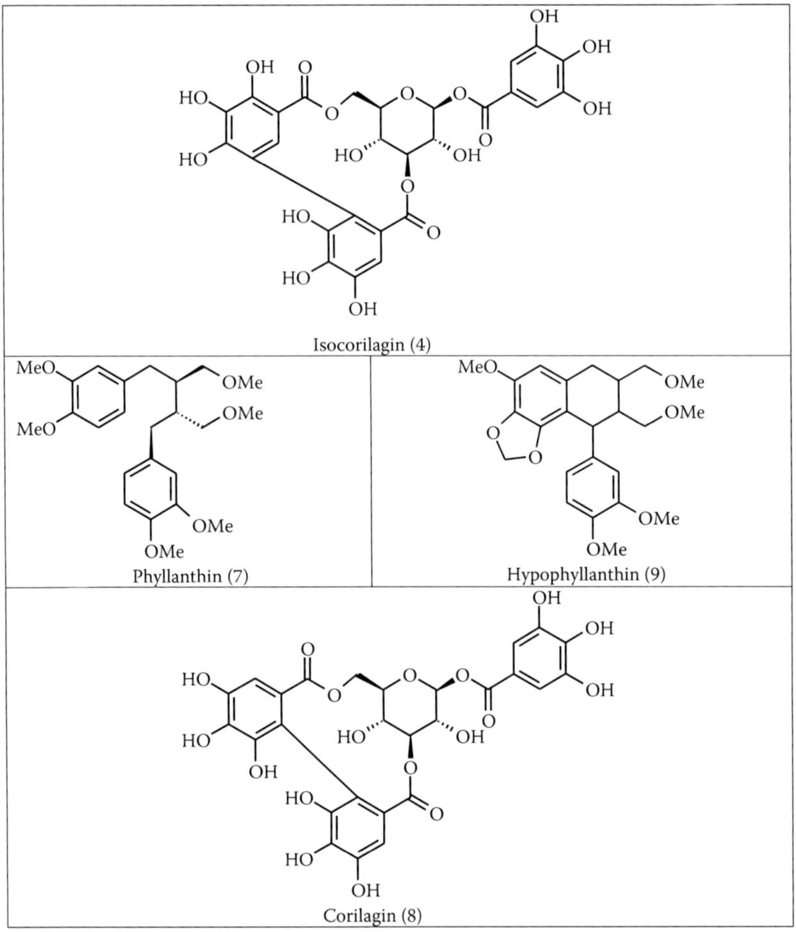

FIGURE 7.1 (continued) Examples of some compounds isolated from the *Phyllanthus* species by hyphenated techniques.

(**8**) (Colombo et al., 2009), and for fingerprint analyses of *Phyllanthus* species (Liu et al., 2007). A number of Taiwanese *Phyllanthus* species (e.g., *P. myrtifolius, P. multiflorus, P. amarus, P. debilis, P. embergeri, P. tenellus,* and *P. urinaria* subsp. *Urinaria*) were successfully analyzed for gallic acid (**12**) content by this hyphenated technique (Lee et al., 2005).

Like the UV-Vis detection method, fluorescence detection can be coupled with an HPLC system. Murugaiyah and Chan (2007) utilized this hyphenated technique for the simultaneous determination of four lignans—phyllanthin (**7**), hypophyllanthin (**9**), phyltetralin (**10**), and niranthin (**11**)—in *Phyllanthus niruri* L. plant samples (Figure 7.1). An isocratic mobile phase consisting of acetonitrile-water (55:45 v/v) was used. The method successfully recorded limits of detection (S/N = 5) for **7** at 0.61 ng/ml, **9** at 6.02 ng/ml, **10** at 0.61 ng/ml, and **11** at 1.22 ng/ml. The limits of

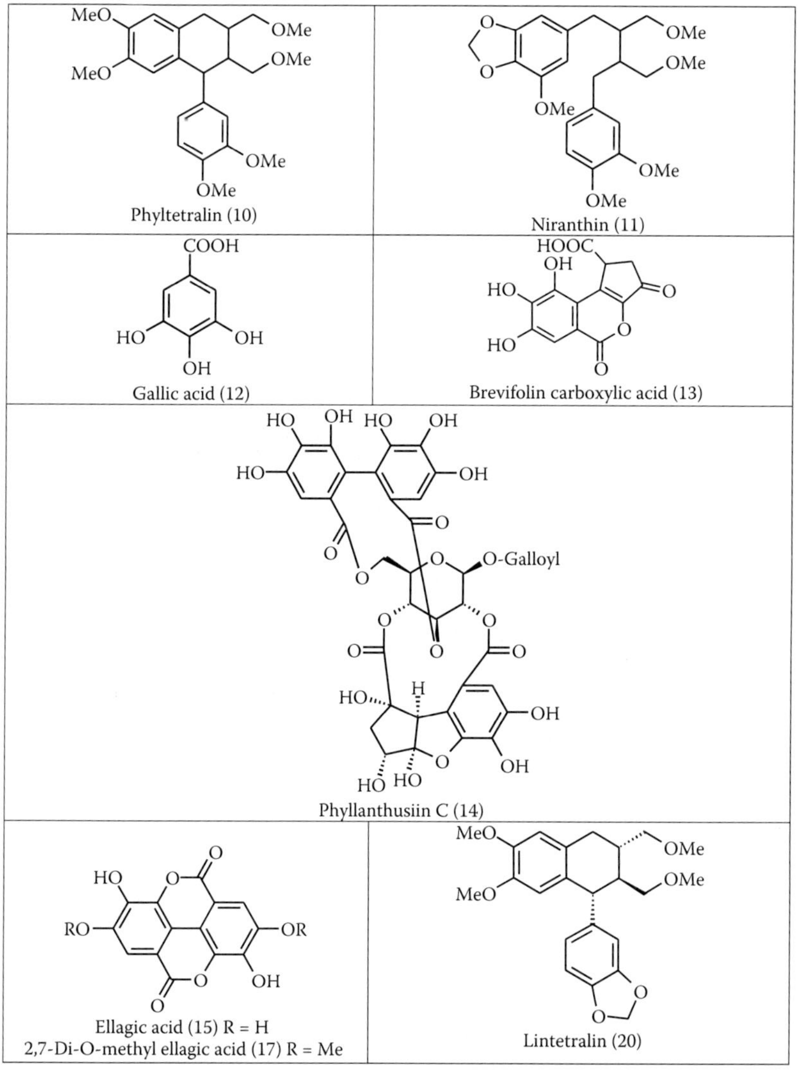

FIGURE 7.1 (continued) Examples of some compounds isolated from the *Phyllanthus* species by hyphenated techniques.

quantification (S/N =12) were 4.88 ng/ml for **7** and **10**, 9.76 ng/ml for **11**, and 24.4 ng/ml for **9**.

LC-MS or HPLC-MS stands for the coupling between an LC with a mass spectrometer. The separated sample coming out of the column can be characterized by its mass spectral data analysis (Sarker et al., 2005). Mass spectroscopy is one of the most sensitive and highly selective methods of molecular analysis and provides information on the molecular weight as well as the fragmentation pattern of the analyte molecule and thus can provide fruitful analysis of the plant extracts or fractions.

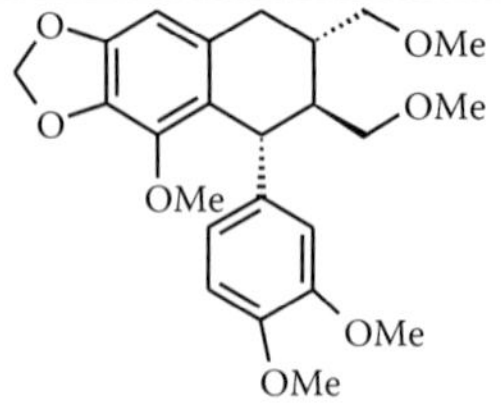

FIGURE 7.1 (continued) Examples of some compounds isolated from the *Phyllanthus* species by hyphenated techniques.

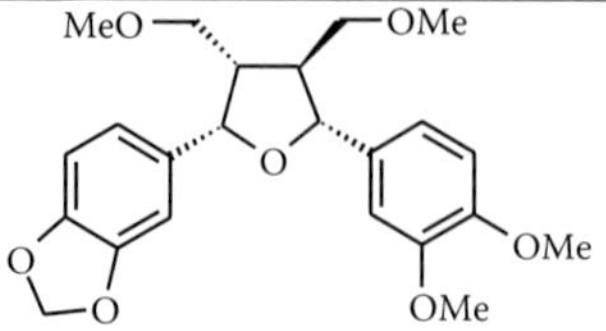

HPLC coupled to UV-Vis and MS, called LC-UV-MS, is extremely useful in combination with biological screening for rapid screening of natural products. For example, the polyphenols in *P. urinaria* have been studied by HPLC-UV-MS (Lam et al., 2007; Huang et al., 2009). Gallic acid (**12**), brevifolin carboxylic acid (**13**), corilagin (**8**), phyllanthusiin C (**14**), and ellagic acid (**15**) were identified successfully as a plant fingerprint by this method (Figure 7.1).

Among the spectroscopic techniques available to date, NMR is probably the least sensitive, yet it provides the most useful structural elucidation of natural products (Sarker et al., 2005). The direct parallel coupling of HPLC systems to NMR introduced the relatively new practical technique HPLC-NMR or LC-NMR, which has been widely known for more than 15 years. A UV-Vis detector or an MS detector is often used as the primary detector in an LC-NMR system. Lam et al. (2007) studied the polyphenol-rich fraction, obtained by Sephadex LH-20 fractionation of the extract of the leaves of *Phyllanthus reticulatus* by an HPLC-SPE-NMR technique, and characterized six compounds, including three flavonoid glycosides. The chromatographic separation was performed on a Merck Purospher Star RP-18e column (250 × 4 mm; 5-μm particle size) eluted with a mixture of water and acetonitrile (9:1 to 7:3 in 25 min, 0.7 ml/min flow rate; primary detection by PDA). The eluate containing each separate analyte was passed through an individual resin cartridge and flushed with dry nitrogen to remove the residual solvents, and the trapped analyte was washed with deuterated acetonitrile and subjected to the Bruker Avance 400-MHz NMR spectrometer's inverse LC probe (120 ml). The ^{1}H-NMR spectrum of each compound was obtained, which helped identification of rutin (**16**), isoquercitrin (**2**), 2,7-di-*O*-methyl ellagic acid (**17**), kaempferol 3-*O*-rutinoside (**18**), kaempferol 3-*O*-glucoside (astragalin, **3**), quercetin 3-*O*-rhamnoside (quercitrin, **19**) (Figure 7.1). This technique only required 1 mg of sample, theoretically equivalent to 0.3 g of dry leaves. This study demonstrated that HPLC-SPE-NMR could be extremely useful for thorough chemical investigation of *Phyllanthus* species. In fact, this method was really based on a previous publication; Wang and Lee (2005) described a similar methodology for the analysis of structurally similar lignans in *P. urinaria* using a reverse-phase HPLC system (a mobile phase consisting of tetrahydrofuran-water-methanol and detection at 225 nm). Coupling of this system with SPE-NMR provided clean ^{1}H-NMR spectra for seven lignans—including hypophyllanthin (**9**), lintetralin (**20**), niranthin (**11**), nirtetralin (**21**), phyltetralin (**10**), phyllanthin (**7**) and virgatusin (**22**)—present in 4 mg of a lignan-rich fraction, equivalent to about 1.0 g of dry plant material (Figure 7.1).

7.3.2 GAS CHROMATOGRAPHY AND MASS SPECTROMETRY

Both LC-MS and GC-MS are the most popular hyphenated techniques in use today (Sarker et al., 2005). GC-MS is a hyphenated technique developed from the coupling between GC and MS. Adequately volatile small compounds, stable in high temperature in GC conditions, can easily be analyzed by GC-MS. Polar compounds are generally derivatized for GC-MS analysis. The most common derivatization technique is the conversion of the analyte to its trimethylsilyl derivative. Currently, GC-MS integrated with different online MS databases for reference compounds is available, and it helps spectra match for the identification of separated components. The use of GC-MS in the analysis of *Phyllanthus* species is almost exclusively confined to volatile components (Quijano et al., 2007; Villarreal et al., 2008; Moronkola et al., 2009). The oils obtained from the aerial parts of *Phyllanthus amarus* have been analyzed by GC-MS, resulting in the identification of 82 compounds representing 87.6% of the total oils (Moronkola et al., 2009). The essential oil from the leaves of *Phyllanthus salviaefolius* was analyzed by GC/MS, and 16 components were identified, which represented 94.6% of the oil. Phytol (21.5%), β-citronellol (17.7%), *trans*-geraniol (13.5%), *cis*-3-hexenol (12.6%), and 1-hexanol (11.3%) were found to be the major components (Villarreal et al., 2008). The volatile compounds in *P. acidus* were analyzed by capillary GC-FID and GC-MS (Quijano et al., 2007). The 46 identified components were dominated by the presence of acids and alcohols.

7.3.3 HYPHENATED CAPILLARY ELECTROPHORESIS

Capillary electrophoresis is an automated separation technique introduced in the early 1990s; it is driven by an electric field, performed in narrow tubes, and can result in the rapid separation of hundreds of different compounds (Sarker et al., 2005). Separation by CE can be monitored or detected by a number of detection techniques (e.g., UV-Vis, MS). Hyphenation of CE with these detection techniques gives rise to hyphenated CE (e.g., CE-MS). While hyphenated CE is a powerful technique for analyzing complex matrices, its use in the analysis of *Phyllanthus* samples was discussed in only one work, published by Durham et al. (2002); the extracts of the roots *Phyllanthus acidus* were examined by hyphenated free-zone CE.

7.4 CONCLUSION

Although various hyphenated techniques have now been around for decades and routinely used in the analysis of crude plant extracts for fingerprinting or chemical profiling, isolation of compounds, and metabolomic studies, the use of hyphenated techniques in the study of the genus *Phyllanthus* is surprisingly limited to HPLC-UV-Vis methods. It can be envisaged, however, that the use of other hyphenated techniques (e.g., HPLC-NMR-MS or CE-MS) in the analyses of *Phyllanthus* species will increase considerably in the future.

REFERENCES

Calixto, J.B., Santos, A.R.S., Cechinel, V., and Yunes, R.A. 1998. A review of the plants of the genus *Phyllanthus*: their chemistry, pharmacology, and therapeutic potential. *Med. Res. Rev.* 18: 225–258.

Colombo, R., Batista, and Teles, H.L., et al. 2009. Validated HPLC method for the standardization of *Phyllanthus niruri* (herb and commercial extracts) using corilagin as a phytochemical marker. *Biomed. Chromatogr.* 23: 573–580.

De Souza, T.P., Holzschuh, M.H., Lionco, M.I., et al. 2002. Validation of a LC method for the analysis of phenolic compounds from aqueous extract of *Phyllanthus niruri* aerial parts. *J. Pharm. Biomed. Anal.* 30: 351–356.

Duke's Phytochemical and Ethnobotanical Databases. 2011. Available at http://www.ars-grin.gov/cgi-bin/duke/ethnobot.pl.

Durham, D.G., Reid, R.G., Wangboonskul, J., and Daodee, S. 2002. Extraction of phyllanthusols A and B from *Phyllanthus acidus* and analysis by capillary electrophoresis. *Phytochem. Anal.* 13: 358–362.

GRIN Taxonomy Database. (2011). USDA, ARS, National Genetic Resources Program. Germplasm Resources Information Network (GRIN) [Online database]. National Germplasm Resources Laboratory, Beltsville, Maryland. Available at http://www.ars-grin.gov/cgi-bin/npgs/html/taxdump.pl?phyllanthus

Huang, S.T., Wang, C.Y., Yang, R.C., et al. 2009. *Phyllanthus urinaria* increases apoptosis and reduces telomerase activity in human nasopharyngeal carcinoma cells. *Forsch. Komplementarmed.* 16: 34–40.

ISI Web of Knowledge. (2009). Thomson Reuters, UK. Available at http://apps.isiknowledge.com/WOS_GeneralSearch_input.do?highlighted_tab=WOS&product=WOS&last_prod=WOS&SID=P26hC5KmD6k98AN@1F@&search_mode=GeneralSearch

Krithika, R., Ramasamy, M., Vrema, R.J., et al. 2009. Isolation, characterization and antioxidative effect of phyllanthin against CCl4-induced toxicity in HepG2 cell line. *Chem. Biol. Interact.* 181: 351–358.

Kumaran, A., and Karunakaran, R.J. 2006. Nitric oxide radical scavenging active components from *Phyllanthus emblica* L. *Plant Food Hum. Nutr.* 61: 1–5.

Lam, S.H., Wang, C.Y., Chen, C.K., and Lee, S.S. 2007. Chemical investigation of *Phyllanthus reticulatus* by HPLC-SPE-NMR and conventional methods. *Phytochem. Anal.* 18: 251–255.

Lee, C.Y., Chiu, T.H., and Tsai, S.W. 2005. Quantitative HPLC methods for gallic acids of *Phyllanthus* (Euphorbiaceae). *J. Liq. Chromatogr. Related Tech.* 28: 2965–2977.

Liu, H., Sheng, W.G., and Peng, S. 2007. Study on HPLC fingerprints of *Phyllanthus urinaria*. *Zhong Yao Cai*. 30: 150–153.

Moronkola, D.O., Ogunwande, I.A., Oyewole, I.O., et al. 2009. Studies on the volatile oils of *Momordica charantia* L. (Cucurbitaceae) and *Phyllanthus amarus* Sch et Thonn (Euphorbiaceae). *J. Essential Oil Res.* 21: 393–399.

Murugaiyah, V., and Chan, K.L 2007. Determination of four lignans in *Phyllanthus niruri* L. by a simple high-performance liquid chromatography method with fluorescence detection. *J. Chromatogr. A* 1154: 198–204.

Nozeran, R., Rossignolbancilhon, L., and Mangenot, G. 1984. Studies on the genus *Phyllanthus* (Euphorbiaceae)—latest developments and perspectives. *Bot. Helv.* 94: 199–233.

Quijano, C.E., Linares, D., and Pino, J.A. 2007. Changes in volatile compounds of fermented cereza agria [*Phylianthus acidus* (L.) Skeels] fruit. *Flavour Fragr. J.* 22: 392–394.

Sarker, S.D., Latif, Z., and Gray, A.I. 2005. Hyphenated techniques. In *Natural products isolation* (2nd ed.). Totowa, NJ: Humana Press.
Sarker, S.D., and Nahar, L. 2005. Hyphenated techniques. In *Natural products isolation* (Eds. Sasker, S.D., Latif, Z., and Gray, A.T.). 2nd Ed., Totowa, Humana Press.
Ueda, M., Asano, M., Sawai, Y., and Yamamura, S. 2000. The chemical control on the leaf-movement of *Phyllanthus urinaria* L. by the internal balance of concentration between two bioactive substances. *Nat. Prod. Lett.* 14: 225–232.
Villarreal, S., Rojas, L.B., Usubillaga, A., Ramirez, I., and Solorzano, M. 2008. Volatile constituents from the leaves of *Phyllanthus salviaefolius* H. B. K. from the Venezuelan Andes. *Nat. Prod. Commun.* 3: 275–277.
Wang, C.Y., and Lee S.S. 2005. Analysis and identification of lignans in *Phyllanthus urinaria* by HPLC-SPE-NMR. *Phytochem. Anal.* 16: 120–126.
Yadav, N.P., Pal, A., Shanker, K., et al. 2008. Synergistic effect of silymarin and standardized extract of *Phyllanthus amarus* against CCl4-induced hepatotoxicity in *Rattus norvegicus*. *Phytomedicine* 15: 1053–1061.

FIGURE 1.2 Morphology of some of the common species of *Phyllanthus* (a–n): (a) *Phyllanthus amarus* (inset: a twig showing flowers); (b) *P. acidus*; (c) *P. gardnerianus*; (d) *P. pinnatus*; (e) *P. polyphyllus*; (f) *P. rheedi*; (g) *P. rotundifolius*; (h) *P. urinaria*; (i) *P. indofisheri*; (j) *P. virgatus*; (k) *P. emblica* flower; (l) *P. simplex*; (m) *P. deblis* male flower; (n) *P. deblis* female flower. (o) Fruit of *P. emblica*; (p) seeds of *P. urinaria*.

FIGURE 3.1 Morphology of some of the common species of *Phyllanthus* (a) *Phyllanthus emblica* tree (insert: a fruiting branch); (b) Emblica orchard; (c) *P. maderaspatensis*; (d) *P. reticulatus* (insert: a twig with fruit).

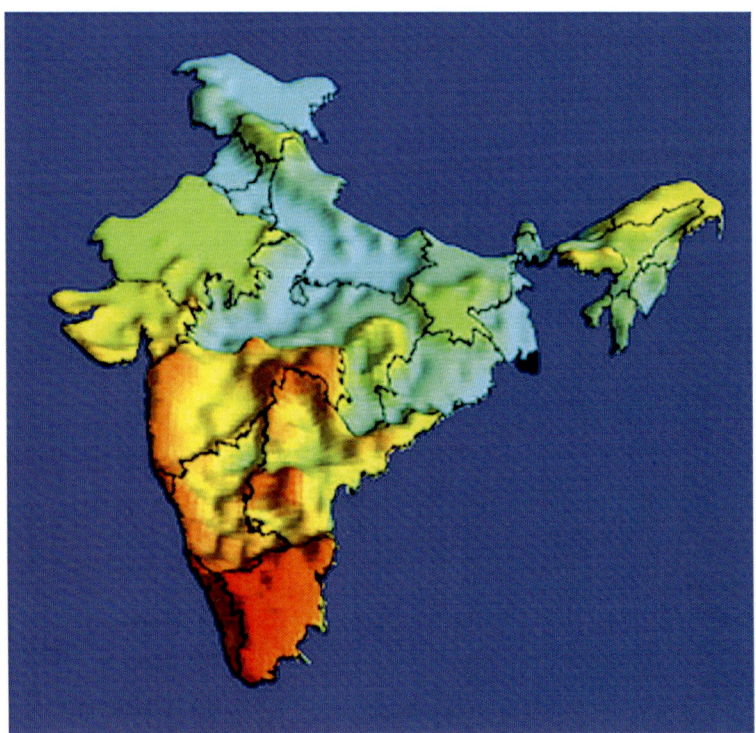

FIGURE 5.1 Hypsographic view of *Phyllanthus* species richness in India. The data on the distribution of the species were obtained from diverse sources (monographs, etc.), and the latitude and longitude were assigned for each record and mapped. The density of the species in each grid of the size 10 km × 10 km was computed and the contours for the density obtained. Based on the contour data, the three-dimensional view was constructed using suitable GIS software.

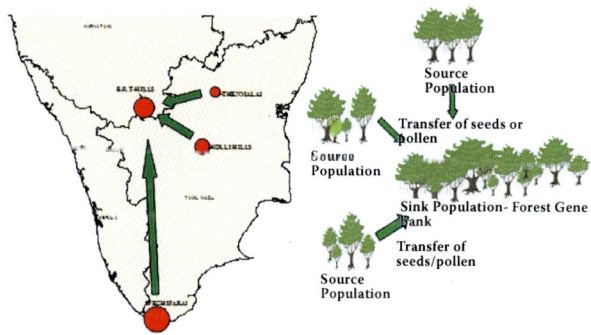

FIGURE 5.5 Genetic diversity map of *Phyllanthus emblica*. The relative size of the circle indicates the levels of gene diversity. The arrows indicate the possible flow of genes either through seeds or pollen to the sink population from the source population (see text for further explanation). (Adapted from Uma Shaanker, R. and Ganeshaiah, K.N. 1997. *Curr. Sci.* 73: 163–168.)

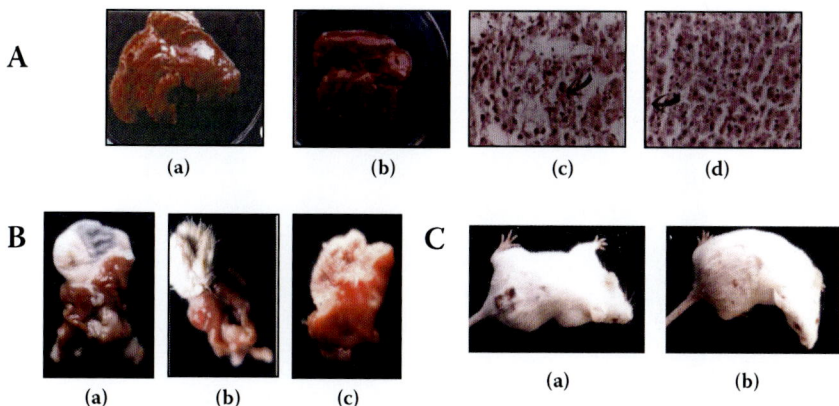

FIGURE 10.1 (A) Inhibition of NDEA-induced hepatocarcinogenesis by PAE. (a) Gross morphology of the liver treated with NDEA at 20 weeks shows proliferation of tumor nodules; (b) the rat liver administered with NDEA and PAE extract at 20 weeks; (c) histopathology of the rat liver treated with NDEA at 20 weeks (×40 magnification); (d) histopathology of the rat liver (×40 magnification) treated with NDEA and PAE at 20 weeks. (B) Inhibition of MNNG-induced gastric cancer by PAE. Morphology of the stomach: (a) untreated; (b) treated with MNNG alone; (c) treated with MNNG plus *P. amarus* 750 mg/kg body weight. (C) Inhibition of DMBA-induced skin cancer by PAE: (a) representative picture from DMBA-alone-treated group; (b) DMBA plus *P. amarus* (5 mg/mouse).

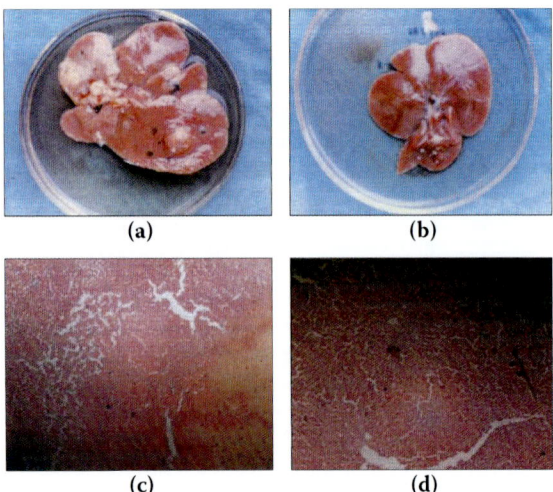

FIGURE 11.2 (a) and (b) Effect of EO on the morphology of liver tumor induced by NDEA. (c) and (d) Effect of EO on the histopathology of liver tumor induced by NDEA. (From Jose, J. K., Kuttan, R., and Bhattacharya, R. K. 1998. Effect of *Embica officinalis* on hepatocarcinogenesis and carcinogene metabolism. *J. Clin. Biochem. Nutr.* 25: 31–39. With permission.)

FIGURE 14.1 *Phyllanthus orbicularis,* a Cuban endemic species. Insert: A twig with flowers.

8 Anti-inflammatory Activity of Various Species of *Phyllanthus*

K. B. Harikumar and Ramadasan Kuttan

CONTENTS

8.1 Introduction 149
8.2 *Phyllanthus Amarus* 150
8.3 *Phyllanthus Emblica* 151
8.4 *Phyllanthus Debilis* 152
8.5 *Phyllanthus Polyphyllus* 152
8.6 *Phyllanthus Reticulatus* 153
8.7 *Phyllanthus Urinaria* 153
8.8 *Phyllanthus Corcovadensis* 153
8.9 *Phyllanthus Singampattiyana* 154
8.10 *Phyllanthus Tenellus* 154
8.11 Formulations with Different *Phyllanthus* Sp. 154
        8.11.1 Entox® 154
        8.11.2 Septilin 154
8.12 Conclusion 155
References 155

8.1 INTRODUCTION

Inflammation (derived from the Latin word *inflammatio*, "to set on fire") in general is a complex biological response of vascular tissue to a variety of external and internal stimuli, which include but are not limited to various microbes, immune reaction, ultraviolet (UV) rays, physical damage, and the like. Inflammatory responses can be either beneficial or detrimental. Inflammation can be either acute or chronic. The acute inflammatory responses are generally regarded as beneficial, while chronic inflammation plays a major role in the etiology of different diseases, like cancer, asthma, cardiovascular diseases, arthritis, diabetes, and neurological diseases (Aggarwal et al., 2006).

Currently, several classes of anti-inflammatory agents are available, such as corticosteroids and nonsteroidal anti-inflammatory drugs (NSAIDs); but, these are often associated with several side effects. So, there is always a demand for better agents with fewer side effects. Throughout the world, plants have been the basis of numerous traditional systems of medicine for many years and continue to provide humankind with new remedies. The natural products derived from the plants become the basis for developing new therapeutic compounds. There are several reports available to support the use of plants in the treatment of inflammation. In this chapter, the anti-inflammatory potential of different species of *Phyllanthus* is described in detail.

8.2 *PHYLLANTHUS AMARUS*

Phyllanthus amarus (PA) has been reported to possess significant anti-inflammatory activity using various *in vitro*, *in vivo*, and *ex vivo* strategies. The oral administration (100 mg/kg) of the hexane fraction of PA inhibited the paw edema formation in mice induced by carrageenan (CGN), bradykinin, platelet-activating factor (PAF), and endothelin 1 (Kassuya et al., 2005). CGN-induced edema is associated with the release of inflammatory mediators such as histamine, kinins, and prostaglandins. PA also decreased the neutrophil infiltration induced by the inflammatory agents mentioned; this action was measured by the activity of myeloperoxidase. Moreover, PA has been shown to inhibit the production of proinflammatory cytokine interleukin 1β (IL-1β). The lignan-rich fraction isolated from PA significantly decreased the CGN-mediated inflammation in paw edema, indicating that PA possesses marked antiedematogenic activity. The authors further characterized the lignan fraction and identified nirtetralin, niranthin, hypophyllanthin, phyltetralin, and phyllanthin as the major lignans (Kassuya et al., 2003) and further characterized the mechanism of action of these lignans (Kassuya et al., 2006). The treatment of mice nirtetralin, phyltetralin, or niranthin significantly decreased PAF-induced paw edema. WEB2170 (a PAF receptor antagonist) was used as the positive control. Niranthin was found to be more effective than the others. It also inhibited the elevated myeloperoxidase activity after PAF challenge. It was concluded that niranthin probably acted as an antagonist to PAF receptor, thereby inhibiting the inflammatory responses.

Our laboratory reported that the water extract of PA (100, 250, and 500 mg/kg orally) decreased the CGN-mediated paw edema in mice. A 1-h prior challenge of extract at these doses produced inhibition of 26%, 33%, and 39%, respectively, at 3 h. Similarly, a methanol extract of PA also was found to inhibit paw edema formation (Raphael and Kuttan, 2003). Intraperitoneal administration of PA (45 mg/kg) significantly attenuated the tumor necrosis factor (TNF) production in endotoxemia in mice induced by lipopolysaccharide (LPS)-galactosamine (Kiemer et al., 2003). Mahat and coworkers evaluated the anti-inflammatory potential of PA in rat models of inflammation (Mahat and Patil, 2007). They found that a methanolic fraction of PA significantly inhibited paw edema formation induced by CGN, bradykinin, serotonin, and prostaglandin E1 (PGE1). However, PA failed to inhibit histamine-induced edema formation. It might be possible that PA does not possess any antihistaminic activity. The authors also tested the efficacy of PA in a CGN-induced air-pouch model of inflammation and observed that PA decreased the leukocyte infiltration and exudate volume. However, PA was less efficient than indomethacin (Mahat and Patil, 2007).

Cotton pellet-induced granulation is yet another widely used *in vivo* system to test the anti-inflammatory activity of different compounds. The inflammatory responses generated after the subcutaneous implantation of the cotton pellet is divided into three phases: transudative, exudative, and proliferative. In a cotton pellet-induced granuloma model, PA produced an inhibition of 23% compared to the standard drug indomethacin, which produced 44% inhibition.

PA inhibited the production of NO (nitric oxide), PGE2, and TNF-α *in vitro* induced by LPS in RAW264.7 cells or rat Kupffer cells (KCs). Moreover, in human whole blood PA significantly decreased the LPS-mediated TNF, IL-1β, IL-10, and interferon γ (IFN-γ) secretion in culture. Western blot analysis indicated that PA attenuated the LPS-mediated expression of iNOS (inducible nitric oxide synthese) and cyclo-oxygenase 2 (COX-2) protein levels. The authors also found that PA inhibited LPS-mediated activation of transcription factors nuclear factor kappa B (NF-κB) and AP-1 (Activator protein-1) (Kiemer et al., 2003).

8.3 PHYLLANTHUS EMBLICA

Asmawi et al. (1993) reported for the first time the anti-inflammatory potential of leaf extracts of *Phyllanthus emblica* (PE) in a rat model of inflammation. The rats were orally treated with either extract (2 g/kg) or phenylbutazone (15 mg/kg as a positive control). After 1 h, animals were challenged with CGN, and edema was measured. The water fraction of the methanolic extract was found to inhibit the edema formation ($p = .057$). Similar results were observed in a dextran-induced inflammatory model (Asmawi et al., 1993). It was observed that PE extract did not inhibit synthesis of thromboxane, leukotriene, or PAF in *in vitro* conditions using isolated human polymorphonuclear leukocytes (PMNs) (Ihantola-Vormisto et al., 1997). But, PE extract inhibited the migration of PMNs induced by leukotriene B4, a mediator of inflammation. This suggested that the antimigration activity might be responsible for the anti-inflammatory potential of PE (Ihantola-Vormisto et al., 1997).

Dried fruit extract of PE was also tested for its potential anti-inflammatory activity in rats. Topical application of either ethyl phenylpropiolate (EPP) or arachidonic acid (AA) was known to develop ear edema in rats, which is associated with fluid accumulation and acute inflammatory responses. In an EPP-induced acute model, release of inflammatory mediators such as kinin, serotonin, and prostaglandins are known to play the major role, while in AA-induced edema activation of the lipoxygenase pathway is the crucial factor. PE extract the edema induced by EPP but failed to afford protection against AA-induced responses. PE extract (at doses of 150, 300, and 600 mg/kg body weight) produced significant inhibition of CGN-induced paw edema. The known COX-2 inhibitor aspirin was used as a positive control in the experiment. In a cotton pellet-induced granuloma model, PE extract as well as aspirin failed to protect the rat against the inflammatory response. In the same experiment, prednisolone, a synthetic adrenal corticosteroid, was used as a positive control and protected the rat against granuloma formation, indicating that PE may not possess any steroidal-like activity (Jaijoy et al., 2010).

Muthuraman et al. (2010) evaluated the anti-inflammatory activity of free and bounded phenolic fraction isolated from PE in rats. In a CGN-induced paw edema model, higher doses (40 mg/kg orally), both phenolic fractions inhibited the edema formation, which was comparable to the standard drug diclofenac sodium (12.5 mg/kg orally). The authors also tested the efficacy of these phenolic fractions in cotton pellet-induced granuloma. Pretreatment with each fraction significantly reduced the granuloma formation in rats. Moreover, the elevated TBARS (thio barbituric acid reactive substances), myeloperoxidase, and plasma extravasation levels after cotton pellet implantation were reduced after treatment with the phenolic fractions (Muthuraman et al., 2010).

Nicolis and coworkers (2008) identified pyrogallol as one of the major active ingredients present in PE with significant anti-inflammatory activity. They compared pyrogallol and ethanolic extract of dried fruits of PE for anti-inflammatory activity. In patients with cystic fibrosis (CF), the major cause of morbidity is infection with *Pseudomonas aeruginosa*. The lung pathology of these patients is characterized by inflammation and chronic infection with *P. aeruginosa*. So, agents that can inhibit the inflammatory reactions have great potential in the treatment of patients with CF. IB3-1 cells were used in the experiment, and these cells were derived from a patient with CF. Ethanolic extract of dried fruits (500 μg/ml) was found to inhibit *P. aeruginosa*-mediated transcription of ICAM-1 (inter-cellular adhesion molecule), IL-8, IL-6, GRO-α (GRO-growth-related oncogene), and GRO-γ in IB3-1 cells. The gas chromatography/mass spectrometry (GC/MS) analysis identified pyrogallol and 5-hydroxy-isoquinoline as the active major components present in the extract. Further studies demonstrated that pyrogallol also inhibited the expression of these genes in a concentration-dependent manner (0–200 μM), while 5-hydroxy-isoquinoline did not influence the transcription of the genes. This confirms pyrogallol as the major active component of the ethanolic fraction. The authors did not provide a clear explanation for the mechanism of action of pyrogallol, and extension of this study to an *in vivo* model is required to completely understand the action of this bioactive phenolic compound.

8.4 PHYLLANTHUS DEBILIS

Phyllanthus deblis (hereafter PD) is a closely related species to *P. amarus*. The extract of *P. deblis* has been reported to inhibit CCl4-induced rat liver dysfunction, and compounds such as β-sitosterol, phyllanthin, hypophyllanthis, and others were isolated from this plant. Petroleum ether extracts of dried plants were used in this study (Chandrashekar et al., 2005). The anti-inflammatory potential was analyzed in two different models. In a CGN-induced paw edema model, PD extract (50–400 mg/kg) significantly reduced the edema formation after CGN challenge. In cotton pellet-induced granuloma model, treatment with PD extract produced a statistically significant decrease in granuloma weight (Chandrashekar et al., 2005).

8.5 PHYLLANTHUS POLYPHYLLUS

Rao et al. (2006) isolated the following four compounds from whole plant extracts of *P. polyphyllus* and evaluated their anti-inflammatory potential *in vitro*:

4-O-methylgallic acid; phyllamyricin C (2-(hydroxymethyl)-6,7,8-trimethoxy-4-(3,4-methylenedioxy-phenyl)-3-naphthoic acid-lactone); Justicidin B (2-(hydroxymethyl)-6,7-dimethoxy-4-(3,4-methylenedioxy-phenyl)-3-naphthoic acid-lactone); and diphyllin (1-hydroxy-2-(hydroxymethyl)-6,7-dimethoxy-4-(3,4-methylenedioxy-phenyl)-3-naphthoic acid-lactone). Murine peritoneal macrophages were stimulated with LPS (2 µg/ml) and IFN-γ (10 U/ml) for 24 h without and with the presence of these compounds. All four compounds decreased the production of NO stimulated by LPS/IFN. The compounds also inhibited the production of TNF and IL-12, the two major cytokines secreted during the early phase of both acute and chronic inflammation. The reported anti-inflammatory activity of *P. polyphyllus* in folk medicine can be attributed to these isolated compounds (Rao et al., 2006).

8.6 PHYLLANTHUS RETICULATUS

Phyllanthus reticulatus is used traditionally in different parts of the world for the treatment of many types of pain and inflammatory conditions. The anti-inflammatory capability of the extract of aerial parts of *P. reticulatus* was evaluated in a CGN-induced paw edema model. The petroleum ether, methanol, and ethyl acetate fractions were used in the study and were administered at doses of 150 and 300 mg/kg orally (Saha et al., 2007). All three fractions were able to decrease the paw edema induced by CGN; the possible mechanism could be the inhibition of prostaglandin synthesis.

8.7 PHYLLANTHUS URINARIA

The anti-inflammatory potential of *Phyllanthus urinaria* (PU) was evaluated *in vitro* in *Helicobacter pylori*-induced inflammation in human gastric epithelial cells. Methanolic and chloroform extracts of PU were tested in human AGS (human gastric epithelial cell line) cells. Both the extracts inhibited the *H. pylori* adhesion and invasion to AGS cells (Lai et al., 2008).

Infection with *H. pylori* resulted in the secretion of inflammatory mediators, especially IL-8, which was mediated through the activation of the transcription factor NF-κB. The chloroform extract of PU significantly reduced the *H. pylori*-induced NF-κB-dependent reporter activity. The secretion of IL-8 was also inhibited by the chloroform extract. These studies clearly demonstrated the potential of PU in the treatment of *H. pylori* infection (Lai et al., 2008). More studies are needed to understand the mechanism of action of the extract.

Kang et al. (2008) also reported the protective effect of PU on paw edema in adjuvant- (heat killed *Mycobacterium tuberculosis*) induced arthritis in mice. PU extract inhibited the myeloperoxidase activity and inhibited the neutrophil migration into inflamed tissue (Kang et al., 2008).

8.8 PHYLLANTHUS CORCOVADENSIS

Phyllanthus corcovadensis is widely used in traditional medicine, especially in Latin American countries. There is not much known about its anti-inflammatory potential. The ethanolic extract of the plant was evaluated for its anti-inflammatory activity in a rat paw edema model induced by formalin, CGN, or dextran. The pretreatment of extract did not influence the paw edema volume induced by CGN, dextran, or formalin, while indomethacin (used as a positive control) significantly reduced the paw edema (Gorski et al., 1993).

8.9 PHYLLANTHUS SINGAMPATTIYANA

Phyllanthus singampattiyana is one of the less-known species of *Phyllanthus* that has not been explored much scientifically. The plant is endemic to Tirunelveli Hills in India. Maridass et al. (2005) reported its anti-inflammatory potential. The extracts of leaves were used in the study. Topical application of the extract reduced rat ear edema induced by TPA (12-O-Tetradecnoylphorbot-13-acetate), and oral administration of the extract significantly reduced the CGN-mediated paw edema. β-Sitosterol has been identified as one of the major components of the extract, and the anti-inflammatory potential of β-sitosterol was already reported in the literature.

8.10 PHYLLANTHUS TENELLUS

Ignácio et al (2001) reported the effect of *P. tenellus* on the production of NO from macrophages. NO, which is a second messenger, plays a major role in the immune system. NO regulates the function of many immune cells, which plays a role in controlling a pathogen attack and associated inflammatory responses. Extracts of fresh and dried parts of *Phyllanthus tenellus* as well as an acetone/water extract of the dried plant were found to stimulate the production of NO both *in vivo* and *in vitro* (Ignácio et al., 2001). Active components of *P. tenellus* are not known yet.

8.11 FORMULATIONS WITH DIFFERENT *PHYLLANTHUS* SP.

8.11.1 Entox®

Deorukhakar et al. (2008) reported the anti-inflammatory potential of a polyherbal formulation named Entox®. One of the major ingredients is the fruits of *Phyllanthus emblica*. These authors tested the anti-inflammatory activity in two *in vivo* models. The first model was of CGN-induced paw edema in rats. Oral administration of the formulation (300 and 600 mg/kg) significantly reduced the CGN-induced paw edema. The cotton pellet-induced granuloma model was the second model. The formulation was found to reduce the granular tissue formation. The formulation at doses of 300 and 600 mg/kg produced inhibition of 27.91% and 53.17%, respectively, while indomethacin (5 mg/kg orally) showed a 57% reduction. These studies clearly demonstrated the anti-inflammatory potential of this formulation (Deorukhakar et al., 2008).

8.11.2 Septilin

One of the major components of Septilin is *Phyllanthus emblica*. Previously, we had reported the immunomodulatory and chemoprotective activity of this formulation (Praveen Kumar et al., 1995; Praveenkumar et al., 1997). The anti-inflammatory activity of Septilin was evaluated in three different *in vivo* models: CGN-induced paw edema, cotton pellet-induced granuloma, and Freund's adjuvant-induced arthritis. Septilin (500 mg/kg orally) showed consistent anti-inflammatory activity in all three models tested (Khanna and Sharma, 2001), but the mechanism of action of this formulation was not studied in detail.

8.12 CONCLUSION

The studies illustrated clearly demonstrated the *in vitro* and *in vivo* anti-inflammatory potential of various species of the genus *Phyllanthus*. Despite these promising results, there is a serious lack of knowledge concerning the mechanism of action of these extracts. More studies are desired to elucidate the active potential of these fractions.

REFERENCES

Aggarwal, B.B., Shishodia, S., Sandur, S.K., et al. 2006. Inflammation and cancer: how hot is the link? *Biochem. Pharmacol.* 72: 1605–1621.

Asmawi, M.Z., Kankaanranta, H., Moilanen, E., and Vapaatalo, H. 1993. Anti-inflammatory activities of *Emblica officinalis* Gaertn leaf extracts. *J. Pharm. Pharmacol.* 45: 581.

Chandrashekar, K.S., Joshi, A.B., Satyanarayana, D., and Pai, P. 2005. Analgesic and anti-inflammatory activities of *Phyllanthus debilis* whole plant. *Pharm. Biol.* 43: 586–588.

Deorukhakar, S.R., Dethe, A., Vohra, R.R., et al. 2008. Antiinflammatory activity of a polyherbal formulation. *Indian J. Pharm. Sci.* 70: 785.

Gorski, F., Correa, C.R., Filho, V.C., et al. 1993. Potent antinociceptive activity of a hydroalcoholic extract of *Phyllanthus corcovadensis*. *J. Pharm. Pharmacol.* 45: 1046–1049.

Ignácio, S., Ferreira, J.L.P., Almeida, M.B., and Kubelka, C.F. 2001. Nitric oxide production by murine peritoneal macrophages *in vitro* and *in vivo* treated with *Phyllanthus tenellus* extracts. *J. Ethnopharmacol.* 74:181–187.

Ihantola-Vormisto, A., Summanen, J., Kankaanranta, H., et al. 1997. Anti-inflammatory activity of extracts from leaves of *Phyllanthus emblica*. *Planta Med.* 63: 518–524.

Jaijoy, K., Soonthornchareonnon, N., Panthong, A., and Sireeratawong, S. 2010. Anti-inflammatory and analgesic activities of the water extract from the fruit of *Phyllanthus emblica* Linn. *Int. J. Appl. Res. Nat. Prod.* 3: 28–35.

Kang, M.H., Kim, T.S., Park, C.G., et al. 2008. Preventive effects of *Phyllanthus urinaria* extract on paw edema in adjuvant-induced arthritis rats. *FASEB J.* 22: 889.5.

Kassuya, C.A., Leite, D.F., de Melo, L.V., et al. 2005. Anti-inflammatory properties of extracts, fractions and lignans isolated from *Phyllanthus amarus*. *Planta Med.* 71: 721–726.

Kassuya, C.A., Silvestre, A., Menezes-de-Lima, O., Jr., et al. 2006. Antiinflammatory and antiallodynic actions of the lignan niranthin isolated from *Phyllanthus amarus*. Evidence for interaction with platelet activating factor receptor. *Eur. J. Pharmacol.* 546: 182–188.

Kassuya, C.A., Silvestre, A.A., Rehder, V.L., and Calixto, J.B. 2003. Anti-allodynic and anti-oedematogenic properties of the extract and lignans from *Phyllanthus amarus* in models of persistent inflammatory and neuropathic pain. *Eur. J. Pharmacol.* 478:145–153.

Khanna, N., and Sharma, S.B. 2001. Anti-inflammatory and analgesic effect of herbal preparation: Septilin. *Indian J. Med. Sci.* 55, 195.

Kiemer, A.K., Hartung, T., Huber, C., and Vollmar, A.M. 2003. *Phyllanthus amarus* has anti-inflammatory potential by inhibition of iNOS, COX-2, and cytokines via the NF-kappaB pathway. *J. Hepatol.* 38: 289–297.

Lai, C.H., Fang, S.H., Rao, Y.K., et al. 2008. Inhibition of *Helicobacter pylori*-induced inflammation in human gastric epithelial AGS cells by *Phyllanthus urinaria* extracts. *J. Ethnopharmacol.* 118: 522–526.

Mahat, M.A., and Patil, B.M. 2007. Evaluation of antiinflammatory activity of methanol extract of *Phyllanthus amarus* in experimental animal models. *Indian J. Pharm. Sci.* 69: 33.

Maridass, M., Victor, B., Benniamin, A., Mannan, M.M., and De Britto, A.J. 2005. Antiinflammatory activity of *Phyllanthus singampattiyana* leaf extract. *Pharm. Biol.* 43: 296–298.

Muthuraman, A., Sood, S., and Singla, S.K. 2010. The antiinflammatory potential of phenolic compounds from *Emblica officinalis* L. in rat. *Inflammopharmacology* July 2 [Epub ahead of print].

Nicolis, E., Lampronti, I., Dechecchi, M.C., et al. 2008. Pyrogallol, an active compound from the medicinal plant *Emblica officinalis*, regulates expression of pro-inflammatory genes in bronchial epithelial cells. *Int. Immunopharmacol.* 8: 1672–1680.

Praveen Kumar, V., Kuttan, G., and Kuttan, R. 1995. Chemoprotective action of septilin against cyclophosphamide toxicity. *Indian J. Pharm. Sci.* 57: 215–217.

Praveenkumar, V., Kuttan, R., and Kuttan, G. 1997. Immunopotentiating activity of Septilin. *Indian J. Exp. Biol.* 35: 1319–1323.

Rao, Y.K., Fang, S.H., and Tzeng, Y.M. 2006. Anti-inflammatory activities of constituents isolated from *Phyllanthus polyphyllus*. *J. Ethnopharmacol.* 103: 181–186.

Raphael, K.R., and Kuttan, R. 2003. Inhibition of experimental gastric lesion and inflammation by *Phyllanthus amarus* extract. *J. Ethnopharmacol.* 87: 193–197.

Saha, A., Masud, M.A., Bachar, S.C., et al. 2007. The analgesic and anti-inflammatory activities of the extracts of *Phyllanthus reticulatus* in mice model. *Pharm. Biol.* 45: 355–359.

9 Hepatoprotective Effects of Plants in the Family Phyllanthaceae

V. V. Asha

CONTENTS

9.1 Introduction 157
9.2 Phyllanthaceae 158
9.3 Hepatoprotection and Phyllanthaceae 159
    9.3.1 *Phyllanthus amarus* Schum & Thonn. (Syn: *Phyllanthus niruri* Sensu Hook F.) 159
    9.3.2 *Phyllanthus debilis* Klein Ex Willd. 161
    9.3.3 *Phyllanthus emblica* L. 161
    9.3.4 *Phyllanthus kozhikodianus* Sivarajan & Manilal 162
    9.3.5 *Phyllanthus maderaspatensis* L 162
    9.3.6 *Phyllanthus niruri* L 163
    9.3.7 *Phyllanthus reticulatus* Poir 164
    9.3.8 *Phyllanthus urinaria* L 164
9.4 Mechanism of Antihepatotoxic Activities of *Phyllanthus* Species 165
    9.4.1 Antioxidative Stress and Hepatoprotection 165
    9.4.2 Anti-inflammatory Activity of Hepatoprotective *Phyllanthus* Species 166
9.5 Conclusion 166
Acknowledgments 167
References 167

9.1 INTRODUCTION

Plants have been used in ethnomedicine and traditional medicine from time immemorial to treat various diseases. The remarkable advances in conventional medicine replaced to a considerable extent the use of crude plant drugs with drugs with a pure chemical entity. Although safe and effective treatments are available in conventional medicine for most diseases, satisfactory treatments are not available in conventional medicine to treat severe liver damage; certain viral diseases, including viral hepatitis; arthritis; and others. However, the use of plant preparations as medicines continues for treatment of liver and other diseases by the mainstream population in developed countries. In addition, in the most recent decade, plants and plant products have played an important role in the management of various liver disorders (Subramoniam and Pushpangadan, 1999). Although plant parts and preparations belonging to different families are used as hepatoprotective medicines, many members of the family Phyllanthaceae (particularly the genus *Phyllanthus*) have unique medicinal properties, including antihepatotoxicity and antihepatitis viral activities (Calixto et al., 1998).

Since ancient times, plants of the genus *Phyllanthus* have commonly been used in the treatment of jaundice and other liver diseases (Thyagarajan and Jayaram, 1992; Thyagarajan et al., 2002). Jaundice and other liver abnormalities are being treated with *Phyllanthus* administration in India, China, Burma, Pakistan, Philippines, Guam, West Indies, South America, East and West Africa, and elsewhere (Calixto et al., 1998; Blumberg, 1998). Plants belonging to the genus *Phyllanthus* are also used in Ayurvedic medicine in combination with other herbals to combat many disorders, including liver diseases (Wealth of India, 2003).

The hepatoprotective effects of some of the traditional hepatoprotective plants belonging to Phyllanthaceae have been proved by scientific studies. The presence of phylogenically specific phytochemicals (chemotaxonomy) could be responsible for the specific pharmacological effects of taxonomically related plants.

The hepatoprotective activity, based on clinical trials of the genus *Phyllanthus* against hepatitis B viral infection, has been reviewed (Liu et al., 2001). Further, this area forms the subject matter of another chapter in this book. Therefore, this review focuses on the hepatoprotective effects of members of Phyllanthaceae against liver disorders arising from induction of toxic chemicals, excess alcohol, and the like. Here, *Phyllanthus* species with scientifically proven hepatoprotective activity are the focus.

9.2 PHYLLANTHACEAE

Phyllanthaceae is one of the largest families, with more than 53 genera and 2,191 known species. Among the genera, *Phyllanthus* L. is the largest one, containing more than 700 species; this is followed by *Lingelsheimia,* which contains at least 200 species. Thirty genera are represented by one species each (Hoffman et al., 2006; APG, 1998). New species are being discovered (Sivarajan and Manilal, 1977). Many hepatoprotective plants belong to the genus *Phyllanthus*. In this genus, there are at least eight important species used as medicine to treat liver diseases in traditional medicine that have been verified as hepatoprotective plants based on pharmacological studies (Table 9.1). These plants are medicinally important, with several therapeutically valuable pharmacological properties (Calixto et al., 1998; Lee et al., 2006).

TABLE 9.1
Some of the Hepatoprotective *Phyllanthus* Species

S. No.	Name of Plants	References
1.	*Phyllanthus amarus* Schum. & Thonn. (Syn: *P. niruri* sensu. Hook f.; *P. fraternus* Webster)	Prakash et al., 1995; Joy and Kuttan, 1998, Raphael, 2004; Sane et al., 1995; Sebastian and Setty, 1999; Padma and Setty, 1999; Sailaja and Setty, 2006; Ahmed et al., 2002
2.	*P. debilis* Klein ex. Willd.	Sane et al.,1995
3.	*P. emblica* L. (Syn: *Emblica officinalis* Gaertn.)	Gulati et al., 1995; Jose and Kuttan, 2000
4.	*P. kozhikodianus* Sivarajan & Manilal (Syn: *P. rheedi* Wight.)	Asha and Pushpangadan, 1998; Suresh and Asha, 2008; Gangopadhyay et al., 2004
5.	*P. maderaspatensis* L.	Asha et al., 2004, 2007
6.	*P. niruri* L.	Syamasundar et al., 1985; Reddy et al., 1993; Bhattacharjee and Sil, 2006a, 2006b; Chatterjee and Sil, 2006
7.	*P. reticulatus* Poir	Das et al., 2008
8.	*P. urinaria* L.	Prakash et al., 1995; Shen et al., 2008

9.3 HEPATOPROTECTION AND PHYLLANTHACEAE

9.3.1 PHYLLANTHUS AMARUS SCHUM & THONN. (SYN: PHYLLANTHUS NIRURI SENSU HOOK F.)

Phyllanthus amarus Schum & Thonn. (syn: *Phyllanthus niruri* sensu Hook F.) is an annual herb found in degraded moist deciduous forests, plantations up to 900 m above sea level, and plains. The plant species is originally from South America and naturalized in the tropics. It is common in South India and distributed in all districts of Kerala State. Flowering and fruiting of this plant in Kerala is during July to October (Sasidharan, 2004). It is a medicinally important species of *Phyllanthus* and is used in traditional medicine for many diseases as a single herbal drug remedy against jaundice in many parts of South India (Wealth of India, 2003).

The antihepatitis viral properties of this plant are reviewed elsewhere in this book. Several studies of *P. amarus* have confirmed that preparations of this plant are hepatoprotective, anti-inflammatory, immunomodulating, and so on in addition to its well-known antihepatitis viral properties (Thyagarajan et al., 1988, 2002).

Phyllanthus amarus is known to protect the liver from toxic chemicals (Raphael, 2004). Standardized extracts of *P. amarus* protected *Ratus norvigicus* from CCl_4-induced hepatotoxicity. Also, synergistic effects of silymarin and the extract of *P. amarus* have been reported (Yadav et al., 2008). Further, the plant is shown to protect the liver of rats from excess alcohol consumption; it also normalizes fatty liver (Khanna et al., 2002). The hepatoprotective effects of the plant are attributed to a large extent to the phytochemicals phyllanthin and hypophyllanthin (Syamasundar et al., 1985). The antihepatotoxic effect of the plant extract may partly be due to its antioxidant activity. *Phyllanthus amarus* was found to have significant antioxidant activity (Karuna et al., 2009).

The plant also exhibits antigenotoxic properties. *Phyllanthus amarus* extract has been reported to inhibit chromosomal damage induced by genotoxic agents (Gowrishankar and Vivekanandan, 1994).

The plant extract has been shown to inhibit mutagenesis caused by various carcinogens or chemicals in *Salmonella typhimurium*. Further, the urinary mutagenesis induced in rats by the administration of benzo[α]pyrene is significantly inhibited by oral administration of *P. amarus* extracts. Mutagens such as aflatoxin B_1 and acetamidofluorene need activation by liver microsomal enzymes for their carcinogenic activity. *Phyllanthus amarus* inhibits the activation of carcinogens and thereby inhibits carcinogenesis (Raphael, 2004).

Phyllanthus amarus has been shown to inhibit liver carcinogenesis caused by N-nitrosodiethylamine (NDEA) in rats (Joy and Kuttan, 1998; Raphael, 2004). *Phyllanthus amarus* extract is reported to increase the life span of rats with hepatocarcinoma (Rajeshkumar and Kuttan, 2000). The extract has antitumor, anticarcinogenic, and radioprotective activities (Rajeshkumar et al., 2002; Harikumar and Kuttan, 2007). *Phyllanthus amarus* was found to inhibit topoisomerase I and II and cdc25 tyrosine phosphatase, and this could be the mechanism of its anticancer activity (Raphael, 2004). Hydrolyzable tannins such as amarin and geraniin, corilagin, and 1,6-digalloyl fucopyranoside as well as rutin and quercetin-3-O-glucopyranoside are reported to be present in *P. amarus* extracts. Some of the hydrolyzable tannins present in *P. amarus* are shown to inhibit protein kinases such as CDPK (calcium dependent protein kinase) and protein kinase C (PKC) (Polya et al., 1995).

Sasidharan (2004) indicated that *P. amarus* Schum. & Thonn. and *P. fraternus* Webstar are the same species; both are found in South India. However, pharmacognostic studies have shown that *P. fraternus* is different from *P. amarus*. For example, phyllanthin and hypophyllanthin are present in *P. amarus* and not in *P. fraternus* (Khatoon, 2006). Therefore, *P. fraternus* could be a genetic variant or a different ecotype of *P. amarus*. In the literature, the hepatoprotective studies discussed next were reported under the name *Phyllanthus fraternus* Webster.

In traditional medicine, this plant is used to treat viral hepatitis and many other ailments (Rao, 2000). Experiments on mice and rats have shown that the extracts of this plant could protect the liver from hepatotoxic chemicals (Padma and Setty, 1999). The plant extracts protected rats from carbon tetrachloride-induced liver injury. When different fractions of alcoholic extracts of aerial parts and roots of *P. fraternus* were tested for antihepatotoxic activity on carbon tetrachloride-induced liver damage in albino rats, the methanol extract was found to be the most active. The degree of protection was measured using biochemical parameters like serum glutamate oxaloacetate transaminase (SGOT), serum glutamate pyruvate transaminase (SGPT), total protein, and total albumin. The hepatoprotective effects of the methanol fraction were further supported by significant recovery of hepatocytes in the histopathological study of the liver showing almost complete normalization of the tissues as neither the fat accumulation nor the necrosis was observed in the extract-treated rats (Ahmed et al., 2002). It has been shown that the plant can protect liver mitochondria from allyl alcohol-induced oxidative stress. Administration of rats with an aqueous extract (100 mg/kg) of *P. fraternus*, prior to allyl alcohol administration, protected rats to a large extent from the decrease in mitochondrial respiration and increase in the generation of superoxide radicals caused by allyl alcohol (Sailaja and Setty, 2006).

Realizing the importance of this plant material as a hepatoprotective agent, *in vitro* micropropagation methods were developed for it (Rajasubramoniam and Saradhi, 1997).

9.3.2 *Phyllanthus debilis* Klein Ex Willd.

The *Phyllanthus debilis* Klein ex Willd. plant species is a herb that is seen in degraded moist and dry deciduous forests and forest plantations 900–1,220 m above sea level in fallow fields and riverbanks. It is distributed in India (including Andamon and Nicobar Islands), Burma, China, Malaysia, Sri Lanka, and elsewhere. This is a common herb (Sasidharan, 2004).

Phyllanthus debilis is used in certain remote villages in India as a hepatoprotective folk medicine. This plant is reported to have a hepatoprotective effect against carbon tetrachloride-induced hepatotoxicity in rats (Sane et al., 1995). However, these reports remain to be confirmed (Subramoniam, 1995).

9.3.3 *Phyllanthus emblica* L

Phyllanthus emblica is a tree generally found throughout the tropics in dry and moist deciduous forests above 800 m sea level. This plant is commonly seen and is also cultivated. The fruit of this plant is edible (used to prepare pickles, etc.) and an ingredient of several Ayurvedic hepatoprotective formulations and preparations (Warrier et al., 2002).

Studies have shown the hepatoprotective activity and several other pharmacological properties of fruits of this plant. The fruit is well known for its rich vitamin C and polyphenol contents. Polyphenols present in the fruit of this plant possess very high antioxidant activity (Khopede, 2001). *Phyllanthus emblica* fruit and quercetin isolated from it exhibit hepatoprotective activity against toxic chemical-induced hepatic damage in rodents. Oral administration of a 50% alcohol extract (100 mg/100 g) of the fruit and quercetin (15 mg/100 g) isolated from it protected rats from hepatotoxic effects of the country-made liquor and mice from paracetamol overdose-induced hepatotoxicity (Gulati et al., 1995). *Phyllanthus emblica* fruit extract was shown to inhibit hepatotoxicity produced by acute and chronic administration of carbon tetrachloride in rats, as seen from the decreased levels of serum and liver lipid peroxides, glutamate pyruvate transaminase, and alkaline phosphatase. The study also suggested that the extract could inhibit liver fibrosis induced by chronic administration of carbon tetrachloride in rats (Jose and Kuttan, 2000). It was also shown to inhibit hepatotoxic and renotoxic effects of metals *in vivo* (Roy et al., 1991). Studies have shown that *P. emblica* fruit extracts could inhibit liver carcinogenesis induced by carcinogenic chemicals (Jeena et al., 1999). *Phyllanthus emblica* fruit extracts also influence carcinogen metabolism (Jose et al., 1998).

9.3.4 Phyllanthus kozhikodianus Sivarajan & Manilal

Phyllanthus kozhikodianus Sivarajan & Manilal is synonymous to *P. rheedii* Wight (Gangopadhyay et al., 2004). This plant species was earlier considered different from *P. rheedei* Wight. However, they are now reported as the same species (Gangopadhyay et al., 2004).

This plant species is a herb, generally found in moist deciduous, semievergreen forests and forest plantations. This is distributed in Western Ghats and Eastern Himalayas (Sasidharan, 2004). *Phyllanthus kozhikodianus* also showed antihepatotoxic activity in experimental animals (Asha and Pushpangadan, 1998; Suresh and Asha, 2008). Plant extracts were shown to have hepatotoxic activity against both paracetamol and D-galactosamine. The alcohol extract of the plant was shown to normalize D-galactosamine-induced histological changes in the liver and alterations in serum marker enzymes for liver damage. The extract also showed marked antioxidant and choleretic activity in rats. The extract also decreased the levels of proinflammatory cytokines. These observations also suggest the possible mechanism of action of this herb as antioxidant activity and the downregulation of proinflammatory cytokines. *Phyllanthus kozhikodianus* is used in ethnomedicine in many parts of Kerala State as a hepatoprotective herb. Scientific studies validated the ethnomedical claim. Phytochemical studies have shown the presence of tannins, flavonoids, and phenolics as major classes of compounds in the extract (Asha et al., 2007).

9.3.5 Phyllanthus maderaspatensis L

The *Phyllanthus maderaspatensis* L. plant species is a herb or small undershrub, generally found in moist as well as dry deciduous forests and plains from the coasts up to 800–1,000 m above sea level. It is seen in Deccan and Carnatic on dry lands, especially black cotton soils and near seacoasts. It is distributed in tropical Africa, Arabia, Sri Lanka, India, China, Java, and Australia (Sasidharan, 2004).

Asha and coworkers established the antihepatotoxic activity of *P. maderspatensis* against experimentally induced liver injury in rats (Asha and Pushpangadan, 1998; Asha et al., 2004, 2007). The whole-plant extracts (n-hexane, ethyl alcohol, and water) showed significant hepatoprotective activity against acetaminophen-induced hepatotoxicity as judged from serum levels of glutamate oxaloacetae transaminase (GOT), glutamate pyruvate transaminase (GPT), and alkaline phosphatase. The water and ethyl alcohol extracts showed moderate activity, whereas the n-hexane extract was very active even at a low dose of 1.5 mg/kg. The antihepatotoxic activity of the hexane extract was found to be better than that of silymarin, a standard hepatoprotective herbal drug. The extract also exhibited choleretic activity in normal rats and *in vitro* hydroxyl radical scavenging activity and inhibition of lipid peroxidation (Asha et al., 2004). The hexane extract of *P. maderaspatensis* also showed antihepatotoxicity against carbon tetrachloride-induced liver damage in rats. The protective effect was evident from the serum biochemical parameters and histopathological analysis. The extract treatment to rats remarkably prevented the elevation of the serum transaminases, lactic dehydrogenase, and liver lipid peroxides in the toxin-treated rats. Glutathione (GSH) levels in the liver were significantly increased by the treatment with the extracts. Histopathological changes induced by carbon tetrachloride and thioacetamide were also significantly reduced by the extract treatment (Asha et al., 2007).

9.3.6 PHYLLANTHUS NIRURI L

The *Phyllanthus niruri* L. plant species originally belonged to the West Indies and is distributed in the Malay Peninsula and elsewhere (Ridley, 1967). This plant species is often confused with *P. amarus* Schum. & Thonn. (Sasidharan, 2004). *Phyllanthus amarus* is a commonly occurring species in India with a striking similarity to this species. It is possible that some of the pharmacological studies carried out in India on *P. amarus* could have been reported erroneously as *P. niruri* L. (voucher specimens were not preserved in the studies).

Experiments on mice and rats have shown that the extracts of *Phyllanthus niruri* could protect the liver from liver damage induced by various toxic chemicals (Reddy et al., 1993; Prakash et al., 1995). *Phyllanthus niruri* protected the liver from experimentally induced damage (Sreenivasa, 1985). Liver-protective effects of alcohol extract of this plant have been shown against carbon tetrachloride-induced liver injury in rats as judged from significant reversal of the elevated serum levels of transaminase (GPT and GOT). The hepatoprotective effect of this plant extract against CCl_4 was found to be better than that of *P. urinaria* (Prakash et al., 1995). Among phyllanthin (a diaryl butane lignan), hypophyllanthin, triacontanal, and tricontanol isolated from a hexane extract of *P. niruri*, phyllanthin and hypophyllanthin protected primary cultured rat hepatocytes against carbon tetrachloride- and galactosamine-induced hepatotoxicity; triacontanal protected only against galactosamine-induced toxicity (Syamasundar et al., 1985).

However, a protein isolate from the herb, when injected intraperitoneally in mice either prior to (preventive) or after the induction of toxicity (curative), has also been shown to protect the liver from carbon tetrachloride intoxication (Bhattacharjee and Sil, 2006b). Administration of CCl_4 increases activities of serum GPT and alkaline phosphatase along with increases in lipid peroxidation and reduces the levels of antioxidant enzymes (superoxide dismutase and catalase) in the liver. Treatment with the protein extract almost restored these changes to normal levels. The hepatoprotective effect of the protein isolate was also evident from histological studies of liver (Bhattacharjee and Sil, 2006b). Also, the protein isolate showed hepatoprotective activity against acetaminophen- and nimesulide-induced liver damages in mice (Bhattacharjee and Sil, 2006a; Chatterjee and Sil, 2006). The protective effects of the protein isolate from *P. niruri* are attributed to a large extent to its antioxidant activity, as judged from a marked reduction in the lipid peroxide levels and increase in the levels of GSH and activities of antioxidant enzymes in the protein-treated intoxicated mice (Bhattacharjee and Sil, 2006b; Chatterjee and Sil, 2006). The aqueous leaf extracts of *P. niruri* and *P. emblica* inhibited the cytotoxic action in the liver caused by lead nitrate and aluminum sulfate in mice (Dhir et al., 1990). This plant extract has antihepatitis viral activity also (Venkateswaran et al., 1987).

9.3.7 PHYLLANTHUS RETICULATUS POIR

Phyllanthus reticulatus is distributed throughout India. It is a large glabrous subscandent shrub. *Phyllanthus reticulatus* is used in traditional medicine to treat various ailments, including liver disorders, and as an astringent, diuretic, stomachic, constipating agent, and so on in India and Pakistan; it is considered useful for treatment of gastropathy, strangury, sores, burns, diarrhea, and the like (Warrier et al., 2002).

It has been shown that *P. reticulatus* has hepatoprotective activity. Fractions obtained from the ethanol extract of aerial parts of this plant have been shown to protect rats from carbon tetrachloride-induced liver damage. The carbon tetrachloride-intoxicated rats receiving the fractions showed promising hepatoprotective activity, as evident from significant changes of pentabarbitone-induced sleeping time, changes in serum marker enzymes for liver damage (glutamate oxaloacetate transaminase, glutamate pyruvate transaminase and alkaline phosphatase), and bilirubin and from histological changes (Das et al., 2008).

9.3.8 PHYLLANTHUS URINARIA L

Phyllanthus urinaria is a herb found in hills 500–1,400 m above sea level, frequently on a thin layer of soil on exposed rocks. This is commonly distributed in pantropic areas and found as a weed in cultivated areas in India (Wealth of India, 2003).

It is a traditional hepatoprotective herb. The plant is used in traditional medicine in the same way as *P. fraternus*, often as a substitute for it (Wealth of India, 2003). This herb is shown to ameliorate the severity of nutritional steatohepatitis both *in vitro* and *in vivo* (Shen et al., 2008). The alcohol extract of this plant has been shown to prevent carbon tetrachloride-induced liver damage in rats, as judged from a reduction in serum marker enzymes for liver damage (Prakash et al., 1995). *Phyllanthus urinaria* attenuates acetaminophen-induced hepatotoxicity. Oral administration of *P. urinaria* extract to mice challenged with a fatal dose of acetaminophen resulted in prevention of mortality and rendered hepatoprotection. Histopathological analysis showed that the extract may protect the hepatocytes from acetaminophen-induced necrosis. The therapeutic dose of the plant extract did not show any toxicological phenomena in mice. Immunohistochemical staining with the cytochrome P450 CYP2E1 antibody revealed that the extract reduced the cytochrome P450 CYP2E1 protein level in mice pretreated with a lethal dose of acetaminophen. *Phyllanthus urinaria* extract also inhibited cytochrome P450 CYP2E1 enzymatic activity *in vitro* (Hau et al., 2009). The major components of the extract were identified as corilagin and gallic acid (Hau et al., 2009). The plant has strong phenolic antioxidants. The antioxidant activities and mushroom-tyrosinase inhibitory activities of phenolic compounds isolated from this plant have been shown (Xu et al., 2007). These properties may contribute to the hepatoprotective property of this plant.

9.4 MECHANISM OF ANTIHEPATOTOXIC ACTIVITIES OF *PHYLLANTHUS* SPECIES

The antihepatotoxic activity of *Phyllanthus* species, to a large extent, could be due to their antioxidant and anti-inflammatory activities. Almost all of the hepatotoxic chemicals cause oxidative stress and inflammation of the liver. Most of the antihepatotoxic *Phyllanthus* species are reported to have anti-inflammatory or antioxidant properties. Therefore, these activities could be involved to a major extent in protecting the liver from toxic chemicals.

9.4.1 ANTIOXIDATIVE STRESS AND HEPATOPROTECTION

One of the mechanisms of action of *Phyllanthus* species is the antioxidant activity of certain phytochemicals present in these plants. In addition to this, these plants may have specific hepatoprotective actions.

The involvement of free radicals in the pathogenesis of liver disorders has been established (Poli, 1993). Increased lipid peroxidation by OH radical and hydroperoxides in experimental acute and chronic alcoholic liver diseases has been shown by several studies (Farzaneh and Moore, 2001; Lieber, 2000). Reactive oxygen species and lipid peroxidation may play a role in the pathogenesis of liver fibrosis. Plasma levels of antioxidant vitamins are generally low in patients with chronic cholestatic liver diseases (Floreani et al., 2000).

Prolonged consumption of excess amounts of alcohol causes severe damage to many tissues, especially the liver, which may become cirrhotic in humans. Excess alcohol intake is an important cause of hepatic cirrhosis, which is sometimes complicated by hepatocellular carcinoma. Alcohol may also have an independent effect on the risk of this cancer. Hepatic iron overload is common in patients with alcoholic liver diseases. Ethanol increases lipid peroxidation in the livers of rats and baboons. Large doses of ethanol decrease GSH levels in liver. A fall in GSH leads to increased lipid peroxidation. Ethanol toxicity to liver, to a large extent, is attributed to an increase in the levels of acetaldehyde. *Phyllanthus amarus* inhibits lipid peroxidation, superoxide radical generation, and hydroxide radical production *in vitro*. The antioxidant property of this plant extract is attributed to polyphenols present in the plant extract. Proven antioxidant phenolic compounds isolated from *P. amarus* include catechin, epi-catachin, gallo-catechin, and epigallo catechin gallate (Hara, 1990). Antioxidant flavonoids such as quercetin and otsragalin are also present in *Phyllanthus* species (Nara et al., 1997). The plant contains amarinic acid and related elagic tannins with antioxidant activity (Foo, 1995). *Phyllanthus urinaria* also inhibited cellular lipid peroxidation and caspase 3 activation by a toxic level of doxorubicin. Endogenous antioxidants defenses, such as by total GSH, catalase, and superoxide dismutase, were favorably modulated by *P. urinaria* (Chularojmontri et al., 2005). *Phyllanthus emblica* fruit acts as a good antioxidant against radiation-induced lipid peroxidation (Khopede, 2001). It is also shown that *P. emblica* fruit polyphenol is a more potent antioxidant than vitamin C. Almost all the reported hepatoprotective *Phyllanthus* species are known to have at least some level of antioxidant property.

9.4.2 ANTI-INFLAMMATORY ACTIVITY OF HEPATOPROTECTIVE PHYLLANTHUS SPECIES

Toxic chemicals, viruses, and so on induce inflammatory diseases of the liver. *Phyllanthus* species such as *P. niruri, P. emblica,* and *P. amarus* contain rutin and other compounds with anti-inflammatory activity (Calixto et al., 1998; Kiemer et al., 2003; Kassuya et al., 2005, 2006; Asmawi and Moilanen, 1997). The extracts of antihepatotoxic *P. reticulatus* are also reported to have anti-inflammatory activity. In the carrageenan-induced mice paw edema model, the methanol extract of this plant at a dose of 300 mg/kg showed 40% inhibition of edema at the end of 4 h (Saha et al., 2007).

Phyllanthus urinaria, P. niruri, and other species are known to have phytochemicals such as quercetin, which is an inhibitor of phospholipase A_2 (Calixto et al., 1998). These inhibitors limit the production of arachidonic acid (AA) from phospholipids. AA is a precursor for the production of inflammatory metabolites. These compounds could help to reduce inflammation in the liver.

Hepatotoxins such as overdose of paracetamol and D-galactoseamine stimulate the production of proinflammatory cytokines. Interestingly, hepatotoxin-induced upregulation of the messenger RNA (mRNA) levels of proinflammatory cytokines, tumor necrosis factor α, and transforming growth factor β (TGF-β) was normalized by *P. rheedi* extract treatment of rats (Suresh and Asha, 2008). The anti-inflammatory mechanism of *P. amarus* appears to be its inhibitory effects on inducible nitric oxide synthase (iNOS), cyclo-oxygenase 2 (COX-2) and cytokines via the nuclear factor kappa B (NF-κB; an important transcription factor) pathway (Kiemer et al., 2003). The hepatoprotective *P. urinaria* also significantly inhibited NF-κB activation induced by doxorubicin (Chularojmontri et al., 2005).

9.5 CONCLUSION

Although the hepatoprotective properties of *Phyllanthus* species such as *P. amarus, P. niruri,* and *P. maderaspatensis* are well established by scientific studies, safe and effective hepatoprotective drugs have not been developed in light of modern science. Further systematic studies involving mechanism of action and detailed toxicity evaluation in experimental animals and follow-up controlled clinical studies are urgently required to develop internationally acceptable and commercially viable medicines for liver diseases. Most of the species in the genus *Phyllanthus* are not screened against liver damage induced by various toxic chemicals. A systematic screening and follow-up studies could bring to light many more liver-protective *Phyllanthus* species. The interesting feature is that the hepatoprotective *Phyllanthus* species are endowed with several pharmacological properties and many active compounds. The interactions between these chemicals in bringing out special pharmacological properties remain to be studied in detail. These plant materials are attractive for developing safe and effective medicine against liver diseases induced by toxic chemicals and hepatitis viruses; they may also prevent chemical-induced carcinogenesis to a large extent. Comparative studies of various hepatoprotective *Phyllanthus* species are required to select the best based on efficacy and safety. *Phyllanthus* species, which act through different mechanisms, can be combined rationally to develop potential polyherbal (many hepatoprotective phytochemical-containing) formulations. These formulations not only could be more effective but also could combat liver damage caused by diverse agents. Focused and integrated studies of potential species are urgently required to develop lifesaving drugs to treat liver diseases.

ACKNOWLEDGMENTS

I express my deep sense of gratitude to Prof. M. Radhakrishna Pillai, director, Rajiv Gandhi Center for Biotechnology, for his encouragement and support. I also would like to thank Dr. A. Subramonium, director, Tropical Botanical Garden and Research Institute, for his critical evaluation of the manuscript. The assistance of Mr. V. Suresh, Miss N. Krishna Radhika, Mr. K. A. Krishnakumar, Mr. P. S. Sreejith, and Mrs. Ciji Varghese is also gratefully acknowledged.

REFERENCES

Ahmed, B., Al-howiriny, T.A., and Mathew, R. 2002. Antihepatotoxic activity of *Phyllanthus fraternus*. *Pharmazie* 57: 855–856.

Asha, V.V., and Pushpangadan, P. 1998. Preliminary evaluation of the anti-hepatotoxic activity of *Phyllanthus kozhikodianus, Phyllanthus maderaspatensis* and *Solanum indicum*. *Fitoterapia* 59: 255–259.

Asha, V.V., Akhila, S., Wills, P.J., and Subramoniam, A. 2004. Further studies on the anti-hepatotoxic activity of *Phyllantus maderaspatensis* Linn. *J. Ethnopharmacol.* 92: 67–70.

Asha, V.V., Sheeba, M.S., Suresh, V., and Wills, P.J. 2007. Hepato-protection of *Phyllanthus maderaspatensis* against experimentally induced liver injury in rats. *Fitoterapia* 78: 134–141.

Asmawi, M.Z., and Moilanen, E.1997. Anti-inflammatory activity of extracts from the leaves of *Phyllanthus emblica*. *Planta Med.* 63: 518–524.

APG (Angiosperm Phylogeny Group). 1998. An ordinal classification for the families of flowering plants. *Ann. Mo. Bot. Gard.* 85: 531–553.

Bhattacharjee, R., and Sil, P.C. 2006a. The protein fraction of *Phyllanthus niruri* plays a protective role against acetaminophen induced hepatic disorder vis its anti-oxidant properties. *Phytother. Res.* 20: 595–601.

Bhattacharjee, R., and Sil, P.C. 2006b. Protein isolate from the herb, *Phyllanthus niruri* L. (Euphorbiaceae), plays hepatoprotective role against carbon tetrachloride induced liver damage via its anti-oxidant properties. *Food Chem. Toxicol.* 45: 817–826.

Blumberg, B.S. 1998. Hepatitis B virus: search for plant derived antiviral. In *Medicinal Plants: their role in health and biodiversity*, ed. T.R. Tomlinson and D. Akerala, 7–12. Philadelphia: University of Pennsylvania Press.

Calixto, J.B., Santos, A.R., Cechinel Filho, V., et al. 1998. A review of the plants of the genus *Phyllanthus*: their chemistry, pharmacology, and therapeutic potential. *Med. Res. Rev.* 18: 225–258.

Chatterjee, M., and Sil, P.C. 2006. Hepatoprotective effects of aqueous extract of *Phyllanthus niruri* on nimesulide-induced oxidative stress in in vivo. *Indian J. Biochem. Biophys.* 43: 299–305.

Chularojmontri, L., Wattanapitayakul, S.K., Herunsalee, A., et al. 2005. Anti-oxidant and cardioprotective effects of *Phyllanthus urinaria* l. on doxorubicin-induced cardiotoxicity. *Biol. Pharm. Bull.* 28: 1169–1171.

Das, B.K., Bepary, S., Datta, B.K., et al. 2008. Hepatoprotective activity of *Phyllanthus reticulatus*. *Pakistan J. Pharm. Sci.* 21: 333–337.

Dhir, H., Roy, A.K., Sharma, A., et al. 1990. Protection offered by aqueous extracts of *Phyllanthus* species against cytotoxicity induced by lead and aluminium salts. *Phytother. Res.* 4: 172–176.

Farzaneh, F.R., and Moore, K. 2001. Nitric oxide and the Liver. *Liver* 21:161–174.

Floreani, A, Baragiotta, A., Martines, D., et al. 2000. Plasma antioxidant levels in chronic cholestatic liver diseases. *Aliment Pharmacol. Ther.* 14: 353–358.

Foo, L.Y. 1995. Amarinic acid and related elagi-tannins from *Phyllanthus amarus*. *Phytochemicals* 39: 217–224.

Gangopadhyay, M., Chakrabarty, T., and Balakrishnan, N.P. 2004. On the status of *Phyllanthus airyshawii* and *P. kozhikodianus* (Euphorbiaceae). *J. Econ. Taxon. Bot.* 28: 585–590.

Gowrishankar, B., and Vivekanandan, O.H. 1994. *In vivo* studies of a crude extract of *Phyllanthus amarus* L. in modifying the genotoxicity induced in *Vicia faba* L. by tannery effluents. *Mutat. Res.* 322: 185–192.

Gulati, R.K., Agarwal, S., and Agrawal, S.S. 1995. Hepatoprotective studies on *Phyllantyus emblica* Linn. and quercetin. *Indian J. Exp. Biol.* 33: 261–281.

Harikumar, K.B., and Kuttan, R. 2007. An extract of *Phyllanthus amarus* protects mouse chromosomes and intestine from radiation-induced damages. *J. Radiat. Res.* 48: 469–476.

Hara, Y. 1990. *Advances in food science and technology*. Tokyo: Nippon Shokuhin Kogyo, Gakkai Korin.

Hau, D.K., Gambari, R., Wong, R.S., et al. 2009. *Phyllanthus urinaria* extract attenuates acetaminophen induced hepatotoxicity: involvement of cytochrome P450 CYP2E1. *Phytomedicine* 16: 751–760.

Hoffman, P., Kathriarachchi, H., and Wurdack K.J. 2006. A phylogenetic classification of Phyllanthaceae (Malpighiales; Euphorbeaceae sensua lato). *Kew Bull.* 61: 37–53.

Jeena, K.J., Joy, K.L., and Kuttan, R., 1999. Effect of *Emblica officinalis, Phyllanthus amarus* and *Picrorrhiza kurroa* on n-nitrosodiethylamine induced hepatocarcinogenesis. *Cancer Lett.* 136: 11–16.

Jose, J.K., Bhattacharya, R.K., and Kuttan R. 1998. Effect of *Emblica officinalis* extract on hepato-carcinogenesis and carcinogen metabolism. *J. Clin. Biochem. Nutr.* 24: 133–138.

Jose, J.K., and Kuttan R. 2000. Hepatoprotective activity of *Emblica officinalis* and chyvanaprakash. *J. Ethnopharmacol.* 72: 135–140.

Joy, K.L., and Kuttan R. 1998. Inhibition by *Phyllanthus amarus* on hepatocarcinogenesis induced by N-nitrosodiethylamine. *J. Clin. Biochem. Nutr.* 24: 133–139.

Karuna, R., Reddy, S.S., Baskar, R., et al. 2009. Anti-oxidant potential of aqueous extract of *Phyllanthus amarus* in rats. *Indian J. Pharmacol.* 41: 64–67.

Kassuya, C.A., Leite, D.F., de Melo, L.V., et al. 2005. Anti-inflammatory properties of extracts, fractions and lignans isolated from *Phyllanthus amarus*. *Planta Med.* 7: 721–726.

Kassuya, C. A., Leite, D.F., de Melo, L.V., et al. 2006. Anti-inflammatory and anti-allodynic actions of the lignan niranthin isolated from *Phyllanthus amarus*. Evidence for interaction with platelet activating factor receptor. *Eur. J. Pharmacol.* 546: 182–188.

Khanna, A.K, Rizvi, F., and Chander, R. 2002. Lipid lowering activity of *Phyllanthus niruri* in hyperlipidemic rats. *J. Ethnopharmacol.* 82: 19–22.

Khatoon, S. 2006. Comparative pharmacognostic studies of three Phyllanthus species. *J. Ethnopharmacol.* 104: 79–86.

Khopede, S.M. 2001. Characterizing the anti-oxidant activity of amla (*Phyllanthus emblica*) extract. *Curr. Sci.* 81: 185–190.

Kiemer, A. K., Hartung, T., Huber, C., et al. 2003. *Phyllanthus amarus* has anti-inflammatory potential by inhibition of iNOS, COX-2, and cytokines via the NF-kappaB pathway. *J. Hepatol.* 38: 289–297.

Lee, C.Y., Peng, W.H., Cheng, H.Y., et al. 2006. Hepatoprotective effects of *Phyllanthus* in Taiwan on acute liver damage induced by carbon tetrachloride. *Am. J. Chin. Med.* 34: 471–482.

Lieber, C.S. 2000. Alcoholic liver disease: new insights in pathogenesis lead to new treatment. *J. Hepatol.* 32. 113–126.

Liu, J., Lin, H., and McIntosh, H. 2001. Genus *Phyllanthus* for chronic hepatitis B virus injection: a systematic review. *J. Viral Hepatitis* 8: 358–366.

Nara, T.K., Glyeye, J., Cerval, E.L., et al. 1997. Flavonoids of *Phyllanthus niruri, Phyllanthus urinaria* and *Phyllanthus orbiculatus*. *Plant Med. Phytother.* 11: 82–89.

Padma, P., and Setty, O.H. 1999. Protective effect of *Phyllanthus fraternus* against carbon tetrachloride induced mitochondrial dysfunction. *Life Sci.* 64: 2411–2417.

Poli, G. 1993. Liver damage due to free radicals. *Br. Med. Bull.* 49: 604–620.

Polya, G.M., Wang, B.H., and Foo, L.Y. 1995. Inhibition of signal regulated protein kinases by plant-derived hydrolysable tannins. *Phytochemistry* 38: 307–314.

Prakash, A., Satyan, K.S., and Wahi, S.P. 1995. Comparative hepato-protective activity of three *Phyllanthus* species, *P. urinaria, P. niruri* and *P. simplex* on carbon tetrachloride induced liver injury in the rat. *Phytother. Res.* 9: 594–596.

Rajasubramoniam, S., and Saradhi P.P. 1997. Rapid multiplication of *Phyllanthus fraternus*: a plant with anti-hepatitis viral activity. *Ind. Crops Prod.* 6: 35–40.

Rajeshkumar, N. V., Joy, K. L., Kuttan G., et al. 2002. Anti-tumor and anti-carcinogenic activity of *Phyllanthus amarus* extract. *J. Ethnopharmacol.* 81: 17.

Rajeshkumar, N.V., and Kuttan, R. 2000. *Phyllanthus amarus* extract administration increases the life span of rats with hepato-carcinoma. *J. Ethnopharmacol.* 73: 215–219.

Rao, K. C. 2000. *Materials for the Database of Medicinal Plants* 226–227. Bangalore, India: Karnataka State Council for Science and Technology.

Raphael R.K. 2004. Pharmacological activity of plant derived drugs and the mechanism of action (with special reference to *Phyllanthus amarus* Schum & Thonn). PhD diss., Mahatma Gandhi University, Kottayam, 79–84.

Reddy, B.P., Murthy, V.N., and Venkateshwarlu, V. 1993. Anti-hepato-toxic activity of *Phyllanthus niruri, Tinospora cordifolia* and *Ricinus communis*. *Indian Drugs* 30: 338–341.

Ridley, H.N. 1967. *The flora of Malay Peninsula*, Vol. 3. London: Reeve.

Roy, A.K., Dhir, H., and Sharma, A. 1991. Comparative efficacy of *Phyllanthus emblica* fruit extract and ascorbic acid in modifying hepatotoxic and renotoxic effects induced by metals *in vivo*. *Int. J. Crude Drug Res.* 29: 117–126.

Saha, A., Masud, M.A., Bachar, S.C., et al. 2007. The analgesic and anti-inflammatory activities of the extracts of *Phyllanthus reticulatus* in mice model. *Pharm. Biol.* 45: 355–359.

Sailaja, R., and Setty, O.H. 2006. Protective effect of *Phyllanthus fraternus* against allyl alcohol-induced oxidative stress in liver mitochondria. *J. Ethnopharmacol.* 105: 201–209.

Sane, R.T., Kuher, V.V., Mary, S.C., et al. 1995. Hepatoprotection by *Phyllanthus amarus* and *P. debilis* on CCl_4 induced liver dysfunction. *Curr. Sci.* 68:1242–1246.

Sasidharan, N. 2004. *Biodiversity documentation for Kerala, part 6: flowering plants*. Peechi, India: Kerala Forest Research Institute

Sebastian, T., and Setty, O.H. 1999. Protective effect of *Phyllanthus fraternus* against ethanol induced mitochondrial dysfunction. *Alcohol* 17: 29–34.

Shen, B., Yu, J., Wang, S., et al. 2008. *Phyllanthus urinaria* ameliorates the severity of nutritional steatohepatitis both *in vitro* and *in vivo*. *Hepatology* 47: 473–483.

Sivarajan, V.V., and Manilal, K.S. 1977. New species of *Phyllanthus* from Kerala. *Indian Botanic Soc.* 56: 165–168.

Sreenivasa, R.Y. 1985. Experimental production of liver damage and its protection with *Phyllanthus niruri* and *Capparis spinosa* (both ingredients of Liv52) in white albino rats. *Probe* 24: 117–119.

Subramoniam, A., 1995. Too many hepatoprotective drugs and a little hepato-protection. *Curr. Sci.* 69: 898–899.

Subramoniam, A., and Pushpangadan, P. 1999. Development of phytomedicines for liver diseases. *Indian J. Pharmacol.* 31: 166–175.

Suresh, V., and Asha, V.V. 2008. Preventive effect of ethanol extract of *Phyllanthus rheedii* Wight on D-galactoseamine induced hepatic damage in Winstar rats. *J. Ethnopharmacol.* 116: 447–453.

Syamasundar, K.V., Singh, B., Thakur, R.S., et al. 1985. Anti-hepatotoxic principles of *Phyllanthus niruri* herb. *J. Ethnopharmacol.* 14: 41–44.

Thyagarajan, S.P., and Jayaram, S. 1992. Natural history of *Phyllanthus amarus* in the treatment of hepatitis B. *Indian J. Med. Microbiol.* 10: 64.

Thyagarajan, S.P., Jayaram, S., Gopalakrishnan, V., et al. 2002. Herbal medicines for liver diseases in India. *J. Gastroentrol. Hepatol.* 17: S370–S376.

Thyagarajan, S.P., Subramoniam, S., and Thirunalasundari, T. 1988. Effect of *Phyllanthus amarus* on chronic carriers of hepatitic virus. *Lancet* 2: 764–766.

Venkateswaran, P.S., Millman, I., and Blumberg, B.S., 1987. Effects of an extract from *Phyllanthus niruri* on hepatitis B virus and wood chuck hepatitis viruses: *in vitro* and *in vivo* studies. *Proc. Natl. Acad. Sci. U S A* 84: 274–278.

Warrier, P.K., Nambiar, V.P.K., and Ramankutty, C. 2002. *Indian medicinal plants: a compendium of 500 species,* Vol. 4. Hyderabad, India: Orient Longman.

Wealth of India. 2003. *A dictionary of Indian raw materials and industrial products. Raw materials,* Vol. 8 [reprint], ed. A. Krishnamurthy. New Delhi, India: CSIR.

Xu, M., Zha, Z.J., Qin, X.L., et al. 2007. Phenolic anti-oxidants from whole plant of *Phyllanthus urinaria. Chem. Bio-diversity* 4: 2246–2250.

Yadav, N.P., Pal, A., Shanker, K., et al. 2008. Synergistic effect of silymarin and standardized extract of *Phyllanthus amarus* against CCl_4-induced hepato-toxicity in *Rattus norvigicus. Phytomedicine* 15: 1053–1061.

10 Anticancer Studies of *Phyllanthus amarus*

K. B. Harikumar and Ramadasan Kuttan

CONTENTS

10.1 Introduction 171
10.2 Effect of *P. Amarus* on Chemically Induced Carcinogenesis 172
10.3 Effect of *P. Amarus* on Transplanted Tumors 174
10.4 Effect of *P. Amarus* on Virally Induced Cancers 174
10.5 Mechanism of Action of *P. Amarus* Extract 178
    10.5.1 Inhibition of Phase I Enzymes 178
    10.5.2 Effect on Phase II Enzymes 178
    10.5.3 Inhibition of Mutagenicity 179
    10.5.4 Inhibition of Clastogenicity 179
    10.5.5 Inhibition of Adduct Formation with Cellular Macromolecules 179
    10.5.6 Induction of Apoptosis 179
    10.5.7 Effect on Transcription Factors 179
    10.5.8 Other Mechanisms 180
Acknowledgment 180
References 180

10.1 INTRODUCTION

Phyllanthus amarus Schumach and Thonn (family Phyllanthaceae) is known for its hepatoprotective (Naaz et al., 2007) and antiviral effects (Thyagarajan et al., 1988). The plant is widely used in different traditional systems of medicine to combat injuries related to the liver. It also exhibits a wide spectrum of pharmacological activities (Calixto et al., 1998; Bagalkotkar et al., 2006). Studies from our laboratory and by others have shown that the plant possesses significant antioxidant (Kumar and Kuttan, 2004), anti-inflammatory (Kassuya et al., 2006), antidiabetic (Raphael et al., 2002b), urolytic, gastroprotective (Raphael and Kuttan, 2003), antifungal, and antibacterial activities. In this chapter, we describe the anticancer effects of *P. amarus*.

10.2 EFFECT OF *P. AMARUS* ON CHEMICALLY INDUCED CARCINOGENESIS

Different plants of *Phyllanthus* species have been reported to inhibit carcinogenesis in various animal models. N-Nitrosodiethylamine- (NDEA) induced hepatic cancer is one of the best-known models to study the process of carcinogenesis in rodent models. In Wistar rats, treatment with NDEA resulted in 100% tumor incidence within 18–20 weeks. *Phyllanthus amarus* extract (PAE) was administered in a simultaneous manner as well as after the development of tumor. As evident from the biochemical and histopathological evaluations, we found that simultaneous treatment of PAE significantly reduced tumor formation in rats (Figure 10.1). Liver weight of normal animals showed 2.86 ± 0.21/100 g body weight, which was increased by NDEA treatment to 4.58 ± 0.88/100 g body weight; in the group treated with PAE (750 mg/kg body weight), it was reduced 3.22 ± 0.66/100 g body weight ($p < .01$). The activity of gamma glutamyl transpeptidase (GGT), which is a marker of neoplasm, was significantly elevated after NDEA treatment and was inhibited significantly by PAE treatment (Table 10.1). In those groups for which treatment started after the development of tumor, PAE significantly elevated the life span of hepatic tumor-bearing animals (Jeena et al., 1999; Rajeshkumar et al., 2002; Rajeshkumar and Kuttan, 2000).

PAE treatment was also found to inhibit the N-methyl-N-nitro-N-nitroso guanidine- (MNNG) induced forestomach carcinoma in rats. MNNG (1 mg/dose) was administered 5 days a week for 28 weeks consecutively to produce forestomach

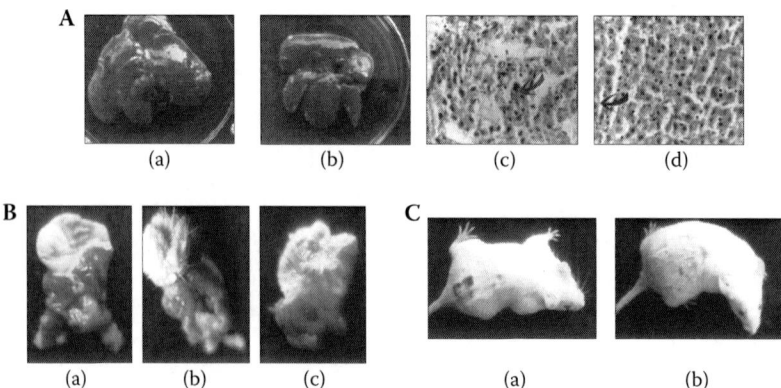

FIGURE 10.1 See color insert. (A) Inhibition of NDEA-induced hepatocarcinogenesis by PAE. (a) Gross morphology of the liver treated with NDEA at 20 weeks shows proliferation of tumor nodules; (b) the rat liver administered with NDEA and PAE extract at 20 weeks; (c) histopathology of the rat liver treated with NDEA at 20 weeks (×40 magnification); (d) histopathology of the rat liver (×40 magnification) treated with NDEA and PAE at 20 weeks. (B) Inhibition of MNNG-induced gastric cancer by PAE. Morphology of the stomach: (a) untreated; (b) treated with MNNG alone; (c) treated with MNNG plus *P. amarus* 750 mg/kg body weight. (C) Inhibition of DMBA-induced skin cancer by PAE: (a) representative picture from DMBA-alone-treated group; (b) DMBA plus *P. amarus* (5 mg/mouse).

TABLE 10.1
Effect of PAE Treatment along with NDEA on Serum Gamma Glutamyl Transpeptidase (GGT) Activity and Hepatic GSH Levels

Group	GGT Activity (U/L at 30°C)	Hepatic GSH (nmol/mg protein)
Untreated	35.3 ± 4.8	8.4 ± 0.6
NDEA alone	151.5 ± 21.4	16.2 ± 3.2
NDEA + *P. amarus* 750 mg/kg body weight	68.8 ± 14.2	10.6 ± 2.1

carcinoma in rats. Simultaneous administration of PAE at 150 and 750 mg/kg body weight was found to reduce the tumor burden (Figure 10.1B). Treatment with PAE decreased the elevated levels of GGT (Table 10.2). Treatment with MNNG suppressed the level of reduced glutathione (GSH) in the forestomach mucosa, and PAE treatment increased significantly the levels of GSH. The count of nucleolar organizer region-associated protein staining (AgNOR staining) is an indicator of the rate of protein synthesis in the cell. The counts were significantly elevated in the vehicle-treated group. PAE treatment also produced a significant decrease in AgNOR counts, demonstrating that the PAE treatment decreased proliferative activity within the cell (Raphael et al., 2006).

We also tested the efficacy of PAE treatment to inhibit sarcoma formation induced by 20-methylcholanthere (20-MC) in mice. 20-MC, which belongs to the class of polyaromatic hydrocarbons, and oxidized products of 20-MC form an adduct with DNA, which leads to tumor formation. PAE treatment inhibited the sarcoma formation as well as increased the life span of sarcoma-bearing animals.

The topical application of 7,12-dimethyl benz(a)nthrecene (DMBA) and croton oil (which act as a promoter) resulted in the formation of skin neoplasms, preferentially papillomas in mice skin. Topical application of PAE (methanolic fraction) at doses of 1 and 5 mg/mouse was found to significantly inhibit the skin papilloma formation (Figure 10.1). All the animals in the vehicle-treated group developed tumors by 20 weeks, while only 56% of the group treated with 5 mg PAE developed tumors at the 20th week. The average number of tumors per mouse in the vehicle-treated group at the 20th week was 6.2 ± 2.6, while in the groups treated with 1 and 5 mg PAE there were 3.4 ± 1.5 and 2.6 ±1.8 tumors, respectively (Raphael, 2004). Bharali et al. also

TABLE 10.2
Effect of PAE Treatment on MNNG-Induced Gastric Carcinoma

Group	GGT Activity (mmol/min/mg protein)	GSH (nmol/min/mg protein)
Untreated	1.6 ± 0.46	9.8 ± 1.2
MNNG alone	20.3 ± 6.7	4.6 ± 0.9
MNNG + *P. amarus* 750 mg/kg body weight	2.8 ± 0.9	8.5 ± 1.4

reported that oral administration of *P. niruri* extract at 1,000 mg/kg body weight inhibited the DMBA-induced papilloma formation (Sharma et al., 2009). Another species, *P. urinaria,* also decreased the papilloma formation (Bharali et al., 2003).

10.3 EFFECT OF *P. AMARUS* ON TRANSPLANTED TUMORS

Administration of PAE inhibited the ascites tumor development in mice induced by either DLA (Dalton's lymphoma ascites) or EAC (Ehrilch's ascites carcinoma) cell lines. In a DLA-induced ascites model, oral administration of PAE at doses of 60, 300, and 1,500 mg/kg body weight increased the survival of tumor-bearing animals to 23.00 ± 2.2, 25.33 ± 3.19, and 29.33 ± 4.20 days, respectively, and the percentage increase in survival was 8.69%, 20.61%, and 39.66%, respectively. Similarly, treatment with PAE also reduced the tumor volume in a solid tumor model in a dose-dependent manner. We also reported that administration of PAE along with cyclophosphamide (CTX) did not reduce the tumor-reducing potential of CTX (Kumar and Kuttan, 2005). In fact, there was a synergistic action of CTX and PAE in reducing the solid tumor burden in mice.

Phyllanthin and hypophyllanthin, which belong to the class of lignans, have been isolated from *P. amarus*. Both these compounds have been shown to inhibit the ascites tumor induced by EAC cells in mice. Treatment of mice with 50 and 100 mg/kg (intraperitoneally) of these compounds inhibited the tumor volume significantly and enhanced the life span of tumor-bearing animals as compared to untreated animals (Islam et al., 2008).

10.4 EFFECT OF *P. AMARUS* ON VIRALLY INDUCED CANCERS

The effect of different species of *Phyllanthus* was studied against several viruses. The antiviral activity of *Phyllanthus* species is described in detail elsewhere in this book.

We evaluated the effect of administration of PAE in a virally induced cancer model, specifically an erythroleukemia induced by Friend murine leukemia virus (FMuLv). This is one of the best- and well-studied animal model available to study the stepwise events that lead to initiation and progression of leukemia due to the reproducibility of the sequential genetic mutations leading to transformation of infected erythroblasts (Lee et al., 2003; Moreau-Gachelin, 2008). Moreover, this retrovirus infection model is also used in the search for the potential of medicinal preparations or establishing new treatment strategies. BALB/c mice were infected with FMuLv in the neonatal stage. Starting from the 14th day after viral challenge, PAE (250 and 750 mg/kg body weight) was administered orally (once daily) until day 70; after that, the animals were kept under observation. On every 30th day, we recorded changes in body weights of all the animals. The blood was collected from the caudal vein, and total white blood cell (WBC) count, red blood cell (RBC) count, and hemoglobin (Hb) levels were determined on day 90. The blood smear was stained by the periodic acid-Schiff method (PAS staining), which is an established method for differential staining of acute leukemia. Starting from the 30th day after the viral challenge, the vehicle-treated animals exhibited symptoms of anemia. The Hb levels (Figure 10.2C) were also decreased, indicating the persistence of anemia

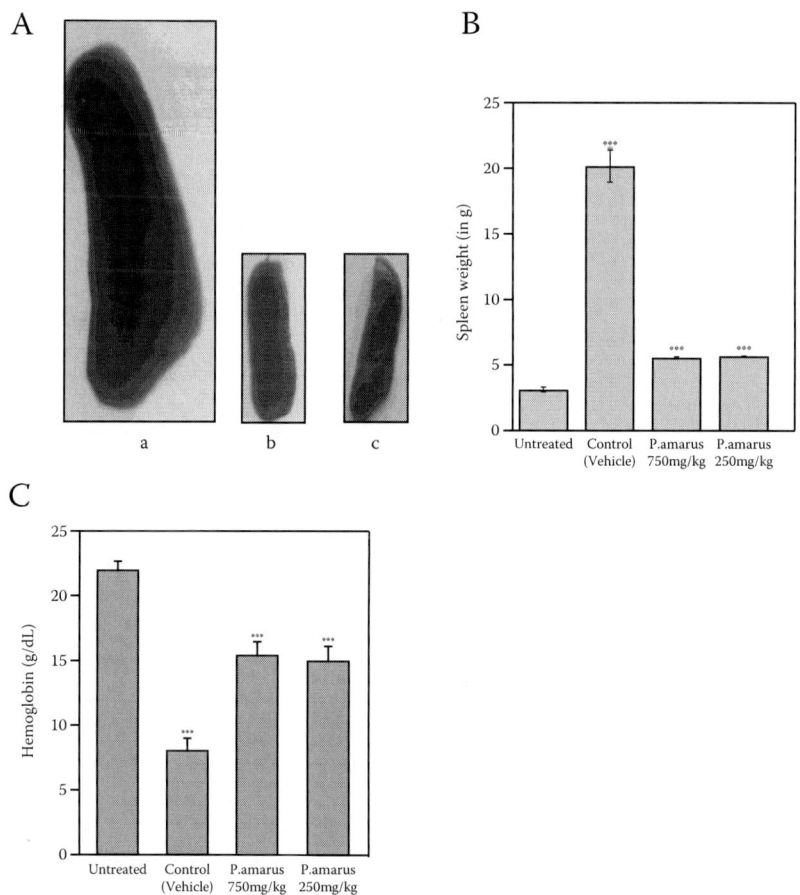

FIGURE 10.2 The effect of administration of PAE on the progression of erythroleukemia induced by FMuLv. (A) Spleen from mice treated with vehicle (a) showing massive enlargement in size; (b) the spleen of an untreated control animal of the same age group; and (c) animal treated with PAE 750 mg/kg body weight. (B) The effect of administration of PAE on spleen weights of leukemia-harboring mice. The vehicle-only-treated group was compared with the normal untreated group. The drug-treated groups were compared with the vehicle group. Data were expressed as mean plus or minus the standard deviation. ***$p < .001$ (analysis of variance followed by Dunnet's test). (C) The effect of administration of PAE on hemoglobin levels of leukemia-harboring mice. The vehicle-only-treated group was compared with the normal untreated group. The drug-treated groups were compared with vehicle group. Data were expressed as mean plus or minus the standard deviation. ***$p < .001$ (analysis of variance followed by Dunnet's test).

in FMuLv-treated animals. There was a decrease in the body weight of all the animals injected with FMuLv when compared to the normal untreated animals. The administration of PAE was found to significantly enhance the life span (Table 10.3) and body weight of erythroleukemia-harboring mice. The treatment of PAE has been shown to increase the Hb levels. The slides prepared from all the animals after PAS

TABLE 10.3
Effect of Administration of *P. amarus* on Survival Rate of Animals Harboring Erythroleukemia Induced by FMuLv

Group	Number of Days						
	30	60	90	105	120	150	180
Normal	5/5	5/5	5/5	5/5	5/5	5/5	5/5
Control	10/10	10/10	8/10	3/10	1/10	0/10	0/10
P. amarus 750 mg	10/10	10/10	10/10	9/10	8/10	6/10	2/10
P. amarus 250 mg	10/10	10/10	10/10	8/10	7/10	6/10	1/10

staining showed typical red color leukemic cells. The positive PAS staining confirmed the induction of leukemia by FMuLv. Moreover, we found that the administration of PAE was effective in decreasing the number of PAS stain-positive cells when compared to the untreated group.

A set of animals from each group was sacrificed on day 90 after viral challenge, and gross necropsy of each animal was performed. Massive enlargement in size and shape of the spleen was observed. The treatment with PAE was effective in decreasing the spleen weight (Figure 10.2B).

Histopathological evaluations of liver sections showed normal architecture in all the groups. The portal areas, sinusoidal spaces, endothelial cell population in the sinusoidal wall, Kupffer cells, and central venous system appeared to be normal. The staining of the spleen from the vehicle-treated group showed hyperplastic lymphoid follicles. Subcapsular sinus and sinusoidal spaces were filled with pleomorphic round or oval cells having the features of blast cells. The presence of degenerating cells and mitotic cells was also observed along with focal areas of hemorrhage and necrosis. In the PAE-administered group compared with that of the vehicle-treated group, these pathological changes were not significant. There were few leukemic cells infiltrating the sinusoidal space. Moreover, the number of mitotic cells was low, indicating that PAE treatment was effective in decreasing the progression of leukemia.

We also performed gene expression analysis of some of the genes involved in progression of leukemia in the spleen of treated and untreated animals. Bcl-2, which is an antiapoptotic gene (Howard et al., 2001), was overexpressed in vehicle-treated animals, and the PAE-administered group did not show any expression, indicating that it is capable of inhibiting the expression of Bcl-2. The tumor suppressor gene p53 was completely lost in the vehicle-treated animals, and PAE treatment has been shown to induce the expression of p53 (Figure 10.3). $p45^{NFE2}$, considered a negative regulator of erythroid differentiation (Howard et al., 2001), showed a complete loss in the vehicle-treated group, and PAE treatment induced expression of $p45^{NFE2}$. Raf-1 and ERK-1 were expressed in all the groups, but the rate of expression was high in the vehicle-treated group. Raf-1/ERK-1 pathway has been reported to play a major role in the FMuLv-induced transformation. The PAE-treated group showed expression of various intensities. The study clearly demonstrated the effectiveness of PAE in the inhibition of progression of FMuLv-induced leukemia (Harikumar et al., 2009).

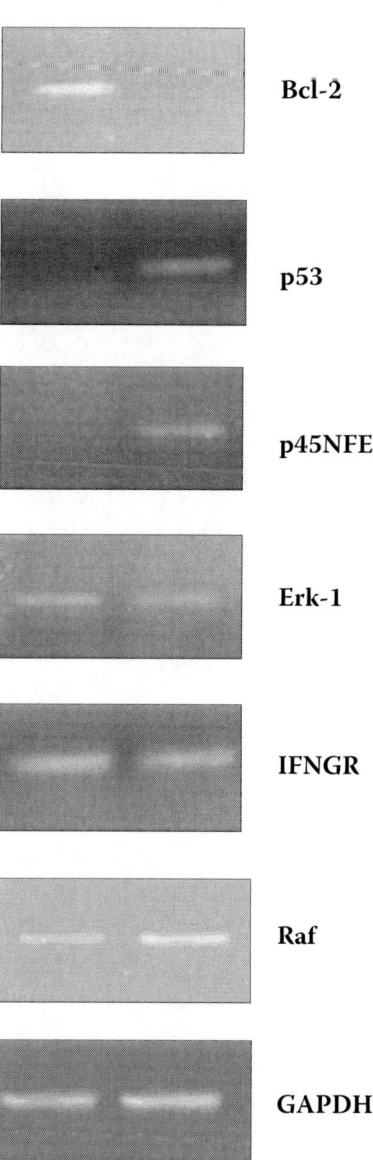

FIGURE 10.3 Gene expression profile of selected panel of genes in the spleen of leukemia-harboring mice at 90th day. The RNA was isolated from the spleen cells, and complementary DNA (cDNA) was prepared. The cDNA was amplified using RT (real time) polymerase chain reaction (PCR) and the amplified products visualized in agarose gels. GAPDH (glyceraldehyde phosphate dehydrogenase) was used as the housekeeping gene.

10.5 MECHANISM OF ACTION OF *P. AMARUS* EXTRACT

The process of carcinogenesis is complex and includes three major steps: initiation, progression, and promotion. In this section, we summarize the possible mechanism through which different species of *Phyllanthus*, in particular *P. amarus*, modulate different events associated with carcinogenesis, thereby inhibiting the process of carcinogenesis.

10.5.1 INHIBITION OF PHASE I ENZYMES

Phase I enzymes belong to the class of cytochrome P450 enzymes. These enzymes play a major role in the biotransformation of procarcinogens, drugs, steroid hormones, various environmental toxicants, and so on in the body. During this process, the procarcinogens become active carcinogens, and some of them directly form an adduct with DNA, thereby causing errors in DNA replication, which eventually leads to mutations. Also, during the conversion there is generation of highly reactive, electrophilic species that can directly react with the nucleophilic centers of macromolecules such as DNA to produce a series of damaging events that leads to initiation of a cancer phenotype. We have shown that PAE was effective in decreasing phase I enzymes both *in vitro* and *in vivo*. The IC_{50} (half-maximal inhibitory concentration) values of the PAE for CYP1A1, CYP1A2, and CYP2B1/2 were found to be 4.65 ± 0.05, 7.79 ± 0.06, and 4.22 ± 0.06 μg/ml, respectively, while that of aniline hydroxylase was 62.38 μg/ml and of aminopyrine-N-demethylase was more than 1,000 μg/ml (Hari Kumar and Kuttan, 2006). Using recombinant cytochrome proteins, it has been shown that the whole plant extract inhibits CYP1A2, CYP3A4, CYP2C9, and CYP2D6 enzymes, and IC_{50} values were 47.5 ± 1.27, 97.0 ± 18.74, 86.0 ± 5.26, and 48.5 ± 0.95 μg/ml, respectively (Appiah-Opong et al., 2008). Another member of the family *P. urinaria* also inhibited CYP2E1, with an IC_{50} value of 500 μg/ml (Hau et al., 2009). Oral administration of the PAE at a dose of 250 mg/kg body weight inhibited the phenobarbitone-induced Cyt P450 activity in liver by 50% (Hari Kumar and Kuttan, 2006). Based on these results, we are hypothesizing that the inhibition of enzyme activity *in vitro* could be due to the direct inactivation of the enzyme by the extract, and *in vivo* inhibition could be due to either direct inhibition of the activity of the enzyme or the inhibition of expression of genes responsible for coding these enzymes.

10.5.2 EFFECT ON PHASE II ENZYMES

Phase II enzymes, often termed carcinogen defense enzymes, are responsible for the detoxification of active carcinogen metabolites and reactive electrophiles in the body. A class of enzymes termed glutathione-S-transferases (GSTs) play an important role in the phase detoxification process. Different species of *Phyllanthus* have been shown to modulate the activity of GST enzymes. In aflatoxin B1-induced hepatic damage, treatment with PAE was shown to elevate the activity of GST (Naaz et al., 2007). We found that in several carcinogenic models treatment with PAE increased the activity of these enzymes. Some other members of the family, like *P. polyphyllus* (Rajkapoor et al., 2007), have increased the activity of GST in different *in vivo* carcinogenic models.

10.5.3 Inhibition of Mutagenicity

All the known carcinogens are well-established mutagens capable of inducing chromosomal aberrations and mutations in DNA, which eventually lead to carcinogenesis. PAE has been shown to inhibit the mutagenicity induced by both direct-acting and indirect-acting mutagens in a dose-dependent manner (Raphael et al., 2002a). Also, PAE treatment significantly inhibited the urinary mutagenicity produced in rats by administration of benzo(a)pyrene. A more detailed account of this subject is described elsewhere in the book.

10.5.4 Inhibition of Clastogenicity

Clastogens are agents that can cause structural and functional changes in the chromosome. Administration of PAE in mice has been shown to afford protection against radiation-induced damages in bone marrow cells. Moreover, PAE treatment decreased the percentage of micronucleated cells and aberrated cells in the bone marrow of irradiated animals (Harikumar and Kuttan, 2007).

10.5.5 Inhibition of Adduct Formation with Cellular Macromolecules

The metabolites of carcinogens form an adduct with cellular macromolecules such as DNA and proteins. Adduct formation with DNA resulted in the initiation of carcinogenesis. Several experimental and epidemiological studies showed that compounds with antioxidant potential were able to inhibit the carcinogen-DNA adduct formation, thereby inhibiting the initiation of carcinogenesis. *Phyllanthus* species are known for their potential antioxidant activity. Different species of *Phyllanthus* have been reported to inhibit the adduct formation. Moreover, some of the isolated compounds from *P. amarus,* such as ellagic acid and gallic acid (Baer-Dubowska et al., 1997), also inhibited the adduct formation.

10.5.6 Induction of Apoptosis

Induction of apoptosis can be another potential mechanism for the anticancer potential of *P. amarus*. PAE was found to induce apoptosis in DLA cells *in vitro.* This was evident from the morphology of the cells and the DNA ladder obtained after treatment with PAE (Harikumar et al., 2009). Treatment with the extract induces the expression of caspase 3 and inhibition of bcl-2. The extract could also alter the mitochondrial membrane potential in HepG2 (hepatocellular carcinoma) cells.

10.5.7 Effect on Transcription Factors

The transcription factor nuclear factor kappa B (NF-κB) has been shown to play an important role in cancer. It has been shown that PAE inhibited NF-κB signaling, NF-κB-dependent gene expression, and proinflammatory cytokine production in HepG2 cells. *Phyllanthus amarus* and related species *P. urinaria* and *P. deblis* were shown to inhibit the activation of NF-κB and COX-2, while the expression of proapoptotic Bax was elevated on treatment. The extracts also inhibited the production of the chemokine interleukin 8 (IL-8) in HepG2 cells (Sureban et al., 2006). The apoptosis induced by tumor necrosis factor (TNF) was synergistically increased by treatment with these extracts. These observations concluded there was potent apoptotic activity of *Phyllanthus* extract in tumor cells.

10.5.8 OTHER MECHANISMS

PAE (phyllanthus amarus extract) has been shown to inhibit cdc25 phosphatase activity, an enzyme that activates cyclin-dependent kinases and thereby enhances cell proliferation and oncogenesis. The IC_{50} was 25 µg/ml. However, *P. amarus* also inhibited the cdc2 kinase activity only at a higher concentration (IC_{50} > 1,000 µg/ml) (Rajeshkumar et al., 2002). Some hydrolyzable tannins purified from *P. amarus* are reported to be potent inhibitors of wheat embryo Ca^{2+}-dependent protein kinase (CDPK), rat brain Ca^{2+} protein kinase, phospholipid-dependent protein kinase C (PKC), and Ca^{2+} calmodulin-dependent myosin light chain kinase (Polya et al., 1995). Cancer cells have high levels of topoisomerase II activity. This enzyme catalyzes the interconversion of DNA topoisomers and plays a crucial role during DNA replication and transcription. The aqueous extract of *P. amarus* inhibited topoisomerase II activity in *Saccaromyces cereviceae* at a concentration of 250 µg/ml.

In conclusion, *P. amarus* exhibits significant anticancer activity either alone or in combination with existing chemotherapy agents. Several compounds have been isolated from *P. amarus,* and some of them (quercetin, rutin, kaempherol, ellagic acid, gallic acid, and lupeol) are known chemopreventive agents. *Phyllanthus amarus* also inhibited (1) the activation of phase I enzymes, (2) adduct formation of carcinogens with cellular macromolecules, (3) topoisomerase activity, and (4) NFκB and NFκB-linked gene expression and proinflammatory cytokine production. It also induced apoptosis in tumor cells, mainly via activation of caspase 3. All these studies clearly demonstrated the significant anticancer potential of the plants, which needs further evaluation in clinical settings.

ACKNOWLEDGMENT

One of the authors (K. B. H.) acknowledges ICMR, New Delhi, for financial support in the form of a senior research fellowship.

REFERENCES

Appiah-Opong, R., Commandeur, J.N., Axson, C., et al., 2008. Interactions between cytochromes P450, glutathione S-transferases and Ghanaian medicinal plants. *Food Chem. Toxicol.* 46: 3598–3603.

Baer-Dubowska, W., Gnojkowski, J., and Fenrych, W. 1997. Effect of tannic acid on benzo[a]pyrene-DNA adduct formation in mouse epidermis: comparison with synthetic gallic acid esters. *Nutr. Cancer* 29: 42–47.

Bagalkotkar, G., Sagineedu, S.R., Saad, M.S., et al., 2006. Phytochemicals from *Phyllanthus niruri* Linn. and their pharmacological properties: a review. *J. Pharm. Pharmacol.* 58: 1559–1570.

Bharali, R., Tabassum, J., and Azad, M.R. 2003. Chemopreventive action of *Phyllanthus urinaria* Linn on DMBA-induced skin carcinogenesis in mice. *Indian J. Exp. Biol.* 41: 1325–1328.

Calixto, J.B., Santos, A.R., Cechinel Filho, V., et al. 1998. A review of the plants of the genus *Phyllanthus*: their chemistry, pharmacology, and therapeutic potential. *Med. Res. Rev.* 18: 225–258.

Harikumar, K.B., and Kuttan, R. 2006. Inhibition of drug metabolizing enzymes (cytochrome P450) *in vitro* as well as *in vivo* by *Phyllanthus amarus* Schum & Thonn. *Biol. Pharm. Bull.* 29: 1310–1313.

Harikumar, K.B., and Kuttan, R. 2007. An extract of *Phyllanthus amarus* protects mouse chromosomes and intestine from radiation induced damages. *J. Radiat. Res.* 48: 469–476.

Harikumar, K.B., Kuttan, G., and Kuttan, R. 2009a. Inhibition of viral carcinogenesis by *Phyllanthus amarus*. *Integr. Cancer Ther.* 8, 254–260.

Hau, D.K., Gambari, R., Wong, R.S., et al. 2009. *Phyllanthus urinaria* extract attenuates acetaminophen induced hepatotoxicity: involvement of cytochrome P450 CYP2E1. *Phytomedicine* 16: 751–760.

Howard, J.C., Li, Q., Chu, W., et al. 2001. Bcl-2 expression in F-MuLV-induced erythroleukemias: a role for the anti-apoptotic action of Bcl-2 during tumor progression. *Oncogene* 20: 2291–2300.

Islam, A., Selvan, T., Mazumdar, U.K., et al. 2008. Antitumor effect of phyllanthin and hypophyllanthin from *Phyllanthus amarus* against Ehrlich ascites carcinoma in mice. *Pharmacology Online*, 2: 796–807.

Jeena, K.J., Joy, K.L., and Kuttan, R. 1999. Effect of *Emblica officinalis*, *Phyllanthus amarus* and *Picrorrhiza kurroa* on N-nitrosodiethylamine induced hepatocarcinogenesis. *Cancer Lett.* 136: 11–16.

Kassuya, C.A., Silvestre, A., Menezes-de-Lima, O., et al. 2006. Antiinflammatory and antiallodynic actions of the lignan niranthin isolated from *Phyllanthus amarus*. Evidence for interaction with platelet activating factor receptor. *Eur. J. Pharmacol.* 546: 182–188.

Kumar, K.B., and Kuttan, R. 2004. Protective effect of an extract of *Phyllanthus amarus* against radiation-induced damage in mice. *J. Radiat. Res.* 45: 133–139.

Kumar, K.B., and Kuttan, R. 2005. Chemoprotective activity of an extract of *Phyllanthus amarus* against cyclophosphamide induced toxicity in mice. *Phytomedicine* 12: 494–500.

Lee, C.R., Cervi, D., Truong, A.H., et al. 2003. Friend virus-induced erythroleukemias: a unique and well-defined mouse model for the development of leukemia. *Anticancer Res.* 23: 2159–2166.

Moreau-Gachelin, F. 2008. Multi-stage Friend murine erythroleukemia: molecular insights into oncogenic cooperation. *Retrovirology* 5, 99.

Naaz, F., Javed, S., and Abdin, M.Z., 2007. Hepatoprotective effect of ethanolic extract of *Phyllanthus amarus* Schum. et Thonn. on aflatoxin B1-induced liver damage in mice. *J. Ethnopharmacol.* 113: 503–509.

Polya, G.M., Wang, B.H., and Foo, L.Y. 1995. Inhibition of signal-regulated protein kinases by plant-derived hydrolysable tannins. *Phytochemistry* 38: 307–314.

Rajeshkumar, N.V., Joy, K.L., Kuttan, G., et al. 2002. Antitumour and anticarcinogenic activity of *Phyllanthus amarus* extract. *J. Ethnopharmacol.* 81: 17–22.

Rajeshkumar, N.V., and Kuttan, R. 2000. *Phyllanthus amarus* extract administration increases the life span of rats with hepatocellular carcinoma. *J. Ethnopharmacol.* 73: 215–219.

Rajkapoor, B., Sankari, M., Sumithra, M., et al. 2007. Antitumor and cytotoxic effects of *Phyllanthus polyphyllus* on Ehrlich ascites carcinoma and human cancer cell lines. *Biosci. Biotechnol. Biochem.* 71: 2177–2183.

Raphael, K.R., Ajith, T.A., Joseph, S., et al. 2002a. Anti-mutagenic activity of *Phyllanthus amarus* Schum & Thonn *in vitro* as well as *in vivo*. *Teratog. Carcinog. Mutagen.* 22: 285–291.

Raphael, K.R., and Kuttan, R. 2003. Inhibition of experimental gastric lesion and inflammation by *Phyllanthus amarus* extract. *J. Ethnopharmacol.* 87: 193–197.

Raphael, K.R., Sabu, M.C., Kumar, K.H., et al. 2006. Inhibition of N-methyl N'-nitro-N-nitrosoguanidine (MNNG) induced gastric carcinogenesis by *Phyllanthus amarus* extract. *Asian Pac. J. Cancer Prev.* 7: 299–302.

Raphael, K.R., Sabu, M.C., and Kuttan, R. 2002b. Hypoglycemic effect of methanol extract of *Phyllanthus amarus* Schum & Thonn on alloxan induced diabetes mellitus in rats and its relation with antioxidant potential. *Indian J. Exp. Biol.* 40: 905–909.

Raphael R.K. 2004. Pharmacological activity of plant derived drugs and the mechanism of action (with special reference to *Phyllanthus amarus* Schum & Thonn). PhD diss., Mahatma Gandhi University, Kottayam, India, 79–84.

Sharma, P., Parmar, J., Verma, P., et al. 2009 Anti-tumor activity of *Phyllanthus niruri* (a medicinal plant) on chemical-induced skin carcinogenesis in mice. *Asian Pac. J. Cancer Prev.* 10: 1089–1094.

Sureban, S.M., et al. 2006. Therapeutic effects of *Phyllanthus* species: induction of TNF-α-mediated apoptosis in HepG2 hepatocellular carcinoma cells. *Am. J. Pharmacol. Toxicol.* 1: 65–71.

Thyagarajan, S.P., Subramanian, S., Thirunalasundari, T., et al. 1988. Effect of *Phyllanthus amarus* on chronic carriers of hepatitis B virus. *Lancet* 2: 764–766.

11 Anticancer Activity of *Phyllanthus emblica*

Jeena Joseph and Ramadasan Kuttan

CONTENTS

11.1 Introduction 183
11.2 Inhibition of Chemical Carcinogenesis by EO 183
11.3 Inhibition of Transplanted Tumors by EO 186
11.4 Inhibition of Tumor Cell Proliferation by EO 186
11.5 Radioprotective Effect of EO 187
11.6 Mechanism of Action of EO Against Cancer 187
      11.6.1 Inhibition of Carcinogen-Metabolizing Enzymes 187
      11.6.2 Inhibition of Mutagenesis 188
      11.6.3 Inhibition of DNA Adduct Formation 188
      11.6.4 Inhibition of Clastogenicity 189
11.7 Effect of EO on Cell Growth and Multiplication 189
11.8 Effect of EO on Induction of Apoptosis 190
11.9 EO in Medicinal Preparations 190
References 190

11.1 INTRODUCTION

Emblica officinalis (EO) (syn *Phyllanthus emblica* L.) (Phyllanthaceae) has been used therapeutically in traditional Indian medicine. Fruits of EO have been used for the treatment of a wide variety of diseases, including diabetes, atherosclerosis, and cancer, and are well known for rejuvenating properties. Scientific evaluations have attributed antioxidant, anti-inflammatory, antiulcer, hepatoprotective, antipyretic, antibacterial, antiparasitic, and antiviral properties to this plant. EO is also proven to possess strong anticancer properties, described in detail in this chapter.

11.2 INHIBITION OF CHEMICAL CARCINOGENESIS BY EO

Carcinogenesis is a multistage process comprising initiation, promotion, and progression. Exposure to environmental and occupational chemicals is an important cause of chemical carcinogenesis. N-Nitrosodiethylamine (NDEA) has been considered

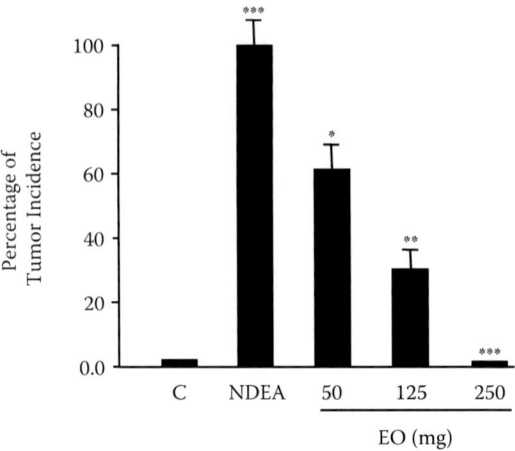

FIGURE 11.1 Effect of EO on NDEA-induced liver tumor. (From Jose, J.K., Kuttan, R., and Bhattacharya, R.K. 1998. Effect of *Embica officinalis* on hepatocarcinogenesis and carcinogen metabolism. *J. Clin. Biochem. Nutr.* 25: 31–39. With permission.)

a potent hepatocarcinogen, and its activity depends on its conversion of guanine to 8-hydroxyguanine by oxidative stress and formation of alkyl DNA adduct. Oral administration of 0.02% NDEA in water 5 days per week for 20 weeks resulted in 100% tumor incidence. Oral administration of EO extract dose dependently reduced the liver tumor incidence, and there were no tumors in the group treated with 250 mg EO extract (Figure 11.1). Biochemical and histopathological analysis indicated a strong inhibitory effect of EO extract on NDEA-induced hepatocarcinogenesis. The morphological pictures of a rat liver treated with- NDEA with- and without EO treatment were given in Figure 11.2 (a and b). The size and weight of the NDEA treated liver were significantly higher than those of the liver from EO treated rats. Histopathological analysis of the liver indicated that it was hepatic adrenocarcinoma of the cystic type. Normal architecture of the liver was preserved in EO treated tissues (Figures 11.2 c and d). Serum and liver gamma glutamyl transpeptidase (γ-GTP) activities were significantly reduced by the administration of EO extract. The level of normal serum γ-GTP was 34.28 ± 2.0 U/l, and this was increased to 135.5 ± 27.9 U/l in the case of control animals treated with NDEA only. The EO-treated group had a lower serum γ-GTP level, which was lowered dose dependently (99.8–44.8 U/l) (Table 11.1). This was also reflected in the liver γ-GTP profile (Jeena et al., 1999; Jose et al., 1997). An isolated polyphenolic fraction from EO was also found to inhibit NDEA-induced hepatocarcinogenesis in rats (Rajeshkumar et al., 2003).

In another report, tumors were induced by initiation with diethylnitrosamine (DEN) (200 mg/kg body weight intraperitoneally) followed by promotion with 2-acetylaminofluorine (2-AAF) (0.02% w/w in diet) for a continuous 6 weeks. EO treatment ameliorated the carcinogenic responses as it reversed the histopathological changes and reduced the number of γ-GT-positive foci and tumor formation in the liver (Sultana et al., 2008).

Subcutaneous injection of a single dose of 20-methylcholanthrene (20-MC; 20 μg/0.1 ml DMSO [dimethylsulfoxide]/mouse) resulted in development of sarcoma

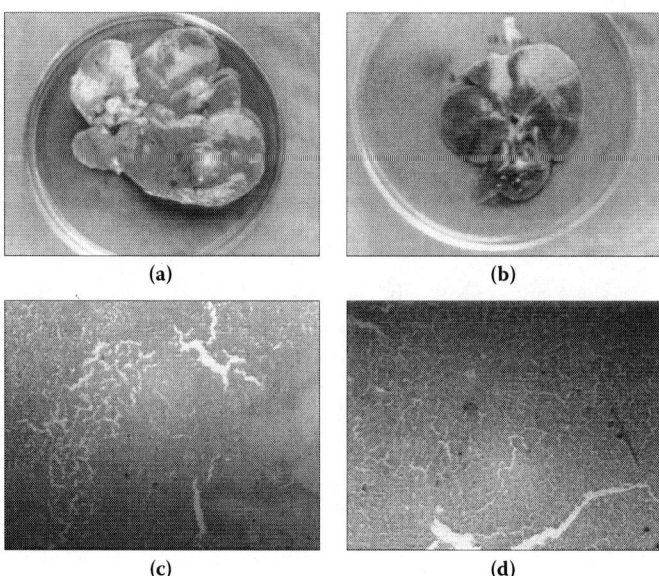

FIGURE 11.2 See color insert. (a) and (b) Effect of EO on the morphology of liver tumor induced by NDEA. (c) and (d) Effect of EO on the histopathology of liver tumor induced by NDEA. (From Jose, J.K., Kuttan, R., and Bhattacharya, R.K. 1998. Effect of *Embica officinalis* on hepatocarcinogenesis and carcinogen metabolism. *J. Clin. Biochem. Nutr.* 25: 31–39. With permission.)

TABLE 11.1
Effect of EO on NDEA-Induced Carcinogen-Metabolizing Enzymes

Group	Aniline Hydroxylase	GST (nmol/min/mg protein)	GHS-Glutathione (nmol/mg/ protein)	γ-GTP Serum (U/l)	γ-GTP Liver (nmol/min/mg)
Normal	0.21 ± 0.18	588 ± 55	8.5 ± 0.1	34.3 ± 2.0	0.08 ± 0.02
Control (NDEA alone)	0.47 ± 0.06	1,414 ± 137	20.4 ± 1.3	135.5 ± 27.9	2.90 ± 0.08
NDEA ± EO (50 mg)	0.41 ± 0.03	886 ± 29**	20.3 ± 1.4	99.8 ± 4.9	2.25 ± 0.05
NDEA + EO (125 mg)	0.35 ± 0.05*	648 ± 21**	17.5 ± 1.6	80.4 ± 6.6*	1.85 ± 0.28***
NDEA + EO) (250 mg)	0.29 ± 0.02**	532 ± 18**	15.2 ± 0.5**	44.8 ± 7.8***	0.38 ± 0.10***

Source: Jose, J.K., Kuttan, R., and Bhattacharya, R.K. 1998. *J. Clin. Biochem. Nutr.* 25: 31–39.
* $p<0.05$, ** $p<0.005$, *** $p<0.001$

in Swiss albino mice. 20-MC is a polyaromatic hydrocarbon that requires oxygen-dependent activation to attain its ultimate form. Oral administration of EO extract was found to inhibit sarcoma development in a dose-dependent manner. EO was also found to increase the life span of animals. Control animals started dying of their tumors after 75 days of carcinogen treatment, and all of the animals were dead by 160 days, whereas in the group treated with EO (10 mg/dose), only 5 animals were dead at 180 days (Jose et al., 1997).

Oral administration of EO extract was found to inhibit the two-stage process of skin carcinogenesis in Swiss albino mice induced by a single application of 7,12-dimethylbenzanthrazene (DMBA) (100 µg/100 µl) and 2 weeks later promoted by repeated application of croton oil (three times a week for 16 weeks). EO extract was administered simultaneously as well as after croton oil treatment. The tumor incidence, tumor yield, tumor burden, and cumulative number of papillomas were higher in the control group compared to the EO-treated group, thus indicating the chemopreventive potential of EO extract on DMBA-induced skin carcinogenesis (Sancheti et al., 2005). In a separate report, Kalpaamruthaa (in which EO is a major component) resulted in inhibition of DMBA-induced experimental mammary carcinoma. Female Sprague-Dawley rats were administered DMBA (25 mg/rat orally) dissolved in olive oil to induce mammary carcinoma. EO showed synergistic activity to mediate the inhibitory effect on mammary carcinogenesis (Arulkumaran et al., 2006).

11.3 INHIBITION OF TRANSPLANTED TUMORS BY EO

Aqueous extract of EO was cytotoxic to L929 cells in culture. The concentration needed for 50% inhibition was 16.5 µg/ml. EO extract reduced the ascites and solid tumors in mice induced by Dalton's lymphoma ascites (DLA) cells. EO extract also increased the life span of animals (Jose et al., 2001). Pyrogallol, an active ingredient from EO, inhibited the growth of lung cancer cells in a xenograft nude mice model (Yang et al., 2009). *Phyllanthus amarus* was also shown to inhibit the transplanted tumors (Rajeshkumar et al., 2003) in mice.

11.4 INHIBITION OF TUMOR CELL PROLIFERATION BY EO

EO extract showed a growth inhibitory property with a certain degree of selectivity against human hepatocellular carcinoma (HepG2) and lung carcinoma (A549). A synergistic effect of EO/doxorubicin or cisplatin (conventional cytotoxic agents) at different dose levels was found in A549 and HepG2 cells (Pinmai et al., 2008).

Active ingredients isolated from EO, such as norsesquiterpenoids and polyphenols, were also found to exert an antiproliferative effect in MK-1 (human gastric adenocarcinoma), HeLa (human uterine carcinoma), and B16F10 (murine melanoma) cells. Norsesquiterpenoid glycosides phyllaemblicins B and C exhibited significant inhibitory effects on B16F10 (GI50 at 2.0 and 3.5 mg/ml, respectively), HeLa (GI50 at 3.0 and 12.0 mg/ml, respectively), and MK-1 (GI50 at 7.0 mg/ml for both compounds) cell proliferation. The major phenolic compounds corilagin and geraniin as well as hydrolysable tannin elaeocarpusin and condensed tannin prodelphinidins

B1 and B2 showed the highest activity against B16F10 and HeLa cells (Zhang et al., 2004).

Extracts of EO and its active principle pyrogallol showed *in vitro* antiproliferative activity toward human tumor cell lines, including human erythromyeloid K562, B-lymphoid Raji, T-lymphoid Jurkat, and erythroleukemic HEL cell lines (Khan et al., 2002).

11.5 RADIOPROTECTIVE EFFECT OF EO

Radiotherapy is an important modality for cancer cure, and to obtain better tumor control with a higher dose of radiation, normal tissue should be protected against radiation injury. Ionizing radiation generates reactive oxygen radicals, which react with cellular macromolecules such as DNA, RNA, and protein and cause cell dysfunction and mortality. Nontoxic radioprotectors that could potentially protect humans against the deleterious effects of radiation in occupational and therapeutic settings are important in clinical radiotherapy. EO was found to be effective in protecting mice against hematological and biochemical modulation in peripheral blood (Singh et al., 2006). Pretreatment with EO extract offers protection of mouse jejunum against radiation-induced reduction in villus height, total cells, and mitotic figures/cript section (Jindal et al., 2009). EO protected intestinal mucosa against radiation-induced damage and increased the life span of animals (Singh et al., 2005). EO also modulated the hematopoietic system and antioxidant enzymes in irradiated animals (Hari Kumar et al., 2004).

11.6 MECHANISM OF ACTION OF EO AGAINST CANCER

Many potential cancer-protective agents can be broadly categorized as blocking agents, which impede the initiation stage, or suppressing agents, which arrest the promotion and progression of tumor, presumably by affecting or disturbing crucial factors that control cell proliferation, differentiation, or apoptosis. EO has shown strong anticancer properties that are mediated by several modes of action that inhibit the following: carcinogen-metabolizing enzymes, mutagenesis, DNA adduct formation, and clastogenicity.

11.6.1 INHIBITION OF CARCINOGEN-METABOLIZING ENZYMES

Many chemical carcinogens, such as benzo(a)pyrene [B(a)P], aflatoxins, and nitrosamines, have been shown to undergo cytochrome P-450-dependent activation to form the ultimate reactive species. Oxygen radicals are associated with the activation of carcinogens as well as the promotion of an initiated cell. Compounds that can inhibit this initiation step will be highly useful in chemoprevention strategies. We have shown that aqueous extract of EO is a potent antioxidant. EO extract was also shown to inhibit the microsomal P-450 enzymes (phase I enzymes) aniline hydroxylase and aminopyrene-N-demethylase both *in vitro* and *in vivo*. The concentration of EO needed for 50% inhibition of aniline hydroxylase was 0.43 mg/ml and for aminopyrene-N-demethylase was 0.30 mg/ml. Oral administration of EO at a dose of 250 mg/kg body weight inhibited NDEA-induced aniline hydroxylase in the liver by 80% (Jose et al., 1998, 1997). These results showed that EO inhibited the cytochrome P-450-dependent activation of carcinogens, thus preventing the ultimate reactive species and the initiation step, and will be highly useful in chemoprevention strategies.

Modulation of endogenous phase II enzymes such as glutathione-S-transferase (GST) abrogate oxidative stress through the scavenging of reactive oxygen species, and metabolism of reactive chemicals is a major mechanism of elimination of carcinogens. GST is a prominent phase II enzyme involved in detoxification, conjugation, and elimination of carcinogens. GST is used as a marker for evaluating anticarcinogenic potential and catalyzes electrophilic conjugation with GSH, thus counteracting a variety of carcinogens. Prophylactic treatment of the rat with *Emblica officinalis* defatted methanolic fruit extract prior and subsequent to 2-AAF in the diet resulted in significant downregulation of the enhanced GST by 72% compared to the 2-AAF-treated control group. Parallel to these changes, 2-AAF induced depletion of tissue GSH content showed marked recovery by 30% compared to 2-AAF-treated control values (Sultana et al., 2008). Oral administration of EO extract was found to inhibit the GST activity induced by NDEA in a dose-dependent manner (Jose et al., 1998).

11.6.2 Inhibition of Mutagenesis

EO extract inhibited the mutagenicity of direct-acting mutagens MNNG (N-methyl-N-nitro-N-nitroso guanidine) and 4-NPDA (nitro-O-phenyldiamine) and the activation of 2-AAF in a dose-dependent manner (Jose et al., 1997). In another report, aqueous extract of EO inhibited aflatoxin B1- (AFB1) and benzopyrene-induced mutagenicity (Sharma et al., 2000b). Their action is possibly mediated through interaction with microsomal-activating enzymes.

11.6.3 Inhibition of DNA Adduct Formation

The damage to DNA caused by the binding of reactive metabolites, which are generated by biotransformation of carcinogenic compounds, is one of the important mechanisms underlying carcinogenic activity. EO extract inhibited the DNA adduct formation induced by AFB1 and B(a)P (Figure 11.3). The concentration of EO required for 50% inhibition of microsome-catalyzed B(a)P adduct formation was 280 µg/ml, and that of AFB1 was 120 µg/ml (Jose et al., 1998). The extract could inhibit DNA adduct formation, which is a prerequisite for the carcinogen-induced mutagenesis and malignancy formation. Interference with metabolic activation of NDEA through inhibition of the mixed-function oxidase system and DNA adduct formation at early stages of NDEA treatment by the EO extract may be one of the important factors associated with the suppression of NDEA-induced hepatocarcinogenesis.

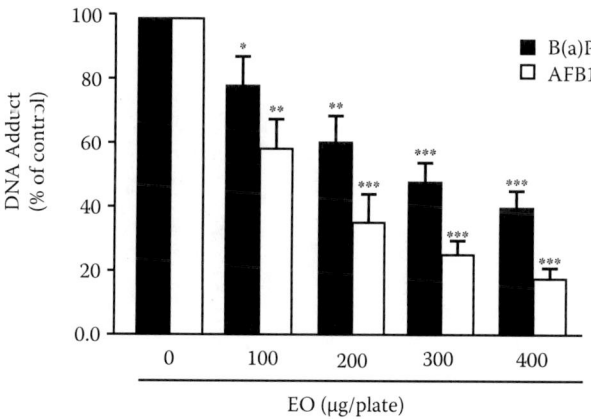

FIGURE 11.3 Effect of EO on microsome-catalyzed carcinogen-DNA adduct formation. (From Jose, J.K., Kuttan, R., and Bhattacharya, R.K. 1998. Effect of *Embica officinalis* on hepatocarcinogenesis and carcinogen metabolism. *J. Clin. Biochem. Nutr.* 25: 31–39. With permission.)

11.6.4 INHIBITION OF CLASTOGENICITY

EO inhibited the *in vivo* clastogenicity induced by B(a)P (Sharma et al., 2000a). Aqueous extract of EO and an equivalent amount of vitamin C protected mice from cesium chloride-induced clastogenicity (Ghosh et al., 1992). *In vivo* administration of EO-extract was alo found to protect mice bone marrow cells from micronuclei induced by metal clastinogen (Dhir et al., 1990, 1991; Roy et al., 1992).

These results suggest that EO can reduce cancer development by reducing the production of carcinogens.

11.7 EFFECT OF EO ON CELL GROWTH AND MULTIPLICATION

Loss of cell cycle control leads to unrestricted proliferation of cancer cells. Inhibition of cell cycle progression by regulating the cell cycle is a major mechanism of preventing oncogenesis. EO extract inhibited the proliferation of different cancer cells. EO extract inhibited cell cycle-regulating enzymes cdc25 phosphatase in a dose-dependent manner. Cdc25 (celldivision cycle) phosphatase controls P34 cdc2/cyclin B activation, which is an absolute requirement for G2/M transition in all dividing cells and is a potential target for antimitotic compounds. The con centration needed for 50% inhibition of cdc25 phosphatase was 5 µg/ml, and that needed for inhibition of cdc2 kinase was 100 µg/ml (Jose et al., 2001). Pyrogallol, an active principle from EO, resulted in G2/M arrest of lung cancer cells. Protein expressions of cyclin B1, cdc2, and cdc25c are decreased and inactive p-cdc25c is decreased on pyrogallol treatment (Yang et al., 2009). The results suggest that antitumor activity of EO extract may partially be due to its interaction with cell cycle regulation.

11.8 EFFECT OF EO ON INDUCTION OF APOPTOSIS

Recent detailed knowledge of molecular carcinogenesis provided the potential for therapeutic intervention in cancer by specifically targeting and sensitizing cancer cells to apoptosis. Polyphenolic fraction isolated from EO induced apoptosis in DLA (Dalton's lymphoma ascites) and CeHa cells. At 200 μg/ml dose, EOP induced membrane blebbing, chromatin condensation, and internucleosomal breaks, as evident from the morphology of the DNA ladder pattern (Rajeshkumar et al., 2003). Pyrogallol, an active constituent from EO, resulted in apoptosis in lung cancer cells (Yang et al., 2009).

11.9 EO IN MEDICINAL PREPARATIONS

EO is a major ingredient in many medicinal preparations, such as Rasayanas (Puri, 2002), Chyavanaprash, Triphala, Kalpaamrutha, and others and is described elsewhere in this book. Chyavanaprash, a drug preparation in which EO is a major component, is widely used as a health tonic and has been used for centuries to rejuvenate the body. It is claimed to reduce aging and age-related ailments such as loss of memory, digestive power, and weakness of sense organs. Preparation of Chyavanaprash involves making a decoction from 35 herbs, many of which are known as immunomodulators and antioxidants. Chyavanaprash was shown to be a potent free radical scavenger (Jose and Kuttan, 1995), thereby preventing carcinogenesis (Jose et al., 1997). It also inhibited the tumor burden (Jose et al., 2001).

Several active ingredients have been reported to be present in EO. Ascorbic acid, tannins, flavonoids, trigalloyl glucose, ellagic acid, and phyllemblic acid have been reported in EO. The protection offered by EO may be attributable to the combined effects of these various plant constituents rather than a single component. EO exerts its anticarcinogenic activity through the combined action via (1) removal of free radicals, (2) modulation of carcinogen-metabolizing enzymes, (3) inhibition of DNA adduct formation, and (4) inhibition of metal-induced clastogenicity. The combined evidence for its capacity to inhibit cancer cell proliferation, regulate cell cycle progression, and induce apoptosis in tumor cells gives strong support for its dietary inclusion as an important strategy in the prevention and treatment of cancer.

REFERENCES

Arulkumaran, S., Ramprasath, V. R., Shanthi, P., and Sachdanandam, P. 2006. Restorative effect of Kalpaamruthaa, an indigenous preparation, on oxidative damage in mammary gland mitochondrial fraction in experimental mammary carcinoma. *Mol. Cell. Biochem.* 291: 77–82.

Dhir, H., Agarwal, K., Sharma, A., and Talukder, G. 1991. Modifying role of *Phyllanthus emblica* and ascorbic acid against nickel clastogenicity in mice. *Cancer Lett.* 59: 9–18.

Dhir, H., Roy, A. K., Sharma, A., and Talukder, G. 1990. Modification of clastogenicity of lead and aluminium in mouse bone marrow cells by dietary ingestion of *Phyllanthus* emblica fruit extract. *Mutat Res.* 241: 305–312.

Ghosh, A., Sharma, A., and Talukder, G. 1992. Relative protection given by extract of *Phyllanthus emblica* fruit and an equivalent amount of vitamin C against a known clastogen—caesium chloride. *Food Chem. Toxicol.* 30: 865–869.

Hari Kumar, K. B., Sabu, M. C., Lima, P. S., and Kuttan, R. 2004. Modulation of haematopoietic system and antioxidant enzymes by *Emblica officinalis* gaertn and its protective role against gamma-radiation induced damages in mice. *J. Radiat. Res. (Tokyo)* 45: 549–555.

Jeena, K. J., Joy, K. L., and Kuttan, R. 1999. Effect of *Emblica officinalis*, *Phyllanthus amarus* and *Picrorrhiza kurroa* on N-nitrosodiethylamine induced hepatocarcinogenesis. *Cancer Lett.* 136: 11–16.

Jindal, A., Soyal, D., Sharma, A., and Goyal, P. K. 2009. Protective effect of an extract of *Emblica officinalis* against radiation-induced damage in mice. *Integr. Cancer. Ther.* 8: 98–105.

Jose, J. K., and Kuttan, R. 1995. Antioxidant activity of *Emblica officinalis*. *J. Clin. Biochem. Nutr.* 19: 63–70.

Jose, J.K., Kuttan, R., and Bhattacharya, R. K. 1998. Effect of *Emblica officinalis* on hepatocarcinogenesis and carcinogen metabolism. *J. Clin. Biochem. Nutr.* 25: 31–39.

Jose, J. K., Kuttan, G., George, J., and Kuttan, R. 1997. Antimutagenic and anticarcinogenic activity of *Emblica officianalis* Gaertn. *J. Clin. Biochem. Nutr.* 22: 171–176.

Jose, J. K., Kuttan, G., and Kuttan, R. 2001. Antitumour activity of *Emblica officinalis*. *J. Ethnopharmacol.* 75: 65–69.

Khan, M. T., Lampronti, I., Martello, D., Bianchi, N., Jabbar, S., Choudhuri, M. S., et al. 2002. Identification of pyrogallol as an antiproliferative compound present in extracts from the medicinal plant *Emblica officinalis*: effects on *in vitro* cell growth of human tumor cell lines. *Int. J. Oncol.* 21: 187–192.

Pinmai, K., Chunlaratthanabhorn, S., Ngamkitidechakul, C., Soonthornchareon, N., and Hahnvajanawong, C. 2008. Synergistic growth inhibitory effects of *Phyllanthus emblica* and *Terminalia bellerica* extracts with conventional cytotoxic agents: doxorubicin and cisplatin against human hepatocellular carcinoma and lung cancer cells. *World J. Gastroenterol.* 14: 1491–1497.

Puri, H. S. *Rasayana: Ayurvedic herbs for longevity and rejuvenation*. Boca Raton, FL: CRC Press, 22–40.

Rajeshkumar, N. V., Pillai, M. R., and Kuttan, R. 2003. Induction of apoptosis in mouse and human carcinoma cell lines by *Emblica officinalis* polyphenols and its effect on chemical carcinogenesis. *J. Exp. Clin. Cancer Res.* 22: 201–212.

Roy, A. K., Dhir, H., and Sharma, A. 1992. Modification of metal-induced micronuclei formation in mouse bone marrow erythrocytes by *Phyllanthus* fruit extract and ascorbic acid. *Toxicol. Lett.* 62: 9–17.

Sancheti, G., Jindal, A., Kumari, R., and Goyal, P. K. 2005. Chemopreventive action of *Emblica officinalis* on skin carcinogenesis in mice. *Asian Pac. J. Cancer Prev.* 6: 197–201.

Sharma, N., Trikha, P., Athar, M., and Raisuddin, S. 2000a. Inhibitory effect of *Emblica officinalis* on the *in vivo* clastogenicity of benzo[a]pyrene and cyclophosphamide in mice. *Hum. Exp. Toxicol.* 19: 377–384.

Sharma, N., Trikha, P., Athar, M., and Raisuddin, S. 2000b. *In vitro* inhibition of carcinogen-induced mutagenicity by *Cassia occidentalis* and *Emblica officinalis*. *Drug Chem. Toxicol.* 23: 477–484.

Singh, I., Sharma, A., Nunia, V., and Goyal, P. K. 2005. Radioprotection of Swiss albino mice by *Emblica officinalis*. *Phytother. Res.* 19: 444–446.

Singh, I., Soyal, D., and Goyal, P. K. 2006. *Emblica officinalis* (Linn.) fruit extract provides protection against radiation-induced hematological and biochemical alterations in mice. *J. Environ. Pathol. Toxicol. Oncol.* 25: 643–654.

Sultana, S., Ahmed, S., and Jahangir, T. 2008. *Emblica officinalis* and hepatocarcinogenesis: a chemopreventive study in Wistar rats. *J. Ethnopharmacol.* 118: 1–6.

Yang, C. J., Wang, C. S., Hung, J. Y., et al., 2009. Pyrogallol induces G2-M arrest in human lung cancer cells and inhibits tumor growth in an animal model. *Lung Cancer* 66: 162–168.

Zhang, Y. J., Nagao, T., Tanaka, T., et al., 2004. Antiproliferative activity of the main constituents from *Phyllanthus emblica*. *Biol. Pharm. Bull.* 27: 251–255.

12 The *In Vivo* and *In Vitro* Proapoptotic and Antiangiogenic Effects of *Phyllanthus urinaria*

Jong-Hwei S. Pang, Sheng-Teng Huang, Rong-Chi Yang, and Hsiao-Ting Wu

CONTENTS

12.1 Traditional Use of *Phyllanthus urinaria* and Related Studies 194
12.2 *Phyllanthus Urinaria* Reduces Viability of Cancer Cells But Not Normal Cells 194
    12.2.1 Induction of Apoptosis—Strategy for Anticancer Drug Development 195
12.3 Mechanism Underlying the Proapoptotic Effect of *P. urinaria* 196
12.4 The *In Vivo* Anticancer Effect of *P. urinaria* 197
    12.4.1 *Phyllanthus urinaria* Decreases *In Vivo* Angiogenesis in Tumor and CAM Assay 198
    12.4.2 *P. urinaria* Inhibited Angiogenesis-Related Functions of Vascular Endothelial Cells 198
    12.4.3 Matrix Metalloproteinase 2 Activity Inhibited by *P. urinaria* Both *In Vivo* and *In Vitro* 199
12.5 Chemical Analysis of *P. Urinaria* 200
    12.5.1 Ellagic Acid: The Major Compound in *P. urinaria* Exerts Antiangiogenic Effect 201
12.6 Conclusion 201
Acknowledgments 202
References 202

12.1 TRADITIONAL USE OF *PHYLLANTHUS URINARIA* AND RELATED STUDIES

Phyllanthus urinaria (*P. urinaria*), one of the herbal plants belonging to the genus *Phyllanthus* (*Euphorbiaceae*), is widely distributed in China, southern India, and South America. It has long been used in folk medicine for the treatment of several diseases, such as kidney and urinary bladder disturbances, intestinal infections, diabetes, and hepatitis B (Calixto et al., 1998). In an animal model, *P. urinaria* has been proven to be effective in protecting acetaminophen- or CCl_4-induced injuries of liver cells (Hau et al. 2009; Lee et al., 2006; Zhou et al., 1997); alleviating the steatohepatitis induced by methionine and choline deficiency (Shen et al., 2008); protecting doxorubicin-induced cardiotoxicity (Chularojmontri et al., 2005); relaxing the histamine-induced contraction of trachea (Paulino et al., 1996); producing pronounced systemic, spinal, and supraspinal antinociception (Santos et al., 1995); inducing the contractile response in the urinary bladder (Dias et al., 1995); and decreasing the blood glucose level in streptozotocin-induced diabetic rats (Higashino et al., 1992). Clinically, *P. urinaria* has been shown to seroconvert hepatitis B e-antibody status in patients from negative to positive (Wang et al., 1995). *Phyllanthus urinaria* has also been demonstrated to exhibit an inhibitory effect on the intracellular HBsAg formation in hepatoma cells (Ji et al., 1993) and the activity of retroviral reverse transcriptase (Suthienkul et al., 1993). In addition, *P. urinaria* was found to diminish HSV-2 (herpes simplex virus) infectivity (Yang et al., 2005). More important, no side effect or toxicity has been reported in any of these studies.

Although the traditional antitumor usage of *P. urinaria* has been described in *Practical Anti-Tumor Herb Medicine* (Charng, 1995), the pharmacological mechanism was not studied. To date, the anticancer effect of the genus *Phyllanthus* has only been reported in no more than 10 articles, as documented by PubMed. The chemopreventive property of *P. urinaria* against 7,12-dimethyl benz(a)nthrecene- (DMBA) induced skin papillomagenesis has been demonstrated in mice (Bharali et al., 2003). *Phyllanthus amarus* protects the liver from hepatocarcinogenesis induced by N-nitrosodiethylamine in animal models (Jeena et al., 1999). The root of *Phyllanthus acuminatus* has inhibited the growth of murine P-388 lymphocytic leukemia and B-16 melanoma cell lines (Powis and Moore, 1985; Pettit et al., 1990). Pure compound isolated from the ethylacetate extract of *P. urinaria* has exhibited anticancer activity by inducing apoptosis through the inhibition of telomerase activity and Bcl-2 expression (Giridharan et al., 2002). Our laboratory has focused our research on investigating the pharmacological mechanism of *P. urinaria*. We have demonstrated the anticancer effect in water extract prepared from *P. urinaria* on cancer cells originated from different tissues and in an animal model. We systematically introduce our works in this chapter.

12.2 *PHYLLANTHUS URINARIA* REDUCES VIABILITY OF CANCER CELLS BUT NOT NORMAL CELLS

At the beginning of our study on *P. urinaria*, we first demonstrated the anticancer effect of a water extract prepared from the whole plant of *P. urinaria* on reducing

TABLE 12.1
Cytotoxic Activity of *P. urinaria* on Human Cancer Cells[a]

Cell Line (Origin)	IC_{50}[b] (mg/ml)[c]
HL-60 (leukemia)	0.12 (0.015)
K562 (leukemia)	1.53 (0.045)
Molt-3 (leukemia)	0.35 (0.045)
HepG2 (liver carcinoma)	1.42 (0.025)
NPC-BM1 (nasopharyngeal carcinoma)	0.81 (0.065)
HT1080 (fibrosarcoma)	1.02 (0.025)

[a] Each cell type was incubated with various doses of *P. urinaria* for 24 hours at 37°C and subjected to trypan blue exclusion assay to measure IC_{50} values.
[b] IC_{50} is the dose at which 50% of cells are no longer viable.
[c] Mean (standard deviation) of triplicate assays.

cell viability of cancer cells. The viability of various cancer cells derived from different tissues, including HL-60, Molt-3, NPC-BM1, HT1080, HepG2, and K562 cells, is decreased dose dependently, with the IC_{50} (half-maximal inhibitory concentration) ranging from 0.12 to 1.53 mg/ml (see details in Table 12.1). More important, *P. urinaria* exerts no cytotoxic effect on normal cells even at higher concentration or for longer exposure. In addition to normal vascular endothelial cells and liver cells, human peripheral mononuclear cells were viable under the same conditions as reported in our previous studies (Huang et al., 2004a, 2004b). These cells are often vulnerable to commonly used cancer drugs that result in the various side effects during chemotherapy in cancer patients. This result confirms that the cytotoxic effect of an aqueous extract of *P. urinaria* is cancer specific. Together with the fact that *P. urinaria* has been used for a long time without any clinical side effects reported, the evidence is provided to support the potential application of *P. urinaria* as a safe anticancer drug.

12.2.1 INDUCTION OF APOPTOSIS—STRATEGY FOR ANTICANCER DRUG DEVELOPMENT

Apoptosis is an active and energy-dependent process of cellular self-destruction to maintain the homeostasis of the human body. This physiological process occurs when a proteolytic system responds to proapoptotic signals and triggers a cascade reaction. The cleavage of a set of proteins will lead to the typical morphological changes of apoptosis, such as cell condensation, membrane blebbing, and formation of apoptotic bodies. Failure or suppression of apoptosis is likely to result in the increase of immortal cells and later the development of cancer. Therefore, it is valuable to find potent agents capable of inducing apoptosis effectively in cancer cells and to use them as therapeutic anticancer drugs.

A variety of stimuli are known to be capable of triggering apoptosis in cancer cells, including irradiation, heat shock, nitric oxide, and some other chemicals. However, most of these stimuli often face the problem of poor specificity since normal cells are also damaged. Therefore, it renders such treatment less appropriate for clinical use. Finding a potential apoptosis-inducing drug with high specificity for cancer cells becomes the focus of related cancer research. Hundreds of natural products have been widely used in traditional and folk medicine for therapeutic anticancer purposes.

The administration of many natural compounds with anticancer effect has been shown to be capable of inducing the apoptotic death of cancer cells. Among them, our study has demonstrated that the water extract of *P. urinaria* could induce the apoptosis of various cancer cells, as demonstrated by the typical change of morphology, induction of DNA fragmentation, and activation of caspsase 3 enzyme. This proapoptotic effect of *P. urinaria* did not affect normal cells as expected.

12.3 MECHANISM UNDERLYING THE PROAPOPTOTIC EFFECT OF *P. URINARIA*

The molecular mechanism underlying the *P. urinaria*-induced apoptosis in cancer cells has been reported to be associated with an increased Bax/Bcl-2 ratio followed by a mitochondria-dependent apoptosis pathway (Giridharan et al., 2002; Huang et al., 2003). In addition, we have demonstrated that the gene expressions of both Fas receptor and Fas ligand are increased by the treatment of *P. urinaria* in HL-60 cells and tightly associated with the apoptosis event. It has been shown that ceramide can act as a downstream effector of the apoptotic pathway initiated by the early Fas signaling system (Furuke and Bloom, 1998; Grullich et al., 2000). Moreover, with doxorubicin treatment, ceramide has been found to induce the gene expression of Fas ligand in human leukemia cells (Herr et al., 1997). Ceramide has been demonstrated to induce mitochondrial activation and increase caspase activity and DNA fragmentation through a Bax/Bcl-2-dependent pathway in different cancer cells (Kim et al., 2001; Von Haefen et al., 2002). The use of fumonisin B1, a specific inhibitor of ceramide synthase, showed the complete inhibition of *P. urinaria*-induced apoptosis, which reveals the critical role of ceramide synthesis in this pathway. The elimination of fumonisin B1 effect by the addition of ceramide further demonstrates that the *P. urinaria*-induced apoptosis in HL-60 cells is mediated through a ceramide-related pathway. Taken together, the apoptosis triggered by *P. urinaria* is mediated through upstream Fas ligand/receptor, followed by ceramide production and Bax/Bcl-2-dependent mitochondrial activation, leading to final cell death (Figure 12.1).

The activation of p53 transcriptional activity is known to be required for the increased expression of Fas receptor (Muller et al., 1998; Fukazawa et al., 1999). However, the *P. urinaria*-induced increase of Fas receptor/ligand gene expression in p53-null HL-60 cells demonstrates the presence of a p53-independent pathway. The fact that several chemotherapeutic agents lack the capacity to induce apoptosis in a large number of tumors appears to be related to a disruption of p53 function. The

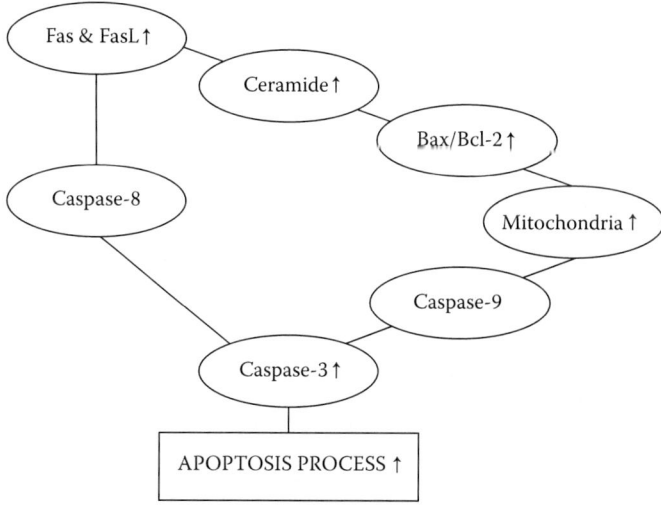

FIGURE 12.1 Mechanisms for the proapoptotic effect of *P. urinaria*.

most striking examples occur in lymphoid malignancies, including non-Hodgkin's lymphoma, acute leukemia, myelodysplastic syndrome, and chronic lymphocytic leukemia (Rouby et al., 1993; Wattel et al., 1994). Although the detailed mechanism for this p53-independent pathway induced by *P. urinaria* is still not clear, our finding suggests that the *P. urinaria*-induced p53-independent apoptotic pathway might be a useful alternative in tumor therapy. In addition, as an herbal medicine, *P. urinaria* has its unique properties of low cost, easy oral consumption, and a long history of use by the human population, all of which are indicative of its potential application as an anticancer agent.

12.4 THE *IN VIVO* ANTICANCER EFFECT OF *P. URINARIA*

In addition to the skin carcinogenesis study using a mice model that showed the *in vivo* anticancer effect of *P. urinaria* in 2003, we demonstrated that the tumor formation of C57BL/6J mice with the implantation of LLC cells is inhibited by the administration of *P. urinaria* extract (Huang et al., 2006). The initial tumor development of C57BL/6J mice was delayed and suppressed in the *P. urinaria* group. The average tumor size was significantly reduced in the *P. urinaria* group compared to that in the control group, clearly proving the *in vivo* anticancer effect of *P. urinaria*. An increased number of apoptotic cancer cells within the tumor in the *P. urinaria* group was also confirmed by TUNEL assay. Like the previous studies of *P. urinaria* in animals or in clinical use, there was no obvious toxicity detected in the *P. urinaria* group, as shown by the comparable mass and histological examination of the major organs, including liver, lung, spleen, kidney, and heart, between the *P. urinaria* group and the control group. Serum drawn from the mice in the *P. urinaria* group also caused the reduced cell viability of cancer cells *in vitro*, indicating that the acting components or metabolites of *P. urinaria* could be absorbed by the gastrointestinal system, enter into the blood circulation after the administration of *P. urinaria* in mice, and exhibit the *in vivo* anticancer effect.

12.4.1 PHYLLANTHUS URINARIA DECREASES IN VIVO ANGIOGENESIS IN TUMOR AND CAM ASSAY

Angiogenesis, the growth of new blood vessels via a well-controlled process, is closely associated with solid tumor formation. The development of capillary blood vessels in the tumor mass for carrying essential nutrients is absolutely required for the rapid growth of solid tumors. A variety of distinct proangiogenic and antiangiogenic factors are associated with modulation of neovascularization related to tumor formation and progression. Since the inhibition of angiogenesis can result in a suppression of tumor growth, we therefore also investigated the potential antiangiogenic effect of *P. urinaria* by examining the neovascularization of the tumor developed in C57BL/6J mice with the implantation of Lewis lung carcinoma cells. We have shown, for the first time, that *P. urinaria* exerts an *in vivo* antiangiogenic effect in tumors developed in this animal model. As we know that the process of angiogenesis in the tumor mass is critically required for the growth of solid tumors to a few millimeters in size, the inhibition of microvessel formation in the tumor therefore was an important contributory cause of reduced tumor size in the *P. urinaria* group, which can be explained by the limited supply of nutrients and oxygen due to the decreased vessel density in the tumor by the administration of *P. urinaria*. Direct evidence was further obtained from using the *in vivo* CAM assay, which again demonstrated that *P. urinaria* efficiently inhibited the *in vivo* process of angiogenesis (Huang et al., 2010).

12.4.2 *P. URINARIA* INHIBITED ANGIOGENESIS-RELATED FUNCTIONS OF VASCULAR ENDOTHELIAL CELLS

There are four distinct phases in the angiogenic cascade involving degradation of the extracellular matrix, cell migration, proliferation, and structure reorganization. Endothelial cells undergoing multiple interactions with different angiogenic factors stimulate proliferation, migration, and remodeling of endothelial cells to form a network of new blood vessels. To further understand the molecular mechanism, we used human umbilical vascular endothelial cells (HUVECs) to study the *in vitro* antiangiogenic effect of *P. urinaria*. This inhibition was unlikely to be caused by the effect of *P. urinaria* on the proliferation of endothelial cells since *P. urinaria* did not interfere with the cell viability of endothelial cells. However, the migration of vascular endothelial cells as determined by the transwell filter assay was clearly inhibited by the treatment with *P. urinaria* in a dose-dependent manner, which is in complete accord with the suppression in gel-induced tube formation of HUVECs by *P. urinaria*, providing a strong clarified mechanism underlying the *in vivo* effect of *P. urinaria* on inhibiting the vessel formation in tumor models (Huang et al., 2006) and CAM assay (Huang et al., 2010).

12.4.3 Matrix Metalloproteinase 2 Activity Inhibited by *P. urinaria* Both *In Vivo* and *In Vitro*

The primary step in the angiogenic process relies on the degradation of subendothelial basement membrane and surrounding extracellular matrix proteins. Matrix metalloproteinases (MMPs), which degrade extracellular matrix proteins, are involved in angiogenesis both *in vitro* and *in vivo*. Therefore, inhibition of the early degradation of extracellular matrix proteins predominantly by MMPs is considered an important strategy to inhibit angiogenesis. Results from the *in vivo* CAM assay showed that *P. urinaria* efficiently inhibited the MMP-2 activity in the protein extract prepared from chick chorioallantoic membrane, suggesting that the inhibition of angiogenesis in chicken embryo by *P. urinaria* could result, in part, from the failure of vascular endothelial cells to degrade the surrounding extracellular matrix, therefore impeding the following migration and proliferation. Among the various MMPs produced by vascular endothelial cells, MMP-2 and MMP-14 (MT1-MMP) have been linked to cell invasion and network formation (Lafleur et al., 2005; Sabeh et al., 2004).

The inhibition of MMP-2 activity is shown to be capable of disrupting the tube formation of endothelial cells (Yang et al., 2006), which may also explain our previous finding that *P. urinaria* could inhibit the matrix-induced tube formation of vascular endothelial cells. The inhibitory effect of *P. urinaria* on MMP-2 activity is not regulated at the transcriptional level or at the posttranslational level. Instead, a novel secretion mechanism is blocked and results in the decrease of MMP-2 activity in the early phase of **angiogenesis** (Richardson et al., 2003). MMP-2 secretion itself is probably regulated by multiple pathways; however, our results suggest the increased expression of RECK (reversion-inducing cysteine-rich protein with kazal motifs) by *P. urinaria* could be involved (Figure 12.2). It is also important to know that RECK

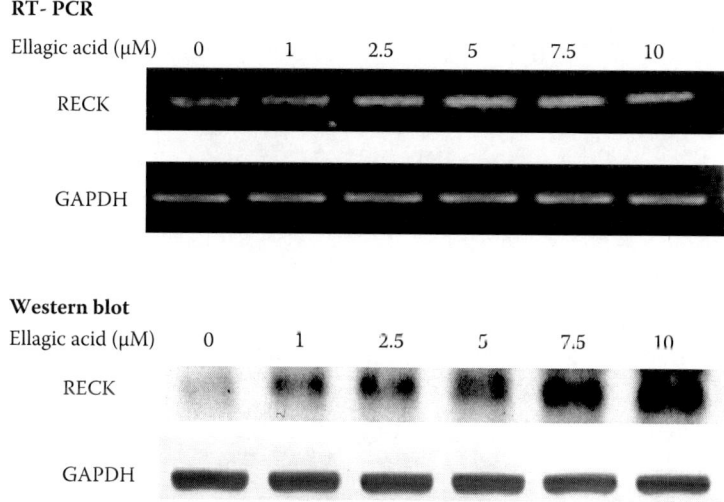

FIGURE 12.2 Ellagic acid dose dependently increases the expression of RECK at both the transcriptional and translational levels.

is thought to be a metastasis suppressor shown to inhibit MMP-2 activation (Clark et al., 2007; Liu et al., 2003).

MMP-2, like other **MMP**s, is a Zn^{2+}-dependent endopeptidase. Since zinc is essential for endopeptidase proteolytic capacity to degrade the ECM (extracellular matrix), compounds with zinc-chelating groups, such as thiol or hydroxamate (Talbot and Brown, 1996), are often used to inhibit MMP activity. We have found that MMP-2 activity can be inhibited directly by incubating with *P. urinaria*, and this inhibition is reversed in the presence of $ZnCl_2$. It is interesting to know that *P. urinaria*, by acting like a zinc chelator, can exert additional non-cell-mediated inhibitory effects on MMP-2 activity. This finding raises the possibility of using *P. urinaria* in clinical therapeutic applications as a zinc chelator in addition use as an antiangiogenic drug.

12.5 CHEMICAL ANALYSIS OF *P. URINARIA*

To further identify which compounds in *P. urinaria* extract exert the anticancer effect, we have analyzed *P. urinaria* by high-performance liquid chromatography (HPLC) and liquid chromatography/mass spectrometry (LC/MS) and characterized 12 polyphenolic compounds (Figure 12.3). This fingerprint profile of *P. urinaria* can then be used for plant identification and as a valuable reference for the following studies. The mass structures of these compounds are tentatively assigned based on mass data mining from existing literature. The major compound in *P. urinaria* is corilagin, followed by gallic acid and ellagic acid.

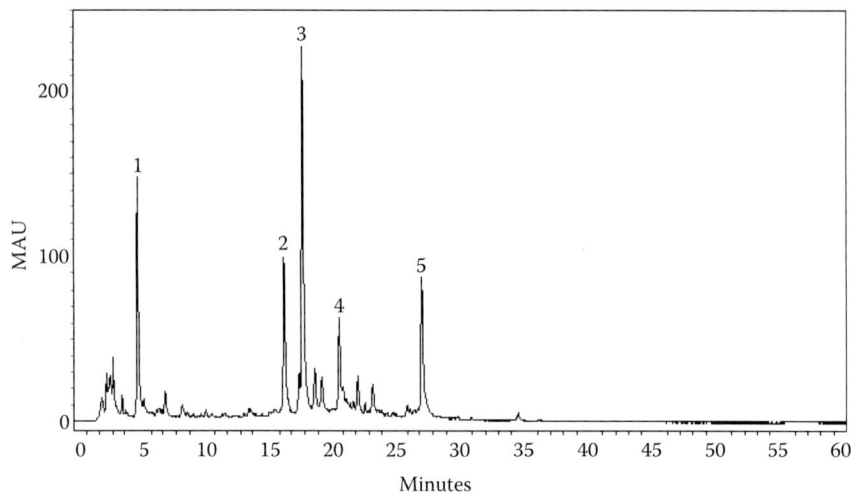

Peak 1. Gallic acid   Peak 2. Brevifolin carboxylic acid

Peak 3. Corilagin   Peak 4. Phyllanthusiin C   Peak 5. Ellagic acid

FIGURE 12.3 Chemical analysis of *P. urinaria*.

12.5.1 ELLAGIC ACID: THE MAJOR COMPOUND IN P. URINARIA EXERTS ANTIANGIOGENIC EFFECT

Ellagic acid (4,4',5,5'6,6'-hexahydroxydiphenic acid 2,6,2'6'dilactone) is a natural phenolic constituent present in woody plants, berries, grapes, and nuts. Ellagic acid has been reported to show many biological properties, including antioxidant, antiproliferative and apoptosis-inducing activities (Han et al., 2006). The anticancer effect of ellagic acid on the angiogenic process was studied *in vitro* and found to inhibit migration and tube formation by downregulation of both platelet-derived growth factor (PDGF) and vascular endothelial growth factor (VEGF) receptors (Labrecque et al., 2005). The inhibition of cancer cell proliferation by ellagic acid is mediated by regulation of MMPs, VEGF, and induction of apoptosis in cancer cells, but not normal cells (Losso et al., 2004). A low concentration of ellagic acid interferes with cell proliferation and induces apoptosis of Molt-4 cancer cells (Mertens-Talcott et al., 2003). Similar results were obtained in our laboratory; in addition, we have confirmed an antiangiogenic effect of ellagic acid in chicken embryos that is also accompanied by the inhibition of MMP-2 activity. Ellagic acid is also found to enhance the expression of genes with antiangiogenic function in vascular endothelial cells as analyzed by a complementary DNA (cDNA) microarray method, further confirming the antiangiogenic effect of ellagic acid. In a bioavailability study using wild-type mice (Seeram et al., 2007), ellagic acid was detectable in the plasma level with oral feeding of ellagitannins to mice. Our study has also shown that digestive-mimicking treatments of HCl (pH 1.0 to 2.0) followed by exposure to β-glucuronidase, effectively changed most compounds except for ellagic acid. This suggests that ellagic acid will survive digestion and be available for absorption into the circulation. From these studies we conclude that the main active constituent of *P. urinara* for its antiangiogenic activity *in vivo* is most likely to be ellagic acid.

12.6 CONCLUSION

Phyllanthus urinaria is documented worldwide to be used for the treatment of various diseases, and no harmful side effect has been reported. Studies of *P. urinaria* on the area of cancer therapy are still limited. However, from both *in vitro* and *in vivo* research results we have obtained, the use of *P. urinaria* as a potential anticancer drug is promising. We have characterized the main compounds of *P. urinaria* and established a fingerprint for quality control. We demonstrated the proapoptotic effect of *P. urinaria* on different cancer cells derived from various origins, but not on normal cells. The proapoptotic effects from four other species in the same genus (*Phyllanthus myrtifolius, Phyllanthus reticulatus, Phyllanthus tenellus,* and *Phyllanthus niruri*) have also been demonstrated; however, their IC_{50} values were higher when compared with that of *P. urinaria*. The apoptosis induced by *P. urinaria* is mediated by both extrinsic and intrinsic pathways. The *in vivo* anticancer effect was also shown in a mice tumor model. More important, we, for the first time, report the *in vivo* and *in vitro* antiangiogenic effect of *P. urinaria*. Underlying mechanisms are likely involved the inhibition of MMP-2 activity by blocking its secretion and chelation of zinc, which is required for MMP-2 activity. In addition, we also proved the inhibition of migration and tube formation of vascular endothelial cells by *P. urinaria* or ellagic acid. Ellagic acid is responsible for almost all the antiangiogenic effects observed for *P. urinaria*. Together, this suggests that the major acting component of *P. urinaria* is likely to be ellagic acid. The anticancer effect of *P. urinaria* is continuously under investigation in our laboratory, and we hope to fully understand its pharmacological composition and associated mechanisms. The aim of all our efforts is the clinical test of *P. urinaria* in the near future for its anticancer efficacy.

ACKNOWLEDGMENTS

We would like to thank Chang Gung Memorial Hospital and Chang Gung University for fully supporting this study, and the technical assistance from Su-Hui Yang and Tzu-Ya Chen is highly appreciated.

REFERENCES

Bharali, R., Tabassum, J., and Azad, M.R. 2003. Chemopreventive action of *Phyllanthus urinaria* Linn on DMBA-induced skin carcinogenesis in mice. *Indian J. Exp. Biol.* 41: 1325–1328.

Calixto, J.B., Santos, A.R., Cechinel Filho, V., and Yunes, R.A. 1998. A review of the plants of the genus *Phyllanthus*: their chemistry, pharmacology, and therapeutic potential. *Med. Res. Rev.* 18: 225–258.

Charng, M.Y. 1995. *Practical anti-tumor herb medicine*, 110–112. Peking: China Medicinal and Technical Publishers.

Chularojmontri, L., Wattanapitayakul, S.K., Herunsalee, A., et al. 2005. Antioxidative and cardioprotective effects of *Phyllanthus urinaria* L. on doxorubicin-induced cardiotoxicity. *Biol. Pharm. Bull.* 28: 1165–1171.

Clark, J.C., Thomas, D.M., Choong, P.F., and Dass, C.R. 2007. RECK—a newly discovered inhibitor of metastasis with prognostic significance in multiple forms of cancer. *Cancer Metastasis Rev.* 26: 675–683.

Dias, M.A., Campos, A.H., Cechinel Filho, V., Yunes, R.A., and Calixto, J.B. 1995. Analysis of the mechanisms underlying the contractile response induced by the hydroalcoholic extract of *Phyllanthus urinaria* in the guinea-pig urinary bladder in-vitro. *J. Pharm. Pharmacol.* 47: 846–851.

el Rouby, S., Thomas, A., Costin, D., et al. 1993. p53 gene mutation in B-cell chronic lymphocytic leukemia is associated with drug resistance and is independent of MDR1/MDR3 gene expression. *Blood* 82: 3452–3459.

Fukazawa, T., Fujiwara, T., Morimoto, Y., et al. 1999. Differential involvement of the CD95 (Fas/APO-1) receptor/ligand system on apoptosis induced by the wild-type p53 gene transfer in human cancer cells. *Oncogene* 18: 2189–2199.

Furuke, K., and Bloom, E.T. 1998. Redox-sensitive events in Fas-induced apoptosis in human NK cells include ceramide generation and protein tyrosine dephosphorylation. *Int. Immunol.* 10: 1261–1272.

Giridharan, P., Somasundaram, S.T., Perumal, K., et al. 2002. Novel substituted methylenedioxy lignan suppresses proliferation of cancer cells by inhibiting telomerase and activation of c-myc and caspases leading to apoptosis. *Br. J. Cancer* 87: 98–105.

Grullich, C., Sullards, M.C., Fuks, Z., et al. 2000. CD95(Fas/APO-1) signals ceramide generation independent of the effector stage of apoptosis. *J. Biol. Chem.* 275: 8650–8656.

Han, D.H., Lee, M.J., and Kim, J.H. 2006. Antioxidant and apoptosis-inducing activities of ellagic acid. *Anticancer Res.* 26: 3601–3606.

Hau, D.K., Gambari, R., Wong, R.S., et al. 2009. *Phyllanthus urinaria* extract attenuates acetaminophen induced hepatotoxicity: involvement of cytochrome P450 CYP2E1. *Phytomedicine* 16: 751–760.

Herr, I., Wilhelm, D., Bohler, T., et al. 1997. Activation of CD95 (APO-1/Fas) signaling by ceramide mediates cancer therapy-induced apoptosis. *EMBO J.* 16: 6200–6208.

Higashino, H., Suzuki, A., Tanaka, Y., and Pootakham K. 1992. [Hypoglycemic effects of Siamese Momordica charantia and *Phyllanthus urinaria* extracts in streptozotocin-induced diabetic rats (the first report)]. *Nippon Yakurigaku Zasshi* 100: 415–421.

Huang, S.T., Wang, C.Y., Yang, R.C., et al. 2010. Ellagic acid, the active compound of *Phyllanthus urinaria*, exerts *in vivo* anti-angiogenic effect and inhibits MMP-2 activity. *Evid. Based Complement. Alternat. Med.* (Epub ahead of print).

Huang, S.T., Yang, R.C., Chen, M.Y., and Pang, J.H. 2004b. *Phyllanthus urinaria* induces the Fas receptor/ligand expression and ceramide-mediated apoptosis in HL-60 cells. *Life Sci.* 75: 339–351.

Huang, S.T., Yang, R.C., Lee, P.N., et al. 2006. Anti-tumor and anti-angiogenic effects of *Phyllanthus urinaria* in mice bearing Lewis lung carcinoma. *Int. Immunopharmacol.* 6: 870–879.

Huang, S.T., Yang, R.C., and Pang, J.H. 2004b. Aqueous extract of *Phyllanthus urinaria* induces apoptosis in human cancer cells. *Am. J. Chin. Med.* 32: 175–183.

Huang, S.T., Yang, R.C., Yang, L.J., et al. 2003. *Phyllanthus urinaria* triggers the apoptosis and Bcl-2 down-regulation in Lewis lung carcinoma cells. *Life Sci.* 72: 1705–1716.

Jeena, K.J., Joy, K.L., and Kuttan, R. 1999. Effect of *Emblica officinalis*, *Phyllanthus amarus* and *Picrorrhiza kurroa* on N-nitrosodiethylamine induced hepatocarcinogenesis. *Cancer Lett.* 136: 11–16.

Ji, X.H., Qin, Y.Z., Wang, W.Y., et al. 1993. [Effects of extracts from *Phyllanthus urinaria* L. on HBsAg production in PLC/PRF/5 cell line]. *Zhongguo Zhong Yao Za Zhi* 18: 496–498, 511.

Kim, H.J., Mun, J.Y., Chun, Y.J., et al. 2001. Bax-dependent apoptosis induced by ceramide in HL-60 cells. *FEBS Lett.* 505: 264–268.

Labrecque, L., Lamy, S., Chapus, A., et al. 2005. Combined inhibition of PDGF and VEGF receptors by ellagic acid, a dietary-derived phenolic compound. *Carcinogenesis* 26: 821–826.

Lafleur, M.A., Drew, A.F., de Sousa, E.L., et al. 2005. Upregulation of matrix metalloproteinases (MMPs) in breast cancer xenografts: a major induction of stromal MMP-13. *Int. J. Cancer* 114: 544–554.

Lee, C.Y., Peng, W.H., Cheng, H.Y., et al. 2006. Hepatoprotective effect of *Phyllanthus* in Taiwan on acute liver damage induced by carbon tetrachloride. *Am. J. Chin. Med.* 34: 471–482.

Liu, L.T., Chang, H.C., Chiang, L.C., and Hung, W.C. 2003. Histone deacetylase inhibitor up-regulates RECK to inhibit MMP-2 activation and cancer cell invasion. *Cancer Res.* 63: 3069–3072.

Losso, J.N., Bansode, R.R., Trappey, A., 2nd, et al. 2004. *In vitro* anti-proliferative activities of ellagic acid. *J. Nutr. Biochem.* 15: 672–678.

Mertens-Talcott, S.U., Talcott, S.T., and Percival, S.S. 2003. Low concentrations of quercetin and ellagic acid synergistically influence proliferation, cytotoxicity and apoptosis in MOLT-4 human leukemia cells. *J. Nutr.* 133: 2669–2674.

Muller, M., Wilder, S., Bannasch, D., et al. 1998. p53 activates the CD95 (APO-1/Fas) gene in response to DNA damage by anticancer drugs. *J. Exp. Med.* 188: 2033–2045.

Paulino, N., Cechinel-Filho, V., Yunes, R.A., and Calixto, J.B. 1996. The relaxant effect of extract of *Phyllanthus urinaria* in the guinea-pig isolated trachea. Evidence for involvement of ATP-sensitive potassium channels. *J. Pharm. Pharmacol.* 48: 1158–1163.

Pettit, G.R., Schaufelberger, D.E., Nieman, R.A., et al. 1990. Antineoplastic agents, 177. Isolation and structure of phyllanthostatin 6. *J. Nat. Prod.* 53: 1406–1413.

Powis, G., and Moore, D.J. 1985. High-performance liquid chromatographic assay for the antitumor glycoside phyllanthoside and its stability in plasma of several species. *J. Chromatogr.* 342: 129–134.

Richardson, S., Neama, G., Phillips, T., et al. 2003. Molecular characterization and partial cDNA cloning of facilitative glucose transporters expressed in human articular chondrocytes; stimulation of 2-deoxyglucose uptake by IGF-I and elevated MMP-2 secretion by glucose deprivation. *Osteoarthritis Cartilage* 11: 92–101.

Sabeh, F., Ota, I., Holmbeck, K., Birkedal-Hansen, H., et al. 2004. Tumor cell traffic through the extracellular matrix is controlled by the membrane-anchored collagenase MT1-MMP. *J. Cell Biol.* 167: 769–781.

Santos, A.R., Filho, V.C., Yunes, R.A., and Calixto, J.B. 1995. Analysis of the mechanisms underlying the antinociceptive effect of the extracts of plants from the genus *Phyllanthus*. *Gen. Pharmacol.* 26: 1499–1506.

Seeram, N.P., Aronson, W.J., Zhang, Y., et al. 2007. Pomegranate ellagitannin-derived metabolites inhibit prostate cancer growth and localize to the mouse prostate gland. *J. Agric. Food Chem.* 55: 7732–7737.

Shen, B., Yu, J., Wang, S., et al. 2008. Phyllanthus urinaria ameliorates the severity of nutritional steatohepatitis both *in vitro* and *in vivo*. *Hepatology* 47: 473–483.

Suthienkul, O., Miyazaki, O., Chulasiri, M., Kositanont, U., and Oishi, K. 1993. Retroviral reverse transcriptase inhibitory activity in Thai herbs and spices: screening with Moloney murine leukemia viral enzyme. *Southeast Asian J. Trop. Med. Public Health* 24: 751–755.

Talbot, D.C., and Brown, P.D. 1996. Experimental and clinical studies on the use of matrix metalloproteinase inhibitors for the treatment of cancer. *Eur. J. Cancer* 32A: 2528–2533.

von Haefen, C., Wieder, T., Gillissen, B., et al. 2002. Ceramide induces mitochondrial activation and apoptosis via a Bax-dependent pathway in human carcinoma cells. *Oncogene* 21: 4009–4019.

Wang, M., Cheng, H., Li, Y., et al. 1995. Herbs of the genus *Phyllanthus* in the treatment of chronic hepatitis B: observations with three preparations from different geographic sites. *J. Lab. Clin. Med.* 126: 350–352.

Wattel, E., Preudhomme, C., Hecquet, B., et al. 1994. p53 mutations are associated with resistance to chemotherapy and short survival in hematologic malignancies. *Blood* 84: 3148–3157.

Yang, C.M., Cheng, H.Y., Lin, T.C., et al. 2005. Acetone, ethanol and methanol extracts of *Phyllanthus urinaria* inhibit HSV-2 infection *in vitro*. *Antiviral Res.* 67: 24–30.

Yang, E.V., Sood, A.K., Chen, M., et al. 2006. Norepinephrine up-regulates the expression of vascular endothelial growth factor, matrix metalloproteinase (MMP)-2, and MMP-9 in nasopharyngeal carcinoma tumor cells. *Cancer Res.* 66: 10357–10364.

Zhou, S., Xu, C., Zhou, N., et al. 1997. Mechanism of protective action of *Phyllanthus urinaria* L. against injuries of liver cells. *Zhongguo Zhong Yao Za Zhi* 22: 109–111.

13 *Phyllanthus* and Hepatitis B, Hepatitis C, and HIV Infections

S. P. Thyagarajan

CONTENTS

13.1 Historical Use of *Phyllanthus* Species in Jaundice 205
13.2 *In Vitro* and *In Vivo* Antihepatitis Property Studies of *P. Niruri* (*P. Amarus*) 206
    13.2.1 Safety Studies of *P. Amarus* 207
    13.2.2 Molecular Mechanism of Anti-HBV Activity of *P. amarus* 208
    13.2.3 Clinical Trials of *P. amarus* in Acute and Chronic Hepatitis B Virus Infections 208
    13.2.4 Clinical Efficacy of *P. amarus* versus Lead Optimization Challenge 209
13.3 *Phyllanthus Amarus* and Hepatitis C 212
13.4 *Phyllanthus* and HIV 212
13.5 Conclusion 214
Acknowledgments 214
References 214

13.1 HISTORICAL USE OF *PHYLLANTHUS* SPECIES IN JAUNDICE

An extensive review of literature in ethnobotany was carried out by Timothy Shepard of Oxford University to identify plants that have been used for treatment of jaundice and liver diseases (Blumberg et al., 1989). Of nearly 1,000 plants identified, 100 had been used in three or more global regions. *Phyllanthus* species were prominent in this list and were frequently used in India, China, Burma, Pakistan, Philippines, Guam, West Indies, South America, East and West Africa, tropics, and subtropics.

In a series of reviews by David Unander (Unander et al., 1990a, 1990b, 1991), it has been shown that about 24 species of *Phyllanthus* are active against clinical jaundice, of which 8 species have been used in India. Botanical Survey of India has observed that *Phyllanthus niruri* (*P. niruri*) in India has been a mixture of three species: *P. amarus, P. debilis,* and *P. fraternus* (Mitra and Jain, 1991).

Of these, *P. niruri* (also including *P. amarus, P. debilis,* and *P. fraternus*) as a treatment modality in jaundice has been extensively reported from several countries all over the world, like Guadeloupe and Martinique (Stehle and Stehle, 1957); South America (Albornoz, 1963); Brazil (Dragendorgg, 1898; Cruz, 1865); West Africa (Oliver-Bever, 1983, 1986); Cuba (Roig and Mesa, 1945); Puerto Rico (Amadeo, 1888); West Indies, including Jamaica, Bahamas, Virgin Islands, Dominican Republic (Morton, 1981; Ayensu, 1981); Thailand (Burnyaphatsara, 1987); Sri Lanka (Ahamed et al., 1984); Pakistan (Ahmad, 1957); Nepal (His Majesty's Government of Nepal, 1984); Philippines (Velazco, 1980); Fiji (Singh, 1986); People's Republic of China, Malay Peninsula, and Indonesia (Burkii et al., 1966; Perry and Metzger, 1980).

India probably has the earliest documentation on the use of *P. niruri* in clinical jaundice in Ayurveda, Siddha, and Unani systems of traditional medicine (Formulatory of Siddha Medicines, 1972; Formulatory of Unani Medicines, 1972; Formulatory of Ayurvedhic Medicines, 1981). While the Indian Council of Scientific and Industrial Research (CSIR) (Wealth of India, 1969), Drury (1873), Dymock (1866), Watt (1892), Dymock et al. (1893), Nadkarni (1954), and Chopra et al. (1958) have outlined the uses of *P. niruri* and the related species (viz. *P. amarus, P. debilis,* and *P. fraterus*) in clinical jaundice, their uses were also reported from Madhya Pradesh (Sahu, 1984); West Bengal (Pal and Jain, 1989); Tamil Nadu (Ainsile, 1984; Krishnakumurthi and Seshadri, 1946); Maharashtra (Mulchandani and Hassaranjani, 1984); central and southern India (Dey and Mair, 1896); Kerala (John, 1984; Sivarajan and Balachandran, 1984); Andhra Pradesh (Hemadri and Rao, 1984; Reddy, 1988); Gujarat (Ramanan and Sainani, 1961); and Uttar Pradesh (Ahuja, 1965).

13.2 *IN VITRO* AND *IN VIVO* ANTIHEPATITIS PROPERTY STUDIES OF *P. NIRURI* (*P. AMARUS*)

The first-ever designed *in vitro* antiviral study of *P. niruri* against any hepatitis virus, with hepatitis B virus (HBV) as the model, was reported by Thyagarajan (1979) from India (Madras). Subsequently, Thyagarajan (Thyagarajan et al., 1982) reported that the whole-plant extract of *P. niruri* using several solvents brought about *in vitro* inactivation of hepatitis B surface antigen (HBsAg). These plants were later identified by Unander as *P. amarus*. Venkateswaran (Venkateswaran et al., 1987) and Blumberg (Blumberg et al., 1989) from the United States showed that the plants collected from Madras, India, whose aqueous extracts bound the surface antigen of HBV *in vitro* had inhibited the viral DNA polymerase (DNAp) of HBV and Woodchuck hepatitis B virus (WHBV) *in vitro* at 50% inhibitory concentrations (IC_{50}) of 59 and 140 µg, respectively; the IC_{50} values for calf thymus DNAp was 115 µg/ml, and the DNAp's of *Escherichia coli* ranged from 120 to 460 µg/ml. When administered intraperitoneally to WHBV-infected woodchucks, both acutely and chronically infected animals, the liver cancer rate in treated chronically infected animals was lower than for the untreated controls.

Unander (Unander et al., 1990a) and Unander and Blumberg (1991) showed that aqueous extracts of *P. niruri* plants (grown in a greenhouse in Puerto Rica) inhibited the viral DNAp of WHBV. Aqueous extracts of plants grown from seeds collected in several countries, including that provided by Thyagarajan from Madras, also inactivated WHBV DNAp. Their activity was relatively uneffected by soil pH or soil calcium but was affected by soil moisture and environmental temperatures. It was also shown that the extracts of *P. niruri* or *P. fraternus* were significantly less inhibitory to WHBV DNAp than that of *P. amarus*.

Mehrotra et al. (1991) have also shown the *in vitro* binding property of *P. amarus* besides showing an effect on HBV DNA. Yanagi et al. (1989) from Japan reported that, from the *Phyllanthus* plants collected from southern India, aqueous extracts of high dilution inhibited HBV DNAp, DNAp I, T4-DNAp, the klenow fragment and reverse transcriptase (RT) of avian myeloblastosis virus. Shead et al. (1990) from Australia (New South Wales) showed the aqueous extracts of *Phyllanthus* sp. inhibited the endogenous DNAp of duck hepatitis B virus (DHBV) at high dilutions. *Phyllanthus amarus*, *P. gasstroemii*, *P. gunnii*, *P. similis*, *P. thymoides,* and *P. tenellus* were tested. *Phyllanthus thymoides* was the most active *in vitro*. However, Niu et al. (1990) from Australia, using *P. amarus* collected from Madras, India, to treat 4- to 5-week-old ducks congenitally infected with DHBV and using suitable controls, showed transient reduction of viral DNA in serum after 10 weeks of treatment but no effects on the level of viral DNA or surface antigen in the liver.

Jayaram (1992) reported *in vitro* inhibition of HBsAg secretion by PLC/PRF/5 (Alexander cell line) for 48 h, when the cell line was treated with 1 mg/ml concentration of *P. amarus* as a single dose. However, HBsAg was detected from the culture medium at a lower concentration after 72 h.

13.2.1 Safety Studies of *P. Amarus*

Mokkhasmit et al. (1971), from Thailand, have used *P. niruri* and reported was nontoxic to mice at 10 g/kg body weight. Krishnakumurthi and Seshadri (1946), from Tamil Nadu, India, showed that the extracted compound phyllanthin from *P. niruri* was toxic to the freshwater fish *Haplochilus malarcia* at 50 ppm and to frogs given intraperitoneally at 100 ppm. Ahamed et al. (1984), from Sudan, found the petroleum ether extract of the roots of *P. niruri* at 25 ppm produced 100% mortality to two species of schistosomiasis-carrying snails. Rao (1985), from Andhra Pradesh (Kurnool), India, reported that a 20% aqueous extract of *P. niruri* leaves was effective as an oral pretreatment at 0.2 ml/100 mg body weight against CCl_4-induced hepatotoxicity in rats. No hepatocellular degeneration was detected compared to extensive degeneration in the controls. Syamasundar et al. (1985), from Lucknow, India, showed the hexane-extracted compounds phyllanthin and hypophyllanthin reduced CCl_4- or galactosamine-induced cytotoxicity to cultured rat hepatocytes. Jayaram et al. (1987), from Tamil Nadu (Madras), India, showed no chronic toxicity in mice using the aqueous extracts of dried whole plant of *P. amarus* at 0.2 mg daily dose per animal for 90 days; no effects on serum glutamic/pyruvic transaminases, serum bilirubin, or histopathology of the liver, spleen, or kidney; and no cytotoxic or cytotonic changes on the vero cell line. Venkateswaran et al. (1987) demonstrated *in vivo* safety using woodchucks as animal models, while Niu et al. (1990) from Australia have shown it to be nontoxic in Pekin ducks chronically infected with duck HBV. Jayaram (1992), from Tamil Nadu (Madras), India, in studying the effect of *P. amarus* on β-galactosamine-induced hepatotoxicity on isolated rat hepatocytes, have shown that (1) *P. amarus* by itself did not bring about any hepatotoxicity on rat hepatocytes; (2) at 1 mg/ml concentration, the aqueous extract protected isolated rat hepatocytes significantly from β-galactosamine-induced hepatotoxicity, thus proving the antihepatotoxic potentials of *P. amarus*.

13.2.2 Molecular Mechanism of Anti-HBV Activity of *P. amarus*

Lee et al. (1996), from Albert Einstein College of Medicine in the United States, using *P. amarus* plant material provided by Thyagarajan, defined the mechanism of action of *P. amarus* in HepG2 2.2.15 cells, which support HBV replication. *Phyllanthus amarus* inhibited HBV polymerase activity, decreased episomal HBV DNA content, and suppressed virus release into culture medium. To examine transcriptional control mechanisms, G26-HBV transgenic mice were used. When *P. amarus* was administered to transgenic mice, hepatic HBsAg messenger RNA (mRNA) levels decreased, indicating transcriptional or posttranscriptional downregulation of the transgene. Further, Ott et al. (1997) from the same group conducted analysis in HuH-7 cells with transfected phasmids using a luciferase reporter and showed that *P. amarus* specifically inhibited HBV enhancer I activity. They predicted that as *P. amarus* downregulates HBV mRNA transcription by a specific mechanism involving interactions between HBV enhancer I and C/EBP transcription factors, purification and further analysis of the active *P. amarus* component will advance insights into its antiviral activity.

13.2.3 Clinical Trials of *P. amarus* in Acute and Chronic Hepatitis B Virus Infections

In all the traditional medicine systems, there have been several formulatory medicines for the treatment of jaundice in general without taking into consideration casewise viral etiology (Formulary of Siddha Medicines, 1972; Formulary of Unani Medicines, 1972; Formulary of Ayurvedhic Medicines, 1981). Even though *P. niruri* (*P. amarus*) was one of the constituents, these were always preparations containing up to 12 medicinal herbs. Most of the treatment evaluations prior to 1970 were based on clinical improvement only.

Sankaran (1977), from Tamil Nadu (Madras), India, using a preparation with nine herbs called "Liver-doks," containing *P. niruri* and *P. emblica*, among others, conducted a clinical trial on unspecified acute hepatitis cases and showed symptomatic recovery of patients. Sundaravalli et al. (1977), using another preparation ("liverite," a 12-herb preparation with *P. niruri*), treated 27 patients and compared results with those obtained with three vitamins. The liverite group showed a rapid decrease in bilirubin and transaminase levels compared to the controls.

Dixit and Achar (1983), from Andhra Pradesh, India, treated acute hepatitis in children using *P. niruri* plant material showed that hepatitis gained clinical and biochemical normalcy within 5 days with no side effects. Jayanthi et al.(1988), from Tamil Nadu (Madras), in a controlled clinical trial in acute viral hepatitis (AVH) using several herbal medicines besides *P. niruri,* showed a significantly greater decrease in transaminases after 2 weeks of treatment with *P. niruri* in both HBsAg-positive and -negative groups. In a virologically characterized AVH clinical trial, Geetha et al. (1992) and Jayaram (1992) showed the following: (1) In acute hepatitis A, biochemical normalcy was achieved significantly faster by both *P. amarus* and "Essentiale," a commercial preparation with essential phospholipids, after 4 weeks of treatment compared to the controls. (2) In acute hepatitis B, *P. amarus* treatment brought about significantly faster biochemical normalcy than Essentiale and the controls. There was also a higher rate of HBsAg clearance in the group treated with *P. amarus* than the others. However, among the non-A, non-B hepatitis group by serological exclusion, better efficacy of Essentiale compared to *P. amarus* treatment was observed. In all the preceding studies mentioned in this section, there were no observed side effects due to *P. amarus* treatment.

In the first-ever double-blind clinical trial concerning healthy chronic carriers of HBV, Thyagarajan et al. (1988) administered 200 mg of *P. amarus* orally in gelatin capsules given three times daily for 30 days to the test group and identical placebo capsules containing lactose to the control group. While 59% of the group treated with *P. amarus* lost HBsAg after treatment, only 4% cleared in the placebo group ($p < .001$). They also reported that HBeAg- (*hepatitis B e-antigen*) positive carriers were less likely to respond; no toxic effects were noticed in any of them. Responding to the criticism of Brook (1988), Thyagarajan et al. (1990) conducted a subsequent open trial on 20 chronic carriers of HBV and showed 20% HBsAg clearance and 63.6% HBeAg seroconversion.

Leelarasamee et al. (1990), from Thailand, in an attempt to reproduce the effect of *P. amarus* treatment in HBV carriers, showed that none of their cases cleared HBsAg by 1 month of treatment with *P. amarus*. Even though the concentration of HBeAg decreased in the group treated with *P. amarus*, it was not clinically significant since HBsAg did not disappear. They thus reported failure of *P. amarus* to eradicate HBsAg from symptomless carriers. In another trial conducted in China by Wang et al. (1991b), using *P. amarus* capsules provided by Thyagarajan, among 11 subjects (3 asymptomatic carriers and 8 subjects with chronic hepatitis), showed that none of them became either HBsAg or HBeAg negative. However, the same group (Wang et al., 1991a), using local species of *Phyllanthus* (i.e., *P. urinaria* and *P. niruri*) treated 77 cases of symptomless HBV carriers, with group A of 42 cases receiving *P. urinaria* and group B of 35 cases receiving *P. niruri*. The HBeAg seroconversion was significant in group A ($p < .025$) and nonsignificant in group B ($p < .05$), while the rate of anti-HBe was significant in both groups ($p < .001$).

13.2.4 CLINICAL EFFICACY OF *P. AMARUS* VERSUS LEAD OPTIMIZATION CHALLENGE

Based on the knowledge drawn from Ayurvedic and Siddha literature, in which it is documented that the efficacy of medicinal plants would depend on the soil in which they are grown and the season during which the plants are used for preparation of traditional medicines, it was hypothesized that the nonreproducibility of uniform treatment efficacy of the drug preparations of *P. amarus* and other species was due to the absence of one or more of the following antiviral and biological properties in the *P. amarus* extract that are essentially required for the efficient treatment of (acute and chronic) hepatitis B:

1. HBsAg binding property facilitating the inactivation of the virus in circulation, ultimately leading to viral clearance
2. HBV-DNA polymerase enzyme-inhibiting potential, thus acting as an antiviral that prevents the multiplication of HBV
3. RT enzyme inhibition, also required for the initiation of HBV replication
4. Inhibition of HBsAg secretion from HBV-transinfected liver cells, thus possessing activity against virus-infected chronic liver disease conditions
5. Hepatoprotective and antihepatotoxic properties against the liver cell toxicity brought about by all hepatitis viruses (A, B, C, D, and E) and other hepatotoxic agents
6. Immunomodulating property to potentiate the immune system of HBV-infected patients toward virus clearance and protective antibody (anti-HBs) responses

To address these scientific requirements, Thyagarajan (1999) developed an agro technology-based standard operating procedure to cultivate the specified biovar of *P. amarus* from the seeds of the wild-grown *P. amarus* from certain regions of Tamil Nadu (India) that showed an optimum *in vitro* activity profile of these six properties.

The drug preparation from this cultivated biovar of *P. amarus* was standardized by a chemobiological fingerprinting methodology through a process that produced an extract with all the biological properties mentioned. Using this "University of Madras *P. amarus* preparation," five clinical trials (one trial in Chennai, India; one in Vellore, India; and one in Glasgow, UK) were conducted on a total of 153 chronic carriers of HBV (E. Walker, Treatment of long-term HBV carriers with *Phyllanthus amarus*: Scottish experience, personal communication, 1999). The consolidated results are presented in Table 13.1.

In summary, these trials have shown a mean HBsAg clearance rate of 25.6% and mean HBeAg seroconversion rate of 55.3% in predetermined chronic carriers of HBV. It was finally decided to recommend a schedule of 500 mg *P. amarus* standardized preparation in capsules given orally three times daily for 6 months.

All this research and drug development work led to the award of the following patents in the name of the University of Madras with Thyagarajan as inventor: Indian patent 405/MAS/99 dated 12.4.1999; U.S. patent 09/719, 486 dated 12.12.2000; North Korean patent 00–1180 dated 11.12.2000; and South African patent 20010148 dated 05.01.2001.

A drug under the brand name Virohep was also marketed in India by Rallis India Pharmaceuticals and Shreya Life Sciences in 2002.

TABLE 13.1
Summary of Seven Clinical Trials Conducted by Thyagarajan and His Collaborators on Human HBV Carriers Using *P. amarus* Grown in Tamil Nadu

Clinical Trial No.	Authors/ Year	Dosage (mg/tds)	Duration (months)	Number Treated Test	Number Treated Placebo	HBsAg Clearance % Test	HBsAg Clearance % Placebo	HBeAg Seroconversion (%) Test	HBeAg Seroconversion (%) Placebo
1	Thyagarajan et al. (1988), Madras	200	1	40	38	59	4	ND	ND
2	Thyagarajan et al. (1990), Madras	250	3	20	Nil	20	—	63.6	—
3	Samuel et al. (1991),* Vellore	250	2	10	12	20	8.3	37.5	0
4	Thyagarajan et al. (1992),* Madras	250	6	72	Nil	25	—	54.0	—
5	Thyagarajan et al. (1993),* Madras	500	3	8	8	25	0	71.4	16.0
6	Walker et al. (1993–1995)* Glasgow	500	4–6	26	Nil	11.6	—	45.4	—
7	Thyagarajan et al. (1996–1997),* Madras	500	6	37	Nil	18.9	—	60.0	—
	Total			213	58	25.1	3.4	55.3	1.7

Note: For convenience, the *P. amarus* grown in Tamil Nadu is termed the "Madras University preparation."

* Clinical trial numbers 3–7 were published by Thyagarajan et al., 2002.

Since several clinical trials have been conducted across the globe to evaluate the efficacy and safety of treatment with plants of the genus *Phyllanthus* for chronic HBV infection, Liu et al.) (2001) performed a systematic review of the randomized clinical trials reported in the literature. Randomized trials comparing the genus *Phyllanthus* versus placebo, no intervention, general nonspecific treatment, other herbal medicine, or interferon treatment for chronic HBV infection were identified by electronic and manual searches. Trials of *Phyllanthus* herb plus interferon (IFN) versus IFN alone were also included. No blinding and language limitations were applied. The methodological quality of trials was assessed by the Jadad scale plus allocation concealment. Twenty-two randomized trials ($n = 1,947$) were thus identified. The methodological quality was high in 5 double-blind trials and low in the 17 remaining trials. The combined results showed that *Phyllanthus* species had a positive effect on clearance of serum HBsAg (relative risk 5.64, 95% confidence interval [CI] 1.85–17.21) compared with placebo or no intervention. There was no significant difference on clearance of serum HBsAg, HBeAg, and HBV DNA between *Phyllanthus* and IFN. *Phyllanthus* species were better than nonspecific treatment or other herbal medicines for the clearance of serum HBsAg, HBeAg, HBV DNA, and liver enzyme normalization. Analyses showed a better effect of the *Phyllanthus* plus IFN combination on clearance of serum HBeAg (relative risk 1.56, 95% CI 1.06–2.32) and HBV DNA (relative risk 1.52, 95% CI 1.05–2.21) than IFN alone. No serious adverse event was reported. Based on this review, *Phyllanthus* species may have a positive effect on antiviral activity and liver biochemistry in chronic HBV infection. However, larger, well-designed, multicentric clinical trials are needed with a chemobiologically fingerprinted and standardized *Phyllanthus* preparation as the optimized drug candidate for hepatitis B.

13.3 *PHYLLANTHUS AMARUS* AND HEPATITIS C

Available literature in the public domain did not reveal any report on the use of *P. amarus* in the treatment of acute and chronic hepatitis C virus (HCV) infection. Hence, the formulation of *Phyllanthus amarus* developed by Thyagarajan (2000b) was utilized to conduct two preliminary clinical trials to treat acute and chronic hepatitis C. The first clinical trial was conducted on two cases by Eric Walker and his coworkers in Glasgow during 1996–1999.

Based on the encouraging clinical efficacy of *P. amarus* treatment shown by Walker and coworkers in Scottish cases of chronic hepatitis C, Thyagarajan et al. at Chennai conducted the second clinical trial on 34 patients with hepatitis C and confirmed the efficacy of the new formulation for the treatment of acute and chronic hepatitis C and other related infections of liver. The results showed normalization of transaminases in 71.6% of cases and HCV RNA negativity in 34.2% of cases. The details are provided in the patents awarded (Thyagarajan, 2000a, 2000b, 2001).

13.4 *PHYLLANTHUS* AND HIV

Extracts of *Phyllanthus* sp. inhibited RT activity. Chang et al. (1995; Mullen, P., personal communication) tested the effect of the aqueous extract of *P. amarus*, provided by Thyagarajan of India, on replication of HIV and reported *P. amarus* to have an effect on the replication of cell free H1RF (HIV) in C8166 cells. Syncytia took longer to appear when *P. amarus* was present in the culture (3–4 days) than when omitted indicating HIV growth-inhibition by *P. amarus*. Similarly, the number of syncytia was lower in cultures with *P. amarus* extract added. A 10-fold decrease in syncytia end point occurred at 100 μg/ml and a 100-fold decrease at 250 μg/ml. The level of P24 antigen in the presence of *P. amarus* extract was also lower at 250 μg/ml.

TABLE 13.2
End-point Dilutions and Antigen Levels for *P. amarus* Extract and Compounds against HIV

Extract/Compound	Syncytia Observed	P24 Antigen (pg ml^{-1})
P. amarus extract	1:160	254
Amarulone	<1:10	77
Amariin	<1:10	254
Elaeocarpusin	<1:10	182
Repundisic acid	1:640	1,515
Corilagin	<1:10	269
Furosin	1:160	5,053
Geraniin	<1:10	196
Untreated control	>1:2,560	2,780

Ogata et al. (1992) independently showed that an aqueous extract of *P. niruri* inhibited HIV-1 RT *in vitro*. The active principle responsible for this anti-HIV property was isolated and identified as repandusinic acid A monosodium salt.

Houghton (personal communication, 1997) used aqueous extract of *P. amarus*, 12 fractions collected from *P. amarus* through reverse-phase silica gel and polyvinylpyrrolidone, and several galloylester polyphenolic compounds provided by Foo (1993) for testing their ability to inhibit HIV-1 by observation of syncytium formation in C8166 cell cultures and quantification of HIV P24 antigen in culture supernatants. The results showed (Table 13.2) that all of the polyphenolic compounds, the aqueous extract of *P. amarus* except repundisic acid, and furosin showed appreciable anti-HIV activity.

In 1995, El-Mekkawy et al. reported that the methanolic extract of *P. emblica* and some of its constituents, such as putranjivain A (IC$_{50}$ 3.9 µM), and other flavonoids and digallic acid, inhibited HIV RT.

Qian-Cutrone et al. (1996) isolated a novel compound from the dried leaf of *P. niruri*; they called it *niruriside* and demonstrated that it has a specific inhibitory activity against the binding of a regulator of expression of the virion (REV) protein to REV response element (RRE) RNA. Since the REV-RRE regulatory mechanism plays a key role in the maintenance of high levels of viral propagation, it was suggested that a partial block of REV function by *P. niruri* may modulate progression in HIV-infected individuals.

Naik and Juvekar (2003) have also shown that the alkaloidal extract of *P. niruri* showed suppressing activity on strains of HIV-1 cultured on MT-4 cell lines. The CC$_{50}$ for the extract was 279.85 µg/ml, whereas the EC$_{50}$ was 20.98 µg/ml, and the selectivity index was 13.34, showing selective toxicity of the *P. niruri* extract for the viral cells, even though inhibition of HIV was observed.

Besides anecdotal treatment of HIV/AIDS cases with *P. amarus* in Tamil Nadu, no case-controlled clinical trial has been conducted.

13.5 CONCLUSION

The inherent problem in the discovery of an effective antiviral drug that is not as successful as an antibacterial drug is the intracellular association of the causative viruses of a disease. For plant-based natural products, there is a much more complicated process of success in view of the "biological synergism" of the bioactive compounds that may be present in the potential medicinal plant to bring about the desired or projected holistic clinical efficacy. Another equally important challenge is the reproducible standardization of the drug preparation using chemobiological fingerprinting protocols. In addition to the reverse pharmacology approach of validation, a system biology approach might make it easier to obtain "blockbusters" from plant sources. *Phyllanthus amarus* is no exception to this modern requirement (Thyagarajan and Jayaram, 1992).

ACKNOWLEDGMENTS

I acknowledge a large number of national and international collaborators and the doctoral students who conducted the laboratory and clinical studies along with me as mentioned in various publications cited in this chapter. The funding support provided by the Indian Council of Medical Research and the British ODA besides the British council and other U.S., Australian, and New Zealand research agencies is gratefully acknowledged.

REFERENCES

Ahamed, E.H.M., Bashir, A.K., and Khier,Y.M. 1984. Investigations of molluscidal activity of certain Sudanese plants used in folk medicine. Part IV. *Planta Medica.* 50: 74–77.

Ahmad, Y.S. *A note on the plants of medicinal value found in Pakistan.* Karachi, Pakistan: Government of Pakistan Press. Abstract A 1908 from NAPRALERT, College of Pharmacy, University of Illinois, Chicago, IL. 1957.

Ahuja, B.S. 1965. *Medicinal plants of Saharanpur,* 56–57. Saharanpur, India: Central Council of Ayurvedic Research.

Ainsile, W. 1984. *Materia indica Vol. II,* 50–152, 436. Delhi, India: Neeraj [reprint].

Albornoz, M.A.R. 1963. *Guia Farmacognosica de Drogas vegetables Y de plantas con Interes Economic de Actualidad. Parts 2, 3, 4 and 5,* 82. Caracas, Venezuela: Central University of Venezuela.

Amadeo, A.J. 1888. The botany and vegetables material medica of the island of Puerto-Rico. *Pharmaceutical J.* 28: 906.

Ayensu, E.S. 1981. *Medicinal plants of West Indies,* 9–100. Alognac, MI; References Publications.

Blumberg, B.S., Millman, I., Venkateswaran, P.S., and Thyagarajan, S.P. 1989. Treatment of HBV carriers with *Phyllanthus amarus. Cancer Detect. Prev.* 14: 195–201.

Brook, M.G. Efficacy of *Phyllanthus amarus* on chronic carriers of hepatitis B virus. 1988. *Lancet* 2: 1017–1018.

Burkii, I.H., Britwistle, W., Foxworthy, F.W., Scrivenor, J.B., and Watson, J.G. 1966. *A dictionary of economic products of the Malay Peninsula Vol. 2,* 1747–1749. Kuala Lumpur, Malaysia: Ministry of Agriculture and Co-operatives, Government of Malaysia.

Burnyaphatsara, N. 1987. Translations provided to Unander from nine books in Thai on uses of *Phyllanthus* species. traditional Thai medicine. Medicinal Plant Information Center, Mahidol University, Bangkok, Thailand.

Chang, C.W., Lin, M.T., Lee, S.S., et al. 1995. Differential inhibition of reverse transcriptase and cellular DNA polymerase-alpha activities by lignans isolated from Chinese herbs, *Phyllanthus myrtifolius* Moon, and tannins from *Lonicera japonica* Thunb and *Castanopsis* hystrix. *Antiviral Res.* 27: 367–374.

Chopra, R.N., Chopra, I.C., Handa, K.L., and Kapur, L.D. 1958. *Chopra's indigenous drugs of India*, 2nd ed., 12–23, 519, 520, 598, 605, 609, 660. Calcutta: Dhar and Sons.

Cruz, G.l. 1865. *Livo verde das Plantas Medicinais e Industrialis do Brasil*, Vol. 2, 438, 708–709. Velloso, Belo Horizonte.

Dey, K.L., and Mair, W. 1896. *The indigenous drugs of India*, 2nd ed., 235. London: W. Thackar and Co.

Dixit, S.P., and Achar, M.P. 1983. Bhumyalaki (*Phyllanthus niruri* Linn) and jaundice in children. *J. Natl. Integr. Med. Assoc.* 25: 269.

Dragendorgg, G. 1898. *Die Helipflanzen der Verschiden Volker and Zeten*, 373–374. Stuttgart, Germany: Verlag Von Ferdinand Enke.

Drury, H. 1873. *The useful plants of India*, 2nd ed., 342. London: Allen.

Dymock, W. 1866. *The vegetable materia medica of western India*, 701–702. Bombay: Education Society's Press.

Dymock, W., Warden, C.J.H., and Hooper, D. 1893. *Pharmacographia indica*, Vol. 3, 265–268. London: Kegan Paul, Trench, Trubner.

el-Mekkawy, S., Meselhy, M.R., Kusumoto, I.T., et al. 1995. Inhibitory effects of Egyptian folk medicines on human immunodeficiency virus (HIV) reverse transcriptase. *Chem. Pharm. Bull. (Tokyo)* 43: 641–648.

Foo, Y.L. 1993. Amariin, a di-dehydrohexahydroxydiphenoyl hydrolysable tannin from *Phyllanthus amarus*. *Phytochemistry* 33: 487–491.

Formulatory of Ayurvedhic medicines. 1981. Madras, India: IMCOPS.

Formulatory of Siddha medicines. 1972. Madras, India: IMCOPS.

Formulatory of Unani medicines. 1972. Madras, India: IMCOPS.

Geetha, J., Manjula, R., Malathi, S., Usha, K., et al. 1992. Efficacy of essential photospholipid substance of soya bean oil and *Phyllanthus niruri* in acute viral hepatitis. *J. Gen. Med.* 4: 53–58.

Hemadri, K., and Rao, S.S. 1984. Jaundice: tribal medicine. *Ancient Sci. Life.* 3: 209–212.

His Majesty's Government of Nepal, Ministry of Forest and Soil Conservation. 1984. Department of Medicinal plants. *Medicinal plants of Nepal* (Suppl. Volume), 98. Katmandu, Nepal: His Majesty's Government Press.

Jayanthi, V., Madanagopalan, N., Thyagarajan, S.P. 1988. Value of herbal medicines, *Phyllanthus niruri, Eclipta alba, Piper longus thippili* (tamil) and combination of *Phyllanthus niruri* and *Ricinus communis* (ictrus-pharm products) in acute viral hepatitis. *J. Gastroenterol. Hepatol.*3: 533–534.

Jayaram, S. 1992. Studies on prevention and control of hepatitis B virus infection. PhD diss., University of Madras, India.

Jayaram, S., Thyagarajan, S.P., Panchanadam, M., et al. 1987. Antihepatitis B virus properties of *Phyllanthus niruri* Linn and *Eclipta alba* Hassk: *in vitro* and *in vivo* safety studies. *Biomedicine* 7: 9–16.

John, D. 1984. One hundred useful raw drugs of the Kani tribes of Trivandrum forest division, Kerala, India. *Int. J. Crude Drug Res.* 22: 17–39.

Krishnakumurthi, G.V., and Seshadri, T.R. 1946. The bitter principle of *Phyllanthus niruri*. *Proc. Indian Acad. Sci. A* 24: 357–364.

Lee, C.D., Ott, M., Thyagarajan, S. P., et al. 1996. *Phyllanthus amarus* down-regulates hepatitis B virus mRNA transcription and replication. *Eur. J. Clin. Invest.* 26: 1069–1076.

Leelarasamee, A., Trakulsomboon, S., Maunwongyathi, P., et al. 1990. Failure of *Phyllanthus amarus* to eradicate hepatitis B surface antigen from symptomless carriers. *Lancet* 1: 1600–1601.

Liu, J., Lin, H., and McIntosh, H. 2001. Genus *Phyllanthus* for chronic hepatitis B virus infection: a systematic review. *J. Viral. Hepat.* 8: 358–366.

Mehrotra, R., Rawat, S., Kulshreshtha, D.K., et al. 1991. In vitro effect of *Phyllanthus amarus* on hepatitis B virus. *Indian J. Med. Res.* 93: 71–73.

Mitra, R.L., and Jain, S.K. 1991. Concept of *Phyllanthus niruri*, Euphorbiaceae in Indian floras. *Bulletin of Botanical Survey of India.* 27: 161–176.

Mokkhasmit, M., Swasdimongkol, K., and Satrawaha, P. 1971. Study on toxicity of Thai medicinal plants. *Bulletin of the Department of Medical Science,* 12: 36–65. Abstract R-001 from NAPRALERT, College of Pharmacy, University of Illinois, Chicago.

Morton, J.F. 1981. *Atlas of medicinal plants of middle America, Bahamas to Yucatan*, 457–463. Springfield, IL; Thomas.

Mulchandani, N.B., and Hassaranjani, S.A. 1984. 4-Methoxy-non-securinine, a new alkaloid from *Phyllanthus niruri. Planta Med.* 50: 104–105.

Nadkarni AK. 1954. *Dr. K. M. Nadkarni's India Materia Medica*, 3rd ed., Vol. 1, 941. Bombay: Popular Book Depot.

Naik, A.D., and Juvekar, A.R. 2003. Effects of alkaloidal extract of *Phyllanthus niruri* on HIV replication. *Indian J. Med. Sci.* 57: 387–393.

Niu, J.Z., Wang, Y.Y., Qiau, M., et al.1990. Effect of *Phyllanthus amarus* on duck hepatitis B virus replication in vivo. *J. Med. Virol.* 32: 212–218.

Ogata, T., Higuchi, H., Mochida, S., et al. 1992. HIV-1 reverse transcriptase inhibitor from *Phyllanthus niruri. AIDS Res. Hum. Retroviruses* 8: 1937–1944.

Oliver-Bever, B. 1983. Medicinal plants in tropical West Africa. III. Anti-infection therapy with higher plants. *J. Ethnopharmacol.* 9: 1–83

Oliver-Bever, B. 1986. *Medicinal plants in tropical West Africa*, 151–163, 168, 258. Cambridge: Cambridge University Press.

Ott, M., Thyagarajan, S.P., and Gupta, S. 1997. *Phyllanthus amarus* suppresses hepatitis B virus by interrupting interactions between HBV enhancer I and cellular transcription factors. *Eur. J. Clin Invest.* 27: 908–915.

Pal, D.C., and Jain, S.K. 1989. Notes on Lodha medicine in Midnapur district, West Bengal, India. *Econ. Bot.* 43: 464–470.

Perry, L.M., and Metzger, J. 1980. *Medicinal plants of East and South-east Asia: attributed properties and uses*, 149–151. Cambridge, MA: MIT Press.

Pio-Correa, M. 1969. *Dicionario das plants Uteis do Brasil e das Exoticas Cultivadas*, Vol. 4, 151–153, 192, 196. Desenvolvimento Florestal, Ministerio da Agricultura Rio de Janeior.

Qian-Cutrone, J., Huang, S., Trimble, J., et al. 1996. Niruriside, a new HIV REV/RRE binding inhibitor from *Phyllanthus niruri. J. Nat. Prod.* 59: 196–199.

Ramanan, M.V., and Sainani, G.S. 1961. Clinical trials with indigenous drugs Kari Manjal Karuppu and *Phyllanthus niruri* in infective hepatitis. *Punjab Med. J.* 10: 667–699.

Rao, Y.S. 1985. Experimental production of liver damage and its protection with *Phyllanthus niruri* and *Capparis spinosa* (both ingredients of LIV52) in white albino rats. *Probe* 24: 117–119.

Reddy, K.R. 1988. Folk medicine from Chittoor district, Andhra Pradesh, India used in the treatment of jaundice. *Int. J. Crude Drug Res.* 26: 137–140.

Roig, Y., and Mesa, J.T. 1945. *Plantas medicinates, aromaticas a venenosas de Cuba. Parte II*, 709–710. Havana, Cuba: Ministerio de Agricultura Servicio de Publicated Y Divulgacion.

Sahu, T.R. 1984. Less known uses of weeds as medicinal plants. *Ancient Sci. Life.* 3: 245–249.

Sankaran, J.R. 1977. Liver-Doks in treatment of viral hepatitis (a clinical trial). *Antiseptic* 74: 621–626.

Shead, A., Vickery, K., Medhurst, R., Freiman, J., and Cossart, Y. 1990. Neutralisation but not cure of duck hepatitis B by Australian *Phyllanthus* extracts. Abstract 602. In: *Scientific program and abstract volume, the 1990 International Symposium on Viral Hepatitis and Liver Disease*, April 4–8, Houston, TX.

Singh, Y.N. 1986. Traditional medicine in Fiji: some herbal folk cures used by Fiji Indians. *J. Ethnopharmacol.* 15: 57–88.

Sivarajan, V.V., and Balachandran, I. 1984. Botanical notes on the identity of certain herbs used in Ayurvedic medicines in Kerala. I. Thamalaki. *Ancient Sci. Life.* 4: 103–105.

Stehle, H., and Stehle, M. F. 1957. *Medicinale Illustree*, Vol. IX, Flore Agronomiique des Antiles Francaises, Anibal Lautric, Pointe-a-Pitre, Guadeloupe, 962: 102, 103.

Sundaravalli, N., Mohan, V.K.K., Ranganathan, G., and Raja, V.B. 1977. Liverite in viral hepatitis. *Antiseptic* 74: 135–142.

Symasundar, K.V., Singh, B., Thakur, R.S., et al. 1985. Antihepatotoxic principles of *Phyllanthus niruri* herbs. *J. Ethnopharmacol.* 14: 41–44.

Thyagarajan, S.P. 1979. Studies on hepatitis B virus infection in Tamilnadu, laying special emphasis on the immunological and biochemical markers with an assessment on the antiviral properties of certain indigenous herbs. PhD diss., University of Madras, India.

Thyagarajan, S.P. 1999. A pharmaceutical formulation from the Indian medicinal plant, *Phyllanthus amarus* for the treatment of hepatitis B and a process for its development. Indian Patent no. 405/MAS/99.12.04.1999.

Thyagarajan, S.P. 2000a. A pharmaceutical formulation from the Indian medicinal plant, *Phyllanthus amarus* for the treatment of hepatitis B and other viral infections with a process for its development. North Korea patent no.00-1180-11.12.2000.

Thyagarajan, S.P. 2000b. A pharmaceutical formulation from the Indian medicinal plant, *Phyllanthus amarus* for the treatment of hepatitis B and other viral infections with a process for its development. USA patent no. 09/719,486-12.12.2000.

Thyagarajan, S.P. 2001. A pharmaceutical formulation from the Indian medicinal plant, *Phyllanthus amarus* for the treatment of hepatitis B and other viral infections with a process for its development. South African patent no. 20010148-05.01.2001.

Thyagarajan, S.P., and Jayaram, S. 1992. Natural history of *Phyllanthus amarus* in the treatment of hepatitis B. *Indian J. Med. Microbiol.* 10: 64–80.

Thyagarajan, S.P., Jayaram, S., Valliammai, T., et al. 1990. *Phyllanthus amarus* and hepatitis B. *Lancet* 2: 949–950.

Thyagarajan, S.P., Jayaram, S., Gopalakrishnan, V. et al., 2002. Herbal medicines for liver diseases in India. *J. Gastroenterol. Hepatol.* 17: 5370–5376.

Thyagarajan, S.P., Subramanian, S., Thirunalasundari. T., Venkateswaran, P.S., and Blumberg, B.S. 1988. Effect of *Phyllanthus amarus* on chronic carriers of hepatitis B virus. *Lancet* 2: 764–766.

Thyagarajan, S.P., Thiruneelakantan, K., Subramanian, S., and Sundaravelu, T. 1982. *In vitro* inactivation of HBsAg by *Eclipta alba* Hassk. and *Phyllanthus niruri* Linn. *Indian J. Med. Res.* 76 Suppl: 124–130.

Unander, D.W., and Blumberg, B.S. 1991. *In vitro* activity of *Phyllanthus* species against the DNA polymerase of hepatitis viruses: effects of growing environment and inter and intra specific differences. *Econ. Bot.* 45: 225–242.

Unander, D.W., Venkateswaran, P.S., Millman, I., Bryan, H.H., and Blumberg, B.S. 1990a. *Phyllanthus* species: sources of new antiviral compounds. In *Advances in new crops*, ed. J. Janick and J.E. Simon, 518–521. Portland, OR: Timber Press.

Unander, D.W., Webster, G.L., and Blumberg, B.S. 1990b. Records of usage or assays in *Phyllanthus* (Euphorbiaceae). I. Subgenera Isocladus, Kirganelia, Cicca and Emblica. *J. Ethnopharmacol.* 30: 233–264.

Unander, D.W., Webster, G.L., and Blumberg, B.S. 1991. Uses and bioassays in *Phyllanthus* (Euphorbiaceae): a compilation. II. The subgenus *Phyllanthus*. *J. Ethnopharmacol.* 34: 97–133.

Velazco, E.A. 1980. Herbal and traditional practices related to maternal and child health care. *Rural Reconstruction Rev.* 2: 27–31.

Venkateswaran, P.S., Millman, I., and Blumberg, B.S. 1987. Effects of an extract from *Phyllanthus niruri* on hepatitis B and woodchuck hepatitis viruses: *in vitro* and *in vivo* studies. *Proc. Natl. Acad. Sci. U S A* 84: 274–278.

Wang, M., Zhou, H.B and Zhao, G.I. 1991a. Effect of *Phyllanthus urinaria* on HBeAg negative conversion. *Hepatol. Rapid Literature Rev.* 21: 25.

Wang, M., Zhou, H.B., Zhao, G.I., Zhao, S., and Mai, K. 1991b. *Phyllanthus amarus* cannot eliminate HBsAg in chronic hepatitis B virus infection. *Hepatol. Rapid Literature Rev.* 21: 22–24.

Watt, G. A. 1892. *Dictionary of economic products of India*, Vol. 6, Part 1, 222–224. London: Allen.

Wealth of India. 1969. *New Delhi, India*: Council of Scientific and Industrial Research, 34–36.

Yanagi, M., Unoura, M., Kobayashi, K., Hattori, N., and Murakami, S. 1989. Inhibitory effect of an extract from *Phyllanthus niruri* on reaction of endogenous HBV—DNA polymerase and other DNA synthetases. In *Abstracts of papers presented at the 1989 meeting on hepatitis B viruses*, 77. Cold Spring Harbor, NY: Cold Spring Harbor Laboratory.

14 Antiviral Activities of *Phyllanthus orbicularis*, an Endemic Cuban Species

Gloria del Barrio and Francisco Parra

CONTENTS

14.1 Introduction 219
14.2 Plant Description, Classification, and Properties 220
    14.2.1 Chemical Constituents 222
14.3 Antiviral Activities 224
    14.3.1 Hepatitis B Virus 225
    14.3.2 Herpesvirus 226
    14.3.3 Enterovirus 230
14.4 Conclusion 230
Acknowledgments 230
References 230

14.1 INTRODUCTION

In recent years, traditional medicine has received increasing attention from the scientific community as an alternative for the development of novel therapeutic approaches. The plants could offer a rich reserve for drug discovery against infectious diseases, particularly when the human population is challenged by new emerging infectious diseases (Mukhtar et al., 2008), the appearance of viral resistance against most of the available drugs, together with their side effects, as well as the ability of some viruses to develop latent or recurrent infections. Hence, there is an urgent need to develop new antivirals of a natural origin.

Phyllanthus L. is a pantropical and subtropical genus of herb and shrub species and is the largest genus (more than 600 species) of the Phyllanthaceae. This family contains about 2,000 species in 59 genera. The Phyllanthaceae is a plant family, previously known as subfamily Phyllantoideae of the Euphorbiaceae. The subfamily

Phyllantoideae was split from Euphorbiacea as a result of lineage analysis based on nuclear ribosomal internal transcribed spacer (ITS) regions and plastid matK DNA sequences (Kathriarachchi et al., 2005, 2006).

The genus *Phyllanthus* has a long history of use in folklore medicine (traditional medicine) for the treatment of many diseases. In recent years, several species have been studied, and their biological properties have been demonstrated (Calixto et al., 1998; Jassim and Naji, 2003) using extracts from different parts of the plant as well as plant-derived products. Several therapeutic properties have been attributed to the genus *Phyllanthus*, such as antibacterial, antiparasitic, antigenotoxic effects as well as chondroprotective potential (Sanchez-Lamar et al., 1999; Ferrer et al., 2004; Lin et al., 2008; Londhe et al., 2008, 2009; Sumantran et al., 2008).

The genus *Phyllanthus* has many growth forms, including annual and perennial herbaceous, arborescent, climbing, floating aquatic, pachycaulous, and phyllocladous. It has a wide variety of floral morphologies and chromosome number and has one of the widest varieties of pollen types.

Cuba is extremely biodiverse, with a rich and vast flora, with a 51% endemism (Fuentes, 2004), representing an excellent natural resource for drug discovery programs. Nevertheless, the plants as a source of antiviral products have scarcely been tested. Of all the identified *Phyllanthus* species, 53 taxa have been reported in Cuba (Hno.León and Hno.Alain, 1953), and only a small number of them have been investigated for the presence of relevant biological activities. Approximately half of these grow on ultramafic (also termed ultrabasic) soils; the rest are not exclusive to these substrates and can develop on other types of soils. Of the 45 taxa analyzed, 24 are nickel hyperaccumulators: 23 from Cuba and 1 from Hispaniola (Dominican Republic) (Berazain et al., 2007). Some species, such as *P. chamacristoides, P. acidus, P. epiphyllanthus, P. discolor,* and *P. xpallidu,* all collected from Cuban ultramafic soils, have been studied for various purposes, looking for different biological activities. *Phyllanthus chamacristoides, P. acidus,* and *P. epiphyllanthus* were screened for hepatitis B surface antigen (HBsAg) inhibition capacity (del Barrio et al., 1995) and *P. discolor* and *P. xpallidu* for nickel accumulation (Berazain et al., 2007). In this chapter, we summarize the main properties and antiviral studies performed on *Phyllanthus orbicularis*.

14.2 PLANT DESCRIPTION, CLASSIFICATION, AND PROPERTIES

Phyllanthus orbicularis Kunth (common local name: Alegría) is a Cuban endemic species growing in xerophytic thorny scrub on serpentine (ultramafic) soils in all Cuban provinces, being locally abundant in some of them. It is a perennial evergreen shrub (Roig y Mesa, 1974) 1–2 m high (Figure 14.1), branched, with cataphylls and reddish young stems. Leaves are simple, alternate, and stipuled; leaf blades are orbicular or nearly round (6–10 mm long, 7–9 mm wide) with a rounded base or slightly cordiform, coriaceous, margin plane, with pinnate venation and veins prominule on

FIGURE 14.1 See color insert. *Phyllanthus orbicularis,* a Cuban endemic species. Insert: A twig with flowers.

both surfaces (Figure 14.1). *Phyllanthus orbicularis* is a monoecious plant with solitary white-pink unisexual pedunculate flowers (Figure 14.1). The male flowers have six obtuse, rounded sepals (2 mm long) and 4–10 stamens joined at the base. The female flowers have six lanceolate acute sepals (3 mm long), a subglobose ovary, and a three-lobed style, with each lobe divided in two. The rounded fruit capsule (5–6 mm diameter) has two small seeds in each locule (Hno.León and Hno.Alain, 1953).

The taxonomy of this family and genus has been confused and only recently, using molecular phylogenetic studies, has the classification of the different species regarding existing taxonomic categories been possible. *Phyllanthus orbicularis* has been placed in clade L: subgenus *xylophylla*, section Orbicularia (Kathriarachchi et al., 2006).

Berazain et al. (2007) reported that this plant was a nickel hyperaccumulator, and significant concentrations of this metal have been found in different leaf and stem tissues. The nickel localization observed for these *Phyllanthus* species raised the question of whether it was a general characteristic of the Euphorbiaceae family or a peculiar feature of the *Phyllanthus* genus. The hypothesis that nickel hyperaccumulation is a protective strategy against microbial pathogenesis or predation seems reasonable (Boyd and Moar, 1999), although the homogeneous distribution of nickel in leaf epidermis of Euphorbiaceae species living in extreme xeric habitats points to alternative functional roles. Berazain et al. (2007) described, in the case of Euphorbiaceae species, that the high concentration of nickel found in their leaves could prevent the oxidative stress that results from exposure to high radiation intensities through the ultraviolet (UV) radiation shielding exerted by this metal at specific regions of the UV spectrum. This shielding has been described for algae growing in high concentrations of ferric iron solutions (Gómez et al., 2004). Studies aiming to evaluate this additional function for nickel in hyperaccumulator plant species grown in xeroedaphic climatic conditions are currently in progress.

14.2.1 Chemical Constituents

Different species of *Phyllanthus* have been phytochemically characterized (see accompanying chapters), and many molecules have been isolated and identified (Calixto et al., 1998; Bagalkotkar et al., 2006). In this context, *P. orbicularis* has been subject to particular attention from our group.

Technical concerns related to which parts of the plants to use in antiviral screenings are frequently discussed by authors. Since most *Phyllanthus* species are commonly small herbs, the whole plant is often used. However, as mentioned, *P. orbicularis* is a herbaceous plant, so we considered it worthwhile to include stems together with leaves in biological evaluations.

The first studies were focused on ascertaining through phytochemical sieving tests the nature of the main families of compounds present in raw plant extracts (Gutierrez-Gaiten et al., 2000). Using these techniques, flavonoids, alkaloids, coumarins, saponins, amino acids, anthocyanidins, mucilages, triterpens, steroids, reducing substances, pyrocathecolic tannins, bitter and astringent metabolites, glycosides and quinones were found (Table 14.1), showing a similar chemical composition to other previously studied species (Yao and Zuo, 1993; Zhang et al., 2000; Bagalkotkar et al., 2006; Obianime and Uche, 2009), such as *P. urinaria*, *P. emblica*, and *P. amarus*.

Despite the fact that *P. orbicularis* grows abundantly everywhere in Cuba, the harvests used in these studies were from a single place in the country (Cajalbana, in the province of Pinar del Rio, the most occidental of the country), and the samples were taken at the same hour of the day and from plants with similar physiological conditions. It is relevant to point out that these compounds were present in all harvests made at different times of the year.

Although the main compound families found in *P. orbicularis* extracts (Table 14.1) were always present, no matter what the climatic conditions at the time of crop were, their concentrations strongly varied depending on the season. This was especially true for tannins and flavonoids, which were studied in more detail due to their reported biological properties (Table 14.2).

On the basis of preliminary findings concerning the biological activity of crude extracts, our group has investigated the main metabolites present in *P. orbicularis*, following a bioactivity-guided protocol, using reverse-phase high performance liquid chromatography (HPLC). The major peaks observed corresponded to phenolic compounds, and the predominant families detected (Table 14.3) were flavanols, flavonol glycosides, condensed tannins (procyanidin dimers and procyanidin trimers), and gallic acid-derivatives. Quercetin-3-O-rutinoside (rutin), kaempferol-3-O-rutinoside, procyanidin B1, procyanidin B2, catechin, epicatechin, and protocatechuic acid were readily identified in the extract (Alvarez et al., 2009a). These compound families have also been found in other species, such as *P. urinaria* (Zhong et al., 1998; Yao and Zuo 1993; Zhang et al., 2000) and *P. emblica* (Habib et al., 2007).

Structure-activity relationships for metabolites found in *P. orbicularis* extracts are now being investigated, and further fractionation procedures are being carried out to investigate whether a unique molecule among these or its synergy contribute to the observed antiviral effects.

TABLE 14.1
Phytochemical Sieving of *Phyllanthus orbicularis*

		Time of Harvest			
Compound	Analytical Method	January	May	August	October
Saponins	Foam	+	+	+	+
Alkaloids	Graguendorf	+	+	+	+
Alkaloids	Mayer	+	+	+	+
Alkaloids	Wagner	+	+	+	+
Flavonoids	Shinoda	+	+	+	+
Anthocyanidins	Anthocyanidin	+	+	+	+
Reducing substances	Fehling	+	+	+	+
Amino acids	Ninhidrin	−	−	−	−
Phenols/tannins	Ferric triclorure	+	+	+	+
Tannins	Gelatin	+	+	+	+
Mucilages	Mucilage	−	−	−	−
Resins	Resin	−	−	−	−
Quinones	Bortrager	+	+	+	+
Cardiotonic glycosides	Kedde	−	−	−	−
Fatty compounds	Sudam	−	−	−	−
Triterpens/steroids	Lieberman-Burchard	+	+	+	+
Glycosides	Molish	+	+	+	+
Coumarins	Baljet	+	+	+	+
Bitter and astringent substances	Flavor	+	+	+	+

+, presence; −, absence (or not detected).

TABLE 14.2
Percentage of Tannins and Flavonoids in *P. orbicularis* Harvests Performed at Different Times of the Year

Time of Harvest	Tannins (% of Mean Values ± SD)	Flavonoids (% of Mean Values ± SD)
January	2.40 ± 0.24	12.11 ± 0.08
May	1.29 ± 0.05	13.08 ± 0.18
August	11.03 ± 0.17	4.75 ± 0.23
October	11.89 ± 0.14	3.2 ± 0.13

TABLE 14.3
Phyllanthus orbicularis Main Families of Compounds and Metabolites

Compound Family	Metabolites
Flavonol glycosides	Quercetin-3-O-rutinoside (rutin)
	Kaempherol-3-O-rutinoside
Flavan-3-ols	Catechin
	Epicatechin
Condensed tannins	Procyanidin B1
	Procyanidin B2
	Procyanidin trimer
	Procyanidin oligomeric and polymeric
Hydrolyzable tannins	Epicatechin gallate
	Procyanidin dimer-gallate
	Procyanidin dimer-digallate
	Procyanidin trimer-gallate
Acidic phenolics	Protocatechuic acid
Flavanone	Not studied

14.3 ANTIVIRAL ACTIVITIES

A wide array of ethnomedicinal plants have shown high levels of antiviral activities, and many of them have complementary and overlapping mechanisms of action, inhibiting either viral propagation or viral genome synthesis (Chattopadhyay and Naik, 2007).

Over the last few years, significant efforts have been made to set up a range of strategies for the identification of potential new antiviral drugs with different mechanisms of action (Greco et al., 2007). In this sense, natural products have the advantage of their amazing structural diversity, and many ethnomedicinal plant extracts represent true cocktails that inhibit several replication steps of many viruses.

The study of the antiviral activity of the genus *Phyllanthus* began about two decades ago when *in vitro* inactivation of HBsAg by *P. niruri* was reported (Thyagarajan et al., 1982).

Following this initial study, several publications showed the *in vitro* and *in vivo* activities of several *Phyllanthus* species, such as *P. niruri*, *P. amarus*, *P. urinaria*, *P. chamacristoides,* and others, in destroying or interfering with antigen detection (Thyagarajan et al., 1988; Thamlikitkul et al., 1991; Yeh et al., 1993; Milne et al., 1994; Unander et al., 1995; del Barrio et al., 1995; Liu et al., 2001) and showing their mechanisms of action in cells (Lee et al., 1996; Ott et al., 1997; Yeh et al., 1993). In recent years, the interest in the plants of the *Phyllanthus* genus has increased considerably, especially regarding their therapeutic potential for the management of many diseases. Viral infections, particularly those associated with sexual transmission and newly emerging infectious viruses, have a crucial importance (Notka et al., 2003, 2004; Bagalkotkar et al., 2006; Lam et al., 2006). The data from ethnobotany, *in vitro* assays, and clinical trials with different species of *Phyllanthus* have fostered further studies of the antiviral activity of *P. orbicularis*, a previously uninvestigated Cuban endemic species. In the following sections, we summarize some of the data obtained against different types of viruses.

14.3.1 Hepatitis B Virus

Hepatitis B virus (HBV) is the prototype of the family *Hepadnaviridae*. HBV infection, a major health concern worldwide, can be acquired by sexual contact or through body fluid transmission. Only a few inhibitors, such as interferonα, nucleoside inhibitors of the viral polymerase, 3TC (Lamivudine), and adefovir, are approved for HBV therapy. However, these treatments are limited by their side effects and the substantial resistance of the virus. Therefore, new antiviral compounds suitable for monotherapy or combination therapy are needed. HBV cannot be propagated in culture systems, and several *in vitro* assays using transfected cultured cells expressing selected viral antigens or the complete replicating genome provide an alternative means to study anti-HBV activities (Yeh et al., 1993; Lee et al., 1996; Lam et al., 2006; Liu et al., 2007).

HBsAg has proven to be a significant risk factor in HBV-induced liver diseases, and an increasing number of mutations in HBsAg are known to enhance the difficulties for therapeutic intervention. From this viewpoint, the early studies were focused on assessing the inhibitory capacity of crude aqueous extracts against HBsAg *in vitro* (del Barrio et al., 1995; del Barrio, 1999). Using similar methods as those previously described (Thyagarajan et al., 1982), the effects of *P. orbicularis* crude aqueous extracts on the detection of HBsAg were investigated after treatment *in vitro* of the sera samples with the plant extract. These studies showed a significant decrease of the amounts of HBsAg detected when the positive sera were incubated with the plant extract *in vitro* (del Barrio et al., 1995). This activity was confirmed from different samples of *P. orbicularis* harvested at different times of the year from the same place (Cajalbana) and plants in the same physiological state (del Barrio et al., 2001). The data showed that these extracts were capable of inactivating or somehow interfering with HBsAg detection, independently of the *Phyllanthus* harvest time. The number of sera that reverted to antigen negative, as measured by enzyme-linked immunosorbent assay (ELISA), was dependent on the concentrations of the extract used. We have only found minor differences between the harvests on August 1993 and October 1994 when smaller concentrations of extracts were used. The data from the phytochemical sieving of these extracts showed the presence of identical metabolites in the different harvests or collections of *P. orbicularis*, but their quantities varied, as can be seen with the value of tannin and flavonoids (Table 14.2).

Further studies have evaluated the effects of *P. orbicularis* extracts on HBsAg particle production by the human hepatoma cell line PLC/PRF/5 (Alexander hepatoma cell line). The aqueous, butanolic, and acetic *P. orbicularis* extracts inhibited the production of HBsAg at 24 and 48 h after treatment, and this activity was plant extract concentration dependent. The selectivity indexes obtained in all cases were above 10 (González et al., 2006). These results were similar to the ones found for

P. urinaria and *P. amarus* using the same cell model (Ji et al., 1993; Jayaram and Thyagarajan, 1996).

The *P. orbicularis* extracts were also evaluated for their anti-HBV activities *in vitro* using the HBV transfected hepatocellular carcinoma HepG2.2.15 cell line, able to continuously produce infectious human HBV particles (Sells et al. 1987). Our preliminary studies indicated that *P. orbicularis* aqueous extract was able to inhibit *in vitro* the expression of HBsAg and *hepatitis B e antigen* (HBeAg) (in culture supernatants measured by ELISA), and HBV DNA replication (evaluated by real-time polymerase chain reaction [PCR]) in a dose-dependent manner.

These findings are consistent with the notion that *P. orbicularis* extracts have inhibitory effects against HBV. The mechanisms involved are currently being studied using several of the experimental systems discussed. Clinical trials have not yet been conducted.

14.3.2 Herpesvirus

The family Herpesviridae includes at least eight viral species pathogenic for humans and several animal species responsible for a wide variety of clinical symptoms. *Herpes simplex virus* (HSV) infection is a major opportunistic infection, and human cytomegalovirus (HCMV) is an important cause of morbidity, in both immunosuppressed patients, often associated with other viral infections, and as a result of solid organ or stem cell transplants. These are therefore serious diseases in areas with high HIV/AIDS prevalence (Whitley et al., 1998; Razonable and Paya, 2003). The treatment of HSV diseases is an important goal because herpetic infections are not controlled by vaccination and due to the emergence of resistant strains against the available drugs currently used (Superti et al., 2008). For these reasons, there is an urgent need of alternative antiherpetic treatments, and for this purpose the natural plant products represent an affordable strategy.

A large number of synthetic and plant-derived anti-HSV drugs have been described, including some used in alternative medicines such as Unani, Chinese, and Ayurvedic (Yoosook et al., 2000; Khan et al., 2005; Mukhtar et al., 2008). Our early data (del Barrio and Parra, 2000) showed the relevance of the genus *Phyllanthus* as a source of effective antivirals based on the antiviral activity found against human herpes simplex type 2 and bovine herpesvirus type 1 in tissue cultures. This activity was further confirmed in other species, such as *P. urinaria* (Yang et al., 2005, 2007). The aqueous extracts obtained from leaves and stems of this plant were evaluated against bovine herpesvirus (BHV) type 1 and herpes simplex type 2, adenovirus and mengovirus in MDBK (Madin Darby bovine kidney), HeLa, Hep 2, and HFF (human foreskin fibroblast) cells. The results demonstrated that these extracts exhibited significant antiviral activity on the alphaherpesviruses, with mean EC_{50} (median effective concentration to produce a 50% effect) values of 21.1 and 25.7 μg/ml against BHV-1 and HSV2, respectively. The average selective indexes (SIs) of the extracts were 12.3 against BHV-1 and 26.01 against HSV-2, while no inhibitory activity was detected against adenovirus type 7 and mengovirus (del Barrio and Parra, 2000). Complementary studies indicated virucidal extracellular effects against herpesviruses but not on the nonenveloped adenovirus (Ad7) and mengovirus assayed (del Barrio and Parra, 2000).

The investigation of antiherpesvirus activity in *P. orbicularis* extracts against HSV, including acyclovir- (ACV) resistant and sensitive clinical isolates, together with laboratory strains of HSV-1 and HSV-2, confirmed the presence of anti-HSV compounds in this plant species. Our laboratory has reported (Alvarez et al., 2009a) that *P. orbicularis* water extract was capable of reducing the virus-induced cytopathic effect (CPE) on Vero (African green monkey) cells by more than 50%. The SI values ranged from 13.3 to 37.6, and no significant differences were found between the assayed HSV strains, except for HSV-1 reference strains F, 8WT, and 8ACV, whose sensitivities to the plant extract were higher than those of clinical isolates. The virucidal activity of the extracts was also demonstrated. These data confirmed previous results (Fernandez-Romero et al., 2003) using butanolic and acetic acid-soluble fractions prepared from the leaves and stems of *P. orbicularis* against acyclovir-sensitive or -resistant HSV-1 strains using HFFs and Vero cells. The results showed antiviral selectivity indexes from 10.3 to 22.8 and very high extracellular virucidal activities on HSV-1 virions, with SIs ranging from 371 to 1,040. These data also highlighted that the observed anti-HSV-1 and -HSV-2 activities were not related to virus resistance or sensitivity against ACV, further supporting the relevance of *P. orbicularis* compounds against herpesvirus infection.

Studies of the mechanism of action of *P. orbicularis* fractions suggested that the strongest inhibition occurred when the extract was added before or during the initial stages of infection. The incubation of virus suspensions with different concentrations of the *P. orbicularis* extract drastically diminished viral infectivity, independent of the HSV isolate, HSV type, or its ACV sensitivity. Although the virucidal effect appeared to be the main activity of the *P. orbicularis* fractions, we cannot rule out that other early events in the HSV replication cycle could also be affected.

Since medicinal plants have been used for the treatment of many diseases without detailed knowledge of their active compounds and mechanisms of action, it is nevertheless advisable to investigate the scientific reasons supporting their use. To this end, we have used an antiviral-guided separation protocol to identify *P. orbicularis* compounds, or families of compounds, responsible for the observed antiviral activity (Alvarez et al., 2009b). The crude methanol extract (Table 14.4) was the most active fraction (SI > 112.9), and it was selected for further size exclusion chromatography analyses. The highest antiviral activity was found in fractions F7 (containing substantial amounts of flavanols, procyanidins, and their gallates) and F8 (containing minor amounts of flavanol and two major compounds with procyanidin gallate-type spectra) (Figure 14.2). Both fractions (F7 and F8) were analyzed using reverse-phase semipreparative HPLC, indicating that their strong capacity for inhibiting HSV-2 *in vitro* replication could be related to the presence of epicatechin gallate and procyanidins, such as B1 and B2, in *P. orbicularis* extracts.

We have also investigated *P. orbicularis* activity against other Herpesviridae members, such as HCMV, belonging to the Betaherpesvirinae subfamily, using primary assays based on the inhibition of the CPE in infected cultures of human embryo lung cells (HEL fibroblasts). A virucidal extracellular activity on HCMV virions

TABLE 14.4
Anti-HSV-2 Activity of *P. orbicularis* Samples

Sample	Anti HSV-2 Activity		
	CC_{50} (µg/ml)	EC_{50} (µg/ml)	SI
ME	>2,500	22.15 ± 4.2	**>112.9**
AE	1,288.2 ± 16.7	71.7 ± 12.9	17.9
MEBu	>2,500	51.96 ± 2.99	>48.1
AEBu	>2,500	60.73 ± 12.4	>41.2
F1	2,356.7 ± 94.3	>100	<23.6[a]
F2	923.18 ± 52.9	>100	<9.23[a]
F3	786.78[b]	>100	<7.87[a]
F4	1,057.8[b]	>100	<10.6[a]
F5	768.6 ± 26.7	44.84 ± 1.9	17.1
F6	532.2 ± 67.4	38.27 ± 0.72	13.9
F7	370.49 ± 40.7	15.3 ± 1.22	**24.2**
F7-A[c]	1:4.103	1:214.1	**52.2**
F7-B[c]	1:4.18	1:144.9	**34.7**
F7-C[c]	NT	1:2.38	>2.38
F8	252.16[b]	12.7 ± 2.1	**19.9**
F8-A[c]	NT	1:4.88	>4.88
F8-B[c]	NT	1:27.58	**>27.5**
F8-C[c]	1:3.16	1:84.2	**26.6**
F9	246.02[b]	21.16 ± 2.28	11.6
F10	385.7 ± 72.9	49.08 ± 1.6	7.86
Rutin[d]	>1,000	>1,500	ND
Catechin[d]	973.8[b]	>1,000	<0.97
B1[d]	327.4 ± 4.49[e]	56.77 ± 3.43[e]	5.77
B2[d]	559.4 ± 57.4[e]	45.0[b,e]	12.4
ACV	>100[b,e]	0.744[b,e]	>134.4

ACV, acyclovir; NT, nontoxic; ND, not analyzed; SI, CC_{50}/EC_{50}.

[a] Since the maximum concentration used in antiviral assays did not allow the quantification of the EC_{50} value, anti-HSV-2 activity could not be confirmed.

[b] Value from an individual experiment.

[c] Cytotoxic and antiviral activities of F7- and F8-derived subfractions were expressed as CD_{50} (dilution reducing cell viability by 50%) and ED_{50} (dilution reducing viral-induced cytopathic effect by 50%), respectively.

[d] Commercial standard used as reference.

[e] Concentration expressed as micromoles.

[f] Items in bold indicate highest antiviral activity.

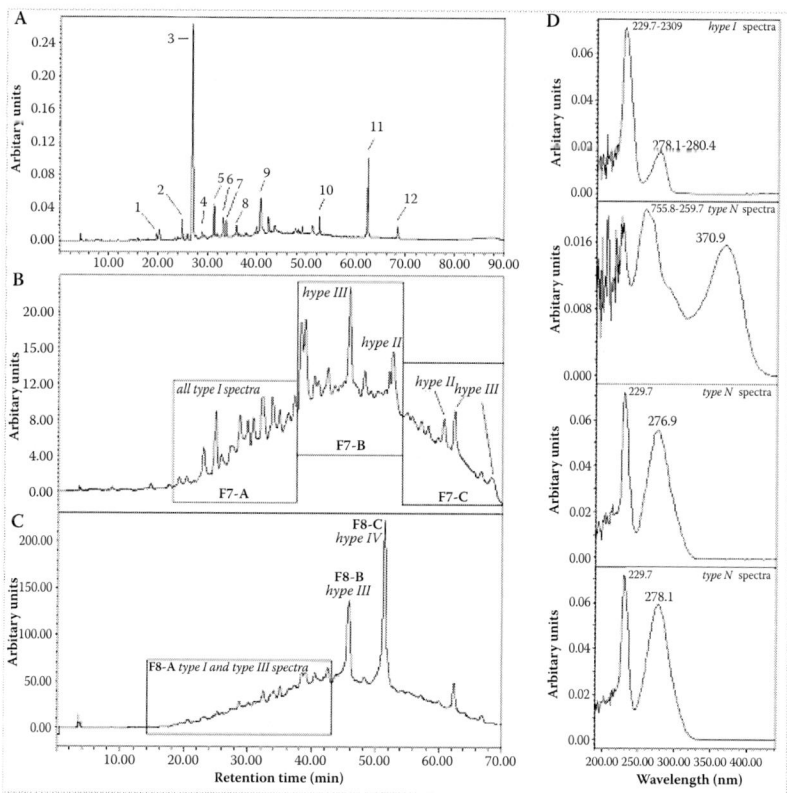

FIGURE 14.2 Reverse-phase HPLC analyses analyzed at 280 nm: (A) *P. orbicularis* methanol extract (ME): 1, protocatechuic acid; 2, 3, 6, and 7, unknown flavonoids; 4, procyanidin B1; 5, catechin; 8, procyanidin B2; 9, epicatechin; 10, unknown hydrolyzable tannin; 11, quercetin-3-O-rutinoside; 12, kaempherol-3-O-rutinoside. (B) and (C) F7 and F8 fractions in semipreparative conditions. Rectangles indicate the regions collected in subfractions. (D) UV-visible spectra: type I spectra correspond to flavan-3-ol-like compounds and procyanidins; type II spectra are typical of flavonol family; type III and IV spectra are characteristic of compounds containing gallic groups.

was also found, although to a lesser extent in comparison with the effects on the HSV strains. Follow-up studies conducted with HCMV demonstrated a decrease in HCMV DNA concentrations, measured by a real-time PCR, in treated cultures with respect to virus controls in which the plant extract was added after virus adsorption. From these data, it can be concluded that *P. orbicularis* extracts have mainly virucidal activity but could also inhibit some early stages of the HSV cycle prior to viral DNA synthesis (immediate-early and early replication stages).

The viral envelope plays a major role in the infectivity of permissive cells and mediates interactions with cell receptors and viral coreceptors. For these reasons, a compound capable of blocking this interaction by either virucidal activity or blocking the adsorption to or penetration into the cell is a potential antiviral drug candidate.

14.3.3 ENTEROVIRUS

Human enteroviruses infect millions of people worldwide each year, resulting in a wide range of clinical outcomes, from unapparent infection to mild respiratory illness (common cold); hand, foot, and mouth disease; acute hemorrhagic conjunctivitis; aseptic meningitis; myocarditis; severe neonatal sepsis-like disease; and acute flaccid paralysis (Melnick, 1990).

The preliminary studies of *P. orbicularis* antiviral activity performed in our laboratory (unpublished results), including *in vitro* assays against several strains of enterovirus, such as the vaccine strain Sabin 1; poliovirus strain Ls-c 2ab (PV1); coxsackievirus B5 reference strain Faulkner (CVB5); coxsackievirus A16 reference strain G10 (CVA16); and echovirus 9 reference strain Hill (E9). The SI of the plant extract against these viruses ranged from 7.1 to 9.9. However, the virucidal extracellular activity showed significant differences between the different assayed strains. In particular, coxsackievirus A16 was not inhibited in the virucidal assay. No previous studies are available in literature describing antienteroviral activity from this plant species.

14.4 CONCLUSION

Based on the data summarized in this review, it can be concluded that *P. orbicularis* offers a variety of antiviral compounds. Thus, screening programs aimed at the identification of potential antiviral agents from this species have great potential in the field of pharmaceutical development.

ACKNOWLEDGMENTS

We are very grateful to Kevin P. Dalton for critical revising of the manuscript. This work was partially funded by the Ayuntamiento de Gijón, the University of Oviedo (Asturias, Spain), the University of Havana (Havana, Cuba), and AECID PCI grant D/023290/09.

REFERENCES

Alvarez, A. L., del Barrio, G., Kouri, V., et al. 2009a. *In vitro* anti-herpetic activity of an aqueous extract from the plant *Phyllanthus orbicularis*. *Phytomedicine* 16: 960–966.

Alvarez, A. L., Diñeiro, Y., del Barrio, G., et al. 2009b. Bioactivity-guided separation of anti HSV-2 and antioxidant metabolites from the plant *Phyllanthus orbicularis*. *Planta Med.* 75: 990.

Bagalkotkar, G., Sagineedu, S. R., Saad, M. S., and Stanslas, J. 2006. Phytochemicals from *Phyllanthus niruri* Linn. and their pharmacological properties: a review. *J. Pharm. Pharmacol.* 58: 1559–1570.

Berazain, R., de la Fuente, V., Sanchez-Mata, D., et al. 2007. Nickel localization on tissues of hyperaccumulator species of *Phyllanthus* L. (Euphorbiaceae) from ultramafic areas of Cuba. *Biol. Trace Elem. Res.* 115: 67–86.

Boyd, R. S., and Moar, W. J. 1999. The defensive function of Ni in plants: response of the polyphagos herbivore *Spodoptera exigua* (lepidoptera: Noctuidae) to hyperaccumulator and accumulator species of *Streptanthus* (Brassicacecae). *Oecologia* 118: 218–224.

Calixto, J. B., Santos, A. R., Cechinel, F., et al. 1998. A review of the plants of the genus *Phyllanthus*: their chemistry, pharmacology, and therapeutic potential. *Med. Res. Rev.* 18: 225–258.

Chattopadhyay, D., and Naik, T. N. 2007. Antivirals of ethnomedicinal origin: structure-activity relationship and scope. *Mini. Rev Med. Chem.* 7: 275–301.

Cimanga, R. K., Tona, L., Luyindula, N., et al. 2004. *In vitro* antiplasmodial activity of callus culture extracts and fractions from fresh apical stems of *Phyllanthus niruri* L. (*Euphorbiaceae*): part 2. *J. Ethnopharmacol.* 95: 399–404.

del Barrio, G. 1999. Actividad antiviral *in vitro* del extracto acuoso de *Phyllanthus orbicularis* HBK. PhD diss., Facultad de Biología, Havana, Cuba.

del Barrio, G., Caballero, O., and Chevalier, P. 1995. Inactivación *in vitro* del AgsHB por extractos de plantas del género *Phyllanthus*. *Rev. Cubana Med. Trop.* 47: 127–130.

del Barrio, G., and Parra, F. 2000. Evaluation of the antiviral activity of an aqueous extract from *Phyllanthus orbicularis*. *J. Ethnopharmacol.* 72: 317–322.

del Barrio, G., Roque, A., and Arias, M. 2001. Estudio comparativo de la capacidad inactivante de distintas colectas de la planta *Phyllanthus orbicularis*. *Rev. Cubana Plant Med.* 3: 83–86.

Fernandez-Romero, J. A., del Barrio, G., Romeu, B., et al. 2003. *In vitro* antiviral activity of *Phyllanthus orbicularis* extracts against herpes simplex virus type 1. *Phytother. Res.* 17: 980–982.

Ferrer, M., Cristofol, C., Sanchez-Lamar, A., et al. 2004. Modulation of rat and human cytochromes P450 involved in PhIP and 4-ABP activation by an aqueous extract of *Phyllanthus orbicularis*. *J. Ethnopharmacol.* 90: 273–277.

Fuentes, V. 2004. Biodiversidad de las especies medicinales. *Rev. Cubana Plant Med.* No. 3, 9.

Gómez, F., Grau, A., Vázquez, L., and Amils, R. 2004. UV radiation effects over microorganisms and study of protective agents. *ESA SP.* 545: 21–25.

González, R., Quintero, A., Morier, L., and Rodríguez, L. 2006. Evaluación de la actividad antiviral de plantas medicinales frente al virus de la hepatitis B (VHB) en células PLC/PRF/5. *Rev. Cubana Med. Trop.* 58: 103–08.

Greco, A., Diaz, J. J., Thouvenot, D., and Morfin, F. 2007. Novel targets for the development of anti-herpes compounds. *Infect. Disord. Drug Targets* 7: 11–18.

Gutierrez-Gaiten, Y., Miranda, M., Hernández, S. T., and Del, B. G. 2000. Estudio de algunos fitoconstituyentes de *Phyllanthus orbicularis* HBK. *Rev. Cubana Farm.* 34: 299–300.

Habib, U. R., Yasin, K. A., Choudhary, M. A., et al. 2007. Studies on the chemical constituents of *Phyllanthus emblica*. *Nat. Prod. Res.* 21: 775–781.

Hno.León, F. S. C., and Hno.Alain, F. S. C. 1953. Flora de Cuba. t. III. Dicotiledóneas: *Malpighiaceae* a *Myrtaceae*, 44–59. Havana, Cuba: Imp. P. Fernández y Cía.,.

Jassim, S. A., and Naji, M. A. 2003. Novel antiviral agents: a medicinal plant perspective. *J. Appl. Microbiol.* 95: 412–427.

Jayaram, S., and Thyagarajan, S. P. 1996. Inhibition of HBsAg secretion from Alexander cell line by *Phyllanthus amarus*. *Indian J. Pathol. Microbiol.* 39: 211–215.

Ji, X. H., Qin, Y. Z., Wang, W. Y., et al. 1993. [Effects of extracts from *Phyllanthus urinaria* L. on HBsAg production in PLC/PRF/5 cell line]. *Zhongguo Zhong. Yao Za Zhi* 18: 496–498, 511.

Kathriarachchi, H., Hoffmann, P., Samuel, R., et al. 2005. Molecular phylogenetics of Phyllanthaceae inferred from five genes (plastid atpB, matK, 3, ndhF, rbcL, and nuclear PHYC). *Mol. Phylogenet. Evol.* 36: 112–134.

Kathriarachchi, H., Samuel, R., Hofmann, P., et al. 2006. Phylogenetics of tribe Phyllantheae (Phyllanthaceae; Euphorbiaceae sensu lato) based on nrITS and plastid matK DNA sequence data. *Am. J. Bot.* 93: 637–655.

Khan, M. T., Ather, A., Thompson, K. D., and Gambari, R. 2005. Extracts and molecules from medicinal plants against herpes simplex viruses. *Antiviral Res.* 67: 107–119.

Lam, W. Y., Leung, K. T., Law, P. T., et al. 2006. Antiviral effect of *Phyllanthus nanus* ethanolic extract against hepatitis B virus (HBV) by expression microarray analysis. *J. Cell Biochem.* 97: 795–812.
Lee, C. D., Ott, M., Thyagarajan, S. P., et al. 1996. *Phyllanthus amarus* down-regulates hepatitis B virus mRNA transcription and replication. *Eur. J. Clin. Invest.* 26: 1069–1076.
Lin, S. Y., Wang, C. C., Lu, Y. L., et al. 2008. Antioxidant, anti-semicarbazide-sensitive amine oxidase, and anti-hypertensive activities of geraniin isolated from *Phyllanthus urinaria*. *Food Chem. Toxicol.* 46: 2485–2492.
Liu, H., Luan, F., Ju, Y., et al. 2007. *In vitro* transfection of the hepatitis B virus PreS2 gene into the human hepatocarcinoma cell line HepG2 induces upregulation of human telomerase reverse transcriptase. *Biochem. Biophys. Res. Commun.* 355: 379–384.
Liu, J., Lin, H., and McIntosh, H. 2001. Genus *Phyllanthus* for chronic hepatitis B virus infection: a systematic review. *J. Viral Hepat.* 8: 358–366.
Londhe, J. S., Devasagayam, T. P., Foo, L. Y., and Ghaskadbi, S. S. 2008. Antioxidant activity of some polyphenol constituents of the medicinal plant *Phyllanthus amarus* Linn. *Redox. Rep.* 13: 199–207.
Londhe, J. S., Devasagayam, T. P., Foo, L. Y., and Ghaskadbi, S. S. 2009. Radioprotective properties of polyphenols from *Phyllanthus amarus* Linn. *J. Radiat. Res.(Tokyo).* 50: 303–309.
Melendez, P. A., and Capriles, V. A. 2006. Antibacterial properties of tropical plants from Puerto Rico. *Phytomedicine* 13: 272–276.
Melmick, J. L. (1990). Enteroviruses: polioviruses, coxsackieviruses, and newer enteroviruses. In *Fields Virology* (Ed., B. N. Fields, D. M. Knipe, R. M. Chanock, M. S. Hirsch, J. L. Melnick, T. P. Monath, and B. Roizman). 2nd Ed. p. 549, New York: Raven Press.
Milne, A., Hopkirk, N., Lucas, C. R., et al. 1994. Failure of New Zealand hepatitis B carriers to respond to *Phyllanthus amarus*. *N. Z. Med. J.* 107: 243.
Mukhtar, M., Arshad, M., Ahmad, M., et al. 2008. Antiviral potentials of medicinal plants. *Virus Res.* 131: 111–120.
Notka, F., Meier, G. R., and Wagner, R. 2003. Inhibition of wild-type human immunodeficiency virus and reverse transcriptase inhibitor-resistant variants by *Phyllanthus amarus*. *Antiviral Res.* 58: 175–186.
Notka, F., Meier, G., and Wagner, R. 2004. Concerted inhibitory activities of *Phyllanthus amarus* on HIV replication *in vitro* and *ex vivo*. *Antiviral Res.* 64: 93–102.
Obianime, A. W., and Uche, F. I. 2009. The phytochemical constituents and the effects of methanol extracts of *Phyllanthus amarus* leaves (kidney stone plant) on the hormonal parameters of male guinea pigs. *J. Appl. Sci. Environ. Manage.* 13: 5–9.
Ott, M., Thyagarajan, S. P., and Gupta, S. 1997. *Phyllanthus amarus* suppresses hepatitis B virus by interrupting interactions between HBV enhancer I and cellular transcription factors. *Eur. J. Clin. Invest.* 27: 908–915.
Rao, Y. K., Fang, S. H., and Tzeng, Y. M. 2006. Anti-inflammatory activities of constituents isolated from *Phyllanthus polyphyllus*. *J. Ethnopharmacol.* 103: 181–186.
Raphael, K. R., and Kuttan, R. 2003. Inhibition of experimental gastric lesion and inflammation by *Phyllanthus amarus* extract. *J. Ethnopharmacol.* 87: 193–197.
Razonable, R. R., and Paya, C. V. 2003. Herpesvirus infections in transplant recipients: current challenges in the clinical management of cytomegalovirus and Epstein-Barr virus infections. *Herpes* 10: 60–65.
Roig y Mesa, J. T. 1974. *Plantas medicinales, aromáticas y venenosas en Cuba,* 85–88. Havana, Cuba: Ciencia y Técnica.
Sanchez-Lamar, A., Fiore, M., Cundari, E., et al. 1999. *Phyllanthus orbicularis* aqueous extract: cytotoxic, genotoxic, and antimutagenic effects in the CHO cell line. *Toxicol. Appl. Pharmacol.* 161: 231–239.

Sells, M. A., Chen, M. L., and Acs, G. 1987. Production of hepatitis B virus particles in Hep G2 cells transfected with cloned hepatitis B virus DNA. *Proc. Natl. Acad. Sci. U. S. A.* 84: 1005–1009.
Sumantran, V. N., Kulkarni, A., Chandwaskar, R., et al. 2008. Chondroprotective potential of fruit extracts of *Phyllanthus emblica* in osteoarthritis. *Evid. Based Complement. Alternat. Med.* 5: 329–335.
Superti, F., Ammendolia, M. G., and Marchetti, M. 2008. New advances in anti-HSV chemotherapy. *Curr. Med. Chem.* 15: 900–911.
Thamlikitkul, V., Wasuwat, S., and Kanchanapee, P. 1991. Efficacy of *Phyllanthus amarus* for eradication of hepatitis B virus in chronic carriers. *J. Med. Assoc. Thai.* 74: 381–385.
Thyagarajan, S. P., Subramanian, S., Thirunalasundari, T., et al. 1988. Effect of *Phyllanthus amarus* on chronic carriers of hepatitis B virus. *Lancet* 2: 764–766.
Thyagarajan, S. P., Thiruneelakantan, K., Subramanian, S., and Sundaravelu, T. 1982. In vitro inactivation of HBsAg by *Eclipta alba* Hassk and *Phyllanthus niruri* Linn. *Indian J. Med. Res.* 76(suppl): 124–130.
Unander, D. W., Webster, G. L., and Blumberg, B. S. 1995. Usage and bioassays in *Phyllanthus* (Euphorbiaceae). IV. Clustering of antiviral uses and other effects. *J. Ethnopharmacol.* 45: 1–18.
Whitley, R. J., Kimberlin, D. W., and Roizman, B. 1998. Herpes simplex viruses. *Clin. Infect. Dis.* 26: 97–109.
Yang, C. M., Cheng, H. Y., Lin, T. C., Chiang, L. C., and Lin, C. C. 2005. Acetone, ethanol and methanol extracts of *Phyllanthus urinaria* inhibit HSV-2 infection *in vitro*. *Antiviral Res.* 67: 24–30.
Yang, C. M., Cheng, H. Y., Lin, T. C., Chiang, L. C., and Lin, C. C. 2007. Hippomanin A from acetone extract of *Phyllanthus urinaria* inhibited HSV-2 but not HSV-1 infection *in vitro*. *Phytother. Res.* 21: 1182–1186.
Yao, Q. Q. & Zuo, C. X. (1993). [Chemical studies on the constituents of *Phyllanthus urinaria* L.]. *Yao Xue. Xue. Bao.* 28: 829–835.
Yeh, S. F., Hong, C. Y., Huang, Y. L., et al. 1993. Effect of an extract from *Phyllanthus amarus* on hepatitis B surface antigen gene expression in human hepatoma cells. *Antiviral Res.* 20: 185–192.
Yoosook, C., Bunyapraphatsara, N., Boonyakiat, Y., and Kantasuk, C. 2000. Anti-herpes simplex virus activities of crude water extracts of Thai medicinal plants. *Phytomedicine* 6: 411–419.
Zhang, L. Z., Guo, Y. J., Tu, G. Z., Miao, F., and Guo, W. B. 2000. [Isolation and identification of a novel polyphenolic compound from *Phyllanthus urinaria* L.]. *Zhongguo Zhong. Yao Za Zhi* 25: 724–725.
Zhang, Y. J., Nagao, T., Tanaka, T., et al. 2004. Antiproliferative activity of the main constituents from *Phyllanthus emblica*. *Biol. Pharm. Bull.* 27: 251–255.
Zhong, Y., Zuo, C., Li, F., et al. 1998. [Chemical constituents of *Phyllanthus urinaria* L. and its antiviral activity against hepatitis B virus]. *Zhongguo Zhong. Yao Za Zhi* 23: 363–364, 384.

15 Diabetes and Diabetic Complications and *Phyllanthus* species

*Geereddy Bhanuprakash Reddy
and Palla Suryanarayana*

CONTENTS

15.1 Diabetes 235
    15.1.1 Diabetic Complications 236
    15.1.2 Biochemical/Molecular Mechanisms and Pathophysiology 236
    15.1.3 Pharmacological Interventions and Their Limitations 237
15.2 Medicinal Plants and Herbal Drugs as Antidiabetic Agents 237
15.3 Antidiabetic Effect of *Phyllanthus* Species 238
    15.3.1 Hypoglycemic or Antidiabetic Effects 238
        15.3.1.1 *Phyllanthus emblica* or *Emblica officinalis* Gaertn 238
        15.3.1.2 *Phyllanthus niruri* Linn 241
        15.3.1.3 *Phyllanthus amarus* Schum & Thunn 242
        15.3.1.4 Other Species of *Phyllanthus* 242
    15.3.2 Experimental Studies with Type 2 Diabetic Models 244
15.4 Human Studies 245
15.5 *Phyllanthus* and Diabetic Complications 245
    15.5.1 Inhibition of Aldose Reductase by *E. officinalis* 246
    15.5.2 Effect of *E. officinalis* on Sorbitol Accumulation under
        High-Glucose Conditions 248
    15.5.3 Delay of STZ-Induced Diabetic Cataract in Rats by
        E. officinalis 248
Acknowledgments 251
References 251

15.1 DIABETES

Diabetes is defined as a clinical condition in which homeostasis of carbohydrate and lipid metabolism is improperly regulated by insulin. This results primarily in elevated fasting and postprandial blood glucose levels. Several distinct forms of diabetes exist that are caused by a complex interaction of genetics, environmental factors, and lifestyle choices. There are three major forms of diabetes: type 1 diabetes (T1D), type 2 diabetes (T2D), and gestational diabetes (GD). The noticeable manifested symptoms of diabetes include increased thirst (polydipsia), increased urination (polyuria), increased appetite (polyphagia), excessive fatigue, unexplained weight loss, and body irritation. Diabetes is the most common form of noncommunicable disease globally and the fourth leading cause of death in developed countries (Amos et al., 1997). Present statistics indicate that there are around 200 million diabetic people worldwide.

15.1.1 Diabetic Complications

Several changes occur due to high blood sugar level, and when these changes become permanent in the body, they develop into serious diabetic complications. While acute complications include diabetic ketoacidosis and nonketotic hyperosmolar state, prolonged exposure to hyperglycemia, without proper management, can lead to various short- and long-term chronic complications. Chronic complications of diabetes affect many organ systems and are responsible for the majority of morbidity and mortality. Macrovascular complications of diabetes are conditions of the large blood vessels that develop consequent to the influence of determinants such as dyslipidemia, obesity, hypertension, and microalbuminaria. Macrovascular complications manifest themselves as accelerated atherosclerosis, clinically resulting in premature ischemic heart disease, increased risk of cerebrovascular disease, and severe peripheral vascular disease. Microvascular complications of diabetes encompass long-term complications of diabetes affecting small blood vessels. The major mechanism of microvascular disease is the toxic effect of prolonged hyperglycemia on insulin-independent tissues like retina, kidney, peripheral nerve, and lens, which results in development of secondary complications of diabetes: retinopathy, nephropathy, neuropathy, and cataract, respectively. Microvascular complications are the major risk in T1D. Although macrovascular diseases are predominant in T2D, microvascular complications are also present (retinopathy, 20%; neuropathy, 9%; and overt diabetic nephropathy, up to 10%). As a consequence of microvascular pathology, diabetes is a leading cause of blindness, end-stage renal failure, and a variety of debilitating neuropathies. Diabetic retinopathy (DR), which affects the microvasculature of the retina, is a leading cause of adult blindness and is the most common complication of diabetes. Diabetic cataract, a nonvascular complication characterized by cloudiness or opacification of the eye lens, is the leading cause of blindness all over the world.

15.1.2 Biochemical/Molecular Mechanisms and Pathophysiology

The hormone insulin is responsible for maintaining the glucose level in the blood. However, in the diabetic condition, due to abnormal insulin metabolism, the body cells and tissues do not make use of glucose from the blood, resulting in an elevated level of blood glucose or hyperglycemia. In T1D, the pancreas cannot synthesize insulin hormone according to the requirement of the body. The pathophysiology of T1D suggests that it is an autoimmune disease in which the body's own immune system generates secretion of substances that attack the beta cells of the pancreas.

Consequently, the pancreas secretes little or no insulin. Therefore, patients with T1D generally require exogenous insulin for survival.

In case of T2D, there is normal production of insulin hormone, but the body cells are resistant to insulin; as a result, glucose remains in the bloodstream. It is commonly manifested by middle-aged adults (those above 40 years). Usually, the first thing that happens in the course of T2D is insulin resistance, an impaired biological response to insulin. Thus, the beta cells of the pancreas have to make more insulin to maintain the glucose levels, which is secondary hyperinsulinemia. As insulin is not usually necessary for treatment of T2D, insulin sensitizers and a variety of other molecules are used for the treatment of T2D. Generally, monotherapy alone is not sufficient to achieve a euglycemic state in a significant number of patients with T2D.

Several molecular mechanisms have been proposed for the development of chronic complications of diabetes (Brownlee, 2001). Glucose-induced damage occurs through four apparently different pathways: increased flux through the polyol pathway (due to increased aldose reductase activity), increased accumulation of advanced glycation end products (AGEs), activation of the protein kinase C (PKC) pathway, and increased flux through the hexosamine pathway. Although there have been major advances in the control of hyperglycemia (diabetes) through dietary changes, hypoglycemic agents, insulin, and islet transplantation, the long-term complications of diabetes remain serious problems to be handled. Therefore, the mentioned biochemical mechanisms serve as the drug targets for the prevention of secondary complications of diabetes.

15.1.3 Pharmacological Interventions and Their Limitations

Although there have been major advances in the control of hyperglycemia and diabetic complications by various pharmacological means, such as hypoglycemic agents, insulin, insulin mimetics, insulin sensitizers, islet transplantation, amylase inhibitors, glucosidase inhibitors, antiglucagon agents, and so on (Campbell, 2003; Krentz and Bailey, 2005; García-Vicente et al., 2007; DeLeon et al., 2002), there is a large gap between the number of diabetic patients and meeting the needs of maintaining strict glycemic control in diabetic people. Furthermore, the long-term complications of diabetes remain serious problems to be dealt with that require treatment or intervention in addition to antidiabetic agents. A number of synthetic aldose reductase inhibitors (ARIs) to block the polyol pathway and a variety of antiglycating agents to prevent the accumulation of AGE, PKC-β inhibitors have been investigated for their efficacy in several *in vitro* and *in vivo* studies against diabetic complications (Kyselova et al., 2004), but few could make it to the clinical trial stage.

15.2 MEDICINAL PLANTS AND HERBAL DRUGS AS ANTIDIABETIC AGENTS

There has been an exponential growth in the field of functional foods, nutraceuticals, and traditional or herbal medicine throughout the world. Extensive research is under way to identify the functional constituents, elucidate the biochemical structures, and determine the mechanisms behind the use of traditional foods and medicines. A number of medicinal plants have been used for over 1,000 years in Indian systems of medicine, and according to the World Health Organization (WHO), 21,000 plants are listed for medicinal purposes around the world. Among these, 2,500 species are present in India. Dietary intervention, particularly the use of traditional foods and medicines, is the mainstay in the management of diabetes (Baily and Day, 1989; Swanston-Flatt et al., 1991). A large number of plants and spices are now well recognized to possess hypoglycemic potential. Thus, plants continue to play an important role in the treatment of diabetes, particularly in the developing countries due to limited resources and availability of modern health care. Simultaneously, there is also an increase in demand in developed countries for the use of traditional medicines or dietary supplements due to decreased side effects and economics associated with plant-based medicines compared to synthetic or semisynthetic drugs. Approximately 450 medicinal plants, including *Physllanthus*, are reported to be used for the treatment of diabetes. However, only a small number of these plant products have been scientifically validated for medical evaluation for their efficacy to treat diabetes. Therefore, the WHO Expert Committee on Diabetes has recommended that traditional medicinal herbs be further investigated.

15.3 ANTIDIABETIC EFFECT OF *PHYLLANTHUS* SPECIES

Phyllanthus is the largest genus in the family Phyllanthaceae. Estimates of the number of species in this genus vary widely, from 750 to 1,200. *Phyllanthus* has a remarkable diversity of growth forms, including annual and perennial herbs, shrubs, climbers, floating aquatics, and pachycaulous succulents. Many species of *Phyllanthus* have a long tradition of use in the Indian Ayurvedic system of medicine and have long been used as medicinal agents in cultures around the world. Traditionally, *Phyllanthus* has been used to treat a variety of complications, including diabetes. This chapter describes the scientific studies of *Phyllanthus* species for utility in the treatment of diabetes and complications associated with diabetes. Taxonomy, phylogenetic aspects, cultivation, other medicinal properties, toxicity, and formulation of *Phyllanthus* are described elsewhere in this book.

15.3.1 Hypoglycemic or Antidiabetic Effects

Many *Phyllanthus* species, mainly *Phyllanthus emblica, Phyllanthus niruri,* and *Phyllanthus amarus,* either individually or in combination with other medicinal herbs, have been reported for their hypoglycemic or antidiabetic effects (Table 15.1).

15.3.1.1 *Phyllanthus emblica* or *Emblica officinalis* Gaertn

Phyllanthus emblica or *Emblica officinalis* Gaertn, also known as Amla (amalaki in Sanskrit), is a member of the small genus of Emblica. It is an important dietary source of vitamin C, minerals, and amino acids and contains phenolic compounds, tannins, phyllembelic acid, phyllemblin, rutin, curcuminoides, and emblicol. All parts of the plant are used for medicinal purposes. Especially, the fruit has been used in Ayurveda as a potent rasayana (Thakur, 1985). A study showed that aqueous

TABLE 15.1
Summary of Antidiabetic Activities of *Phyllanthus* Species and Their Formulations

Plant/ Formulation	Pharmacological Effect	Experiment Model	Reference
Phyllantus emblica	Antidiabetic	Alloxan-diabetic rats	Qureshi et al., 2009
Phyllanthus amarus	Hypoglycemic, diuretic, and hypotensive	Human study	Srividya and Periwal, 1995
Phyllanthus amarus	Hypoglycemic	Alloxan-diabetic rats	Raphael et al., 2002
Phyllanthus amarus	Hypoglycemic and hypocholesterolemic	Normal mice	Adeneye et al., 2006
Phyllanthus amarus	Hypoglycemic	T2D subjects	Moshi et al., 2001
Phyllanthus niruri	Antihyperglycemic and antioxidant	STZ-diabetic rats	Mazunder et al., 2005
Phyllanthus niruri	Antidiabetic	STZ-Wistar rats	Nwanjo, 2007
Phyllanthus fraternus	Antidiabetic and antioxidant	Alloxan-diabetic rats	Garg et al., 2008
Phyllanthus sellowianus	Hypoglycemic	STZ-diabetic mice	Hnatyszyn et al., 2002
Phyllanthus reticulates	Hypoglycemic	Alloxan-diabetic mice	Kumar et al., 2008
Phyllanthus simplex	Antidiabetic and antioxidant	Alloxan-diabetic rats	Shabeer et al., 2009
Phyllanthus debilis	Antihyperglycemic and hypoglycemic	Normoglycemic mice	Wanniarachchi et al., 2009
Phyllanthus rheedii	Antihyperglycemic and antihyperlipidemic	STZ- diabetic rats	Sivajothi et al., 2008
Triphala	Antidiabetic and antioxidant	Alloxan diabetic rats	Sabu and Kuttan, 2002
Rajanyamalakadi	Antidiabetic and hypolipidemic	T2D patients	Faizal et al., 2009
Hyponidd	Antihyperglycemic and antioxidant	STZ-diabetic rats	Babu and Prince, 2004
Dihar	Antihyperglycemic, antihyperlipidimic, and antioxidant	STZ-diabetic rats	Patel et al., 2009
Diarun plus	Antidiabetic	STZ-diabetic mice	Senthilvel et al., 2006
Diasulin	Antidiabetic, antihyperlipidemic, and antiperoxidative	Alloxan-diabetic rats	Saravanan and Pari, 2005

fruit extract in a dose of 200 mg/kg body weight significantly decreased blood glucose levels in alloxan-induced diabetic rats. The effect was almost similar to that of a known antidiabetic drug, chlorpropamide, at a dose of 84 mg/kg. The aqueous extract also induced hypotriglyceridemia by decreasing triglyceride levels at 1, 2, and 4 h in diabetic rats (Qureshi et al., 2009).

In most cases, various species of *Phyllanthus*, *E. officinalis* in particular, either individually or in combination with other medicinal plants, were shown to have antioxidant potential. Diarun plus is a polyherbal formula containing *E. officinalis* along with *Curucma longa, Momordica charantia, Eugenia jambolona, Trigonella, Gymnema,* and *Salacia*; it was shown to be effective in controlling hyperglycemic status in rats and mice. The drug at 500 mg/kg body weight showed significant antidiabetic activity, as evaluated by serum glucose levels. However, it did not significantly influence the levels of serum insulin in both diabetic and normoglycemic rats (Senthilvel et al., 2006).

Diasulin is another polyherbal drug composed of an ethanolic extract of 10 medicinal plants: *Cassia auriculata, Coccinia indica, Curcuma longa, Emblica officinalis, Gymnema sylvestre, Momordica charanti, Scoparia dulcis, Syzigium cumini, Tinospora cardifolia,* and *Trigonella foenum graecum*. Oral administration of ethanolic extract of Diasulin (200 mg/kg body weight) to alloxan-treated diabetic rats for 30 days resulted in a significant reduction of blood glucose and increase in plasma insulin. Diasulin also resulted in a significant decrease in tissue lipids and lipid peroxide formation. The effect produced by Diasulin was comparable with that of glibenclamide. The antidiabetic and antihyperlipidemic effect of Diasulin may be due to the effect of active constituents (alkaloid and pectins from *Coccinia indica*, alkaloids from *Tinospora cordifolia*, emblicanin A and B from *Emblica officinalis*; Saravanan and Pari, 2005).

Oral administration of a methanolic extract (75%) of *E. officinalis* alone and in combination with two other plant extracts (*Terminalia chebula* and *Terminalia belerica*) (named Triphala) reduced the blood sugar level in normal and in alloxan-induced diabetic rats significantly within 4 h at a single dose of 100 mg/kg body weight. Daily administration of the drug produced a sustained effect from the 7th to the 11th day. However, the reduction in glucose levels in diabetic rats was not nearer to control levels. Although the mechanism of action of these extracts in diabetic animals is not known, it is thought to be mediated by reducing the effect of inflammatory cytokines released during diabetes through antioxidant or free radical scavenging effect (Sabu and Kuttan, 2002).

Dihar is a polyherbal formulation composed of *Syzygium cumin, Momordica charantia, Emblica officinalis, Gymnema sylvestre, Entcostemma littorale, Azadiracta indica, Tinospora cordifolia,* and *Curcuma longa*. Treatment with Dihar at a dose of 100 mg/kg for 6 weeks to streptozotocin (STZ)-induced diabetic rats has shown significant antihyperglycemic activity along with antihyperlipidimic and antioxidant activities, as shown by a reduction in serum glucose and lipids and increase in insulin levels (Patel et al., 2009).

Hyponidd is a herbomineral formulation composed of the extracts of 10 medicinal plants, including *E. officinalis*. Other medicinal plants that constitute hyponid are *Momordica charantia, Melia azadirachta, Pterocarpus marsupium, Tinospora cordifolia, Gymnema sylvestre, Enicostemma littorale, Eugenia jambolana, Cassia auriculata,* and *Curcuma longa.* Oral administration of hyponidd at a dose of 100 and 200 mg/kg body weight for 45 days to STZ-induced diabetic rats resulted in significantly lowered levels of blood glucose (Babu and Prince, 2004). It also decreased the levels of glycosylated hemoglobin and lipid peroxidation and improved the insulin levels. The results suggest that hyponid exhibits antihyperglycemic activity through the antioxidant phytochemicals present in its various constituents.

There is increasing evidence to show that diabetes and its complications are associated with increased oxidative stress induced by the generation of free radicals by means of glucose autoxidation, increased formation of glucose-derived AGE, and enhanced glucose flux through the polyol pathway. Elevated generation of free radicals resulting in the consumption of antioxidant defense components may lead to disruption of cellular functions and oxidative damage to membranes and may enhance susceptibility to lipid peroxidation. Under physiological conditions, a variety of antioxidants, like glutathione (GSH), vitamin C, vitamin E, and endogenous antioxidant enzymes like catalase (CAT), superoxide dismutase (SOD), glutathione-S-transferase (GST), and glutathione peroxidase (GPx), protect the body against the adverse effects of free radicals. Diabetes-induced oxidative stress has been found to affect different parts of the body directly or indirectly, leading to various secondary complications (Oberley, 1988). Dietary intervention, particularly the use of traditional foods and medicines derived from natural sources, employed in the management of diabetes has mostly been targeted to contain the oxidative-induced damage.

Therefore, it is logical to understand that the effect of natural agents with antioxidant potential in addition to hypoglycemic activity could exert antidiabetic activity. A study reported that Amla in the form of either the commercial enzymatic extract (Sun Amla, Taiyo Kagaky Co. Ltd., Japan) at 20 or 40 mg/kg body weight per day or a polyphenol-rich fraction of ethyl acetate extract at 10 or 20 mg/kg body weight per day was effective as a strong free radical scavenging agent when given orally for 20 days to the STZ-induced diabetic rats (Rao et al., 2005). *Emblica officinalis* is used in many Ayurvedic preparations and polyherbal drugs, and most of these drugs have shown antidiabetic activity in experimental animals; some of these studies are discussed in the rest of this chapter.

15.3.1.2 *Phyllanthus niruri* Linn

Phyllanthus niruri Linn is commonly known as "chanca piedra" and belongs to the family Phyllanthaceae; it is an indigenous antidiabetic plant popularly used in South India for diabetes mellitus. It can be found in all tropical regions of the world, including southern India and China. There are several studies of the antidiabetic effect of *P. niruri*. For example, a methanol extract of *P. niruri* (MEPN) has shown significant antihyperglycemic activity along with antioxidant activity at 125 and 250 mg/kg body weight in the STZ-induced diabetic rat model (Mazunder et al., 2005). In subsequent studies, Nwanjo (2007) showed the hypoglycemic effect of an aqueous extract of *P. niruri* at a dose of 120 and 240 mg/kg body weight in STZ-induced diabetic rats in a dose-dependent manner. Other researchers confirmed the hypoglycemic properties of *P. niruri* (Ramakrishnan et al., 1982; Sivaprakasam et al., 1995).

TABLE 15.2
Hypoglycemic Effect of Methanolic Extract of *P. amarus* on Blood Sugar Level in Alloxan-Induced Diabetic Rats (Multidose Long-Term Study)

Treatment (dose/kg body weight)	Blood Sugar Level (mg/dl) Days					
	3	6	9	12	15	18
Normal	75.6 ± 2.8	75.6 ± 2.9	76.2 ± 3.2	75.7 ± 3.3	75.4 ± 2.9	76.3 ± 2.1
Diabetic control	479 ± 16.5	436 ± 4.6	391 ± 5.1	357 ± 10.2	307 ± 11.4	266 ± 22.3
P. amarus (200 mg)	489 ± 27.4	285 ± 5.3*	200 ± 9.2*	185 ± 11.5*	146 ± 11.2*	117 ± 6.1*
P. amarus (1,000 mg)	501 ± 20.2	232 ± 22.1*	168 ± 8.5	144 ± 8.4*	120 ± 7.2*	94.5 ± 6.4*

Source: Adapted from Raphael, K.R., Sabu, M.C., and Kuttan, R. 2002. *Indian J. Exp. Biol.* 40: 905–909.

Note: Values are mean plus or minus standard deviation (SD) (*n* = 6) and represent random glucose level in alloxan-treated animals with or without extract from 3rd day to 18th day.

* $p < .001$ (compared to values on 0 h of first day of the same group).

15.3.1.3 Phyllanthus amarus Schum & Thunn

A methanol extract of *P. amarus* reduced the blood sugar in alloxan-induced diabetic rats by 6% at a dose of 200 mg/kg body weight and 19% at 1,000 mg/kg body weight. Continued administration of the extract for 15 days produced a significant reduction in blood sugar by bringing down the glucose values almost close to control rats after 18 days at 1,000 mg/kg dose (Raphael et al., 2002) (Table 15.2). In addition, the extract exhibited antioxidant potential *in vitro* as shown by its ability to scavenge superoxide and hydroxyl radicals and inhibition of lipid peroxidation in rat liver homogenate. Aqueous leaf and seed extracts of *P. amarus* at oral doses of 150, 300, and 600 mg/kg have shown antidiabetic and antilipidemic activity in normal mice in a dose-dependent manner (Adeneye et al., 2006). It is suggested that the *P. amarus* extract could be enhancing peripheral utilization of glucose, although the mechanism is unclear. Interestingly, α-amylase inhibitory activity was also reported for the hexane extract of *P. amarus* (Ali et al., 2006) (Figure 15.1). On extraction and fractionation, it was found that a mixture of oleanolic acid and ursolic acid (2:1) was responsible for α-amylase inhibition, with an IC_{50} (half-maximal inhibitory concentration) of 4.4 μ*M*.

15.3.1.4 Other Species of *Phyllanthus*

The petroleum ether and ethanolic extracts of leaves of *Phyllanthus reticulatus* at 500 and 1,000 mg/kg were shown to have a hypoglycemic effect in alloxan-induced diabetic mice. At 1,000 mg/kg, *P. reticulates* showed a sustained decrease in blood glucose levels on subchronic administration for 21 days. The phytochemical screening of extracts revealed the presence of terpenoid glycosides, protein, and carbohydrates and the absence of alkaloids and steroids (Kumar et al., 2008).

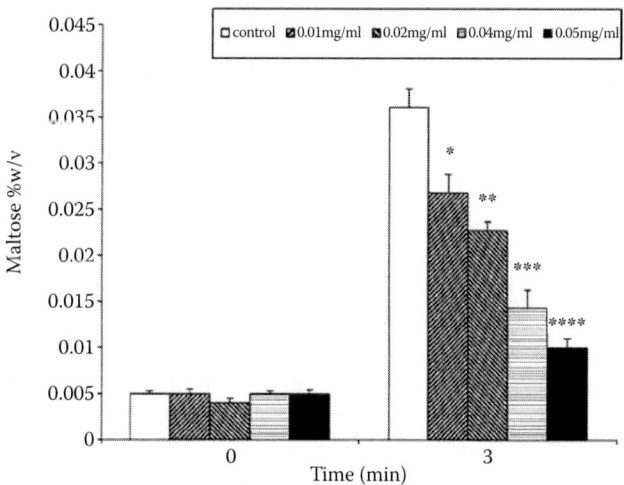

FIGURE 15.1 Amylase inhibitory activity of *Phyllanthus amarus*. The α-amylase inhibition assay was performed by maltose formation after 3 min in the presence of *Phyllanthus amarus*. Results are expressed as mean plus or minus the standard error ($n = 4$). (*Source:* Adapted from Ali, H., Houghton, P.J., Soumyanath, A. 2006. *J. Ethnopharmacol.* 107: 449–455.)

Phyllanthus simplex is a fibrous perennial herb widely distributed in tropical and subtropical regions. Antidiabetic or hypoglycemic and antioxidant effects of various fractions of *P. simplex* were studied on normal and alloxan-induced diabetic Charles Foster rats. In the normoglycemic rats, methanol extract (125 and 250 mg/kg) and aqueous fractions (150 and 300 mg/kg) showed a significant hypoglycemic effect on day 21. In diabetic rats, methanolic and aqueous extracts showed a significant antihyperglycemic effect. The active fractions were able to normalize the marked alterations in antioxidant enzymes and antioxidant parameters in liver and kidney. Treatment with the active fractions also normalized the diabetes-induced hyperlipidemia and liver glycogen (Shabeer et al., 2009).

Antidiabetic and antioxidant potential of petroleum ether and ethanolic and aqueous extracts of the whole plant of *Phyllanthus fraternus* was studied in alloxan-induced diabetic albino rats. Administration of three different plant extracts (0.5 g/kg) of *P. fraternus* remarkably improved the elevated levels of blood glucose. Ethanolic extract reduced the blood sugar levels in a significant and sustained manner throughout the study. Aqueous extract also showed good activity during the second study week but could not sustain this to the third week. Petroleum ether extract could not produce any significant results (Garg et al., 2008).

The whole plant of *Phyllanthu debilis* Linn. is used in Sri Lanka for the treatment of diabetes mellitus. The antidiabetic potential of aqueous extract of *P. debilis* was studied on normoglycemic mice. Mice treated orally with three doses (497, 995, or 1,990 mg/kg) of aqueous extract lowered the fasting blood glucose level in a dose-dependent manner. Further, an aqueous extract of *P. debilis* markedly improved the oral glucose and sucrose tolerance tests up to 5 h posttreatment. The improvement of the glucose tolerance was dose dependent. In addition, it significantly inhibited glucose absorption from the small intestine (Wanniarachchi et al., 2009).

Phyllanthus rheedii is a slender, branching, erect herb with the calyx lobes usually white margined and is found throughout India. It is used as an oriental folk medicine in diabetes mellitus (Rajan et al., 2002). Ethanolic extracts of the whole plant of *P. rheedii* were studied for antihyperglycemic, antihyperlipidemic, and antioxidant effects in STZ-induced diabetic rats. Oral administration of *P. rheedii* for 21 days resulted in a significant reduction in blood glucose level, lipid metabolism, and enzyme level and significant improvement in SOD and catalase in liver tissues of STZ-induced diabetic rats (Sivajothi et al., 2008).

Phyllanthus sellowianus is a plant used in folk medicine as a hypoglycemic and diuretic agent. The hypoglycemic effect of various fractions of the stem barks of *P. sellowianus* was studied using bioassay-guided fractionation protocols and an STZ-induced hyperglycemic mice model. The aqueous extract was partitioned between dichloromethane and butanol. Aqueous and butanol fractions at a dose of 200 mg/kg orally caused a significant reduction in blood glucose concentration at 6 and 9 h, while the same dose of dichloromethane fraction was ineffective. The reduction in blood glucose levels obtained with the *P. sellowianus* fractions was similar to that observed with glibenclamide (10 mg/kg). Phytochemical analysis of butanol and aqueous fractions revealed the presence of flavonoid compounds, of which rutin and isoquercitin were the major constituents, respectively, suggesting the involvement of these flavonoids in the observed hypoglycemic effect of the fractions (Hnatyszyn et al., 2002).

15.3.2 Experimental Studies with Type 2 Diabetic Models

Most of the studies mentioned with regard to the hypoglycemic and antidiabetic potential of various *Phyllanthus* species were conducted using normal rats or STZ- or alloxan-induced diabetic rats. In principle, STZ- or alloxan-treated rat or mice models resemble human T1D as these chemicals destroy pancreatic beta-cells, leading to insulin deficiency. T1D accounts for about 5–10% of all people with diabetes mellitus, whereas T2D results from the combined action of defects in insulin secretion and insulin action, or both, either of which may predominate. Insulin resistance is the characteristic feature of patients with T2D. T2D may account for as much as 90% of all people with diabetes. Although hyperglycemia is the end state of both T1D and T2D, molecular events in the insulin-deficient and insulin-resistant state could be entirely different, so are the effect and mechanism of antidiabetic agents or drugs in T1D and T2D. Thus, it may not be indisputable to extend the mechanisms of action based on T1D models to T2D. However, there are not many studies, barring one or two, to demonstrate the antidiabetic potential of *Phyllanthus* using T2D models. A study showed that an aqueous extract of *P. amarus* promoted glucose uptake, as assessed by a glucose tolerance test in the Wistar albino rat. In addition, the extract showed a cholesterol and low-density lipoprotein (LDL) lowering effect at 50, 100, and 200 mg/kg body weight (James et al., 2009), and these results may have implications for the treatment of T2D. However, there is a need for extensive studies to evaluate the potential of

Phyllanthus as an antidiabetic or hypoglycemic agent using T2D or insulin resistance animal models.

15.4 HUMAN STUDIES

Although a few drug formulations of *Phyllanthus* in combination with other medicinal plants, such as Triphala, Diasulin, Diarun, and Dihar, have been in use for human consumption as antidiabetic drugs, there are not many scientific studies that validated the antidiabetic potential of *Phyllanthus* on humans. A study evaluated the hypoglycemic and hypolipidemic effects of an Ayurvedic medicine, Rajanyamalakadi, containing *E. officinalis, Curcuma,* and *Salacia* in T2D patients over a period of 3 months. Rajanyamalakadi showed a significant antidiabetic and hypolipidemic effect (Faizal et al., 2009). It was thought that the mechanism of action might be through stimulation of insulin secretion and antioxidant actions.

Another study reported that *P. amarus* is a potential diuretic, hypotensive, and hypoglycemic agent in humans (Srividya and Periwal, 1995). Results of this controlled human study suggest that administration of dried powder of whole plant (5 g in three equal doses per day) for 10 days increased 24 h urine volume and urine and serum sodium levels. Significant reductions in systolic blood pressure in nondiabetic hypertensive patients and blood glucose levels were significantly decreased in both diabetic and nondiabetic subjects compared to a control group. Thus, *P. amarus* seems to have a significant potential therapeutic value as a diuretic, hypotensive, and in particular hypoglycemic agent with no detrimental side effects. However, in another study of non-insulin-dependent diabetic patients, it was demonstrated that an aqueous extract of *P. amarus* administered for 1 week was not effective in lowering blood glucose in both a fasting and a postprandial state (Moshi et al., 2001). Nevertheless, many controlled human studies and clinical trials are very much needed to promote *Phyllanthus* as an antidiabetic agent in an appropriate formulation.

15.5 *PHYLLANTHUS* AND DIABETIC COMPLICATIONS

Although strict glycemic control is expected to prevent diabetic complications, perfect glycemic control is not always possible. Further, persistence of progression of diabetic complications during the subsequent period of normal glucose homeostasis (called *metabolic memory*) suggests that exclusive management of glucose can no longer be viewed as sufficient for the control of long-term complications. Hence, agents that can prevent diabetic complications, in addition to or irrespective of glycemic control, need critical evaluation in the management of secondary complications. Even though a large number of plants and spices are now well recognized to possess hypoglycemic potential, studies of the beneficial effects of medicinal plants for the treatment of secondary complications of diabetes like cataract, retinopathy, nephropathy, and neuropathy are scant. In this context, studies were conducted to investigate the therapeutic evaluation of traditional medicines, including *Phyllanthus,* for their potential to treat or prevent diabetic complications (Suryanarayana et al., 2004, 2007; Saraswat et al., 2008, 2009).

15.5.1 INHIBITION OF ALDOSE REDUCTASE BY *E. OFFICINALIS*

Aldose reductase (ALR2; EC: 1.1.1.21) is the rate-limiting enzyme of the polyol pathway and reduces glucose to sorbitol utilizing NADPH as a cofactor (Kinoshita, 1990). Normally, the polyol pathway represents a minor route of glucose utilization, accounting for less than 3% of glucose consumption. However, in the presence of high glucose, the activity of the polyol pathway is substantially increased, and up to 30% of total glucose is metabolized by this pathway (Bhatnagar and Srivastava, 1992). Accumulation of osmotically active sorbitol leads to osmotic and oxidative stress, culminating in tissue injury. Evidence for the involvement of ALR2 in diabetic neuropathy, retinopathy, nephropathy, and cataract emerged from several independent studies (Kinoshita, 1990; Bhatnagar and Srivastava, 1992). Studies of experimental animal models suggested that the compounds that inhibit ALR2 could be effective in the prevention of certain complications (Kador et al., 1985). Although a wide variety of compounds has been synthesized to inhibit ALR2 and studied in experimental models, only a limited number of drugs reached clinical trials (Raskin and Rosenstock, 1987).

Since *E. officinalis* is extensively used against many chronic ailments, including diabetes, inhibition of ALR2 by *E. officinalis* was assessed. Aqueous extract of *E. officinalis* inhibited rat lens and recombinant human ALR2, with IC_{50} values of 0.72 and 0.88 mg/ml, respectively (Table 15.3). To identify the chemical entities responsible for ALR2 inhibition, investigations were carried out with ascorbic acid for ALR2 inhibition as *E. officinalis* is considered to be a rich source of ascorbic acid. However, ascorbic acid did not inhibit ALR2 at concentrations as high as 5 mM (Suryanarayana et al., 2004). Further, it is interesting to note the observations of some studies indicating that the biological actions, particularly antioxidant activities, of *E. officinalis* cannot be attributed to ascorbic acid alone (Khopde et al., 2001; Bhattacharya et al., 1999). A study attributed the potent vitamin C-like and antioxidative activity of *E. officinalis* to low molecular weight hydrolyzable tannoids present in it, which were identified as emblicanin A, emblicanin B, penigluconin, and pedunculagin (Ghosal et al., 1996).

Therefore, we investigated whether these hydrolyzable tannoids have ALR2 inhibitory potential. As shown in Figure 15.2, the standardized extract of *E. officinalis* with enriched tannoids exhibited remarkable inhibition against both rat lens

TABLE 15.3
IC_{50} Values of Aqueous Extract and Tannoid Principles of *E. officinalis* and Quercetin with Aldose Reductase (ALR2)

Inhibitor	Rat Lens ALR2	Human Recombinant ALR2
Emblica aqueous extract (mg/ml)	0.72	0.88
Emblica tannoids (µg/ml)	6.1	9.8
Quercetin (µg/ml)	9.2	13.5

Source: Adapted from Suryanarayana, P., Kumar, P.A., Saraswat, M., et al. 2004. *Mol. Vis.* 10:148–154.

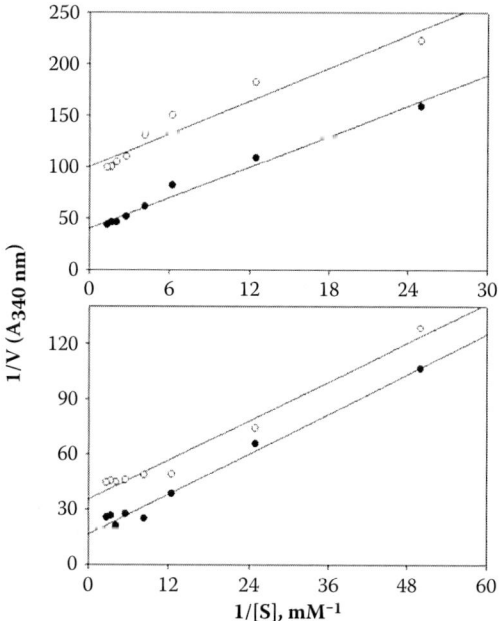

FIGURE 15.2 Inhibition of aldose reductase (ALR2) by *E. officinalis* tannoids. Double-reciprocal plots of rat lens ALR2 (upper) and human recombinant ALR2 (lower) in the presence (open circles) and absence (closed circles) of tannoids of *E. officinalis*. (*Source:* Adapted from Suryanarayana, P., Kumar, P.A., Saraswat, M., et al. 2004. *Mol. Vis.* 10: 148–154.)

ALR2 and human recombinant ALR2 (Suryanarayana et al., 2004). The inhibitory potential of isolated tannoids was about 100 times higher compared to the aqueous extract of *E. officinalis,* suggesting their potential application in the prevention of diabetic cataract (Table 15.3). On comparison with IC_{50} values of quercetin, a natural flavonoid with ALR2 inhibitory potential, tannoids of Amla appear to be more potent (Table 15.3). Decreased V_{max} and K_m with glyceraldehyde as substrate indicated that tannoids of *Emblica* inhibited ALR2 in an uncompetitive manner (Figure 15.2; Table 15.4).

TABLE 15.4
Kinetic Parameters of Rat Lens and Human Recombinant ALR2 in the Absence and Presence of Tannoids of *E. officinalis*

Kinetic Paramete	Rat Lens ALR2		Human Recombinant ALR2	
	Tannoids (–)	Tannoids (+)	Tannoids (–)	Tannoids (+)
K_m (mM)	0.279 ± 0.032	0.161 ± 0.022	0.085 ± 0.017	0.037 ± 0.013
V_{max}	24.2 ± 1.57	10.6 ± 1.42	3.88 ± 0.219	1.72 ± 0.23

Source: Adapted from Suryanarayana, P., Kumar, P.A., Saraswat, M., et al. 2004. *Mol. Vis.* 10:148–154.

15.5.2 Effect of *E. officinalis* on Sorbitol Accumulation under High-Glucose Conditions

Accumulation of high concentrations of polyols in target tissues leads to excessive hydration, gain of sodium, and loss of potassium ions due to an increase in intracellular ionic strength. Finally, there is a loss of membrane permeability and leakage of free amino acids, glutathione, myoinositol, and other low molecular weight substances. The resulting hyperosmotic stress-associated oxidative insult is postulated to be the primary cause for the development of diabetic complications such as cataract, retinopathy, neuropathy, and nephropathy (Bhatnagar and Srivastava, 1992).

To understand the significance of *in vitro* inhibition of ALR2 by tannoids of *E. officinalis*, the effect of an enriched tannoid fraction against osmotic stress was investigated in a lens organ culture system. Rat lenses incubated with 55 mM glucose for 24 h developed vacuoles and Y sutures and showed significantly increased ALR2 activity compared to lenses incubated with 5.5 mM glucose (36.2 ± 3.51 vs. 26.4 ± 2.45 µmol/h/100 mg; $n = 3$). However, the morphology of the lenses incubated with 55 mM glucose along with 50 µg/ml of tannoid mixture of *E. officinalis* appeared similar to control lenses (Figure 15.3). More importantly, activation of ALR2 due to hyperglycemic stress was prevented when lenses were incubated with 55 mM glucose in the presence of 50 µg/ml tannoids *of E. officinalis* in the medium (36.2 ± 3.51 vs. 29.8 ± 2.82 µmol/h/100 mg; $n = 3$). Together, these results imply that *E. officinalis* ingredients may be explored for the treatment of diabetic complications.

15.5.3 Delay of STZ-Induced Diabetic Cataract in Rats by *E. officinalis*

A number of studies with experimental animals suggested that the compounds that inhibit ALR2 could be effective in the prevention of certain diabetic complications (Bhatnagar and Srivastava, 1992; Kador et al., 1985; Raskin and Rosenstock, 1987; Banditelli et al., 1997). Nonetheless, clinical trials of synthetic ALR2 inhibitors against neuropathy and retinopathy have met with limited success, and some of them were associated with deleterious side effects and poor penetration of target tissues, such as nerve and retina (Pfeifer et al., 1997). Since an aqueous extract

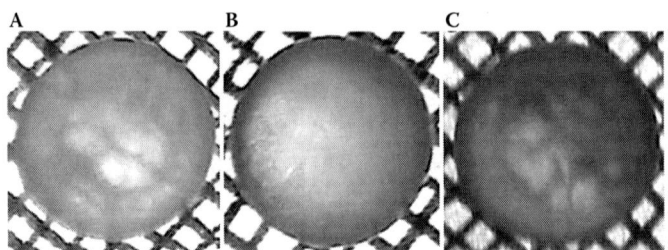

FIGURE 15.3 Effect of *E. officinalis* tannoids on lens transparency under osmotic conditions. Rat lenses were cultured in modified TC-199 in the presence of 5.5 mM glucose (A), 55 mM glucose (B), and 55 mM glucose along with 50 µg/ml tannoid mixture of amla (C). (*Source:* Adapted from Suryanarayana, P., Kumar, P.A., Saraswat, M., et al. 2004. *Mol. Vis.* 10: 148–154.)

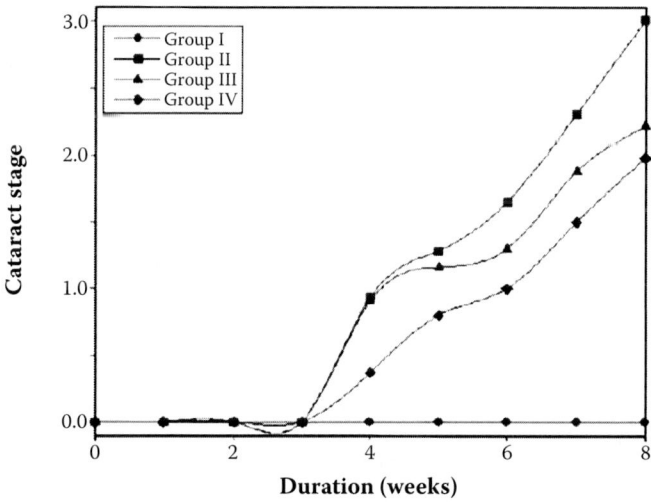

FIGURE 15.4 Delay of diabetic cataract in rats by *Emblica* and its tannoid-enriched fraction. Cataract formation was monitored weekly by slit-lamp microscope, and the stage of cataract was scored. Stages of cataract in each group were averaged at the given time, and the average stage of cataract was plotted as a function of time. *Emblica* (group IV) and its constituent tannoids (group III) delayed the maturation of diabetic cataract due to slow progression compared to untreated diabetic rats (group II). Lenses in control rats (group I) were clear during the experimental period. (Source: Adapted from Suryanarayana, P., Saraswat, M., Petrash, J.M., and Reddy, G.B. 2007. *Mol. Vis.* 13: 1291–1297.)

of *E. officinalis* and hydrolyzable tannoids inhibited ALR2 and the tannoids of *Emblica* prevented diabetic cataract in a lens organ culture system (Suryanarayana et al., 2004), we evaluated the efficacy of whole *Emblica* pericarp and the enriched tannoid mixture for effectiveness in prevention or delay of the onset and progression of cataracts in the STZ-induced diabetic rat model. The onset of cataract due to hyperglycemia was observed in diabetic animals after 4 weeks of STZ injection (Figure 15.4). Interestingly, there was a delay in the onset of cataract in diabetic animals fed with *E. officinalis* when compared to untreated diabetic animals. At the end of 8 weeks, the severity of cataracts was significantly lower in diabetic animals fed with tannoids or whole extract of *E. officinalis* compared to untreated diabetic animals, indicating that the *Emblica* and its constituent tannoids delayed the maturation of diabetic cataract due to slow progression (Figure 15.4). These data suggest that progression of lens abnormalities and maturation of cataracts due to STZ-induced hyperglycemia were significantly delayed with feeding of *Emblica* and its tannoids (Suryanarayana et al., 2007). The whole *Emblica* was more effective than its tannoid fraction in delaying the onset and maturation of diabetic cataract.

As expected, blood glucose levels were elevated, and insulin levels were decreased significantly in untreated diabetic rats compared to control animals. However, treatment with either *Emblica* or its tannoids did not reverse the changes in blood glucose and insulin levels, indicating that *Emblica* and tannoid treatment had no effect on the hyperglycemia (Suryanarayana et al., 2007). These findings may appear contradictory to the studies that reported antidiabetic or hypoglycemic activity for *Phyllanthus* species. Most of the studies with *E. officinalis* and other *Phyllanthus* species that showed hypoglycemic and antidiabetic activity were in combination with other medicinal plants. Further, the doses used in those studies were many-fold (50- to 100-fold) higher compared to the dose (2% in the diet) used in our study. Thus, it is not surprising that the dose used in this study lacked influence on the hyperglycemic state of the animal. However, it is interesting to note that, despite unaltered hyperglycemic state, the *Emblica* and tannoid fraction of *Emblica* were effective in arresting the progression of diabetic complications (in this case, diabetic cataract) at very low levels, suggesting that *Emblica* could act downstream of glucose-mediated events that lead to complications.

Consistent with observations of *in vitro* inhibition of ALR2 by *Emblica* tannoids, the specific activity of ALR2 in lenses from diabetic animals fed with *Emblica* and its tannoids was decreased significantly compared to untreated diabetic rats (Suryanarayana et al., 2007). Further, as expected, there was an increase of sorbitol levels in the eye lens of diabetic animals due to activation of the polyol pathway. However, feeding of tannoids and *Emblica* resulted in a lower but incomplete normalization of diabetes-induced sorbitol accumulation. Based on these results, it appears that *Emblica* and tannoids at the doses used in the study are partially effective against osmotic stress caused by hyperglycemia. Moreover, partial inhibition of ALR2 and sorbitol accumulation brought about by *Emblica* and its tannoids is in agreement with a delay but not complete prevention of diabetic cataract. Nonetheless, it may be feasible to observe more pronounced effects with higher doses as used in other studies. If that is achieved, delay of diabetic complications by treatment with *Emblica* tannoids merits further attention.

Recent studies indicated that the polyol pathway may be related to hyperglycemia-induced oxidative stress, and there may be a metabolic connection between the polyol pathway and oxidative stress (Chung et al., 2003; Srivastava et al., 2005). Further, it was reported that ALR2 inhibitors reduce oxidative damage (Chung et al., 2003; Srivastava et al., 2005). On the other hand, many studies have reported antioxidant activity for *Phyllanthus* species. The antioxidant effect of a tannoid-rich fraction of *Emblica* has also been shown in a stress-induced oxidative damage model in rat brain (Bhattacharya et al., 1999). Hence, the effect of *E. officinalis* on diabetes-induced oxidative stress in rat lens was assessed by monitoring TBARS, protein carbonyls, and important antioxidant enzymes. Interestingly, treatment with *Emblica* and its tannoids prevented the alterations not only in TBARS, but also in protein carbonyls, despite elevated glucose levels. *Emblica* and tannoids treatment partially prevented the altered activities of antioxidant enzymes. These data clearly demonstrated that *Emblica* and tannoids not only inhibited osmotic stress but also prevented hyperglycemia-induced lenticular oxidative stress (Suryanarayana et al., 2007). Thus, it may be stated that *Emblica* or its constituents, such as tannoids, could be effective in delaying the progression of diabetic complications. Although multiple mechanisms may contribute to these effects, the antiosmotic and antioxidant effects of *Emblica* appear to be the predominant mechanism of action.

In summary, although various species of *Phyllanthus* or their active constituents appear to have great potential in the prevention and treatment of diabetes and its complications, further studies are needed to provide a strong scientific support for its antidiabetic effect.

ACKNOWLEDGMENTS

The authors were supported by grants from the Department of Science & Technology, Department of Biotechnology, and the Life Sciences Research Board, Government of India. The Indian Herbs Research & Supply Company, Saharanpur, India is acknowledged for the generous supply of standardized extract of *Emblica* tannoids.

REFERENCES

Adeneye, A.A., Amole, O.O., and Adeneye, A.K. 2006. Hypoglycemic and hypocholesterolemic activities of the aqueous leaf and seed extract of *Phyllanthus amarus* in mice. *Fitoterapia* 77: 511–514.

Ali, H., Houghton, P.J., and Soumyanath, A. 2006. alpha-Amylase inhibitory activity of some Malaysian plants used to treat diabetes; with particular reference to *Phyllanthus amarus*. *J. Ethnopharmacol.* 107: 449–455.

Amos, A.F., McCart, D.J., and Zimmet, P. 1997. The rising global burden of diabetes and its complications: estimates and projections to the year 2010. *Diabet. Med.* 14(suppl 5): S1–S85.

Babu, P.S., and Ignacimuthu, S 2007. Antihyperlipidemic and antioxidant effect of hyponidd in the brain of streptozotocin induced diabetic rat. *Int. J. Biol. Chem.* 1: 196–204.

Babu, P.S., and Prince, P.S.M. 2004. Antihyperglycaemic and antioxidant effect of hyponidd, an Ayurvedic herbomineral formulation in streptozotocin induced diabetic rats. *J. Pharm. Pharmacol.* 56: 1435–1442.

Bailey, C.J., and Day, C. 1989. Traditional plant medicines as treatments for diabetes. *Diabetes Care* 12: 553–564.

Banditelli, S., Boldrini, E., Vilardo, P.G., et al. 1997. A new approach against sugar cataract through aldose reductase inhibitors. *Exp. Eye Res.* 69: 533–538.

Bhatnagar, A., and Srivastava, S.K. 1992. Aldose reductase: congenial and injurious profiles of an enigmatic enzyme. *Biochem. Med. Metab. Biol.* 48: 91–121.

Bhattacharya, A., Chatterjee, A., Ghosal, S., and Bhattacharya, S.K. 2000. Antioxidant activity of active tannoid principles of *Emblica officinalis* (Amla). *Indian J. Exp. Biol.* 38: 877–880.

Brownlee, M. 2001. Biochemistry and molecular cell biology of diabetic complications. *Nature* 414: 813–820.

Campbell, R.K. 2003. Type 2 diabetes: where we are today: an overview of disease burden, current treatments, and treatment strategies. *J. Am. Pharm. Assoc.* 49(suppl 1): S3–S9.

Chung, S.S., Ho, E.C., Lam, K.S., and Chung, S.K. 2003. Contribution of polyol pathway to diabetes-induced oxidative stress. *J. Am. Soc. Nephrol.* 14: S233–S236.

DeLeon, M.J., Chandurkar, V., Albert, S.G., and Mooradian, A.D. 2002. Glucagon-like peptide-1 response to acarbose in elderly type 2 diabetic subjects. *Diabetes Res. Clin. Pract.*

Faizal, P., Suresh, S., Satheesh Kumar, R., and Augusti, K.T. 2009. A study on the hypoglycemic and hypolipidemic effects of an Ayurvedic drug Rajanyamalakadi in diabetic patients. *Indian J. Clin. Biochem.* 24: 82–87.

García-Vicente, S., Yraola, F., Marti, L., et al. 2007. Oral insulin-mimetic compounds that act independently of insulin. *Diabetes*, 56: 486–493.

Garg, M., Dhar, V.J., and Kalia, A.N. 2008. Antidiabetic and antioxidant potential of *Phyllanthus fraternus* in alloxan induced diabetic animals. *Pharmacog. Mag.* 4: 138–143.
Ghosal, S., Tripathi, V.K., and Chauhan, S. 1996. Active constituents of *Emblica officinalis*: Part 1. The chemistry and antioxidative effects of two new hydrolysable tannins, Emblicanin A and B. *Indian J. Chem.* 35B: 941–948.
Hnatyszyn, O., Miño, J., Ferraro, G., and Acevedo, C. 2002. The hypoglycemic effect of *Phyllanthus sellowianus* fractions in streptozotocin-induced diabetic mice. *Phytomedicine* 9: 556–559.
James, D.B., Owolabi, O.A., Elebo, N., et al. 2009. Glucose tolerance test and some biochemical effect of *Phyllanthus amarus* aqueous extracts on normoglycemic albino rats. *African J. Biotech.* 8: 1637–1642.
Kador, P.F., Kinoshita, J.H., and Sharpless, N.E. 1985. Aldose reductase inhibitors: a potential new class of agents for the pharmacological control of certain diabetic complications. *J. Med. Chem.* 28(7) 841–849.
Kasuni, K., Wanniarachchi, L., Dinithi, C., and Peiris, W.D. 2009. Ratnasooriya antihyperglycemic and hypoglycemic activities of *Phyllanthus debilis* aqueous plant extract in mice. *Pharm. Biol.* 47: 260–265.
Khopde, S.M., Priyadarshini, K.I., Mohan, H., et al. 2001. Characterizing the antioxidant activity of amla (*Phylanthus emblica*) extract. *Curr. Sci.* 81: 185–190.
Kinoshita, J.H. 1990. A thirty year journey in the polyol pathway. *Exp. Eye Res.* 50: 567–573.
Krentz, A.J., and Bailey, C.J. 2005. Oral antidiabetic agents: current role in type 2 diabetes mellitus. *Drugs* 65: 385–411.
Kumar, S., Kumar, D., Deshmukh, R.R., et al. 2008. Antidiabetic potential of *Phyllanthus reticulatus* in alloxan-induced diabetic mice. *Fitoterapia* 79: 21–23.
Kyselova, Z., Stefek, M., and Bauer, V. 2004. Pharmacological prevention of diabetic cataract. *J. Diabetes Complications* 18: 129–140.
Mazunder, U.K., Gupta, M., and Rajeshwar, Y. 2005. Antihyperglycemic effect and antioxidant potential of *Phyllanthus niruri* (Euphorbiaceae) in streptozotocin-induced diabetic rats. *Eur. Bull. Drug Res.* 13: 15–23.
Moshi, M.J., Lutale, J.J., Rimoy, G.H., et al. 2001. The effect of *Phyllanthus amarus* aqueous extract on blood glucose in non-insulin dependent diabetic patients. *Phytother. Res.* 15: 577–580.
Nwanjo, H.U. 2007. Studies on the effect of aqueous extract of *Phyllanthus niruri* leaf on plasma glucose level and some hepatospecific markers in diabetic wistar rats. *Int. J. Lab. Med.* 2 (2). Available at http://www.aboutus.org/ISPUB.COM.
Oberley, L.W. 1988. Free radicals and diabetes. *Free Rad. Biol. Med.* 5: 113–124.
Patel, S.S., Shah, R.S., and Goyal, R.K. 2009. Antihyperglycemic, antihyperlipidemic and antioxidant effects of Dihar, a polyherbal Ayurvedic formulation in streptozotocin induced diabetic rats. *Indian J. Exp. Biol.* 47: 564–570.
Pfeifer, M.A., Schumer, M.P., and Gelber, D.A. 1997. Aldose reductase inhibitors: the end of an era or the need for different trial designs? *Diabetes* 46: S82–S89.
Qureshi, S.A., Asad, W., and Sultana, V .2009. The effect of *Phyllantus emblica* Linn on type-2 diabetes, triglycerides and liver-specific enzymes. *Pak. J. Nutr.* 8: 125–128.
Rajan, S., Sethuraman, M., and Mukherjee, P.K. 2002. Ethnobiology of the Nilgiri Hills, India. *Phytother. Res.* 16: 98–116.
Ramakrishnan, P.N., Murugesan, R., Palanichamy, S., and Murugesh, N. 1982. Oral hypoglycaemic effect of *Phyllanthus niruri* Linn. *Indian. J. Pharm. Sci.* 44: 10–12.
Rao, T.P., Sakaguchi, N., Juneja, L.R., et al. 2005. Amla (*Emblica officinalis* Gaertn.) extracts reduce oxidative stress in streptozotocin-induced diabetic rats. *J. Med. Food.* 8: 362–368.

Raphael, K.R., Sabu, M.C., and Kuttan, R. 2002. Hypoglycemic effect of methanol extract of *Phyllanthus amarus* Schum & Thonn on alloxan induced diabetes mellitus in rats and its relation with antioxidant potential. *Indian J. Exp. Biol.* 40: 905–909.

Raskin, P., and Rosenstock, J. 1987. Aldose-reductase inhibitors and diabetic complications. *Am. J. Med.* 83: 298–296.

Sabu, M.C., and Kuttan, R. 2002. Anti-diabetic activity of medicinal plants and its relationship with their antioxidant property. *J. Ethnopharmacol.* 81: 155–160.

Saraswat, M., Muthenna, P., Suryanarayana, P., et al. 2008. Dietary sources of aldose reductase inhibitors: prospects for alleviating diabetic complications. *Asian Pac. J. Clin. Nutr.* 17: 558–565.

Saraswat, M., Reddy, P.Y., Muthenna, P., and Reddy, G.B. 2009. Prevention of non-enzymic glycation of proteins by dietary agents: prospects for alleviating diabetic complications. *Br. J. Nutr.* 101: 1714–1721.

Saravanan, R., and Pari, L. 2005. Antihyperlipidemic and antiperoxidative effect of Diasulin, a polyherbal formulation in alloxan induced hyperglycemic rats. *BMC Compl. Alter. Med.* 22: 5–14.

Senthilvel, G., Jegadeesan, M., Anoop, A., et al. 2006. Effect of a polyherbal formulation (Diarun plus) on streptozotocin induced experimental diabetes. *Int. J. Trop. Med.* 1: 88–92.

Shabeer, J., Srivastava, R.S., and Singh, S.K. 2009. Antidiabetic and antioxidant effect of various fractions of *Phyllanthus simplex* in alloxan diabetic rats. *J. Ethnopharmacol.* 124: 34–38.

Sivajothi, V., Dey, A., and Jayakar, B. 2008. Antihyperglycemic, antihyperlipidemic and antioxidant effect of *Phyllanthus rheedii* on streptozotocin induced diabetic rats. *Iranian J. Pharm. Res.* 7: 53–59.

Sivaprakasam, K., Yasodha, R., Sivandam, G., and Veluchamy, G. 1995. Clinical evaluation of *Phyllanthus amarus* Schum & Thonn in diabetes mellitus. In *Seminar on research in Ayurveda and Siddha,* 7. New Delhi, India: CCRAS.

Srivastava, S.K., Ramana, K.V., and Bhatnagar, A. 2005. Role of aldose reductase and oxidative damage in diabetes and the consequent potential for therapeutic options. *Endocr. Rev.* 26: 380–392.

Srividya, N., and Periwal, S. 1995. Diuretic, hypotensive and hypoglycaemic effect of *Phyllanthus amarus*. *Indian J. Exp. Biol.* 33: 861–864.

Suryanarayana, P., Kumar, P.A., Saraswat, M., et al. 2004. Inhibition of aldose reductase by tannoid principles of *Emblica officinalis*: implications for the prevention of sugar cataract. *Mol. Vis.* 10: 148–154.

Suryanarayana, P., Saraswat, M., Petrash, J.M., and Reddy, G.B. 2007. *Emblica officinalis* and its enriched tannoids delay streptozotocin-induced diabetic cataract in rats. *Mol. Vis.* 13: 1291–1297.

Swanston-Flatt, S.K., Flatt, P.R., Day, C., and Bailey, C.J.1991. Traditional dietary adjuncts for the treatment of diabetes mellitus. *Proc. Nutr. Soc.* 50: 641–651.

Thakur, C.P. 1985. *Emblica officinalis* reduces serum, aortic and hepatic cholesterol in rabbits. *Experentia* 41: 423–424.

Wanniarachchi, K.K., Dinithi, L., Peiris, C., et al. 2009. Antihyperglycemic and hypoglycemic activities of *Phyllanthus* debilis aqueous plant extract in mice. *Phrmaceutical Biol.* 47: 260–265.

16 Chemoprotective, Genotoxic, and Antigenotoxic Effects of *Phyllanthus* Sp.

Rakesh K. Johri

CONTENTS

16.1 Introduction 255
    16.1.1 Background 256
16.2 Effect of *P. Emblica* Fruit Against Known Mutagens or Genotoxic Agents 257
16.3 Effect of *P. Amarus* Against Known Mutagens or Genotoxic Agents 257
16.4 Effects of *Phyllanthus* Species Against Radiation-Induced Mutagenicity or Genotoxicity 258
16.5 A Mechanistic Understanding of The Antimutagenic or Antigenotoxic Profile of *Phyllanthus* Sp. 259
16.6 Concluding Remarks 261
References 262

16.1 INTRODUCTION

DNA repair systems protect the cell from genomic instability that results from accumulation of critical mutagenic base lesions. Certain chemical compounds cause failure of the DNA repair mechanism and thus could be potent mutagenic or genotoxic in nature, causing cell death. Considerable evidence exists for a variety of such agents and their possible mechanisms of action. Among the chemical agents most notable are the heavy metals, such as aluminum, arsenic, cadmium, chromium, lead, and nickel. Some metals, like arsenic and nickel, are also comutagenic: they augment the effects of radiation and other DNA-damaging agents. In this context, aromatic compounds (polycyclic hydrocarbons and heterocyclic amines), and some other agents like aflatoxin B1 (AFB1) (a dietary genotoxin), benzopyrene (BP) (an environmental genotoxin), cyclophosphamide (CP), and sodium azide are also of potential relevance. Ionizing radiations induce damage to DNA and other macromolecules and are known to be mutagenic or genotoxic. Over the past decades, considerable efforts have been made to identify plants or their constituents capable of attenuating mutagenesis or genotoxicity. One such notable herb is *Phyllanthus*, a large genus in the family Phyllanthaceae. Several species of this plant have been used to treat a wide number of diseases in many traditional systems of medicine in diverse geographical areas and are attributed with beneficial pharmacological effects.

This review presents a chemoprotective (antimutagenic/antigenotoxic) profile of *Phyllanthus* sp. Among many of these, the protective effects of two, namely, *P. emblica* (syn. *Emblica officinalis*) and *P. amarus* (PA; syn. *P. niruri*), have been eminently significant. Efficacy of some other *Phyllanthus* species is also discussed.

16.1.1 Background

Chemical mutagens or genotoxic agents as well as ionizing radiation damage cellular DNA, which is of prime biological importance. Their deleterious effects are revealed in the form of chromosomal aberrations, elevated frequencies of DNA strand breaks, sister-chromatid exchanges and DNA base damage, elimination of bases, and sugar damage. Double-strand breaks (DSBs) and DNA-DSB complexation are considered the most lethal events following ionizing radiation and have been found to be the main target of cell killing. Unrepaired DNA contributes to chromosome fragmentation, and following division, chromosomes with a damaged kinetocore appear as micronuclei (Rao et al., 2006; Jagetia, 2007).

Inhibition of such aberrations has therefore served as an important indicator for assessing the potential of a test agent as antimutagenic or antigenotoxic. In experimental studies, various end-point determinations (mutation assays and other relevant parameters), carried out in both animal-derived and *in vitro* models, have provided much evidence for the preventive role of *Phyllanthus*. One widely employed *in vitro* test is an Ames assay utilizing multiple strains of *Salmonella typhimurium* (*S. typhimurium*), such as TA98 (detects frameshift mutation), TA100 (detects basepair substitutions), and others. Since many aromatic amines and amides are promutagens that require metabolic activation, an assay utilizing mammalian S9 preparation. Simple preparations from the liver are called an S9 mix, which contains many of the enzymes necessary to perform such metabolic conversions or activation. In addition, radiation-protective effects have been evaluated on the basis of the ability of a putative antigenotoxic or antimutagenic agent to ameliorate gastrointestinal (GI) injury and consequently to influence the survival rate of animals. The GI tract is, after the bone marrow, the organ most sensitive to the effects of radiation. With this, the intestine becomes conspicuously susceptible to acute morphological changes (Akpolat et al., 2009). Rapidly dividing tissues such as cells of hematopoietic systems are more prone to radiation damage. In this respect, an assessment of protective efficacy of *Phyllanthus* in maintaining intestinal crypt cells and villus topography and other functional changes in hematopoietic systems subsequent to radiation challenge have been relied on. In most of the animal studies, a gamma radiation source has been used. Utilizing these assays, a variety of extracts (aqueous or solvent derived) from aerial parts of *Phyllanthus* have been investigated.

16.2 EFFECT OF *P. EMBLICA* FRUIT AGAINST KNOWN MUTAGENS OR GENOTOXIC AGENTS

The protective effect of *P. emblica* fruit (PEF) against several known mutagens or genotoxic agents is documented. PEF extracts (aqueous as well as nonaqueous) prevented mutagenicity induced by sodium azide, 4-nitro-O-phenylenediamine (4-NPD), 2-acetylamiofluorine (2-AAF), N-methyl-N-nitro-N-nitrosoguanidine (MNNG), and 4-nitro-O-phenylenediamine (4-NPDA) in TA97, TA98, and TA100 strains of *S. typhimurium* (Grover and Kaur, 1989; Jose et al., 1997). A similar inhibitory effect was also found against AFB1 and BP mutagenicity in TA98 and TA100 strains of *S. typhimurium* in the presence of a liver S9 preparation (Sharma et al., 2000b). A 7-day oral treatment of PEF extract inhibited chromasomal aberrations and micronuclei formation in mice bone marrow induced by BP, CP, and 7,12-dimethylbenzanthracene (DMBA) (Sharma et al., 2000a; Haque et al., 2001; Banu et al., 2004). BP mutagenicity was suppressed in mice receiving a PEF extract as dietary supplementation (Nandi et al., 1997). The antimutagenic effects of PEF were reported against several heavy metals. An oral administration of a PEF extract provided prophylactic protection against aluminum-, arsenic-, cesium-, lead-, and nickel-induced chromosomal breaks, micronuclei formation, and sister-chromatid exchanges in mouse bone marrow cells (Agarwal et al., 1992; Dhir et al., 1990a, 1990b, 1991, 1993; Ghosh et al., 1992; Roy et al., 1992; Biswas et al., 1999). Chromium-mediated DNA fragmentation in rat lymphocytes remained inhibited in the presence of this extract (Sai Ram et al., 2003). PEF reduced chromosomal abnormalities induced by metanil yellow, zinc chloride, and ethyl parathion (Giri and Banerjee, 1986). In a recent study, PEF extract orally administered to mice reversed lead-induced sperm head mutagenesis in the germ cells and arsenic-induced karyolysis in hepatocytes (Madhavi et al., 2007; Sharma et al., 2009).

16.3 EFFECT OF *P. AMARUS* AGAINST KNOWN MUTAGENS OR GENOTOXIC AGENTS

The role of PA in the amelioration of genetic damage has also been investigated. In an Ames assay, a methanolic extract of PA attenuated the activation and mutagenicity of AFB1, sodium azide, and several aromatic amines, such as 2-aminofluorene (2-AF), 2-AAF, MNNG, N-ethyl-N-nitrosoquanidine (NENG), and 4-NPDA in a *Salmonella* mutation assay using several strains (TA98, TA100, TA102, TA1535) as well as in an *Escherichia coli* WP2 strain (Raphael et al., 2002). An aqueous extract of the plant protected against mutagenicity induced by 2-AF, 2-nitrofluorene, sodium azide, 2-aminoanthracene (2-AA), 4-nitroquinolone-1-oxide, and NENG in *S. typhimurium* TA98 and TA110, whereas given orally it prevented DNA single-stand breaks caused by dimethylnitrosamine in hamster liver cells (Sripanidkulchai et al., 2002).

A methanolic extract of the plant inhibited urinary mutagenicity of BP (Raphael et al., 2002). Oral administration of a PA aqueous extract was found to reduce the cytotoxic action of lead nitrate and aluminum sulfate by preventing the frequency of chromosomal breakages in mice (Dhir et al., 1990b). In a study using tannery effluents as mutagens, a crude extract of PA was found to reduce chromosomal aberrations in root meristems of *Vicia faba*. This study showed that phyllanthin, a constituent of PA, was antigenotoxic (Gowrishankar and Vivekanandan, 1994). CP-induced genotoxicity was suppressed by PA; oral administration of its methanolic extract significantly decreased the myelosuppression and increased the white blood count, bone marrow cellularity, and the number of maturing monocytes in a dose-dependent manner (Kumar and Kuttan, 2005).

Regarding effects of other *Phyllanthus* species against known mutagens or genotoxic agents, *P. orbicularis* (PO) has shown antimutagenic or antigenotoxic activities in a *Salmonella* assay using different cotreatment, pretreatment, and posttreatment approaches against hydrogen peroxide and several aromatic amines, such as 2-AF, 2-AA, m-phenylenediamine (m-PDA), 1-aminopyrene (1-AP), and 9-aminophenantrene (9-AP) in *S. typhimurium* YG1024, in different cotreatment approaches. It was suggested that the protective effect of PO was due to competitive interactions with S9 enzymes, which could result in the conversion of promutagenic amines into nonmutagenic entities (Ferrer et al., 2001, 2002). In another study, a protective effect of PO was also noted in the presence of 2-amino-1-methyl-6-phenylimidazo [4,5-b] pyridine (PhIP) and 4-aminobiphenyl (4-ABP). In this study, activated liver microsomal fractions from both rats and humans were used to metabolize these amines (Ferrer et al., 2004).

A genotoxic assessment of PO has also been undertaken. Low concentrations of an aqueous extract of the plant were not found to be genotoxic. It did not induce either primary DNA damage or chromosomal aberrations in Chinese hamster ovarian (CHO) cells and in various other *in vitro* and *in vivo* assays. On the other hand, when tested at high concentrations, it was found that certain compounds could be susceptible to activation to genotoxic or mutagenic metabolites (Sanchez-Lamar et al., 1999; Sanchez-Lamar et al., 2002). A chemopreventive effect of *P. maderaspatensis* is reported; an ethanolic extract of the plant prevented the occurrence of micronucleated polychromatic erythrocytes in the bone marrow of mice treated with cisplatin (Chandrasekar et al., 2006). One species, *P. tenellus,* has been associated with some mutagenicity, as indicated by a frameshift mutation in an Ames test using TA98 strain (Rivera et al., 1994).

16.4 EFFECTS OF *PHYLLANTHUS* SPECIES AGAINST RADIATION-INDUCED MUTAGENICITY OR GENOTOXICITY

Several investigations revealed the efficacy of *Phyllanthus* sp. against radiation-induced damage (Jagetia, 2007). A PEF aqueous extract administered before gamma radiation exposure significantly reduced the weight loss and increased the survival rate of mice (Singh et al., 2005). In such animals, a significant increase in red and white blood cells and hemoglobin and hematocrit values was observed compared to untreated irradiated mice (Singh et al., 2006). On further examination of intestine, it was found that crypt cell population, mitotic figures, and the villus length, which were markedly reduced after gamma radiation, tended to return to normal in those mice were treated with PEF extract prior to irradiation (Jindal et al., 2009). In an *in vitro* study, a PEF ethnolic extract inhibited gamma radiation-induced plasmid DNA strand breaks in a concentration-dependent manner (Bhattacharya et al., 2006).

Radiation-protective effects of PA are known. A methanolic extract of PA given prophylactically before the exposure of mice to gamma radiation significantly inhibited myelosuppression and increased white blood cell count, bone marrow cellularity, and α-esterase activity as a marker of maturing monocytes, in a dose-dependent manner, compared to untreated irradiated mice. An intestinal study of irradiated mice revealed damage to intestinal epithelium in the form of distorted villus architecture, edema, and infiltration of lymphocytes. Such pathological alterations were reversed after oral administration of PA methanolic extract to irradiated mice. One other noteworthy feature in such mice was a reduced incidence of chromosomal aberrations and micronuclei formation in bone marrow (Uma Devi et al., 2000; Harikumar et al., 2004; Harikumar and Kuttan, 2007). In a subsequent study, this radiation-protective activity of PA was attributed to the antigenotoxic action of a group of several compounds (mainly ellagitannins and flavonoids), which were shown to prevent gamma radiation-induced single-strand breaks in plasmid DNA (Londhe et al., 2008). The antimutagenic action of PO was reported; in an SOS chromotest assay, an aqueous extract of the plant inhibited gamma radiation-induced DNA damage in *E. coli* PQ37 cells, possibly by stimulation of DNA repair mechanisms (Fuentes et al., 2006).

16.5 A MECHANISTIC UNDERSTANDING OF THE ANTIMUTAGENIC OR ANTIGENOTOXIC PROFILE OF *PHYLLANTHUS* SP.

The mechanisms by which an agent exerts its mutagenic or genotoxic effects are largely unknown. However, certain cellular events that are modulated under the influence of a putative geno- or radiation protector, such as *Phyllanthus*, have causally been linked to their mechanistic profile. In this context, some of the key events that have been shown to play a prominent role are (1) activation of microsomal enzymes and (2) generation of reactive oxygen species (ROS). Therefore, the protective effect of this herb in mitigating mutagenesis or genotoxicity is now widely understood in terms of its ability to deactivate CYP450 enzymes and to quench ROS, although several other mechanisms working independently or in a concerted manner may also be involved because of the presence of many active compounds.

Although a large proportion of carcinogens is chemically inert, documented reports amply suggest that all such chemicals increase the levels of several CYP450-dependent biotransformation enzymes. The metabolic activation of CYPs generates highly reactive and unstable metabolites, which play an important role in the carcinogenicity of most chemical agents, and this has been recognized as a major mechanism for tissue-specific carcinogenic or toxic effects. BP is reported to cause genotoxic damage in liver, bone marrow, and peripheral blood via induction of CYP1A1 (Kim et al., 2003). CYP1A2 is involved in the bioactivation of 2-AA (Jemnitz et al., 2004). A significantly increased genotoxic response of cyclophosphamide in bone marrow cells was observed due to the activation of CYP2B1 (Voskoboinik et al., 1997). CYP3A4 and CYP1A2 were found to be the major isoforms involved in AFB1 activation to induce genotoxic AFB1 oxidation products (Ueng et al., 1995). CYP2E1 has been implicated in the bioactivation of several dialkylnitrosamines and N-nitrosamines (Van Vleet et al., 2001; Cooper and Porter, 2001).

Counteracting the adverse effects of several known mutagens or genotoxins, thus either preventing the conversion of promutagens into mutagenic entities or facilitating their chemical modification into nonmutagenic derivatives via CYP deactivation, has been suggested to be one biochemical event contributing notably in the preventive role of *Phyllanthus* (Jose et al., 1997; Sharma et al., 2000b; Ferrer et al., 2001, 2004; Banu et al., 2004). This has largely been deduced from a number of experimental studies utilizing mammalian cell/tissue and recombinant systems. In several such studies, the detrimental action of many genotoxic agents (alkylating promutagens, aromatic and heterocyclic amines, and polycyclic aromatic hydrocarbons) shown to occur as a result of adverse interactions of an electrophile (generated by CYPs) with cellular DNA and other macromolecules has also been demonstrated in mutagenicity or genotoxicity tester *in vitro* systems (Fujita et al., 2001; Duarte et al., 2005). Thus, it could be demonstrated in *in vitro* and *in vivo* studies that aqueous or alcoholic PA extracts inhibited basal as well as stimulated activity of multiple CYP isoforms (CYP1A1/2, CYP2B1/2, CYP2C9, CYP2D6, CYP3A4/5/7) (Hari Kumar and Kuttan, 2006; Appiah-Opong et al., 2008; Agbonon et al., 2010). A similar deactivation of enzyme activity and deinduction of CYP2E1 by PEF and a *P. urinaria* extract was documented (Jose et al., 1997; Tasduq et al., 2005; Shen et al., 2008; Hau et al., 2009). A significant decrease of the mutagenesis of several aromatic amines by PO has been suggested to be due to effects of some components of the extract with CYP enzymes (Ferrer et al., 2001, 2004). One mechanism that has also been proposed for the observed antigenotoxic profile of *P. emblica* is that it is able to modify the gene expression profile, which is known to be altered by a mutagen (Mischiati et al., 2006).

Mutagenic or genotoxic stimuli suppressed glutathione-S-transferase (GST). Inhibition of GST may increase harmful effects of electrophilic compounds. An augmentation of GST by PEF/PA extract administered orally in the presence of a metal or chemical mutagen such as DMBA, BP, CP, and ionizing radiation has therefore been considered of particular relevance to their protective actions (Sharma et al., 2000a, 2009; Banu et al., 2004; Kumar and Kuttan, 2004).

One of the most deleterious consequences of ionizing radiation with a biological system is generation of ROS, such as superoxide anion radical, hydroxyl radicals, and hydrogen peroxide, which are capable of causing direct oxidative damage to DNA, resulting in single-strand DNA breaks, and inhibition of major DNA repair systems (REF); this has also been recognized as one underlying mechanism in metal-induced pathology (Beyersmann and Hartwig, 2008). Some experimental evidence also exists to show that this damage to plasmid DNA by free radicals generated from radiation was preventable by ellagitannins and flavonoids isolated from PA (Londhe et al., 2008). A similar protective effect of PEF/PA against several metals could also be due predominantly to their antioxidant property since all heavy metals are capable of redox cycling, thus reducing a cell's antioxidant pool. The chemopreventing effect of *P. maderaspatensis* against one chemotherapeutic drug, cisplatin, has also been attributed to its antioxidant potential (Chandrasekar et al., 2006). Cisplatin mutagenicity is believed to be caused by misincorporation of noncomplimentary bases and, if left unpaired, generates point or frameshift mutations. An aqueous extract of PO was also shown to inhibit hydrogen peroxide-mediated chromosomal aberrations in CHO cells (Sanchez-Lamar et al., 1999).

Indirectly, ROS disturb glutathione (GSH) homeostasis. Under such conditions, glutathione reductase (GR) remained inhibited, causing depletion of GSH, a major cellular reductant, and as a cause of or consequence of which a concomitant increase in the lipid peroxidation occurs. A substantially low cellular GSH content along with high lipid peroxidation have been viewed as important contributing factors for oxidative stress in the event of a mutagenic or genotoxic challenge. With the result a decrease in the endogenous enzymic defense systems, such as catalase, glutathione peroxidase (GPx), and superoxide dismutase (SOD), and GST disturb oxidant-antioxidant balance. Ionizing radiation-mediated intestinal pathology is also linked to these events. ROS also damage the hematopoeitic system.

Phyllanthus sp. were found to reverse such ROS-mediated alterations. Many investigations showed the ability of *Phyllanthus* sp. and their constituents to act as potential antioxidants, and this has been considered a major mechanism associated with their observed efficacy. Several studies have shown that PEF effectively maintains a favorable cell/tissue antioxidant defense by lowering the lipid peroxidation and elevating antioxidant enzyme systems (Jose et al., 1997; Tasduq et al., 2005; Reddy et al., 2009). A similar antioxidant profile of *P. urinaria* is also known (Shen et al., 2008). Prevention of chromosomal aberrations and micronuclei formation in the presence of PEF or PA was shown to be related to their actions in restoring hepatic GSH content and GPx and GR activity in mice under the influence of DMBA, BP, CP, and arsenic (Haque et al., 2001; Banu et al., 2004; Sharma et al., 2000a, 2009; Ghosh et al., 1992).

A PEF extract efficiently inhibited lipid peroxidation and restored SOD in gamma-irradiated rat mitochondria (Bhattacharya et al., 2006). The ability of PEF extract and several constituents from PA in preventing plasmid DNA breakage has been associated with their potential free radical scavenging activity (Bhattacharya et al., 2006; Londhe et al., 2008). Antigenotoxic and antimutagenic effects of PO are also attributed to their antioxidative potential (Ferrer et al., 2002; Fuentes et al., 2006). Prevention of GI damage by PEF as well as PA against gamma radiation has been related to the ability of the plant to contain lipid peroxidation and enhance GSH content, catalase, GPx, and SOD activities in the intestine, thus strengthening the antioxidant defense system in irradiated mice (Harikumar and Kuttan, 2004; Singh et al., 2006; Jindal et al., 2009).

16.6 CONCLUDING REMARKS

The uses of medicinal plants in Ayurveda and other traditional medicines are widespread. These herbs still serve as an important reservoir in the quest for pharmacologically active agents. A large number of modern chemotherapeutic drugs have their origin in one or the other natural product derived from such plants. Although evaluation of *Phyllanthus* sp. for their protective effects against chemical mutagens or genotoxins has been going on for over a decade, the radiation-protective profile of this herb is relatively more recent. *Phyllanthus* sp. thus present a new source of a herbal drug or potentially novel natural product against genotoxicity of chemical agents and ionizing radiations. There exists a profound need for such effective non-toxic protectors for therapeutic as well as occupational needs.

REFERENCES

Agarwal, K., Dhir, H., Sharma A., and Talukdar, G. 1992. The efficacy of two species of Phyllanthus in counteracting nickel clastogenicity. *Fitoterapia* 63: 49–54.

Agbonon, A., Eklu-Gadegbeku, K., Aklikokou, K., et al. 2010. In vitro inhibitory effect of West African medicinal and food plants on human cytochrome P450 3A subfamily. *J. Ethnopharmacol.* 128: 390–394.

Akpolat, M., Kanter, M., and Uzal, M.C. 2009. Protective effects of curcumin against gamma radiation-induced ileal mucosal damage. *Arch. Toxicol.* 83: 609–617.

Appiah-Opong, R., Commandeur, J.N.M., Axson, C., and Vermeulen, N.P.E. 2008. Interactions between cytochromes P450, glutathione S-transferase and Ghanaian medicinal plants. *Food Chem. Toxicol.* 46: 3598–3603.

Banu, S.M., Selvendiran, K., Singh, J.P., and Sakthisekaran, D. 2004. Protective effect of *Emblica officinalis* ethanolic extract against 7,12-dimethylbenz(a) anthracene (DMBA) induced genotoxicity in Swiss albino mice, *Hum. Exp. Toxicol.* 23: 527–531.

Beyersmann, D., and Hartwig, A. 2008. Carcinogenic metal compounds: recent insight into molecular and cellular mechanisms. *Arch. Toxicol.* 82: 493–512.

Bhattacharya, S., Subramanian, M., Kamat, J.P., et al. 2006. Radioprotective property of *Emblica officinalis* fruit ethanol extract. *Pharm. Biol.* 44: 682–690.

Biswas, S., Talukdar, G., and Sharma, A. 1999. Protection against cytotoxic effects of arsenic by dietary supplementation with crude extracts of *Emblica officinalis* fruit. *Phytother. Res.* 13: 513–516.

Chandrasekar, M.N., Bommu, P., Nanan, M.J., and Suresh, B. 2006. Chemopreventive effect of *Phyllanthus maderaspatensis* in modulating cisplatin-induced nephrotoxicity and genotoxicity. *Pharm. Biol.* 44: 100–106.

Cooper, M.T., and Porter, T.D. 2001. Cytochrome b(5) coexpression increases the CYP2E1-dependent mutagenicity of dialkylnitrosamines in methyltransferase-deficient strains of *Salmonella typhimurium*. *Mutat. Res.* 484: 61–68.

Dhir, H., Roy, A.K., Sharma, A., and Talukdar, G. 1990a. Modification of clastogenicity of lead and aluminium in mouse bone marrow cells by dietary ingestion of *Phyllanthus emblica* fruit extract. *Mutat. Res.* 241: 305–312.

Dhir, H., Roy, A.K., Sharma, A., and Talukdar, G. 1990b. Protection afforded by aqueous extracts of *Phyllanthus* species against cytotoxicity induced by lead and aluminium salts. *Phytother. Res.* 4: 172–176.

Dhir, H., Agarwal, K., Sharma, A., and Talukdar, G et al. 1991. Modifying role of *Phyllanthus emblica* and ascorbic acid against nickel clastogenicity in mice. *Cancer Lett.* 59: 9–18.

Dhir, H., Roy, A.K., and Sharma, A. 1993. Relative efficiency of Phyllanthus emblica fruit extract and ascorbic acid in modifying lead and aluminium-induced sister-chromatid exchanges in mouse bone marrow. *Environ. Mol. Mutagen.* 21: 229–236.

Duarte, M.P., Palma, B.B., Laires, A., et al. 2005. *Escherichia coli* BTC, a human cytochrome P450 competent tester strain with a high sensitivity towards alkylating agents: involvement of alkyltransferases in the repair of DNA damage induced by aromatic amines. *Mutagenesis* 20: 199–208.

Ferrer, M., Cristofol, C., Sanchez-Lamar, A., et al. 2004. Modulation of rat and human cytochromes P450 involved in PhIP and 4-ABP activation by an aqueous extract of *Phyllanthus orbicularis*. *J. Ethnopharmacol.* 90: 273–277.

Ferrer, M., Sánchez-Lamar, A., Fuentes, J. L., et al. 2001. Studies on the antimutagenesis of *Phyllanthus orbicularis*: mechanisms involved against aromatic amines. *Mutat. Res.* 498: 99–105.

Ferrer, M., Sanchez-Lamar, A., Fuentes, J.L., et al. 2002. Antimutagenic mechanisms of *Phyllanthus orbicularis* when hydrogen peroxide is tested using Salmonella assay. *Mutat. Res.* 517: 251–254.

Fuentes, J.L., Alonso, A., Cuetara, E., et al. 2006. Usefulness of the SOS chemotest in the study of medicinal plants as radioprotectors. *Int. J. Radiat. Biol.* 82: 323–329.

Fujita, K., Nakayama, K., Yamazaki, Y., et al. 2001. Construction of *Salmonella typhimurium* YG7108 strains, each coexpressing a form of human cytochrome P450 with NADPH-cytochrome P450 reductase. *Environ. Mol. Mutagen.* 38: 329–338.

Ghosh, A., Sharma, A., and Talukder, G. 1992. Relative protection given by extract of *Phyllanthus emblica* fruit and an equivalent amount of vitamin C against a known clastogen—caesium chloride. *Food Chem. Toxicol.* 30: 865–869.

Giri, A.K., and Banerjee, T.S. 1986. Antagonistic activity of herbal drug (*Phyllanthus emblica*) on cytological effects of environmental chemicals on mammalian cells. *Cytologia* 52: 375–380.

Gowrishankar, B., and Vivekanandan, O.S. 1994. *In vivo* studies of a crude extract of *Phyllanthus amarus* L. in modifying the genotoxicity induced in *Vicia faba* L. by tannery effluents. *Mutat. Res.* 322: 185–192.

Grover, I.S., and Kaur, S. 1989. Effect of *Emblica officinalis* Gaertn. (Indian gooseberry) fruit extract on sodium azide and 4-nitro-o-phenylenediamine induced mutagenesis in *Salmonella typhimurium*. *Ind. J. Exp. Biol.* 27: 207–209.

Haque, R., Bin-Hafeez, B., Ahmad, I., et al. 2001. Protective effects of *Emblica officinalis* Gaertn. in cyclophosphamide-treated mice. *Hum. Exp. Toxicol.* 20: 643–650.

Hari Kumar, K.B., and Kuttan, R. 2006. Inhibition of drug metabolizing enzymes (cytochrome P450) *in vitro* as well as *in vivo* by *Phyllanthus amarus* SCHUM & THONN. *Biol. Pharm. Bull.* 29: 1310–1313.

Harikumar, K.B., and Kuttan, R. 2007. An extract of *Phyllanthus amarus* protects mouse chromosomes and intestine from radiation induced damages. *J. Radiat. Res.* 48: 469–476.

Harikumar, K.B., Sabu, M.C., Lima, P.S., and Kuttan, R. 2004. Modulation of haematopoetic system and anti-oxidant enzymes by *Emblica officinalis* Gaertn and its protective role against gamma-radiation induced damages in mice. *J. Radiat. Res.* 45: 549–555.

Hau, D.K., Gambari, R., Wong, R.S., et al. 2009. *Phyllanthus urinaria* extract attenuates acetaminophen induced hepatotoxicity:involvement of cytochrome P450 CYP 2E1. *Phytomedicine* 16: 751–760.

Jagetia, G.C. 2007. Radioprotective potential of plants and herbs against the effect of ionizing radiation. *J. Clin. Biochem. Nutr.* 40: 74–81.

Jemnitz, K., Veres, Z., Torok, G., et al. 2004. Comparative study in the Ames test of benzo[a] pyrene and 2-aminoanthracene metabolic activation using rat hepatic S9 and hepatocytes following *in vivo* or *in vitro* induction. *Mutagenesis* 19: 245–250.

Jindal, A., Soyal, D., Sharma, A., and Goyal, P.K. 2009. Protective effect of an extract of *Emblica officinalis* against radiation-induced damage in mice. *Integr. Cancer Ther.* 8: 98–105.

Jose, J.K., Kuttan, G., George, J., and Kuttan, R. 1997. Antimutagenic and anticarcinogenic activity of *Emblica officinalis* Gaern. *J. Clin. Biochem. Nutr.* 22: 171–176.

Kim, J.Y., Lee, S.K., Kim, C.H., et al. 2003. Effects of polycyclic aromatic hydrocarbons on liver and lung cytochrome P450s in mice. *Arch. Pharm. Res.* 26: 394–404.

Kumar, K.B.H., and Kuttan, R. 2004. An extract of *Phyllanthus amarus* protects mouse chrosomes and intestine from radiation induced damages. *J. Radiat. Res.* 45: 133–139.

Kumar, K.B.H., and Kuttan, R. 2005. Chemoprotective activity of an extract of *Phyllanthus amarus* against cyclophosphamide-induced toxicity in mice. *Phytomedicine* 12: 494–500.

Londhe, J.S., Devasagayam, T.P., Foo, L.Y., and Ghaskadbi, S.S. 2008. Antioxidant activity of some polyphenol constituents of the medicinal plant *Phyllanthus amarus* Linn. *Redox Rep.* 13: 199–207.

Madhavi, D., Devil, K.R., Rao, K.K., and Reddy, P.P. 2007. Modulating effect of *Phyllanthus* fruit extract against lead genotoxicity in germ cells of mice. *J. Environ. Biol.* 28: 115–117.

Mischiati C., Sereni, A., Hassan Khan, M. T., et al. 2006. Effects of plant extracts on gene expression profiling: from macroarrays to microarray technology. *Adv. Phytomed.* 2: 21–33.

Nandi, P., Talukdar, G., and Sharma, A. 1997. Dietary chemoprevention of clastogenic effects of 3,4-benzo(a)pyrene by *Emblica officinalis* Gaertn. fruit extract. *Br. J. Cancer.* 76: 1279–1283.

Rao, B.S.S., Shanbhoge, R., Upadhya, D., et al. 2006. Antioxidant, anticlastogenic and radioprotective effect of *Coleus aromaticus* on Chinese hamster fibroblast cells (V79) exposed to gamma radiation, *Mutagenesis* 21: 237–242.

Raphael, K.R., Ajith, T.A., Joseph, S., and Kuttan, R. 2002. Anti-mutagenic activity of *Phyllanthus amarus* Schum & Thonn *in vitro* as well as *in vivo*. *Teratog. Carcinog. Mutagen.* 22: 285–291.

Reddy, V.D., Padvathi, P., and Varadacharyulu, N.Ch. 2009. *Emblica officinalis* protects against alcohol-induced liver mitochondrial dysfunction in rats. *J. Med. Food.* 12: 327–333.

Rivera, I. G., Martins, M.T., Sanchez, P.S., et al. 1994. Genotoxicity assessment through the Ames test of medicinal plants commonly used in Brazil. *Environ. Toxicol. Water Qual.* 9: 87–93.

Roy, A.K., Dhir, H., and Sharma, A. 1992. Modification of metal-induced micronuclei formation in mouse bone marrow erythrocytes by *Phyllanthus* fruit extract and ascorbic acid. *Toxicol. Lett.* 62: 9–17.

Sanchez-Lamar, A., Fiore, M., Cundari, E., et al. 1999. *Phyllanthus orbicularis* aqueous extract: cytotoxic, genotoxic, and antiumutagenic effects in the CHO cell line. *Toxicol. Appl. Pharmacol.* 15: 231–239.

Sanchez-Lamar, A., Fuentas, J.L., Fonseca, G., et al. 2002. Assessment of the potential genotoxic risk of *Phyllanthus orbicularis* HBK aqueous extract using *in vitro* and *in vivo* assays. *Toxicol Lett.* 136: 97–96.

Sai Ram, M., Neetu, D., Deepti, P., et al. 2003. Cytoprotective activity of Amla (*Emblica officinalis*) against chromium (VI) induced oxidative injury in murine macrophages. *Phytother. Res.* 17: 430–433.

Sharma, N., Trikha, P., Athar, M., and Raisuddin, S. 2000a. Inhibitory effect of *Emblica officinalis* on the *in vivo* clastogenicity of benzo[a]pyrene and cyclophosphamide in mice. *Hum. Exp. Toxicol.* 19: 377–384.

Sharma, N., Trikha, P., Athar, M., and Raisuddin, S. 2000b. *In vitro* inhibition of carcinogen-induced mutagenicity by *Cassia occidentalis* and *Emblica officinalis*. *Drug Chem. Toxicol.* 23: 477–484.

Sharma, A., Sharma, M.K., and Kumar, M. 2009. Modulatory role of *Emblica officinalis* fruit extract against arsenic-induced oxidative stress in Swiss albino mice. *Chem. Biol. Interact.* 180: 20–30.

Shen, B., Yu, J., Wang, S., et al. 2008. *Phyllanthus urinaria* ameliorates the severity of nutritional steatohepatitis both *in vitro* and *in vivo*. *Hepatology* 47: 473–483.

Singh, I., Sharma, A., Nunia, V., and Goyal, P.K. 2005. Radioprotection of Swiss albino mice by *Emblica officinalis*. *Phytother. Res.* 19: 444–446.

Singh, I., Soyal, D., and Goyal, P.K. 2006. *Emblica officinalis* (Linn.) fruit extract provides protection against radiation-induced hematological and biochemical alterations in mice. *J. Environ. Pathol. Toxicol. Oncol.* 25: 643–654.

Sripanidkulchai, B., Tattawasart, U., Laupatarakasem, P., et al. 2002. Antimutagenic and anticarcinogenic effects of *Phyllanthus amarus*. *Phytomedicine* 9: 26–32.

Tasduq, S.A., Mondhe, D.M., Gupta, D.K., et al. 2005. Reversal of fibrogenic events in liver by *Emblica officinalis* (fruit), an Indian natural drug. *Biol. Pharm. Bull.* 28: 1304–1306.

Ueng, Y.F., Shimada, T., Yamazaki, H., and Guengerich, F.P. 1995. Oxidation of aflatoxin B1 by bacterial recombinant human cytochrome P450 enzymes. *Chem. Res. Toxicol.* 8: 218–225.

Uma Devi, P.U., Rao, R.K., and Kamath, B.S.S. 2000. Radioprotective effect of *Phyllanthus niruri* on mouse chromosomes. *Curr. Sci.* 78: 1245–1247.

Van Vleet, T.R., Bombick, D.W., and Coulombe, R.A., Jr. 2001. Inhibition of human cytochrome P450 2E1 by nicotine, nicotinine, and aqueous cigarette tar extract *in vitro*. *Toxicol. Sci.* 64: 185–191.

Voskoboinik, I., Drew, R., and Ahokas, J.T. 1997. Peroxisome proliferator nafenopin potentiated cytotoxicity and genotoxicity of cyclophosphamide in the liver and bone marrow cells. *Chem. Biol. Interac.* 105: 81–97.

17 Antiaging Effects of *Phyllanthus* Species

Vasudevan Mani and Shanmugapriya Thulasimani

CONTENTS

17.1 Introduction 267
17.2 Free Radicals and Aging 269
17.3 Dementia and *Phyllanthus*..272
17.4 Diabetes and *Phyllanthus* 272
17.5 Cardiovascular Disease and *Phyllanthus* 272
17.6 Conclusion 273
References..274

17.1 INTRODUCTION

The word *aging* does not give a good feeling to most of us because of problems and diseases associated with aging. The history of the world is replete with tales of individuals trying to stave off aging and death. Many individuals take megadoses of vitamin E, drink kombucha tea, and so on, all in the hope of finding the "fountain of youth." Wealthy people go to private European medical centers for lamb cell injections. According to social, behavioral, physiological, morphological, cellular, and molecular changes, aging can be considered in many different ways. Aging in its broadest sense is the continuous and irreversible decline in the efficiency of various physiological processes once the reproductive phase of life is over (Balcombe and Sinclair, 2001). In recent years, aging has become the social and political agenda. Aging stories, particularly antiaging therapies, are a big attraction in newspapers and magazines (Laura, 2005). It has attracted a high level of attention from all over the world; the research agenda is focused on aging for the 21st century (Giacomoni, 2005).

It is impossible to define the onset of "old age," which is the result of many physiological and sociobehavioral changes. Aging is a complex process that can be modified only by multipronged interventions (Klatz et al., 1996; Evans et al., 1992). Positive lifestyle is pivotal in the promotion of health and well-being. While one can extol the virtues and benefits of a good lifestyle for health maintenance and longevity, the investment in "behavior change today for health tomorrow" is a difficult pathway of intervention. Mental and physical idleness result in loss of vitality. An important promoter of aging is lack of mind or body activity, resulting in the "disuse syndrome," with its hallmark of premature aging (Rabins and Lauber, 2005; Woodruff-Pak, 1997).

One of the most important problems in modern gerontology is the development of means to extend a healthy life span. A number of nutrients and chemicals have been widely advertised as antiaging drugs or supplements. Experimental studies have repeatedly shown the life-extending effect of such substances, often referred to as life-span-prolonging drugs or geroprotectors (Emanuel and Obukhova, 1978). The life-extending capacity was shown for a number of geroprotectors, including antioxidants, chelate agents and lathyrogens, succinate, adaptogens and herbs, neurotropic drugs, inhibitors of monoamine oxidase, glucocorticoids, dehydroepiandrosterone, sex and growth hormones, melatonin, pineal peptide preparations, protein inhibitors, antidiabetic biguanides, thymic hormones and peptides, immunomodulators and enterosorbents (Anisimov, 2001), and mimetics of superoxide dismutase and catalase (Melov et al., 2000). The natural and synthetic dietary supplements and chemicals, including antioxidants, vitamins, and hormones, are among the most popular products on the market, even without solid scientific evidence (Olshansky et al., 2002). However, excessive intake of antioxidants or hormones is known to destroy delicate control mechanisms of homeostatic balance. It is therefore unlikely that they have a long-term beneficial impact (Goto, 2004).

Many theories of aging have been proposed, but no single explanation suffices. Despite uncertainty about aging theories, modern antiaging research has identified several key disorders as processes that promote tissue aging. These processes include immune impairment, sleep deprivation, obesity, adverse lifestyle, genetic programming, poor nutrition, hormonal deficiencies or deregulation, inflammation, oxidative stress to tissues, deficient methylation, and the formation of glycated proteins (advanced glycation end products, or ACEs) (Kyriazis, 2005). A wide range of natural substances has been identified that can provide favorable nutritional or chemical effects on several of these common disorders or processes that accelerate tissue aging. Modern research has identified many natural substances with antiaging properties, but simple interventions or single supplements (or drugs) cannot address efficiently the multifactorial aspects of tissue aging. Furthermore, the biochemical cascades of events involved in aging require a synergistic approach to the formulation of antiaging substances, specifically dietary supplements (Renpwal, 1998; Hendler, 1985).

Phyllanthus is the largest genus in the family Phyllanthaceae. *Phyllanthus emblica* (syn. *Emblica officinalis*), *Phyllanthus niruri* (syn. *Phyllanthus amarus*), *Phyllanthus acidus, Phyllanthus acuminatus, Phyllanthus caroliniensis, Phyllanthus mirabilis, Phyllanthus urinaria,* and *Phyllanthus fluitans* are some selected species with numerous therapeutic applications. *Phyllanthus* species are distributed in all tropical and subtropical regions on earth. Traditionally, *Phyllanthus* species have been used to treat jaundice, gonorrhea, frequent menstruation, dysentery, and diabetes. They have also been used topically as a treatment for skin ulcers, sores, swelling, and itchiness (Mabberley, 2008). *Phyllanthus emblica* is a species of *Phyllanthus* that is considered to be one of the strongest rejuvenative herbs in Ayurvedic medicine. It is the primary ingredient used in one of the renowned Ayurvedic herbal formulas, called *Chayavanprasha,* which has great respect as a tonic. In addition, huge numbers of *Phyllanthus* species possess several antiaging mechanisms. The objective of this chapter is to review the possible role of *Phyllanthus* species in the antiaging mechanism.

17.2 FREE RADICALS AND AGING

The free radical theory of aging was first proposed by Harman (1981). It was explained that the oxygen radicals are responsible for many degenerative changes that come with aging. Free radicals have been involved not only in aging but also in degenerative disorders, including cancer, atherosclerosis, cataracts, and neurodegeneration (Nordberg and Arner, 2001). Several cytokines, growth factors, hormones, and neurotransmitters use reactive oxygen species (ROS) as secondary messengers in intracellular signal transduction (Thannical and Fanburg, 2000). The antioxidant system supplies the main protection against unstable and potentially harmful molecules (Haliwel and Gutterridge, 1990). It includes several enzymes, such as superoxide dismutase, glutathione peroxidase, catalase, and several nonenzymatic molecules, such as vitamins C and E and β-carotene. These molecules prevent mostly oxidative damage but not completely. High levels of antioxidants, which organisms possess lifelong, have been shown to be correlated with life span.

There are many studies related to *Phyllanthus* species and antioxidant activities (Jose and Kuttan, 1995). Some of the studies are interestingly correlated with antioxidant and antiaging mechanisms. The ethyl acetate extract of Amla (*Emblica officinalis* Gaertn.) reduced the inducible nitric oxide synthase (iNOS) and cyclooxygenase 2 (COX-2) expression levels by inhibiting nuclear factor kappa B (NF-κB) activation in aged rats. These results indicated that Amla would be a useful antioxidant for the prevention of age-related renal diseases (Yokozawa et al., 2007a). Also, *Emblica officinalis* controlled the lipid metabolism and protein expression involved in oxidative stress during the aging process through reduction of serum and hepatic cholesterol, triglycerides, and mitochondrial thiobarbituric acid-reactive substance and increased PPARalpha (peroxisome proliferator activated receptor alpha) protein levels in aged rats (Yokozawa et al., 2007b). Moreover, it helped to protect the skin from the damaging effects of free radicals, nonradicals, and transition metal-induced oxidative stress (Chaudhuri, 2002). The oral administration of *Emblica officinalis* aqueous extract along with ochratoxin significantly ameliorated ochratoxin-induced lipid peroxidation in the testis of mice (Verma and Chakraborty, 2008). *Emblica officinalis* has been reported as a rich source of vitamin C, which plays an important role in scavenging free radicals (Scartezzini et al., 2006). Free gallic and ellagic acids as well as emblicanins A and B in the *Emblica officinalis* extract were separated by thin-layer chromatography (TLC) and identified the antioxidant activity by *in vitro* methods (Pozharitskaya et al., 2007). The fruit extracts of *Emblica officinalis* significantly inhibited chromium-induced free radical production and restored the antioxidant status by controlled lipid peroxidation and improved glutathione peroxidase activity as well as glutathione (GSH) levels by *in vitro* study (Ram et al.,2002). Thangaraj et al. (2007) focused on the significant prevention of stress-induced oxidative stress of *Emblica officinalis* by restoring the enzymatic antioxidant status like that of superoxide dismutase, catalase, and glutathione peroxidase in lymphoid organs of thymus and spleen. Chronic administration of tannoid principles, namely, emblicanin A, emblicanin B, punigluconin, and pedunculagin, of *E. officinalis* normalized the activities of superoxide dismutase, catalase, and glutathione peroxidase and reduced lipid peroxidation in rat brain frontal cortex and striatum against chronic stress-induced perturbations (Bhattacharya et al., 2000).

An antioxidant protein 35-kDa molecule was isolated from the leaves of *Phyllanthus niruri* and demonstrated antioxidant as well as cytoprotective activities in hepatocytes (Sarkar and Sil, 2010; Sarkar et al., 2009). An isolated protein from leaves of *Phyllanthus niruri* also normalized the antioxidant marker enzymes like superoxide dismutase and catalase and reduced lipid peroxidation in liver against carbon tetrachloride-induced liver damage (Bhattacharjee and Sil, 2007). Aqueous extract of the herb *Phyllanthus niruri* also extended the antioxidant activity with improved hepatic antioxidant enzymes superoxide dismutase, catalase, and GSH and reduced lipid peroxidation against nimesulide- and acetaminophen-induced oxidative stress in liver (Chatterjee and Sil, 2006; Bhattacharjee and Sil, 2006). The nitric oxide-scavenging activity with *Phyllanthus niruri* also was demonstrated *in vitro* (Jagetia and Baliga, 2004).

Karuna et al. (2009) evidenced the antioxidant activity of *Phyllanthus amarus*. Rats treated with an aqueous extract of *Phyllanthus amarus* showed a significant decline in plasma lipid peroxides and a significant increase in plasma vitamin C, uric acid, glutathione reductase, catalase, glutathione peroxidase, and superoxide dismutase activities. Also, it had a significant protective effect against hydrogen peroxide-, streptozotocin-, and nitric oxide-induced DNA damage in lymphocytes. Londhe et al. (2008) isolated some polyphenol constituents, namely, amariin, 1-galloyl-2,3-dehydrohexahydroxydiphenyl (DHHDP)-glucose, repandusinic acid, geraniin, corilagin, phyllanthusiin D, rutin, and quercetin 3-O-glucoside and examined their free radical scavenging ability using *in vitro* methods. In addition, a few more recent studies also explained the antioxidant mechanism of *Phyllanthus amarus* in various experimental models (Faremi et al., 2008; Adeneye and Benebo, 2008; Naaz et al., 2007; Raphael et al., 2002).

Phyllanthus urinaria Lin, one of the herbal plants belonging to the genus *Phyllanthus* possesses numerous antioxidant principles: geraniin, phyllanthin, phyltetralin, trimethyl-3,4-dehydrochebulate, methylgallate, rhamnocitrin, methyl brevifolincarboxylate, β-sitosterol-3-*O*-β-d-glucopyranoside, quercitrin, and rutin (Fang et al., 2008; Lin et al., 2008). In addition, some of the recent reports strengthen information on the antioxidant mechanisms of *Phyllanthus urinaria* (Kumaran and Karunakaran, 2007; Chularojmontri et al., 2009; 2005; Xu et al., 2007). Furthermore, Kumaran and Karunakaran (2007) highlighted the antioxidant activity of *Phyllanthus debilis, Phyllanthus virgatus,* and *Phyllanthus maderaspatensis* using *in vitro* methods. Another study with ethanol extract of *Phyllanthus maderaspatensis* increased the levels of GSH in heart tissue and protected against adriamycin-induced toxicity in mice (Bommu et al., 2008).

Different fractions from *Phyllanthus simplex* significantly increased the antioxidant enzymes like superoxide dismutase, catalase, and glutathione peroxidase in the liver and kidney of alloxan-induced diabetic rats (Shabeer et al., 2009). *Phyllanthus rheedii* and *Phyllanthus orbicularis* are also rich sources of tannins, flavanoids, and phenolic compounds and have proved to be antioxidant plants (Suresh and Asha, 2008; Ferrer et al., 2002). Antioxidant activity of *Phyllanthus tenellus, Phyllanthus caroliniensis, Phyllanthus ussurensis,* and *Phyllanthus polyphyllus* were also proven using various experimental models (Ignacio et al., 2001; Filho et al., 1996; Chung et al., 2003; Rajkapoor et al., 2007).

Several related theories containing an ROS component have also been proposed (Weinert and Timiras, 2003). One that has been extensively studied is the mitochondrial theory of aging, which hypothesizes that mitochondria are the critical component in control of aging. In support of a mitochondrial theory of aging, evidence suggests that mitochondrial DNA damage is increased with aging (Hamilton et al., 2001). Interestingly, ethyl acetate extract of *Emblica officinalis* significantly reduced thiobarbituric acid-reactive substance levels of mitochondria in aged rats (Yokozawa et al., 2007a). Tertiary butyl hydroperoxide- (TBHP) induced mitochondrial cellular damage in hepatocytes was controlled with a novel antioxidant protein molecule isolated from *Phyllanthus niruri* (Sarkar and Sil, 2010). *Phyllanthus amarus* also protected rat liver mitochondria against oxidative damage (Londhe et al., 2008).

The aqueous extract of *Phyllanthus fraternus* also showed antioxidant activity by reducing mitochondrial dysfunction induced by bromobenzene. Interestingly, it maintained the level of antioxidant enzymes like catalase, glutathione peroxidase, glutathione reductase, and superoxide dismutase and reduced lipid peroxides and protein carbonyl levels (Gopi and Setty, 2010). In addition, numerous studies confirmed the possible protective mechanism of mitochondrial dysfunction with *Phyllanthus fraternus* against different inducing agents (Sebastian, and Setty, 1999; Padma and Setty, 1999).

An additional theory that has gained more attention in recent years is the molecular inflammatory theory of aging, whereby the activation of redox-sensitive transcriptional factors by age-related oxidative stress causes the upregulation of proinflammatory gene expression. As a result, various proinflammatory molecules are generated, leading to inflammation processes in various tissues and organs. This inflammatory cascade is exaggerated during aging and has been linked with many age-associated pathological changes, such as cancer, various cardiovascular diseases, arthritis, and several neurodegenerative diseases (Chung et al., 2006). Interestingly, a common phenomenon in aging-related pathological changes is the discovery of ROS as a potential unifying mechanism contributing to many of these diseases (Fortuno et al., 2005; Madamanchi et al., 2005; Moreira et al., 2005).

Interestingly, some of the species (namely, *Emblica officinalis, Phyllanthus amarus, Phyllanthus debilis, Phyllanthus reticulates, Phyllanthus singampattiyana,* and *Phyllanthus urinaria*) have been demonstrated to have anti-inflammatory action with free radical scavenging, and it may protect the molecular inflammation and control the aging process (Asmawi et al., 1993; Fang et al., 2008; Kassuya et al., 2006; Chandrashekar et al., 2005; Saha et al., 2007; Maridass et al., 2005; Rao et al., 2006).

Some evidence has been reported concerning the role of *Phyllanthus* species in various antiaging mechanisms related to oxidative damage. Further molecular studies will explain the quantitative utility of these plants in the antiaging mechanism.

17.3 DEMENTIA AND *PHYLLANTHUS*

Aging may affect memory by changing the way the brain stores information and by making it harder to recall stored information. Dementia is related to progressive loss of memory and one other cognitive disturbance, like speech disorder and loss of space orientation. Many different mechanisms are hypothesized for age-related brain changes, including apoptosis, telomere loss, neuroendocrine alterations, and autoimmune changes, and oxidative stress causes memory loss. Age-related brain damage may be produced by dysfunction of the neuronal cytoskeleton or damage to mitochondria that diminishes neuronal energy (Hayflick, 2000).

The four common types of dementia in persons over the age of 65 include Alzheimer's disease (AD), vascular dementia, diffuse Lewy body disease (dLbd), and alcohol-induced dementia. There is also extensive evidence linking the central cholinergic system to AD and memory (Parle et al., 2004). *Emblica officinalis* and *Phyllanthus amarus* were shown to be memory enhancers and reversed the drug-induced and aging-induced memory deficit in experimental animals. Moreover, they facilitated the cholinergic activity by inhibiting cholinesterase activity in the brain (Vasudevan and Parle, 2007a, 2007b; Joshi and Parle, 2007).

17.4 DIABETES AND *PHYLLANTHUS*

Diabetes mellitus (DM) is one of the most important and prevalent chronic diseases. It currently affects 250 million people worldwide, with 6 million new cases reported each year. DM has been implicated as a risk factor for dementia not only of the vascular type but also concerning AD. Patients with type 1 or type 2 diabetes mellitus have been found to present cognitive deficits associated with reduced performance on multiple domains of cognitive function (Ott et al., 1999). In addition, diabetes and aging are associated with skin damage, a reduction in skin blood flow, and numerous changes in the structure of the skin, including changes in skin collagen, skin thickness, and the response of the skin vasculature to local and global heat stress (Petrofsky et al., 2008).

Various species of *Phyllanthus* have been proved to facilitate hypoglycemic and antihyperglycemic activity in different experimental models as described in detail elsewhere in this book.

17.5 CARDIOVASCULAR DISEASE AND *PHYLLANTHUS*

Age is a major risk factor for cardiovascular disease, including heart diseases. The incidence of heart disease increases with age for both men and women. Aging is associated with the development of cardiovascular structural and functional alterations, which can explain the age-related increase in cardiovascular risk. Among them, hypertension and atherosclerosis develop largely with the aging process. The development of extensive atherosclerosis of major arteries of the heart, brain, and lower extremities is a particularly frequent problem in elderly individuals and is responsible for the majority of the cardiovascular morbidity and mortality in this population (Costopoulos et al., 2008).

Epidemiological studies revealed that the clinical sequelae of atherosclerosis-related conditions like coronary artery disease, ischemic stroke, and peripheral artery disease are all increased with age (Robert, 1999; Lakatta, 2007; Hori et al., 2010). Autopsy studies also supported these findings and demonstrated that atherosclerosis in the vascular beds supplying blood to the heart, brain, and lower extremities also increased with age.

Elevation of plasma low-density lipoproteins (LDLs) is clearly a major risk factor. In part, the age-related increase in atherosclerosis may be secondary to the rise in LDL cholesterol levels that occurs with increasing age (Lakatta, 2007). Flavonoids from *Emblica officinalis* significantly reduced the LDL and very-low-density lipoprotein (VLDL) cholesterol in rats. Interestingly, they significantly inhibited the activity of 3-hydroxy-3-methylglutaryl (HMG) coenzyme A (CoA) reductase in liver, and there was an increase in plasma LCAT (lecithin cholesterol acyltransferase) in experimental animals (Anila and Vijayalakshmi, 2002).

Moreover, numerous studies focused on the effect of *Emblica officinalis* in controlling hypercholesterolemia and preventing atherosclerosis (Kim et al., 2010; 2005; Yokozawa et al., 2007b; Vasudevan and Parle, 2007b; Mathur et al., 1996). *Phyllanthus niruri* significantly increased the serum high-density lipoprotein (HDL) as well as plasma LCAT levels and reduced the levels of LDL and HDL in triton-induced hyperlipidemic rats (Khanna et al., 2002). Weight loss and decreased plasma cholesterol effects of *Phyllanthus amarus* aqueous extract of leaves and seeds were studied in normal mice (Adeneye et al., 2006). Oral treatment for 21 days with *Phyllanthus rheedii* reduced lipid metabolites such as total cholesterol and triglycerides and maintained the HDLs in streptozotocin-induced diabetic rats (Sivajothi et al., 2008).

Therefore, it seems likely that species of *Phyllanthus* are useful antiaging agents in view of their atherosclerosis seducing property, and they may control age-related AD.

Hypertension generally affects people later in life. Almost two-thirds (62%) of the population with hypertension are age 55 and older. Hypertension can lead to other chronic diseases, such as heart disease, kidney failure, or stroke. Reports are pouring in showing a strong link between aging and hypertension (Taddei et al., 2006; James et al., 2006; Qiao et al., 2008; Umeno et al., 2010). A few reports focused on the role of *Phyllanthus* in control of hypertension. Geraniin, an isolated compound from *Phyllanthus urinaria* lowered systolic and diastolic blood pressure in experimental rats (Lin et al., 2008). *Phyllanthus amarus, Phyllanthus corcovadensis,* and *Phyllanthus sellowianus* were reported as diuretic plants that can also control hypertension (Wright et al., 2007). A significant reduction in systolic blood pressure in nondiabetic hypertensives and female subjects was noted with *Phyllanthus amarus*, and it increased 24-h urine volume and urine and serum sodium levels (Srividya and Periwal, 1995).

17.6 CONCLUSION

Aging is characterized by a progressive accumulation of damaged macromolecules, organelles, and cytomembranes, which may account for the age-associated malfunctioning of many biological processes. The inefficiency and failure of maintenance repair and turnover pathways may be the main cause of damage accumulation during aging. Many theories of aging are based on the concept that damage, due to either normal toxic by-products of metabolism or inefficient repair/defensive systems, accumulates throughout the entire life span and causes aging; it also acts as the biggest risk factor for many diseases, including cardiovascular and neurodegenerative diseases. The present chapter highlighted the possible role of *Phyllanthus* species in reduction of various aging mechanisms, like oxygen free radical, mitochondrial damage, cytotoxicity, and molecular inflammation. Also, we focused on the various aging disorders, namely, dementia, diabetes, hypertension, and atherosclerosis, presenting promising activity of different plants from the *Phyllanthus* species. Most of the reports were established based on animal experiments, but further clinical evaluation is needed to prove the potential of these plants. From the evidence given here, there is a need to establish either useful molecules or polyherbal formulations or extracts from *Phyllanthus* with a focus on several antiaging principles, and a clinical focus is very much essential.

REFERENCES

Adeneye, A.A., Amole, O.O., and Adeneye, A.K. 2006. Hypoglycemic and hypocholesterolemic activities of the aqueous leaf and seed extract of *Phyllanthus amarus* in mice. *Fitoterapia* 77: 511–514.

Adeneye, A.A., and Benebo, A.S. 2008. Protective effect of the aqueous leaf and seed extract of *Phyllanthus amarus* on gentamicin and acetaminophen-induced nephrotoxic rats. *J. Ethnopharmacol.* 118: 318–323.

Anila, L., and Vijayalakshmi, N.R. 2002. Flavonoids from *Emblica officinalis* and *Mangifera indica* effectiveness for dyslipidemia. *J. Ethnopharmacol.* 79: 81–77.

Anisimov, V.N. 2001. Life span extension and cancer risk: myths and reality. *Exp. Gerontol.* 36: 1101–1136.

Asmawi, M.Z., Kankaanranta, H., Moilanen, E., and Vapaatalo, H. 1993. Anti-inflammatory activities of *Emblica officinalis* Gaertn leaf extracts. *J. Pharm. Pharmacol.* 45: 581–584.

Balcombe, N.R., and Sinclair, A. 2001. Ageing: definitions, mechanisms and the magnitude of the problem. *Best Pract. Res. Clin. Gastroenterol.* 15: 835–849.

Bhattacharya, A., Ghosal, S., and Bhattacharya, S.K. 2000. Antioxidant activity of tannoid principles of *Emblica officinalis* (Amla) in chronic stress induced changes in rat brain. *Indian J. Exp. Biol.* 38: 877–880.

Bhattacharjee, R., and Sil, P.C. 2006. The protein fraction of *Phyllanthus niruri* plays a protective role against acetaminophen induced hepatic disorder via its antioxidant properties. *Phytother. Res.* 20: 595–601.

Bhattacharjee, R., and Sil, P.C. (2007) Protein isolate from the herb, *Phyllanthus niruri* L. (Euphorbiaceae), plays hepatoprotective role against carbon tetrachloride induced liver damage via its antioxidant properties. *Food Chem. Toxicol.* 45: 817–826.

Bommu, P., Nanjan, C.M., Joghee, N.M., et al. 2008. *Phyllanthus maderaspatensis*, a dietary supplement for the amelioration of adriamycin-induced toxicity and oxidative stress in mice. *J. Nat. Med.* 62: 149–154.

Chandrashekar, K.S., Joshi, A.B., Satyanarayana, D., and Pai, P. 2005. Analgesic and anti inflammatory activities of *Phyllanthus debilis* whole plant. *Pharm. Biol.* 43: 586–588.

Chatterjee, M., and Sil, P.C. 2006. Hepatoprotective effect of aqueous extract of *Phyllanthus niruri* on nimesulide-induced oxidative stress *in vivo*. *Indian J. Biochem. Biophys.* 43: 299–305.

Chaudhuri, R.K. 2002. *Emblica* cascading antioxidant: a novel natural skin care ingredient. *Skin Pharmacol. Appl. Skin Physiol.* 15: 374–380.
Chularojmontri, L., Ihara, Y., Muroi, E., et al. 2009. Cytoprotective role of *Phyllanthus urinaria* L. and glutathione-S transferase in doxorubicin-induced toxicity in H9c2 cells. *J. Med. Assoc. Thai.* 92: S43–S51.
Chularojmontri, L., Wattanapitayakul, S.K., Herunsalee, A., et al. 2005. Antioxidative and cardioprotective effects of *Phyllanthus urinaria* L. on doxorubicin-induced cardiotoxicity. *Biol. Pharm. Bull.* 28: 1165–1171.
Chung, H.Y., Sung, B., Jung, K.J., et al. 2006. The molecular inflammatory process in aging. *Antioxid. Redox. Signal.* 8: 572–581.
Chung, S.K., Nam, J.A., Jeon, S.Y., et al. 2003. A prolyl endopeptidase-inhibiting antioxidant from *Phyllanthus ussurensis*. *Arch. Pharm. Res.* 26: 1024–1028.
Costopoulos, C., Liew, T.V., and Bennett, M. 2008. Ageing and atherosclerosis: mechanisms and therapeutic options. *Biochem. Pharmacol.* 75: 1251–1261.
Emanuel, L.M., and Obukhova, L.K. 1978. Types of experimental delay in aging patterns. *Exp. Gerontol.* 13: 25–29.
Evans, W., Rosenberg, I., and Thompson, J. 1992. *Biomarkers: the 10 keys to prolonging vitality.* New York: Simon & Schuster.
Fang, S., Rao, Y.K., and Tzeng, Y. 2008. Anti-oxidant and inflammatory mediator's growth inhibitory effects of compounds isolated from *Phyllanthus urinaria*. *J. Ethnopharmacol.* 116: 333–340.
Faremi, T.Y., Suru, S.M., Fafunso, M.A., and Obioha, U.E. 2008. Hepatoprotective potentials of *Phyllanthus amarus* against ethanol-induced oxidative stress in rats. *Food Chem. Toxicol.* 46: 2658–2664.
Ferrer, M., Sanchez-Lamar, A., Luis Fuentes, J., et al. 2002. Antimutagenic mechanisms of *Phyllanthus orbicularis* when hydrogen peroxide is tested using *Salmonella* assay. *Mutat. Res.* 517: 251–254.
Filho, V.C., Santos, A.R., De Campos, R.O., et al. 1996. Chemical and pharmacological studies of *Phyllanthus caroliniensis* in mice. *J. Pharm. Pharmacol.* 48: 1231–1236.
Fortuno, A., Jose, G.S., Moreno, M.U., et al. 2005. Oxidative stress and vascular remodelling. *Exp. Physiol.* 90: 457–462.
Giacomoni, P.U. 2005. Ageing science and the cosmetics industry. *EMBO Rep.* 6: S45–S48.
Gopi, S., and Setty, O.H. 2010. Protective effect of *Phyllanthus fraternus* against bromobenzene induced mitochondrial dysfunction in rat liver mitochondria. *Food Chem. Toxicol.* 48: 2170–2175.
Goto, S. 2004. Hormesis and intervention of aging: an emerging paradigm in gerontology. *Geriatrics Gerontol. Int.* 4: S79–S80.
Haliwel, B., and Gutterridge, J.M.C. 1990. Role of free radicals and catalytic metal ions in disease: an overview. *Methods Enzymol.* 186: 1–85.
Hamilton, M.L., Van Remmen, H., Drake, J.A., et al. 2001. Does oxidative damage to DNA increase with age? *Proc. Natl. Acad. Sci. U. S. A.* 98: 10469–10474.
Harman, D. 1981. The aging process. *Proc. Natl. Acad. Sci. U. S. A.* 78: 7124–7128.
Hayflick, L. 2000. The future of ageing. *Nature* 408: 267–269.
Hendler, S. 1985. *The complete guide to anti-aging nutrients.* New York: Simon & Schuster.
Hori, Y., Funabashi, N., Uehara, M., et al. 2010. Positive influence of aging on the occurrence of fat replacement in the right ventricular myocardium determined by multislice-CT in subjects with atherosclerosis. *Int. J. Cardiol.* 142: 152–158.
Ignacio, S.R., Ferreira, J.L., Almeida, M.B., and Kubelka, C.F. 2001. Nitric oxide production by murine peritoneal macrophages *in vitro* and *in vivo* treated with *Phyllanthus tenellus* extracts. *J. Ethnopharmacol.* 74: 181–187.
Jagetia, G.C., and Baliga, M.S. 2004. The evaluation of nitric oxide scavenging activity of certain Indian medicinal plant *in vitro*: a preliminary study. *J. Med. Food.* 7: 343–348.

James, M.A., Tullett, J., Hemsley, A.G., and Shore, A.C. 2006. Effects of aging and hypertension on the microcirculation. *Hypertension*. 47: 968–974.

Jose, J. K., and Kuttan, R. 1995. Antioxidant activity of *Emblica officinalis*. *J. Clin. Biochem. Nutr.* 19: 63–70.

Joshi, H., and Parle, M. 2007. Evaluation of the antiamnesic effects of *Phyllanthus amarus* in mice. *Colombia Med.* 38: 132–139.

Karuna, R., Reddy, S.S., Baskar, R. et al. 2009. Antioxidant potential of aqueous extract of *Phyllanthus amarus* in rats. *Indian J. Pharmacol.* 41: 64–67.

Kassuya, C.A.L., Silvestre, A., Jr., O.M., et al. 2006. Antiinflammatory and antiallodynic actions of the lignan niranthin isolated from *Phyllanthus amarus*: evidence for interaction with platelet activating factor receptor. *Eur. J. Pharmacol.* 546: 182–188.

Khanna, A. K., Rizvi, F.N., and Chander, R. 2002. Lipid lowering activity of *Phyllanthus niruri* in hyperlipemic rats. *J. Ethnopharmacol.* 82: 19–22.

Kim, H.J., Yokozawa, T., Kim, H.Y., et al. 2005. Influence of amla (*Emblica officinalis* Gaertn.) on hypercholesterolemia and lipid peroxidation in cholesterol-fed rats. *J. Nutr. Sci. Vitaminol.* 51: 413–418.

Kim, H.Y., Okubo, T., Juneja, L.R., and Yokozawa, T. (2010) The protective role of amla (*Emblica officinalis* Gaertn.) against fructose-induced metabolic syndrome in a rat model. *Br. J. Nutr.*, 103, 502–512.

Klatz, R., Kovarik, F., and Goldman, R. 1996. *Advances in anti-aging medicine,* Vol. 1. Larchnront, NY: Liebert.

Kumaran, A., and Karunakaran, R. J. 2007. *In vitro* antioxidant activities of methanol extracts of five *Phyllanthus* species from India. *Food Sci. Technol.* 40: 344–352.

Kyriazis, M. 2005. *Anti-aging medicines.* London: Waikins.

Lakatta, E.G. 2007. Central arterial aging and the epidemic of systolic hypertension and atherosclerosis. *J. Am. Soc. Hypertension.* 1: 302–340.

Laura, H. 2005. Ageing research in the media. *EMBO Rep.* 6: S81–S83.

Lin, S.Y, Wang, C.C., Lu, Y.L., et al. 2008. Antioxidant, anti-semicarbazide-sensitive amine oxidase, and anti-hypertensive activities of geraniin isolated from *Phyllanthus urinaria*. *Food Chem. Toxicol.* 46: 2485–2492.

Londhe, J.S., Devasagayam, T.P., et al. 2008. Antioxidant activity of some polyphenol constituents of the medicinal plant *Phyllanthus amarus* Linn. *Redox Rep.* 13:199–207.

Mabberley, D. J. 2008. *Mabberley's plant-book,* 3rd ed. Cambridge: Cambridge University Press.

Madamanchi, N.R., Vendrov, A., and Runge, M.S. 2005. Oxidative stress and vascular disease. *Arterioscler. Thromb. Vasc. Biol.* 25: 29–38.

Maridass, M., Victor, B., Benniamin, A., et al. 2005. Anti-inflammatory activity of *Phyllanthus singampattiyana* leaf extract. *Pharm. Biol.* 43: 296–298.

Mathur, R., Sharma, A., Dixit, V.P., and Varma, M. 1996. Hypolipidaemic effect of fruit juice of *Emblica officinalis* in cholesterol-fed rabbits. *J. Ethnopharmacol.* 50: 61–68.

Melov, S.J., Ravenscroft, S., Malik, M.S., et al. 2000. Extension of life-span with superoxide dismutase/catalase mimetics. *Science* 289: 1567–1569.

Moreira, P.I., Smith, M.A., Zhu, X., et al. 2005. Oxidative stress and neurodegeneration. *Ann. N. Y. Acad. Sci.* 1043: 545–552.

Naaz, F., Javed, S., and Abdin, M.Z. 2007. Hepatoprotective effect of ethanolic extract of *Phyllanthus amarus* Schum. et Thonn. on aflatoxin B1-induced liver damage in mice. *J. Ethnopharmacol.* 113: 503–509.

Nordberg, J., and Arner, E. S. J. 2001. Reactive oxygen species, antioxidants, and the mammalian thioredoxin system. *Free Radic. Biol. Med.* 31:1287–1312.

Olshansky, S.J., Hayflick, L., and Carnes, B.A. 2002. No truth to the fountain of youth. *Sci. Am.* 286: 92–95.

Ott, A., Stolk, R.P., van Harskamp, F., et al. 1999. Diabetes mellitus and the risk of dementia: the Rotterdam Study. *Neurology* 53: 1937–1942.

Padma, P., and Setty, O.H. 1999. Protective effect of *Phyllanthus fraternus* against carbon tetrachloride-induced mitochondrial dysfunction. *Life Sci.* 64: 2411–2417.

Parle, M., Dhingra, D., and Kulkarni, S.K. 2004. Neurochemical basis of learning and memory. *Indian J. Pharm. Sci.* 66: 371–376.

Petrofsky, J.S., Prowse M., and Lohman, E. 2008. The influence of ageing and diabetes on skin and subcutaneous fat thickness in different regions of the body. *J. Appl. Res.* 8: 55–61.

Pozharitskaya, O.N., Ivanova, S.A., Shikov, A.N., and Makarov, V.G. 2007. Separation and evaluation of free radical-scavenging activity of phenol components of *Emblica officinalis* extract by using an HPTLC-DPPH method. *J. Sep. Sci.* 30: 1250–1254.

Qiao, X., McConnell, K.R., and Khalil R.A. 2008. Sex steroids and vascular responses in hypertension and aging. *Gend. Med.* 5: S46–S64.

Rabins, P., and Lauber, L. 2005. *Getting old without getting anxious.* London: Penguin.

Rajkapoor, B., Sankari, M., Sumithra, M., et al. 2007. Antitumor and cytotoxic effects of *Phyllanthus polyphyllus* on Ehrlich ascites carcinoma and human cancer cell lines. *Biosci. Biotechnol. Biochem.* 71: 2177–2183.

Ram, S.M., Neetu, D., Yogesh, B., et al. 2002. Cyto-protective and immunomodulating properties of Amla (*Emblica officinalis*) on lymphocytes: an in-vitro study. *J. Ethnopharmacol.* 8: 5–10.

Rao, Y.K., Fang, S.H., and Tzeng, Y.M. 2006. Anti-inflammatory activities of constituents isolated from *Phyllanthus polyphyllus*. *J. Ethnopharmacol.* 103: 181–186.

Raphael, K.R., Sabu, M.C., and Kuttan, R. 2002. Hypoglycemic effect of methanol extract of *Phyllanthus amarus* Schum & Thonn on alloxan induced diabetes mellitus in rats and its relation with antioxidant potential. *Indian J. Exp. Biol.* 40: 905–909.

Renpwal, S.T. 1998. *The anti-aging revolution.* Emmaus, PA: Rodale Press.

Robert, L. 1999. Aging of the vascular-wall and atherosclerosis. *Exp. Gerontol.* 34: 491–501.

Sarkar, M.K., Kinter, M., Mazumder, B., and Sil, P.C. 2009. Purification and characterisation of a novel antioxidant protein molecule from *Phyllanthus niruri*. *Food Chem.* 114: 1405–1412.

Sarkar, M.K., and Sil, P.C. 2010. Prevention of tertiary butyl hydroperoxide induced oxidative impairment and cell death by a novel antioxidant protein molecule isolated from the herb, *Phyllanthus niruri*. *Toxicol. In Vitro* 24: 1711–1719.

Scartezzini, P., Antognoni, F., Raggi, M.A., Poli, F., and Sabbioni, C. 2006. Vitamin C content and antioxidant activity of the fruit and of the Ayurvedic preparation of *Emblica officinalis* Gaertn. *J. Ethnopharmacol.* 104: 113–118.

Sebastian, T., and Setty, O.H. 1999 Protective effect of *P. fraternus* against ethanol-Induced mitochondrial dysfunction. *Alcohol* 17: 29–34.

Shabeer, J., Srivastava, R.S., Singh, S.K. 2009. Antidiabetic and antioxidant effect of various fractions of *Phyllanthus simplex* in alloxan diabetic rats. *J. Ethnopharmacol.* 124: 34–38.

Sivajothi, V., Dey, A., Jayakar, B., and Rajkapoor, B. 2008. Antihyperglycemic, antihyperlipidemic and antioxidant effect of *Phyllanthus rheedii* on streptozotocin induced diabetic rats. *Iranian J. Pharm. Res.* 7: 53–59.

Srividya, N., and Periwal, S. 1995. Diuretic, hypotensive and hypoglycaemic effect of *Phyllanthus amarus*. *Indian J. Exp. Biol.* 33: 861–864.

Suresh, V., and Asha, V.V. 2008. Preventive effect of ethanol extract of *Phyllanthus rheedii* Wight. on D-galactosamine induced hepatic damage in Wistar rats. *J. Ethnopharmacol.* 116: 447–453.

Taddei, S., Virdis, A., Ghiadoni, L., Versari, D., and Salvetti, A. 2006. Endothelium, aging, and hypertension. *Curr. Hypertens. Rep.* 8: 84–89.

Thangaraj, R., Ayyappan, SR., Manikandan, P., and Baskaran, J. 2007. Antioxident property of *Emblica officinalis* during experimentally induced restrain stress in rats. *J. Health Sci.* 53: 496–499.

Thannical, V. J., and Fanburg, B. L. 2000. Reactive oxygen species in cell signaling. *Am. J. Physiol. Lung Cell. Mol. Physiol.* 279: L1005–L1028.

Umeno, T., Shimada, T., Tsukihashi, H., et al. 2010. The effect of hypertension, aging and benidipine on arterial elasticity in elderly hypertensives. *CVD Prev. Cont.* 5: 45–50.

Vasudevan, M., and Parle, M. 2007a. Effect of Anwala churna (*Emblica officinalis* Gaertn.): an Ayurvedic preparation on memory deficit rats. *Yakugaku Zasshi* 127: 1701–1707.

Vasudevan, M., and Parle, M. 2007b. Memory enhancing activity of Anwala churna (*Emblica officinalis* Gaertn.): an Ayurvedic preparation. *Physiol. Behav.* 91: 46–54.

Verma, R., and Chakraborty, D. 2008. *Emblica officinalis* aqueous extract ameliorates ochratoxin-induced lipid peroxidation in the testis of mice. *Acta Pol. Pharm.* 65: 187–194.

Weinert, B.T., and Timiras, P.S. 2003. Invited review: theories of aging. *J. Appl. Physiol.* 95: 1706–1716.

Woodruff-Pak, D. 1997. *The neuropathy of aging.* Malden, MA: Blackwell.

Wright, C.I., Van-Buren, L., Kroner, C.I., and Koning, M.M. 2007. Herbal medicines as diuretics: a review of the scientific evidence. *J. Ethnopharmacol.* 114: 1–31.

Xu, M., Zha, Z.J., Qin, X.L., et al. 2007. Phenolic antioxidants from the whole plant of *Phyllanthus urinaria*. *Chem. Biodivers.* 4: 2246–2252.

Yokozawa, T., Kim, H.Y., Kim, H.J., et al. 2007a. Amla (*Emblica officinalis* Gaertn.) attenuates age-related renal dysfunction by oxidative stress. *J. Agric. Food. Chem.* 55: 7744–7752.

Yokozawa, T., Kim, H.Y., Kim, H.J., et al. 2007b. Amla (*Emblica officinalis* Gaertn.) prevents dyslipidaemia and oxidative stress in the ageing process. *Br. J. Nutr.* 97: 1187–1195.

18 Toxicity Studies of *Phyllanthus* Species

K. N. S. Sirajudeen

CONTENTS

18.1 Introduction 279
18.2 Toxicity Studies of *Phyllanthus* SPP. 281
    18.2.1 *Phyllanthus amarus/Phyllanthus niruri* 281
    18.2.2 *Phyllanthus urinaria* 283
    18.2.3 *Phyllanthus emblica* (*Emblica officinalis*) 283
    18.2.4 Other *Phyllanthus* Species 284
        18.2.4.1 *Phyllanthus reticulatus* 284
        18.2.4.2 *Phyllanthus maderaspatensis* 285
        18.2.4.3 *Phyllanthus rheedii/Phyllanthus kozhikodianus* 285
        18.2.4.4 *Phyllanthus tenellus* 285
        18.2.4.5 *Phyllanthus fraternus* 285
18.3 Conclusion 285
References 286

18.1 INTRODUCTION

Plant materials have been used for medicinal purpose since ancient times as the treatment for various diseases. Over the past two decades, interest in drugs derived from plant sources has increased. The World Health Organization (WHO) estimated that 70–80% of World's population use essentially traditional medicine for primary health care. The use of herbal medicine is popular not only in developing countries but also now in developed countries (Calixto, 2000).

Generally, herbal products are perceived by the public as natural, safe, and free from toxic side effects, but this is not necessarily the case. Well controlled clinical trials showed that toxicity really exists. Plant materials contain many constituents, and some are potentially toxic, such as cytotoxic anti-cancer derived drugs, digitalis, the pyrrolizidine alkaloids, ephedrine, phorbol ester, and others (Brown, 1992; Drew and Myers, 1997; Calixto, 2000). Only limited data exist for most plants to ensure their safety, efficacy, and quality. Safety is now a common concern for all kinds of herbal medicines, including Arabic, Ayurvedic, and Kampo medicines (Saad et al., 2006; Ikegami et al, 2004; De Smet, 2004). Herbal toxicity can result from contamination, adulteration, and misidentification and from natural chemical constituents of the herbs (Pillans, 1995).

To assess the safety of herbal products for clinical use, toxicological studies should be carried out in various experimental animals to predict toxicity and to provide guidelines for selecting a safe dose in humans. The toxicological evaluation of repeated-dose (subacute, subchronic, and chronic) studies in experimental animals is relevant in determining the cumulative toxicity of plant material preparations on the target organ and on physiologic metabolic tolerance. To select the ideal dose for a repeated-dose study, an acute toxicity study with a range of doses has to be conducted, and the doses selected should be at and above the suggested human dose (Arnold, 1990; Rhiouani et al., 2008).

A wide variety of toxic effects can be assessed in toxicity studies by monitoring different parameters, including hematological, clinical chemistry, behavioral, and histological evaluation (WHO, 1993).

Blood provides an index of physiological and pathological status in human and animals, and important parameters in a hematological study include hemoglobin, packed cell volume, white blood cell count, and platelet count (Schlam et al., 1975). The liver is an important organ involved in biotransformation of xenobiotics, so it is vulnerable to xenobiotics (Sturgill and Lambert, 1997). The kidneys are principal organs involved in excretion, especially of xenobiotics, so they are easily affected by potentially toxic agents. Some herbal medicines have hepato- and nephrotoxic effects (Torar and Petzel, 2009).Therefore, the liver and kidney functions should be monitored in herbal toxicity studies. The liver contains a host of enzymes, such as alanine transaminase (ALT), aspartate transaminase (AST), alkaline phophatase (ALP), and lactate dehydrogenase (LDH). The activities of these enzymes are used to assess the functional status of the liver and as biochemical markers of liver injury (Moss and Ralph Handerson, 1999). Hepatotoxic agents cause damage to the liver cell membrane, and these enzymes are leaked into serum and show increased activities (Sturgill and Lambert, 1997). Drug-induced nephrotoxicities are often associated with marked elevations in blood urea and serum creatinine (Ferguson et al., 2008)

One of the plant genera widely used traditionally for the treatment of many diseases is *Phyllanthus* (family Phyllanthaceae), and it is distributed in most tropical and subtropical countries, comprising approximately of 550–750 species throughout the world. *Phyllanthus* spp. are used traditionally for the treatment of viral, bacterial, and parasitic infections and other diseases. Among the *Phyllanthus* spp., the most studied are *P. amarus/P. niruri, P. urinaria,* and *P. emblica* (Calixto et al., 1998; Kumar and Kuttan, 2004; Liu et al., 2001). The following section provides information regarding the toxicity or safety studies carried out on these plant species as well as other *Phyllanthus* spp. Since not many experimental studies directly involved proper toxicity studies, the *Phyllanthus* spp. tested for their therapeutic efficacy and with toxicity parameters are included in the section. As mentioned, liver damage and kidney damage are associated with elevation in their injury markers. In efficacy studies, the animals were challenged with hepatotoxic and nephrotoxic agents and assessed for their markers of injury with the administration of *Phyllanthus* spp. If any amelioration of the elevated toxicity markers occurs, this indicates not only the efficacy of the plant extract but also safety of the extract for use at the dosage tested.

18.2 TOXICITY STUDIES OF *PHYLLANTHUS* SPP.

In this section, the *in vivo* studies involving toxicity of different *Phyllanthus* species are described for *P. amarus*/*P. niruri*, *P. urinaria*, *P. emblica*, and other *Phyllanthus* spp. Liu et al. (2001) published a meta-analysis of the efficacy and safety of *Phyllanthus* for chronic hepatitis B virus (HBV) infection. Among the 22 randomized clinical trials included in the meta-analysis, no serious adverse reactions were reported. Most of the adverse effects were monitored through hematological, biochemical, and urinary examinations or questionnaire report.

18.2.1 *Phyllanthus amarus*/*Phyllanthus niruri*

Among the *Phyllanthus* species, *P. amarus* is widely studied in human as well as animal models. In our study, we found that acute administration of *P. amarus* extract even at a dose of 5 g/kg body weight did not produce any signs of toxicity or mortality (Sirajudeen et al., 2006). Toxicologists agree that any test substance that is not lethal on acute administration at a concentration of 5 g/kg body weight is essentially nontoxic (Organization for Economic Cooperation and Development, 1981; Brock et al., 1995) In the repeated-dose study, no significant difference was observed between the control and *P. amarus* extract administered to (male and female) rats (at doses of 100, 400, and 800 mg/kg body weight for 6 weeks) in the total body weight gain as well as the liver marker enzymes analyzed in serum. The nontoxic nature of *P. amarus* extract administration was also confirmed by histological studies, which showed no observable changes between the control rats and those given *P. amarus* extract.

Lawson-Evi et al. (2008) carried out a toxicological assessment of *Phyllanthus amarus* Schum and Thonn aqueous and hydroalcoholic extracts. Acute and subacute toxicity of the extracts was evaluated in Swiss mice and Wistar rats. At an oral dose of 5 g/kg body weight, the mice did not elicit any signs of toxicity or mortality over a 14-day observational period. In the subacute toxicity study, mice orally administered extracts at doses of 1 and 3 g/kg body weight for 28 days did not show any significant difference in body weight gain when compared with control animals, and the clinical biochemistry parameters AST, ALT, GGT (gamma glutamyl transpeptidase), creatinine, and creatine kinase revealed no toxic effect. No gross abnormalities or histopathological changes were observed in liver, kidney, and pancreas. This study indicated that the extracts of *P. amarus* given orally are considered nontoxic in animals since the median lethal dose (LD_{50}) was greater than 5 g/kg body weight.

Oral administration of an aqueous extract of *Phyllanthus amarus* to rats at a dose of 200 mg/kg body weight every day for 8 weeks did not cause any change in body weight gain when compared with untreated control, and no visible side effects were noticed during the experimental period. Genotoxicity of an aqueous extract of *Phyllanthus amarus* was assessed under *in vivo* conditions and did not result in lymphocyte DNA damage, which revealed the plant extract had a nongenotoxic property (Karuna et al., 2009).

Pretreatment of rats with the aqueous and methanolic extracts (100 mg/kg body weight) of *P. niruri* markedly reduced changes induced by carbon tetrachloride (CCl_4) in the serum AST and ALT markers of liver injury. Administration of these extracts alone did not affect the serum enzymes (Harish and Shivanandappa, 2006).

Administration of a crude extract of *Phyllanthus amarus* at a dose of 500 mg/kg body weight for 45 days exhibited reversible antifertility action with little or no side effects in male mice (Rao et al., 1997).

Urinary mutagenicity produced in rats by benzo[a]pyrene was significantly inhibited by the oral administration of *Phyllanthus* methanolic extract (500 mg/kg body weight) for 12 days prior to a single-dose intraperitoneal injection of benzo[a]pyrene. These results indicated significant antimutagenicity of the *P. amarus* extract under *in vivo* conditions (Raphael et al., 2002).

Although most of the studies carried out indicated the nontoxic nature of the plant extract, the following few studies showed toxic side effects on *P. amarus* administration:

Six chromatographic fractions of *P. amarus* were administered orally to the albino rats at doses of 400, 800, and 1,600 mg/kg body weight for 14 days. They showed a toxic effect in the serum markers of liver and kidney injury (Adedapo et al., 2005a).

Oral administration of an aqueous leaf extract of *P. amarus* to rats at doses of 400, 800, and 1,000 mg/kg for 30 days showed toxic effects on hematological and biochemical markers, as evidenced by histological changes in liver, kidneys, and testes (Adedapo et al., 2005b).

An aqueous extract of *Phyllanthus amarus* was administered orally at doses of 400 and 800 mg/kg body weight daily for 30 days to adult Wistar rats. The histological study results showed the toxic effects of chronic administration of *P. amarus* on the microanatomy of the renal cortical structure, but there was no remarkable difference in observed distortion between the treated low-dose and high-dose groups. It has been speculated that toxic effects could be independent of concentration and hypothesized that the function of the kidney may be adversely affected (Adjene and Nwose, 2010). But, these effects have yet to be confirmed with further studies.

Although the few studies mentioned suggest the nephrotoxic nature of *P. amarus* extract, in another study, single daily oral doses of 100–400 mg/kg of the leaf and seed aqueous extract of *Phyllanthus amarus* were studied for their protective effects in acetaminophen- and gentamicin-induced nephrotoxic Wistar rats for 14 days (Adeneye and Benebo, 2008). In the acetaminophen nephrotoxic rats, the extract significantly attenuated elevations in the serum creatinine and blood urea nitrogen levels in a dose-related fashion, and there was attenuation of acetaminophen-induced tubulonephrosis. A gentamicin-induced acute renal injury model also showed similar effects. Histological findings of the effects of the *Phyllanthus* extract on the renal architecture differed with those of Adedapo et al. (2005a, 2005b). The observed variance could be attributed to differences in the administered extract doses, animal models, and duration of drug exposure. However, several independent animal (Rao and Alice, 2001; Kumar and Kuttan, 2004) and human (Srividya and Periwal, 1995; Moshi et al., 2001) studies have reported and confirmed the high safety profile of the plant extracts.

In an acute toxicity study, acetone extracts of *P. amarus* at doses of 250, 500, 1,000, 2,000, 4,000, and 8,000 mg/kg body weight were administered to Swiss male albino mice. There was no mortality at any of the tested doses by the end of 7 days of observation, indicating the nontoxic nature of the plant extract. *Phyllanthus amarus* may also offer protection against toxic effects of alcohol to the liver (Shokunbi and Odetola, 2008).

Thyagarajan et al. (1988) demonstrated that oral administration of *P. amarus* extract had little or no toxic effect on patients (chronic carriers of HBV).

Sixty-five adult asymptomatic chronic carriers of HBV were enrolled to a randomized controlled efficacy study of *Phyllanthus amarus*. Thirty-four received *Phyllanthus amarus* 600 mg per day for 30 days. Adverse effects were not observed in all patients receiving the plant (Thamlikitkul et al., 1991).

Phyllanthus amarus was administered to 30 asymptomatic carriers of hepatitis B surface antigen (HBsAg) in a dose of 250 to 500 mg three times daily for 4 to 8 weeks. Although none of the 30 subjects cleared HBsAg, *Phyllanthus amarus* was well tolerated, with no clinical side effects or changes in the organ profiles for safety evaluation (Doshi et al., 1994).

18.2.2 PHYLLANTHUS URINARIA

In a double-blind, placebo-controlled study of *Phyllanthus urinaria* in which the patients were randomized into groups and received *P. urinaria* at a dose of 1, 2, and 3 g three times daily for 6 months or placebo, no serious adverse event was reported. Incidence of adverse events in the various dosage groups was similar to that for placebo (Chan et al., 2003).

There was no obvious toxicity detected in the mice group administered *P. urinaria* (500 mg/kg body weight/day in drinking water for 4 weeks), as shown by the comparable mass and histological examination of major organs, including liver, lung, spleen, kidney, and heart, between the *P. urinaria* group and the control (Huang et al., 2006).

A therapeutic dose (200 mg/kg) of *Phyllanthus urinaria* extract administered to mice for 3 days did not show any toxicological phenomenon (Hau et al., 2009).

18.2.3 PHYLLANTHUS EMBLICA (EMBLICA OFFICINALIS)

Pretreatment with *E. officinalis* at doses of 100 and 200 mg/kg body weight for 7 days, prior to CCl_4 intoxication showed a significant reduction in the levels of SGOT (serum glutamate oxaloacetate transaminase) and SGPT (scrum glutamate pyruvate transaminase; Sultana et al., 2005).

A hydroalcoholic (50%) extract of *Emblica officinalis* (fruit) administered (100 mg/kg orally) daily for 5 weeks reduced the severity of hepatic fibrosis induced by CCl_4 and thioacetamide (TAA) and reduced the elevated serum levels of AST, ALT, ALP, and bilirubin (Tasduq et al., 2005b).

Administration of a 50% hydroalcoholic extract of *Emblica officinalis* (fruit) (100 mg/kg orally) daily for 30 days reversed the liver toxicity in rats induced by antitubercular drugs, as indicated by the serum markers of liver injury and by reversal of the normal lobular pattern with no evidence of necrotic areas (Tasduq et al., 2005a).

Oral administration of *P. emblica* to rats at a dose of 75 mg/kg for 7 days (posttreatment after daily oral administration of ethanol at 4 g/kg for 7 days) lowered the ethanol-induced levels of AST and ALT. Histopathological studies confirmed the beneficial roles of *P. emblica* against ethanol-induced liver injury in rats (Pramyothin et al., 2006).

Oral administration of ethyl acetate extract of *Emblica officinalis* at a dose 10 mg/kg body weight daily for 100 days reduced the elevated levels of serum creatinine and urea nitrogen (which are indicators of renal dysfunction in clinical practice) in the aged rats compared with aging control rats. This suggests that *Emblica officinalis* extract may ameliorate renal dysfunction in the aging process (Yokozawa et al., 2007).

The protective role of the fruits of *Emblica officinalis* aqueous extract was studied by Sharma et al. (2009) in adult Swiss albino mice against arsenic-induced hepatopathy. In this study, *Emblica* fruit extract (500 mg/kg body weight orally) was administered 10 days before $NaAsO_2$ (4 mg/kg body weight) and continued up to 30 days after arsenic treatment. Combined treatment of *Emblica* and arsenic decreased the serum transaminases, improved ALP activity, and reduced karyolysis, karyorrhexis, necrosis, and cytoplasmic vacuolization induced by $NaAsO_2$ intoxication.

The protective action of crude extracts of *Phyllanthus emblica* fruits against lead- and aluminum-induced sister chromatid exchanges (SCEs) was studied in bone marrow cells of Swiss albino mice (Dhir et al., 1993). Oral administration of PFE (phyllanthus emblica fruit extract) for 7 consecutive days at a dose of 685 mg/kg body weight prior to exposure of mice to the metals by intraperitoneal injections reduced the frequencies of SCEs induced by both metals. No significant changes in SCE frequencies or in proliferation rate index (PRI) values were observed in mice fed PFE without any metal treatment when compared to the negative control might indicate the nongenotoxic nature of the extracts.

18.2.4 OTHER PHYLLANTHUS SPECIES

18.2.4.1 *Phyllanthus reticulatus*

Acute oral toxicity studies of petroleum ether and ethanolic extracts of leaves of the *P. reticulatus* administered to rats indicated safety up to a dose of 5,000 mg/kg body weight, which indicates the nontoxic nature of the plant extracts (Kumar et al., 2008). In addition, oral administration of ethanolic extracts of aerial parts of *P. reticulatus* (200 mg/kg body weight daily) for 15 days to rats (intoxicated with CCl_4, a well-known liver toxicant) decreased the serum AST and ALT and restored serum bilirubin, which was elevated due to CCl_4 toxicity, to the normal range. Histological studies indicated the reversal or prevention of microscopic changes that occurred due to CCl_4 intoxication (Das et al., 2008).

18.2.4.2 Phyllanthus maderaspatensis

Administration of acetaminophen, a hepatotoxicant, elevates the activities of serum marker enzymes for liver damage (AST, ALT, and ALP) in rats. Administration of *P. maderaspatensis* extracts (water, alcohol, or *n*-hexane) at a dose of 200 mg/kg remarkably prevented acetaminophen-induced elevation of serum AST, ALT, and ALP (Asha et al., 2004). In addition, administration of hexane extract of *P. maderaspatensis* (200 mg/kg body weight) also prevented the elevation of serum liver marker enzymes and histological changes induced in rats treated with CCl_4 and thioacetamide (Asha et al., 2007).

18.2.4.3 Phyllanthus rheedii/Phyllanthus kozhikodianus

The ethanolic extract of *Phyllanthus rheedii* was analyzed for its preventive effect in liver damage induced by D-galactosamine (D-GalN) in rats at a dose of 200 mg/kg body weight; it showed a positive effect in liver marker enzymes and prevented the toxin-induced changes in the liver (Suresh and Asha, 2008). In the toxicity study, there was no mortality observed, and animals did not show any significant changes in skin, fur, eyes, body temperature, body weight, internal organs, and behavior. The serum parameters of the toxicity analysis showed that there were no significant changes between control and extract-treated groups. This suggests that the extract is without any significant toxicity at a dose of 2,000 mg/kg and that the lethal dose of the extract is very high (i.e., more than 2 g/kg) (Suresh and Asha, 2008).

18.2.4.4 Phyllanthus tenellus

Acute toxicity of fresh *P. tenellus* extract at doses of 200–450 mg/kg administered intraperitonealy to mice did not show any apparent effect on the animal behavior during the 4 days of experimental observation. At doses of 500 mg/kg and above, mortality was recorded and LD_{50} obtained was 877mg/kg (Ignacio et al., 2001).

18.2.4.5 Phyllanthus fraternus

Acute toxicity properties of the crude (water) plant extract of *Phyllanthus fraternus* in mice with intraperitoneal treatment at a doses of 500–1,500 mg/kg showed that all animals survived beyond 24 h only up to a dose of 600 mg/kg and an LD_{50} calculated to be approximately 700 mg/kg (Matur et al., 2009).

18.3 CONCLUSION

According to the available literature, *Phyllanthus spp.* are considered safe at the dosages tested for the extracts under study, although a few studies indicated some toxicities, but these are yet to be confirmed. So, when the *Phyllanthus* extract is administered in humans higher than the dose and duration of the already studied or recommended clinical doses, it should be carefully monitored for the toxicity markers. Care also should be taken when administering to pregnant or lactating woman or patients with severe organ diseases because safety has not been established for that purpose.

REFERENCES

Adedapo, A.A., Abatan, M.O., Idowu, S.O., and Olorunshogo, O.O. 2005a. Toxic effects of chromatographic fractions of *Phyllanthus amarus* on the serum biochemistry of rats. *Phytother. Res.* 19: 812–815.

Adedapo, A.A., Adegbayibi, A.Y., and Emikpe, B.O. 2005b. Clinicopathological changes associated with the aqueous extract of the leaves of *Phyllanthus amarus* in rats. *Phytother. Res.* 19: 971–976.

Adeneye, A.A., and Benebo, A.S. 2008. Protective effect of the aqueous leaf and seed extract of *Phyllanthus amarus* on gentamicin and acetaminophen-induced nephrotoxic rats. *J. Ethnopharmacol.* 118: 318–323.

Adjene, J.O., and Nwose, E.U. 2010. Histological effects of chronic administration of *Phyllanthus amarus* on the kidney of adult Wistar rat. *North Am. J. Med. Sci.* 2: 193–195.

Arnold, D.L. 1990. Oral ingestion studies. In *Handbook of in vivo toxicity testing*, ed. D.L. Arnold, H.C. Grice, and D.R. Krewski, 167–188. San Diego, CA: Academic Press.

Asha, V.V., Akhila, S., Wills, P.J., and Subramoniam, A. 2004. Further studies on the antihepatotoxic activity of *Phyllanthus maderaspatensis* Linn. *J. Ethnopharmacol.* 92: 67–70.

Asha, V.V., Sheeba, M.S., Suresh, V., and Wills, P.J. 2007. Hepatoprotection of *Phyllanthus maderaspatensis* against experimentally induced liver injury in rats. *Fitoterapia* 78: 134–141.

Brock, W.D., Trochimowicz, H.J., Millischer, R.J.H., et al. 1995. Acute and subchronic toxicity of 1,1-dichloro-1-fluroethane (HCFC-141 b). *Food Chem. Toxicol.* 33: 483–490.

Brown, R.G. 1992. Toxicity of Chinese herbal remedies. *Lancet* 340: 673.

Calixto, J. B. 2000. Efficacy, safety, quality control, marketing and regulatory guidelines for herbal medicines (phytotherapeutic agents). *Braz. J. Med. Biol. Res.* 33: 179–189.

Calixto, J. B., Santos, A.R.S., Cechinel-Filho, V., and Yunes, R. A. 1998. A review of the plants of the genus *Phyllanthus*: their chemistry, pharmacology and therapeutic potential. *Med. Res. Rev.* 18: 225–258.

Chan, H.L., Sung, J.J., Fong, W.F., et al. 2003. Double-blinded placebo-controlled study of *Phyllanthus urinaria* for the treatment of chronic hepatitis B. *Aliment Pharmacol. Ther.* 18: 339–345.

Das, B.K., Bepary, S., Datta, B.K., et al. 2008. Hepatoprotective activity of *Phyllanthus reticulatus*. *Pak. J. Pharm. Sci.* 21: 333–337.

De Smet, P.A.G.M. 2004. Health risks of herbal remedies: an update. *Clin. Pharmacol. Ther.* 76: 1–17.

Dhir, H., Roy, A.K., and Sharma, A. 1993. Relative efficiency of *Phyllanthus emblica* fruit extract and ascorbic acid in modifying lead and aluminium-induced sister-chromatid exchanges in mouse bone marrow. *Environ. Mol. Mutagen.* 21: 229–236.

Doshi, J.C., Vaidya, A.B., Antarkar, D.S., et al. 1994. A two-stage clinical trial of *Phyllanthus amarus* in hepatitis B carriers: failure to eradicate the surface antigen. *Indian J. Gastroenterol.* 13: 7–8.

Drew, A.K., and Myers, S.P. 1997. Safety issues in herbal medicine: implications for the health professions. *Med. J. Australia* 166: 538–541.

Ferguson, M.A., Vaidya, V.S., and Bonventre, J.V. 2008. Biomarkers of nephrotoxic acute kidney injury. *Toxicology* 20: 182–193.

Harish, R., and Shivanandappa, T. 2006. Antioxidant activity and hepatoprotective potential of *Phyllanthus niruri*. *Food Chem.* 95: 180–185.

Hau, D.K., Gambari, R., Wong, R.S., et al. 2009. *Phyllanthus urinaria* extract attenuates acetaminophen induced hepatotoxicity: involvement of cytochrome P450 CYP2E1. *Phytomedicine* 16: 751–760.

Huang, S.T, Yang, R.C., Lee, P.N., et al. 2006. Anti-tumor and anti-angiogenic effects of *Phyllanthus urinaria* in mice bearing Lewis lung carcinoma. *Int. Immunopharmacol.* 6: 870–879.

Ikegami, F., Fujii, Y., and Satoh, T. 2004. Toxicological considerations of Kampo medicines in clinical use. *Toxicology* 198: 221–228.

Karuna, R., Reddy, S.S., Baskar, R., and Saralakumari, D. 2009. Antioxidant potential of aqueous extract of *Phyllanthus amarus* in rats. *Indian J. Pharmacol.* 41: 64–67.

Kumar, K.B., and Kuttan, R. 2004. Protective effect of an extract of *Phyllanthus amarus* against radiation-induced damage in mice. *J. Radiation Res.* 45: 133–139.

Kumar, S., Kumar, D., Deshmukh R.R., et al. 2008. Antidiabetic potential of *Phyllanthus reticulatus* in alloxan-induced diabetic mice *Fitoterapia* 79: 21–23

Lawson-Evi, P., Eklu-Gadegbeku, K., Agbonon, A., et al. 2008 Toxicological assessment on extracts of *Phyllanthus amarus* Schum and Thonn. *Sci. Res. Essays* 3: 410–415.

Liu, J., Lin, H., and McIntosh, H. 2001. Genus *Phyllanthus* for chronic hepatitis B virus infection: a systematic review. *J. Viral Hepatol.* 8: 358–366

Matur, B.M., Matthew, T., and Ifeanyi, C.I.C. 2009. Analysis of the phytochemical and *in vivo* antimalaria properties of *Phyllanthus fraternus* Webster extract. *New York Sci. J.* 2: 12–19.

Moshi, M.J., Lutale, J.J., Rimoy, G.H., et al. 2001. The effect of *Phyllanthus amarus* aqueous extract on blood glucose in non-insulin dependent diabetic patients. *Phytother. Res.* 15: 577–580.

Moss, D.W., and Ralph Handerson, A. 1999. Clinical enzymology. In *Tietz text book of clinical chemistry*, ed. C.A. Burtis and E.R. Ashword, 3rd ed., 651–683. Philadelphia: Saunders.

Organization for Economic Cooperation and Development. 1981. *Guidelines for the testing of chemicals*. Paris: Organization for Economic Cooperation and Development.

Pillans, P.I. 1995. Toxicity of herbal products. *NZ Med. J.* 108: 469–471.

Pramyothin, P., Samosorn, P., Poungshompoo, S., and Chaichantipyuth, C. 2006. The protective effects of *Phyllanthus emblica* Linn. extract on ethanol induced rat hepatic injury. *J. Ethnopharmacol.* 107: 361–364.

Rao, M.V., and Alice, K.M. 2001. Contraceptive effects of *Phyllanthus amarus* in female mice. *Phytother. Res.* 15: 265–267.

Rao, M.V., Shah, K.D., and Rajani, M. 1997. Contraceptive effects of *Phyllanthus amarus* extract in the male mouse (*Mus musculus*). *Phytother. Res.* 11: 594–596.

Raphael, K. R., Ajith, T.A., Joseph S., and Kuttan, R. 2002. Anti-Mutagenic activity of *Phyllanthus amarus* Schum & Thonn *in vitro* as well as *in vivo*. *Teratogen. Carcinogen. Mutagen.* 22: 285–291.

Rhiouani, H., El-Hilaly, J., Israili, Z.H., and Lyoussi B. 2008. Acute and sub-chronic toxicity of an aqueous extract of the leaves of *Herniaria glabra* in rodents. *J. Ethnopharmacol.* 118: 378–386.

Saad, B., Azaizeh, H., Abu-Hijleh, G., and Said, O. 2006. Safety of traditional Arab herbal medicine. *Evid. Based Complement. Alternat. Med.* 3: 433–439.

Schlam, O. W., Iain N, C., and Caroll, E. J. 1975. *Veterinary Hematology*. Philadelphia: Lea and Febiger.

Sharma, A., Sharma, M.K., and Kumar, M. 2009. Modulatory role of *Emblica officinalis* fruit extract against arsenic induced oxidative stress in Swiss albino mice. *Chem. Biol. Interact.* 180: 20–30.

Shokunbi, O.S., and Odetola, A.A. 2008. Gastroprotective and antioxidant activities of *Phyllanthus amarus* extracts on absolute ethanol induced ulcer in albino rats. *J. Med. Plants Res.* 2: 261–267.

Sirajudeen, K.N.S., Suliaman, S.A., Madhava, M., et al. 2006. Safety evaluation of aqueous extract of leaves of a plant *Phyllanthus amarus* in rat liver. *Afr. J. Trad. CAM.* 3: 78–93.

Srividya, N., and Periwal, S. 1995. Diuretic, hypotensive and hypoglycemic effect of *Phyllanthus amarus*. *Indian J. Exp. Biol.* 33: 861–864.

Sturgill, M.G., and Lambert, G.H. 1997. Xenobiotics-induced hepatotoxicity; mechanism of liver injury and method of monitoring hepatic function. *Clin. Chem.* 43: 1512–1526.

Sultana, S., Ahmad, S., Khan, N., and Jahangir, T. 2005. Effect of *Emblica officinalis* (Gaertn) on CCl_4 induced hepatic toxicity and DNA synthesis in Wistar rats. *Indian J. Exp. Biol.* 43: 430–436.

Suresh, V., and Asha, V.V. 2008. Preventive effect of ethanol extract of *Phyllanthus rheedii* Wight. on D-galactosamine induced hepatic damage in Wistar rats. *J. Ethnopharmacol.* 116: 447–453.

Tasduq, S.A., Kaisar, P., Gupta, D.K., et al. 2005a. Protective effect of a 50% hydroalcoholic fruit extract of *Emblica officinalis* against anti-tuberculosis drugs induced liver toxicity. *Phytother. Res.* 19: 193–197.

Tasduq, S.A., Mondhe, D.M., Gupta, D.K., et al. 2005b. Reversal of fibrogenic events in liver by *Emblica officinalis* (fruit), an Indian natural drug. *Biol. Pharm. Bull.* 28: 1304–1306.

Thamlikitkul, V., Wasuwat, S., and Kanchanapee, P. 1991. Efficacy of *Phyllanthus amarus* for eradication of hepatitis B virus in chronic carriers. *J. Med. Assoc. Thai.* 74: 381–385.

Thyagarajan, S.P., Subramanian, S., Thirunalasundari, T., et al. 1988. Effect of *Phyllanthus amarus* in chronic carriers of hepatitis B virus. *Lancet* 2: 764–766.

World Health Organization. 1993. *Research guidelines for evaluating the safety and efficacy of herbal medicine.* Manila: Regional Office for the Western Pacific.

Yokozawa, T., Kim, H.Y., Kim, H.J., et al. 2007. Amla (*Emblica officinalis* Gaertn.) attenuates age-related renal dysfunction by oxidative stress. *J. Agric. Food Chem.* 55: 7744–7752.

19 Clinical Trials Involving *Phyllanthus* Species

Mulyarjo Dirjomuljono and Raymond R. Tjandrawinata

CONTENTS

19.1 Introduction 289
19.2 Immunomodulatory Effects...290
    19.2.1 Pulmonary Tuberculosis 290
    19.2.2 Vaginal Candidiasis 295
    19.2.3 Varicella Zoster Infection 296
    19.2.4 Urolithiasis..297
    19.2.5 Antidiabetes..299
19.3 *Phyllanthus* in Combination with Other Herbs or Medication....................303
    19.3.1 Tonsillopharyngitis 303
    19.3.2 Bronchial Asthma 304
    19.3.3 Others..305
19.4 Safety and Tolerability...306
19.5 Conclusion 308
Acknowledgment 308
References..308

19.1 INTRODUCTION

Herbal medicines have been used for a long time for the treatment of many different diseases. To date, there is a growing interest in research on plants traditionally used as folk medicines; the research is aimed to scientifically evaluate the potential use of a given plant or compound for treating a particular illness as well as to examine its potential benefits in clinical practice. Most plants or compounds need further and larger carefully conducted studies to confirm the preliminarily indicated pharmacological activities of the herbals. The plants of *Phyllanthus*, one of the largest genera in the family Phyllanthaceae, are widely distributed in most tropical and subtropical countries and have long been used in folk remedies to treat kidney and urinary bladder disturbances, intestinal infections, diabetes, and hepatitis B. Supporting such a traditional use of *Phyllanthus,* substantial progress on their chemistry and pharmacological properties has been made (Calixto et al., 1998). Through numerous preclinical studies and a few clinical studies, *Phyllanthus* species have been reported to exert biological activities against hepatitis B virus (HBV) (Thyagarajan et al., 1988, 1990; Venkateswaran et al., 1987; Blumberg et al., 1989) and other viruses, such as human immunodeficiency virus (HIV) (Notka et al., 2003, 2004; Naik and Juvekar, 2003) as well as herpes simplex virus (HSV) (Yang et al., 2005). The genus has also been studied for its anticarcinogenic (Pettit et al., 1983, 1990; Rajeshkumar and Kuttan, 2000; Rajeshkumar et al., 2002); antinociceptive (Santos et al., 1995, 2000); lipid-lowering (Khanna et al., 2002); antilithiasis (Nishiura et al., 2004; Micali et al., 2006); as well as antidiabetic effects (Ramakrishnan et al., 1982; Srividya and Periwal, 1995; Moshi et al., 2001; Raphael et al., 2002; Adeneye et al., 2006; Kumar et al., 2008). Several preclinical and clinical studies have also been done with respect to the immunomodulatory effect of *Phyllanthus* sp. (Ma'at, 1996; Raveinal, 2003; Amin, 2005; Halim and Saleh, 2005; Radityawan, 2005; Pramayanti et al., 2005; Sarisetyaningtyas et al., 2006).

Indeed, of hundreds of *Phyllanthus* species known, only a few certain species have widely and soundly been studied and either confirmed or just indicated to possess any or some of those activities mentioned. More randomized, controlled, and adequately powered clinical studies evaluating *Phyllanthus* in various diseases are yet necessary to establish the potential benefits of the genus in clinical practice. In this chapter, we discuss clinical trials conducted involving *Phyllanthus* species, including a phase 1 study in healthy volunteers. However, to avoid reiterating a similar subject, the effects and potential as well as clinical benefits of *Phyllanthus* against hepadnaviruses, including clinical trials on the herb in hepatitis, are not discussed in this chapter as they are presented in detail elsewhere in this book.

19.2 IMMUNOMODULATORY EFFECTS

A preclinical study of *Phyllanthus niruri* by Ma'at (1996) demonstrated the effect of *Phyllanthus niruri* on enhancing activities and function of immune system components, both humoral and cellular immunities. Such immunomodulatory effect of the herb was further evaluated in several clinical trials involving patients with pulmonary tuberculosis (TB) (Raveinal, 2003; Amin, 2005; Halim and Saleh, 2005; Radityawan, 2005); vaginal candidiasis (Pramayanti et al., 2005); as well as chicken pox (Sarisetyaningtyas et al., 2006). In such diseases, the aid of an effective immune system is critical in determining patients' response to treatment as well as the eradication of the pathogens.

19.2.1 PULMONARY TUBERCULOSIS

The causal pathogen of pulmonary TB in humans, *Mycobacterium tuberculosis*, is a bacillus that lives intracellularly (Barnes and Modlin, 1994; Chan and Kauffman, 1994). To eradicate such intracellular pathogens, the cellular immune response plays an important role; thus, T lymphocytes together with their secreted cytokines, which may activate and promote the phagocytic function of macrophages, become critical determinant factors. Among the cytokines, interferon γ (IFN-γ) appears to be the key effector cytokine in the control of mycobacterial infection (Smith et al., 1997;

Dannenberg, 1993). Depressed type 1 T-helper (Th-1) responses, which also appear in defects in IFN-γ production, are predominant in patients with TB (Lin et al., 1996). An elevated level of IFN-γ, which is required to defend and fight against *M. tuberculosis*, is of critical importance for a gradual improvement in the immunological status of patients with TB. Therefore, the success of TB therapy is largely determined by the capability of such treatment to increase the secretion of IFN-γ. IFN-γ is a cytokine that has antiviral and antiproliferative properties, biologically affects macrophage activation and natural immunity, as well as promotes the antigen-presenting process, phenotyping development of T-helper cells, and humoral immunity (Kauffman, 1993).

Two clinical studies of *Phyllanthus niruri* Linn on pulmonary TB demonstrated the immunomodulatory effect of the herb through elevation of IFN-γ levels during 2 to 6 months of TB therapy (Table 19.1) (Amin, 2005; Radityawan, 2005). Those studies were both designed as randomized, parallel, double-blind, placebo-controlled clinical studies, which used either *Phyllanthus niruri* extract at a dose of 50 mg three times daily or the placebo in addition to World Health Organization (WHO) standardized regimens of TB therapy (rifampisin, isoniazid, pyrazinamide, ethambutol) (WHO, 2003).

Radityawan (2005), in his study involving 40 patients with TB, reported that after 2 months of therapy a significant elevation of plasma IFN-γ level was observed in patients receiving additional treatment with *Phyllanthus*. A trend toward a similar result was reported separately by Amin (2005) in a study involving 67 patients with pulmonary TB. Such elevation was still demonstrated after up to 6 months of therapy, even though it was not statistically significant. Subjects with standard TB regimens alone (or those in the placebo group) also showed an elevated level of IFN-γ after 2 months of treatment. However, in those subjects, the IFN-γ could not be maintained at that level. In fact, it then decreased along the therapeutic course of 2 to 6 months (Amin, 2005).

Amin (2005) also evaluated the effect of *Phyllanthus niruri* on tumor necrosis factor α (TNF-α) (Table 19.1), another cytokine that also plays a critical role in halting the disease progression of TB. In the *Phyllanthus niruri* group, the insignificant decrease of TNF-α level after 2 months of therapy, which was then followed by its elevation at the sixth month, indicated the presence of a favorable immunological response toward the suppression of mycobacterial growth. Synergistic action between TNF-α and IFN-γ in inducing nitric oxide (NO) production is highly important in granuloma formation, which limits the spread of bacterial infection throughout the body, and thus mycobacteria eradication occurs (Fenton and Vermeulen, 1996; Smith et al., 1997). Local production of IFN-γ and TNF-α by leukocytes is also critical for the differentiation and activation of recruited peripheral monocytes involved in killing the mycobacteria (Fenton and Vermeulen, 1996).

Other than secretion of IFN-γ and TNF-α, *Phyllanthus niruri* may also affect the secretion of other cytokines by type 2 T-helper (Th-2) lymphocytes. In a double-blind, randomized, placebo-controlled study involving 39 patients newly diagnosed with TB with moderate-to-severe lesions on radiology and positive sputum AFB (acid-fast bacillus) test, Halim and Saleh (2005) reported a moderate suppression of human interleukin (IL) 10 secretion by Th-2 due to *Phyllanthus niruri* extract

TABLE 19.1
Immunomodulatory Effect of *Phyllanthus* sp. Observed in Clinical Studies

Studies	Treatment	Evaluated Immunological Parameters	Time of Evaluation	Efficacy		
				Standard Regimens + *Phyllanthus niruri*	Standard Regimens Alone	*p* between Groups
Pulmonary Tuberculosis						
Amin (2005): Randomized controlled trial	WHO standardized TB regimens *plus* either • *Phyllanthus niruri* extract: 50 mg thrice daily ($n = 34$), or • Placebo ($n = 33$) Treatment duration: 6 months	IFN-γ (pg/ml): Mean (SD)	Baseline 2 months of treatment 6 months of treatment	2.73 (3.92) 4.32 (8.53) 5.24 (4.73)	2.29 (3.34) 6.63 (13.62) 5.35 (3.79)	NS NS NS
		TNF-α (pg/ml): Mean (SD)	Baseline 2 months of treatment 6 months of treatment	11.39 (19.95) 1.54 (3.03) 27.72 (74.16)	39.82 (119.87) 8.64 (26.14) 6.07 (2.96)	NS NS NS
Radityawan (2005): Randomized controlled trial	WHO standardized TB regimens *plus* either • *Phyllanthus niruri* extract: 50 mg thrice daily ($n = 20$), or • Placebo ($n = 20$) Treatment duration: 2 months	IFN-γ (pg/m;): Mean (SD)	Baseline 2 months of treatment	5.24 (4.30) Δ + 7.65 vs. baseline	7.73 (7.78) Δ + 0.41 vs. baseline	.010
Halim and Saleh (2005): Randomized controlled trial	WHO standardized TB regimens *plus* either • *Phyllanthus niruri* extract: 50 mg thrice daily ($n = 20$), or • Placebo ($n = 19$) Treatment duration: 2 months	IL-10: Mean (SD)	2 months of treatment	↓1.5 × baseline	↑1.25× baseline	NS

Study	Treatment	Parameter	Timepoint	Mean (SD)	p
Raveinal (2003): Randomized controlled trial	WHO standardized TB regimens plus either: • *Phyllanthus niruri* extract: 50 mg thrice daily ($n = 20$), or • Placebo ($n = 20$) Treatment duration: 4 weeks	CD4 (/mm³): Mean (SD)	Baseline	45.55 (6.07)	
			4 weeks of treatment	56.25 (5.95)[a]	
			Baseline	42.70 (5.97)	NS
			4 weeks of treatment	47.15 (5.69)[b]	.001
		CD8 (/mm³): Mean (SD)	Baseline	33.30 (4.47)	
			4 weeks of treatment	33.00 (2.41)	
			Baseline	31.95 (4.11)	NS
			4 weeks of treatment	33.70 (3.70)	.730
		CD4/CD8: Mean (SD)	Baseline	1.39 (0.22)	
			4 weeks of treatment	1.71 (0.21)[a]	
			Baseline	1.36 (0.25)	NS
			4 weeks of treatment	1.41 (0.2)	.010
Vaginal Candidiasis					
Pramayanti et al. (2005): Randomized controlled trial	Oral ketokonazole 200 mg twice daily for 5 days plus either: • *Phyllanthus niruri* extract: 100 mg thrice daily for 7 days ($n = 15$), or • Placebo ($n = 15$)	IFN-γ (pg/ml): Mean (SD)	Baseline	120.14 (44.51)	
			7 days of treatment	138 (34.67)	
			1 month posttreatment	159.10 (58.76)	
			3 months posttreatment	128.48 (24.92)	
			Baseline	105.11 (28.67)	NS
			7 days of treatment	100.76 (28.54)	.004
			1 month posttreatment	96.26 (28.39)	<.001
			3 months posttreatment	91.35 (30.37)	<.001
		IL-12 (pg/m;): Mean (SD)	Baseline	71.68 (68.71)	
			7 days of treatment	118.23 (109.15)[b]	
			1 month posttreatment	128.31 (112.76)[b]	
			3 months posttreatment	97.80 (81.60)[a]	
			Baseline	60.10 (25.20)	NS
			7 days of treatment	60.13 (28.04)	.129
			1 month posttreatment	67.88 (23.95)	.194
			3 months posttreatment	55.47 (20.44)	.061

Note: Evaluated parameters listed in the table are only those related to the immunological system, through which *Phllyanthus* may exert its action. Those studies also evaluated relevant clinical outcomes.

NS, not significant ($p > .05$); Δ, change from baseline.

[a] $p < .01$ versus baseline.
[b] $p < .05$ versus baseline.

treatment (Table 19.1). In the study, *Phyllanthus niruri* extract was given at a dose of 50 mg three times daily in addition to standard TB regimens. After 2 months of therapy, a considerable increase of IL-10 level from baseline was observed in the group receiving the standard TB regimen only, while in the *Phyllanthus niruri* group the level was moderately suppressed. IL-10, also known as human cytokine synthesis-inhibitory factor (CSIF), is an anti-inflammatory cytokine produced primarily by monocytes and to a lesser extent by lymphocytes. It is capable of inhibiting synthesis of proinflammatory cytokines like IFN-γ, IL-2, IL-3, TNF-α, and granulocyte-macrophagee colony-stimulating factor (GM-CSF) produced by macrophages and Th-1 lymphocytes. IL-10 is released by cytotoxic T cells ($CD8^+$) to inhibit the actions of natural killer (NK) cells during the immune response to infection (Moore et al., 2001).

Suppression of IL-10 secretion by *Phyllanthus niruri* facilitates the inflammatory reaction necessary for microbial eradication. In brief, suppression of IL-10 secretion indicated an improvement in immunological status of patients with TB who received *Phyllanthus niruri* in adjunct to the standard TB regimens. Such suppression might also mean optimizing the action of T-helper subsets ($CD4^+$) as demonstrated in another study conducted by Raveinal (2003).

In line with favorable effects of *Phyllanthus niruri* on IFN-γ, TNF-α, and IL-10, Raveinal (2003), in a double-blind, randomized, placebo-controlled clinical study involving 40 patients with TB, reported that addition of *Phyllanthus niruri* at a dose of 50 mg extract three daily in addition to the standard TB regimens, significantly increased peripheral $CD4^+$ percentage as well as the ratio $CD4^+/CD8^+$ compared to their own baseline and with significantly greater values compared to those of the group with standard TB regimens alone (Table 19.1). The increase of $CD4^+$ percentage seen in the group treated with standard TB regimens alone was also significant as compared to baseline, but it was not accompanied by a significant elevation in $CD4^+/CD8^+$ ratio. No significant changes from baseline were observed in $CD8^+$ levels in both groups. Patients in both groups were at a comparable immunological status at baseline.

Patients with advanced TB had a higher percentage of $CD8^+$ lymphocytes and lower $CD4^+/CD8^+$ ratio in peripheral blood than patients with less severe disease (Raveinal, 2003). Previous studies reported a decreased $CD4^+$ percentage and increased $CD8^+$ percentage, causing a decreased $CD4^+/CD8^+$ ratio in the peripheral blood of patients with advanced TB (Jones et al., 1997; Tsao et al., 2002). The increase of $CD4^+$ percentage and thus the ratio of $CD4^+/CD8^+$, which were already observed by Raveinal (2003) after 4 weeks of treatment initiation with the addition of *Phyllanthus niruri*, suggested a positive response to treatment through the recovery of immunity found in TB patients.

We only focus our discussion on the immunological aspects of *Phyllanthus* treatment to correlate them with the positive findings observed in preclinical studies. However, it was reported also in those studies that the improvement in immunological parameters seen with *Phyllanthus* was positively translated into clinical and radiological improvement (Raveinal, 2003; Amin, 2005; Halim and Saleh, 2005; Radityawan, 2005). Addition of *Phyllanthus niruri* extract to the TB standard regimens also accelerated the conversion of sputum AFB, which was already observed within only 1 week after treatment initiation. More subjects (52.9%) in the *Phyllanthus*-treated group than in the placebo group (39.4%) experienced such sputum AFB conversion after 1 week of treatment (Amin, 2005). Even though such a percentage was not statistically significant, it was of considerable clinical importance, particularly at a community level. A faster sputum conversion will benefit the community as it will reduce the risk of TB transmission from patients with TB with positive sputum. This study result suggests that the use of *Phyllanthus* supplementation may potentially lead to reducing the risk of TB transmission.

Despite the small sample size in each pulmonary TB study, in considering all the immunological improvement demonstrated in the studies, *Phyllanthus niruri* with its immunomodulatory property has a promising potency for inclusion in the management of TB therapy, in which it may synergistically act with the standard TB regimens to optimize the therapy.

19.2.2 Vaginal Candidiasis

Vaginitis is the most-often given gynecological diagnosis worldwide, and up to 25% of the cases are candidal (*Candida albicans*) in origin, also called vaginal candidiasis (Sobel, 1997; Egan and Lipsky, 2000). Lack of IFN-γ production due to impaired function of T lymphocytes was a predisposition to recurrent vaginal candidiasis (Corrigan et al., 1998). Promoting the immune response of the host, particularly the specific response of the Th-1 subset to candida, plays an important role in the healing process of such infections. Such approach is also known as Th-1-associated anticandidal protection (Ma'at, 2001). Since the effectiveness and safety of current antifungals are limited, a combination of chemotherapy (using antifungals) and immunotherapy is applied in approaching candidiasis, particularly for those with recurrent cases (Magliani et al., 2002a, 2002b).

Inspired by the immunomodulatory property of *Phyllanthus niruri*, Pramayanti et al. (2006) conducted a randomized, double-blind, controlled study to evaluate its effectiveness in candidiasis. Thirty married female patients who had vaginal candidiasis received oral ketoconazole 200 mg twice daily for 5 days and either *Phyllanthus niruri* or its matching placebo at a dose of 100 mg extract three times daily for 7 days. They were followed up until 3 months after the treatment ceased. There was no significant change in IFN-γ level of vaginal secretions in patients without *Phyllanthus niruri* treatment after 7 days of treatment as compared to baseline. However, at 1 and 3 months after treatment the level dropped even lower than the baseline. In those receiving *Phyllanthus niruri*, the IFN-γ levels at day 7, month 1, and month 3 were all significantly higher than at baseline. The levels in the *Phyllanthus niruri* group at those evaluation points were also significantly higher than those of the placebo group (Table 19.1). Such elevated IFN-γ level indicates increased activity of the cellular immune response (Th-1 subset), which concurrently suppresses the secretion of IL-4 and IL-10 (by Th-2 subset) and activates the macrophages, eradicating the candida from the vaginal area (Witkin et al., 2000).

In the study, the elevation of IFN-γ in the *Phyllanthus niruri* group was also accompanied by a significantly higher level of IL-12 in the vaginal secretions at day 7, month 1, and month 3 than at baseline. There were no significant changes of

IL-12 observed in the placebo group (Pramayanti et al., 2005). In specific cellular immunity, IL-12 plays a role in stimulating IFN-γ production by NK and T cells and in promoting the NK cell cytotoxicity (Watford et al., 2003).

The immunological improvement in the *Phyllanthus niruri* group was consistently followed by a remarkably higher recovery rate (73.33%) after 7 days of treatment but lower recurrence rate in the first and third month (18.2% and 45.5%, respectively) after stopping treatment than those observed in the placebo group (26.67%, 50.0%, and 100%, respectively for those values) (Pramayanti et al., 2005).

Even though further larger clinical studies are needed, the result of the preliminary study with *Phyllanthus niruri* was in line with the approach of chemoimmunotherapy in speeding up the recovery as well as restraining the recurrence of vaginal candidiasis (Magliani et al., 2002b).

19.2.3 Varicella Zoster Infection

Varicella, known as chicken pox, is highly contagious. In tropical countries, it is reported as a benign disease in childhood, yet not in adults. The causal pathogen is varicella-zoster virus, which is classified as a herpes virus and is also the cause of herpes zoster infection (shingles) (Oxman and Alani, 1993). The use of a substance that may exert immunostimulating activity rather than an antiviral is advisable for immunocompetent patients with varicella infection. This is particularly because in such patients the recovery from varicella infection is mostly dependent on the effectiveness of the immune system of the body.

A double-blind, randomized, controlled study was conducted to investigate the clinical benefits of *Phyllanthus* sp. in pediatric subjects with varicella-zoster infection (Sarisetyaningtyas et al., 2006). In the study, children 2–14 years old who experienced varicella without complications were randomly assigned to receive either *Phyllanthus niruri* Linn extract at a dose of 25 mg/5 ml syrup three times daily (treated group) or placebo (control group). The efficacy was measured by calculating the number of papules and crusts after treatment with the extract for 4 days. Evaluation of the efficacy and safety was performed daily by providing each subject's parents daily monitoring sheets.

The healing process of varicella infection is clinically marked with the absence of fever, new papules, and followed by the crust formation in most parts of the body. Evaluation of 101 enrolled study subjects did not show any statistically significant difference between treated and control groups in terms of clinical signs and symptoms. In this study, fever in both treated and control groups was no longer detected after 2 days of treatment. *Phyllanthus niruri* did not seem to have any effects on fever. After 4 days of treatment, more subjects in the treated group (43.1%) than in the control group (30.0%) had more than 50% of their crustic lesions aborted ($p = .053$, with a statistical number needed to treat [NNT] of 7.6). This considerably low NNT value means that for one patient to completely recover, we just need to treat eight varicella-infected patients only. Furthermore, it was also reported that at day 3 of treatment, there were more subjects in the treated group than in the control group with crust formation. The result suggested that, from a clinical point of view, *Phyllanthus niruri* extract was beneficial in shortening the healing process of varicella, particularly because it accelerated the formation (appearance) of the crusts, which was then followed by disease relief (disappearance of the crusts). In the recovery process of varicella infection, the appearance of many crusts is of clinical importance as it means that the infection is no longer contagious (Sarisetyaningtyas et al., 2006).

None of the study subjects had secondary skin lesions. This was in line with many literature studies that stated that immunocompetent patients with varicella infection are rarely inflicted with secondary skin infections (Dunkle et al., 1991; Oxman and Alani, 1993).

The study demonstrated a preliminary finding of the potential benefit of *Phyllanthus niruri* in shortening the overall period of varicella disease progress, particularly the morphological progress of the skin lesions. Further larger clinical studies in pediatric patients are required to confirm this finding. Further preclinical studies are also necessary to explore whether such potential was associated with its immunomodulatory or its antiviral activity. To date, we do not have available data studying the effect of the herb on varicella-zoster virus.

Results of all the immunomodulatory-related studies discussed, considered together, explain that *Phyllanthus niruri* acts as an immunomodulator via activation of the cellular immune system. Specifically, *Phyllanthus niruri* acts to activate neutrophils, macrophages/monocytes, and T and B lymphocytes. The activation of the phagocytic process by neutrophils suggests a quicker eradication process of invading microbes, viruses, or fungi, especially the extracellular foreign pathogens, and their clearance out of the body. On the other hand, the increase of phagocytic profile of monocytes/macrophages by *Phyllanthus niruri* promotes the lysis of cells infected by intracellular microbes such as *Mycobacterium tuberculosis*, rendering the microbes exposed to other components of the immune system in the extracellular compartments. The increased cytokine expression and secretion, such as of IFN-γ, TNF-α, IL-4, IL-6, IL-12, and suppression of IL-10 observed in the clinical studies provide a strong indication that *Phyllanthus niruri* may also influence the defense reactions of the body against foreign pathogens involving cellular immune system.

19.2.4 Urolithiasis

Urolithiasis is a disease highly prevalent throughout the world, carrying significant morbidity and consequent costs. Although considerable efforts have been made to identify effective treatments for the disease, it is yet a goal to be achieved.

In Brazil, *Phyllanthus niruri* ("stone breaker" or *quebra pedra*) has also been used for years in folk medicine to treat kidney and bladder stones (Calixto et al., 1998). Alkaloids from plants of the genus *Phyllanthus* present an antispasmodic activity, leading to smooth muscle relaxation, mostly evidenced in the urinary tract, which would facilitate the elimination of urinary calculi (Calixto et al., 1998). An *in vitro* study using canine distal tubular cell culture described the powerful inhibitory effect of *Phyllanthus niruri* on calcium oxalate (CaOx) crystal adhesion and suggested that *Phyllanthus niruri* intake might reduce urinary calcium in patients with hypercalciuria (Campos and Schor, 1999). Using an *in vitro* model of crystallization of human urine, Barros et al. (2003) confirmed the inhibition on CaOx crystal growth and aggregation by the aqueous extract from *Phyllanthus niruri*. In a rat model, the aqueous extract of the whole plant of *Phyllanthus niruri* was demonstrated to prevent an increase of the size of matrix bladder calculi as well as the size and number of formed satellite crystals (Freitas et al., 2002). Those preclinically positive findings were preliminarily confirmed by Nishiura et al. (2004) and Micali et al. (2006) through their controlled clinical studies.

A randomized, placebo-controlled clinical study involving 69 calcium stone-forming (CSF) patients was conducted by Nishiura et al. (2004) to evaluate the potential therapeutic role of *Phyllanthus niruri* in urolithiasis. The herb was administered at a dose of 450 mg extract (lyophilized 2% aqueous extract of *Phyllanthus niruri*) three times daily for 3 months. Overall, the study result in the short-term follow-up showed that no significant differences in calculi voiding or pain relief between groups taking *Phyllanthus niruri* or the placebo were detected. However, there was an important finding reported in this study. Following a 3-month period of *Phyllanthus niruri* administration, a significant effect of *Phyllanthus niruri* in reducing 24-h urinary calcium was found in a subset of patients presenting with hypercalciuria (i.e., 29% of the total studied subjects), one of the most prevalent biochemical abnormalities found in urolithiasis. While all hypercalciuric patients receiving *Phyllanthus niruri* had their calcium levels normalized, the levels of hypercalciuric patients receiving placebo were not significantly different from the baseline (pretreatment) levels (Table 19.2). The number of calculi as well as calculus size observed by ultrasonography were not modified by *Phyllanthus niruri* or the placebo. Four of 33 patients receiving *Phyllanthus niruri* and 5 of 36 patients receiving placebo passed calculi during the 3 months of the study.

The study identified a potential beneficial effect of *Phyllanthus niruri* on hypercalciuria, an important risk factor for stone formation (Nishiura et al., 2004). Yet, larger clinical studies involving particularly hypercalciuric patients and with a longer-term follow-up period are necessary to confirm these findings and to define whether such biochemical modifications can be translated into clinical benefit.

Also associated with the potential benefit of *Phyllanthus niruri* in urolithiasis, Micali et al. (2006) evaluated further the efficacy of *Phyllanthus niruri* extract treatment after shock-wave lithotripsy (SWL) for renal lithiasis. SWL is widely used in the treatment of patients with renal and ureteral calculi as it is noninvasive and generally well tolerated. Joint European Association of Urology/American Urological Association guidelines on the management of ureteral calculi reviewed the success rates after SWL, indicating a range from 68% to 90% (Preminger et al., 2007). The success rate of SWL may be affected by a number of features connected to both stone and patient characteristics. Stone size (Sorensen and Chandhoke, 2002), location (Elbahnasy et al., 1998; Sumino et al., 2002), and composition (Krishnamurthy et al., 2005) influence the clearance of residual fragments, while a high body mass index compromises the results of stone focusing (Pareek et al., 2005). Other researchers also found the influence of type of lithotripsy used on the outcome (Matin et al., 2001; Obek et al., 2001; Sheir et al., 2003).

Regarding stone location, success rates for renal calculi, in particular for lower-pole stones, are often lower than those of ureteral stones (Albala et al., 2001; Pace et al., 2005; Pearle et al., 2005). Lower-pole stones still represent a difficult problem due to the reduced discharge of the fragments and their consequent recrystallization. The cause may be attributed to morphological features of the renal lower pole, such as infundibulopelvic angle, infundibular length and width (Pearle et al., 2005), as well as the size of caliceal stones (Albala et al., 2001). Supportive medical therapy (expulsive medical treatment) to improve the SWL outcome has been attempted in the past using potassium or magnesium citrates, resulting in an increased clearance rate of calcium oxalate (Cicerello et al., 1994), greater rate of remission, and lower rate of recurrence (Soygur et al., 2002). Selective alpha-adrenergic blockers and calcium channel blockers are also choices to modulate the response of the ureter to a calculus; both provide smooth muscle relaxation (Madeb et al., 2007).

Micali et al., through a randomized controlled study involving 150 patients with renal calculi with size of no less than 25 mm and containing calcium oxalate, evaluated the use of *Phyllanthus niruri* extract as a post-SWL expulsive medical treatment (Micali et al., 2006; Schuler et al., 2009). The extract was given at a dose of 2 g once daily for a minimum of 90 days. There was no significant difference in stone size between groups at baseline. At the end of treatment (180 days), complete stone clearance (stone-free condition with residual fragments less than 3 mm) occurred in 93.5% of patients in the treated group but in only 83.3% of patients in the control group. A stone-free condition without any residual fragments occurred in 88.5% and 76.4% of patients in the treated and control groups, respectively. The differences were not statistically significant. However, of those with lower caliceal stones (41% and 33% of patients in treated and control groups, respectively), 93.7% in the treated group were stone free compared to 70.8% in the nontreated group (Table 19.2) (Micali et al., 2006). In line with other studies (Sorensen and Chandhoke, 2002; Pace et al., 2005), this study also found that stone size and medical therapy were the main predictors of stone clearance.

Also note that in this study patients might have had up to three SWL treatments. Regarding the stone-free patients, 39.7% and 43.3% of patients in each group, respectively, underwent SWL retreatment (Micali et al., 2006).

Regular administration of *Phyllanthus niruri* after the SWL procedure may be suggested to improve the outcome. Nevertheless, such significant benefits are only evident for stones in the lower caliceal location, where the extract may act to prevent recrystallization of stone fragments and thus increase their clearance. Further larger clinical studies involving subjects with lower-pole renal stones are still needed to confirm the findings.

Meanwhile, preliminary findings by Nishiura et al. (2004) and Micali et al. (2006) should be seen as the doorway to the deeper exploration and wider use of *Phyllanthus* in the treatment and prevention of urolithiasis and renal lithiasis, both in larger clinical studies as well as in clinical practice.

19.2.5 Antidiabetes

Traditionally, *Phyllanthus* sp. have been used in Ayurvedic medicine for over 2,000 years for reducing the blood sugar level (Nadkarmi, 1993). Preclinical evidence has also proven the hypoglycemic activity of *Phyllanthus* sp. in diabetic animals (Ramakrishnan et al., 1982; Raphael et al., 2002; Adeneye et al., 2006; Kumar et

TABLE 19.2
Effects of *Phyllanthus* sp. Observed in Clinical Studies

Studies	Treatment	Evaluated Parameters	Time of Evaluation	Efficacy *Phyllanthus* sp. Pre	Post	Control Pre	Post	p between Groups
Antilithiasis								
Urolithiasis: Nishiura et al. (2004): Randomized controlled trial	After shock-wave lithotripsy: Treatment with either • *Phyllanthus niruri* extract: 450 mg extract (lyophilized 2% aqueous extract of *P. niruri*) thrice daily for 3 months (n = 33), or • Placebo (n = 36)	Calcium (mg/urine 24 h): Mean (SD)	3 months of treatment	200 (86)	206 (97)	231 (108)	230 (109)	NS
		Calcium (mg/kg/urine 24 h): Mean (SD)						
		• Those with hypercalciurics		4.8 (1.0)	3.4 (1.1)	4.9 (0.6)	4.5 (1.8)	<.05
		• Nonhypercalciurics		2.4 (0.8)	2.9 (1.5)	2.4 (0.6)	2.8 (1.0)	NS
		Number of calculi: mean (SD)		1.8 (0.9)	1.5 (1.4)	2.0 (1.2)	1.6 (1.4)	NS
		Calculi size (cm): mean (SD)		0.6 (0.2)	0.6 (0.2)	0.6 (0.2)	0.6 (0.3)	NS
Renal lithiasis: Micali et al. (2006): Randomized controlled trial	After shock-wave lithotripsy: Treatment with either • *Phyllanthus niruri* extract: 2 g once daily for a minimum of 90 days (n =78), or • Without any oral medical treatment (n = 72)	Stone-free rates (%): With residual fragments < 3 mm:	180 days of treatment					
		• Overall			93.5		83.3	.10
		• Those with baseline stone size < 10 mm			97.2		84.6	.02
		• Those with lower caliceal			93.7		70.8	<.001
		Without any residual:						
		• Overall			88.5		76.4	.08
		• Those with lower calicea			90.6		62.5	.03

Clinical Trials Involving *Phyllanthus* Species

Antidiabetic Effect

Study	Treatment	Outcome measure	Treatment group	Control group	p
Moshi et al. (2001): Randomized, controlled, crossover trial	Treatment with either • *Phyllanthus amarus* extract 12.5 g twice daily for 1 week, or • Without any oral medical treatment	Fasting blood glucose (mmol/l): Mean (SD), 1 week of treatment	10.59 (4.4)	10.03 (3.9)	NS

Phyllanthus sp. in Combination with Other Herbals

Study	Treatment	Outcome measure	Treatment group	Control group	p
Dirjomuljono et al. (2008): Randomized, controlled, crossover trial	Treatment with either • Combination of *Phyllanthus niruri* 50 mg + *Vigella sativa* 360 mg extract in a formulated preparation thrice daily for 7 days ($n = 97$), or • Placebo ($n = 99$)	• Sore-pain reduction (VAS: mm): 6 h • Reduction of sore pain (%): 2 days Mean (SD) • Swallowing difficulty reduction (VAS: mm): 6 h • Reduction of swallowing difficulty (%): 2 days Mean (SD) • Complete relief rate (%): 7 days of treatment	29.29 (19.46) 58.71 (23.39) 33.81 (18.56) 62.06 (25.27) 60.0	22.72 (19.08) 46.44 (30.44) 26.40 (20.36) 48.10 (32.79) 38.4	.018 .008 .002 .001 .022

Note: Evaluated parameters listed in the table are only some of all outcomes evaluated in the study that are considered critical to indicate the potential benefits of *Phyllanthus* sp. in clinical practice. Complete outcomes can be retrieved from the cited references.
NS, not significant ($p > .05$); VAS, visual analogue scale.

al., 2008). Yet another study documented *Phyllanthus* has aldose reductase inhibitory properties (Shimizu et al., 1989). Aldose reductase is an enzyme involved in carbohydrate metabolism that converts glucose to sorbitol. The enzyme acts on nerve endings exposed to a high concentration of blood glucose and thus may lead to many diabetic complications, such as diabetic neuropathy and macular degeneration. Substances that inhibit the enzyme can prevent some of the chemical imbalances that occur and thus protect the nerves from the risk of developing such complications.

Several studies attempting to prove clinically the antidiabetic potency of *Phyllanthus* have also been conducted, even though the results were not consistent with each other in terms of the hypoglycemic activity of the genus.

A clinical study not specifically aimed to evaluate the hypoglycemic activity of *Phyllanthus* examined several effects (i.e., diuretic, hypotensive, and hypoglycemic effects of *Phyllanthus amarus*) of the herb at once in a study (Srividya and Periwal, 1995). Nine individuals with mild hypertension (four also with diabetes mellitus) were treated with a preparation of the whole plant of *Phyllanthus amarus* for 10 days. Physiological profile and dietary pattern before and after the treatment period were also examined. Significant increase in 24-h urine volume and urine and serum sodium levels was observed. A significant reduction of systolic blood pressure in nondiabetic hypertensives and female subjects was noted. Blood glucose was also significantly reduced in the treated group. Clinical observations revealed no harmful side effects.

Despite all the observations indicating potential diuretic, hypotensive, and hypoglycemic effects of *Phyllanthus amarus* in human subjects, they were basically preliminary. The study sample size was too small to draw an adequately powered conclusion, including regarding its blood glucose-lowering effect. No control group was allocated in the study. Also, for a metabolic disease such as diabetes, the dietary profile during the study has a significant impact on the blood glucose evaluation and thus should be controlled before and after treatment and, importantly, ensuring comparable profiles between all enrolled subjects.

Quite a different finding was reported in another crossover clinical study (Moshi et al., 2001). In the study, Moshi et al. evaluated the glycemic response of 21 patients with non-insulin-dependent diabetes mellitus (NIDDM) to a standard meal in three study phases consecutively: while on oral antihypergylcemic agents, after a 1-week washout period, and after a 1-week, twice-daily treatment with 100 ml of an aqueous extract from 12.5 g of powdered aerial parts of *Phyllanthus amarus*. After the 1-week washout period, the fasting blood glucose (FBG) and postprandial blood glucose (PPBG) significantly increased compared to that while on oral antihyperglycemics (Table 19.2). After a 1-week *Phyllanthus amarus* treatment, no hypoglycemic activity was observed. Both FBG and PPBG remained similar to that observed after the washout period. One week of treatment with the aqueous extract of *Phyllanthus amarus* was incapable of lowering both FBG and PPBG in untreated patients with NIDDM (Moshi et al., 2001).

Inconsistent findings from one study with another may be due to one or more of the following possible reasons: inadequate power, absence of a control group, too loose inclusion-exclusion criteria, short-term follow-up period of the studies, different species of *Phyllanthus* used between studies, or exactly the same species between studies but with different ways of extraction or different formulations. Regardless of the study results, all the studies together should lead to further larger well-designed clinical studies to confirm the antidiabetic potency of *Phyllanthus*.

19.3 *PHYLLANTHUS* IN COMBINATION WITH OTHER HERBS OR MEDICATION

Several clinical studies also evaluated the effects of *Phyllanthus* sp., particularly when it was combined with other herbs in a specifically formulated preparation (a fixed-combination preparation) or used in combination with other proven medication (Dirjomuljono et al., 2008; Sen et al., 2009).

19.3.1 Tonsillopharyngitis

Other than cases of bacterial infection, *Phyllanthus*, with its immunomodulatory activity, was also studied in cases of viral infection, such as acute tonsillopharyngitis. In a randomized, parallel, double-blind, placebo-controlled clinical study, *Phyllanthus niruri* extract was used in combination with another herbal, *Nigella sativa*, which has anti-inflammatory activity (Dirjomuljono et al., 2008). All of 196 patients enrolled in the study were those with moderate-to-severe sore throat, and most of them did not have nasal congestion. The combination containing 50 mg *Phyllanthus niruri* and 360 mg *Nigella sativa* extracts was orally administered three times daily for 7 days. Such a combination readily exerted its efficacy, as shown by markedly alleviated sore throat, measured as a reduction in visual analogue scale of swallowing pain and swallowing difficulty in patients receiving such combination within 5 and 6 h after administration. The reduction was greater than that in the placebo group. In line with the significant pain alleviation, from days 0 to 2 of treatment, patients treated with such a combination also needed significantly less escape (analgesic) therapy (paracetamol tablets) than those in the placebo group. At the end of treatment (day 7), a significantly greater proportion of patients in the *Phyllanthus niruri–Nigella sativa* group than in the placebo group had the sore throat completely relieved (Table 19.2) (Dirjomuljono et al., 2008). In the study, the effectiveness of *Phyllanthus niruri–Nigella sativa* extract in the treatment of acute tonsillopharyngitis was not influenced by baseline severity of the symptoms, indicating that the extract can be used for a wide range of tonsillopharyngitis severity, especially those with moderate-to-severe symptoms.

The effectiveness of such a combination demonstrated in the study might partly be attributable to the antinociceptive activity of *Phyllanthus niruri* (Santos et al., 1995, 2000) as well as the antibacterial property of the *Phyllanthus niruri–Nigella sativa* combination (Sunardi and Tjandrawinata, 2006). In an *in vitro* study, *Phyllanthus niruri–Nigella sativa* extract was proven to be active against gram-positive and gram-negative bacteria, including *Streptococcus pneumoniae*, one of the most common pathogenic bacteria in tonsillopharyngitis cases. The antibacterial activities of the *Phyllanthus niruri–Nigella sativa* extract were even higher than those demonstrated by either herbal extract alone, suggesting that a combination of both herbal extracts provides a synergistic effect in delivering the antibacterial activity (Hanafi and Hatem, 1991; Sunardi and Tjandrawinata, 2006).

This study confirmed the clinical benefits of *Phyllanthus niruri–Nigella sativa* combination in patients with acute tonsillopharyngitis (Dirjomulijono et al., 2008). The irrational use of antibiotics in such a viral-origin infection may thus be minimized, which in turn may reduce the incidence of antibiotic resistance. Based on this study alone, it cannot be confirmed, however, whether *Phyllanthus niruri* or *Nigella sativa* alone would provide such benefits in those subjects. This study alluded to the potential of using *Phyllanthus* in combination with other herbs to give a synergistic efficacy while still retaining their safety profiles. This is expected to be applied in clinical practice.

19.3.2 Bronchial Asthma

To date, bronchial asthma is pharmacologically treated by β-adrenergic receptor agonists and corticosteroids, either as a single therapy or in combination, with various kinds of formulations that enable the drugs to act either locally on the respiratory organ or systemically. Attempts to minimize the adverse drug reactions of such modern (synthetic) medications, particularly due to their prolonged use in chronic disease, have been made by introducing the use of herbal medicines to the formal medical practice. With respect to this issue, *Phyllanthus* has also been studied in India for its efficacy in bronchial asthma.

A randomized, controlled, parallel-design clinical study involving 30 subjects (12–70 years old) with bronchial asthma was conducted (Sen et al., 2009) to compare the efficacy of three groups of treatment in resolving clinical symptoms of bronchial asthma: (1) *Phyllanthus fraternus* Webster; (2) an oral standard fixed combination of 2 mg of salbutamol (a β-adrenergic receptor agonist) and 100 mg of theophylline (α-adrenergic receptor agonist), supported with salbutamol inhaler 100 μg/dose (0.005% w/v); and (3) the combination of both treatments. The symptoms measured included both the subjective (breathlessness, cough, expectoration, wheezing, ronchi, *jaranashakti* [a Sanskrit word that means "energetic feeling" in English], *ruchi* ["appetite" in English]) and objective (forced expiratory volume at 1 second [FEV_1], forced vital capacity [FVC], peak expiratory flow rate [PEFR], differential leukocyte count for eosinophil and absolute eosinophil count [AEC]) parameters. This study was designed to evaluate whether the use of *Phyllanthus fraternus*, either alone or in combination with the modern medicines, could provide a synergistic effect in improving efficacy. Study treatment was given for 45 days, with *Phyllanthus fraternus* at a dose of 500 mg orally three times daily, the standard treatment a dose three times daily, while the salbutamol inhaler was allowed to be taken on an as-needed basis only (Sen et al., 2009). This study was conducted based on a previous report of an experimental study demonstrating the antihistaminic property of *Phyllanthus fraternus* Webster (syn. *Phyllanthus niruri* Linn) (Mishra, 1978).

At the end of the study, all three groups experienced a significant improvement in all subjective symptoms compared to their respective baseline. *Phyllanthus fraternus* provided a significantly better improvement of energetic feeling and appetite compared to that of standard treatment, while a combination of both *Phyllanthus fraternus* and standard treatment provided a significantly greater improvement in all subjective symptoms than that of any other groups. Regarding the objective parameters, all three groups experienced a significant improvement in all parameters measured but the PEFR, of which the improvement from baseline was only observed in the group receiving a combination of both treatments. With comparison to combination treatment, *Phyllanthus fraternus* alone demonstrated an equivalent efficacy. With comparison to the standard treatment, *Phyllanthus fraternus* was better in reducing AEC, while the combination treatment was better in reducing AEC, differential eosinophil, and FVC. Overall study results indicated that *Phyllanthus fraternus* alone was better than the standard salbutamol-theophylline treatment in improving the symptoms of bronchial asthma. A combination of both treatments provided the best efficacy (Sen et al., 2009).

Despite the potential benefits found (Sen et al., 2009), it was not reported in the study whether the use of such a combination might reduce a patient's need for inhaled salbutamol. It also was not reported whether the use of *Phyllanthus fraternus* alone or in combination might significantly minimize the adverse reactions commonly observed with the standard therapy. Such unreported considerably important results and the small sample size together should also be noted to underline that the positive findings in the study were still preliminary and thus need to be confirmed in further studies with adequate power that include the evaluation of critical safety parameters compared to the standard therapy.

19.3.3 Others

Several preliminary clinical trials have been conducted involving *Phyllanthus emblica* (Lakshmipathi and Venugopala, 1962; Misra, 1970; Indira et al., 1973; Misger et al., 1977; Marya and Gulati, 1978; Vijayasarathy and Sharma, 1981). However, all the studies used the specifically formulated preparations in which *Phyllanthus emblica* extract was only one of many other active ingredients. Therefore, it is likely impossible to appraise the individual effect of the herb that contributed to the observed efficacy as well as adverse events. It is also rather beyond the theme of this chapter, so we do not discuss them further but list the cases studied.

Three studies involving first- to third-degree hemorrhoid cases evaluated the efficacy of the same oral and ointment preparations (Misger et al., 1977; Marya and Gulati, 1978; Vijayasarathy and Sharma, 1981). Only the oral preparation contained *Phyllanthus emblica*. One study evaluated the effects of a formulated oral preparation in acute otorhinolaryngologic infections (Lakshmipathi and Venugopala, 1962). One study evaluated the efficacy of a formulated oral preparation in management of diarrhea, both specific and nonspecific types, with mild-to-moderately severe cases (Misra, 1970). Another study evaluated the digestive corrective and anabolic effect of a formulated oral preparation in newborns and infants (Indira et al., 1973). In brief, in all preparations used in the studies, *Phyllanthus emblica* was reasonably used for its richness of ascorbic, ellagic, and gallic acids; thus, it probably exerted its action through its antioxidant as well as anti-inflammatory properties (Ihantola-Vormisto et al., 1997). All the studies reported a good efficacy and safety profile of the oral preparations evaluated. Yet, we cannot claim that such results were attributed to one or more of the components individually as all the components acted together, whether additively or synergistically, to bring about the observed effects.

19.4 SAFETY AND TOLERABILITY

The safety profile and tolerability of *Phyllanthus* are based on available safety data and adverse events recorded in some of the clinical studies discussed in this chapter. Furthermore, a phase 1 clinical study of *Phyllanthus* extract specifically aimed to evaluate the safety and tolerability of a formulated *Phyllanthus* extract in Thai healthy subjects is also available (Tangpukdee et al., 2004).

In pulmonary TB studies, *Phyllanthus niruri* administration to the patients for a long-term period (from 2 months up to 6 months, at a dose of 50 mg extract three times daily) in addition to the WHO standardized TB regimens did not affect liver functions, as measured by serum aspartate and alanin aminotransferase (AST and ALT) (Amin, 2005). The adverse events reported in the group treated with *Phyllanthus niruri* and the control group were itching (59% and 57%, respectively); nausea/vomiting (41% and 42%, respectively); arthralgia (20% and 9%, respectively); headache (9% and 12%, respectively); and rash (6% and 3%, respectively) (Amin, 2005). Radityawan (2005) also reported 1 (of 20) patient who received *Phyllanthus niruri* and experienced dizziness, but it ended in only 1 week. Most of the events were mild in severity, and all had already been resolved by the end of the study (Amin, 2005; Radityawan, 2005). In particular for the itching and nausea/vomiting events, they were probably related to the drugs included in the standard TB regimens, as such events are commonly found in patients taking all four drugs of the regimens at once. No different rates of adverse events were observed between groups, which indicates that all the adverse events were unlikely to be related to *Phyllanthus niruri* administration.

Raveinal (2003) also reported similar adverse events observed during 2 months of study treatment. The adverse events experienced by the group treated with *Phyllanthus niruri* and the control group included nausea/vomiting (30% and 50%, respectively); malaise (20% and 15%, respectively); arthralgia/myalgia (10% and 10%, respectively); and headache (5% and 30%, respectively). Other adverse events, such as icterus, itching, and numbness, were only observed in the control group. They were all mild to moderate in severity and had already been resolved by the end of the study.

Phyllanthus niruri extract was also reported to be safe for and well tolerated by pediatric patients with varicella infection. Regardless of such technical difficulty in obtaining parents' permission that limited patients (60 patients in total or 30 per group) in whom laboratory examinations could be performed, all laboratory parameters (liver and renal function examinations) of both groups were still in the normal ranges. No clinically significant adverse events were found during the study course (Sarisetyaningtyas et al., 2006).

In urolithiasis and renal lithiasis patients, treatment with *Phyllanthus* was associated with lack of clinically significant adverse events (Nishiura et al., 2004; Micali et al., 2006).

Use of *Phyllanthus amarus* in diabetic patients did not significantly affect both liver and renal functions based on ALT and serum creatinine, respectively. Although the lymphocyte and monocyte levels were reported to be significantly decreased and the granulocyte level was significantly increased after treatment, the overall total white blood cell (WBC) count and hemoglobin (Hb) were not significantly affected by *Phyllanthus amarus* treatment (Moshi et al., 2001).

In combination with *Nigella sativa*, the combination *Phyllanthus niruri* (50 mg) and *Nigela sativa* (360 mg) extract administered three times daily for 7 days was also found to be safe and well tolerated in patients with acute tonsillopharyngitis, with only polyuria, which was probably related to the combination. Such event was found in 14.5% patients in the group treated with *Phyllanthus niruri–Nigella sativa* compared to only 5% in the control group (Dirjomuljono et al., 2008). Even though polyuria might result from the increased water intake by the subjects as advised by the investigators to help them restore their health condition from tonsillopharyngitis, such adverse effect was more likely to be attributed to *Phyllanthus niruri* extract. *Phyllanthus niruri* has been reported to have such an effect (Unander et al., 1995). However, through several previous and current studies we have not found any clinically significant consequences of polyuria due to *Phyllanthus niruri* extract, such as hypotension or dehydration. Nevertheless, caution still has to be taken, especially when this extract is used by patients who are hypotensive or dehydrated.

In the Thai phase 1 study, *Phyllanthus* extract was administered orally at two dosage regimens (500 and 1,000 mg/day doses for 14 days) to a total of 16 healthy volunteers (Tangpukdee et al., 2004). Gastrointestinal, central nervous, cardiovascular, and dermatological effects and other changes possibly attributable to *Phyllanthus* administration were evaluated. Routine blood investigations (hematology and biochemistry) and urinalysis were performed prior to and at the end of the study (7 days before initial dose, at follow-up visits, and until 28 days after the last dose, i.e., days 1, 3, 7, 10, 14, and 42). Collected data included serum chemistry profile (electrolytes, blood urea nitrogen, creatinine, total protein, albumin, AST and ALT, alkaline phosphatase, amylase, total and direct bilirubin, and fasting blood sugar); hematology profile (red blood cell count, Hb, WBC with differential, and platelet count); urine analysis (specific gravity, pH, occult blood, glucose, ketone, and protein); and chest X-ray, serum anti-HIV, and urine pregnancy tests (in females), which were performed on the screening day only. The volunteers were physically examined, and adverse reactions during the study were recorded with the date and time at which they occurred and disappeared.

In this study, oral doses of *Phyllanthus* extract up to 1,000 mg/day were well tolerated. No serious clinical or laboratory adverse events were observed in either study group. Laboratory monitoring showed variations between each of the follow-up days, but no significant differences were observed among those days with respect to the baseline data and among doses of treatments. Some volunteers experienced some adverse symptoms during *Phyllanthus* administration in the study, as follows: headache (12%), hunger (6%), anorexia (18%), diarrhea (6%), constipation (12%), insomnia (18%), and drowsiness (18%). The most frequently reported degree of symptom was mild. Nausea and vomiting were not reported by any volunteer throughout the study.

All adverse events reported were considered as unlikely to be related to *Phyllanthus* treatment (Tangpukdee et al., 2004).

Other available published articles on *Phyllanthus* did not report any safety data or adverse events. This may likely be because the administration of *Phyllanthus* in those studies did not cause any adverse events clinically significant enough to be noted or cautioned; thus, they were justifiably omitted from the published reports. Taken together, all currently available safety data for *Phyllanthus* apparently depict the good safety profile of the herb in patients with various diseases.

19.5 CONCLUSION

To date, other than its efficacy in hepatitis, particularly hepatitis B and its direct effect against the pathogenic hepadnavirus, only limited numbers of clinical studies have been published involving *Phyllanthus* species, numerous *in vitro* and animal studies demonstrating various potential benefits of the genus of *Phyllanthus* notwithstanding. Moreover, only a few certain species of such a large genus have already been thoroughly studied, in particular *Phyllanthus niruri, Phyllanthus amarus,* and *Phyllanthus urinaria*. Some of them demonstrated inconsistent results that did not positively correspond with the findings of preceding preclinical studies. Even among those with positive correlation with preclinical findings, most were still preliminary in terms of confirming the clinical potential benefits of the herbs. Further well-designed, randomized, controlled, adequate-power clinical studies are yet to be conducted. Moreover, there are still many other species of *Phyllanthus* queuing to be explored.

In terms of safety, *Phyllanthus* species were proven to lack adverse reactions. Their good safety profile combined with many potential clinical benefits found in various clinical studies thus far constitute promising virtues that make *Phyllanthus* deserve further larger clinical studies to confirm their potential efficacy in various diseases and with specific types of patient. In turn, this will place the herbals in a position for management of various diseases in formal clinical practice.

ACKNOWLEDGMENT

We would like to gratefully acknowledge Liana W. Susanto, MSc, Clinical Research Division, Dexa Medica Group, Indonesia, for her assistance in the preparation of this chapter.

REFERENCES

Adeneye, A.A., Amole, O.O., and Adeneye, A.K. 2006. Hypoglycemic and hypocholesterolemic activities of the aqueous leaf and seed extract of *Phyllanthus amarus* in mice. *Fitoterapia* 77: 511–514.

Albala, D.M., Assimos, D.G., Clayman, R.V., et al. 2001. Lower pole I: a prospective randomized trial of extracorporeal shock wave lithotripsy and percutaneous nephrostolithotomy for lower pole nephrolithiasis—initial results. *J. Urol.* 166: 2072–2080.

Amin, Z. 2005. The effect of *Phyllanthus* extract as an additional treatment in tuberculosis patients with minimal and moderately advanced radiological lesion. Diss., University of Indonesia, Jakarta.

Barnes, P.F., and Modlin, R.L. 1994. T-cell responses and cytokines. In *Tuberculosis pathogenesis, protection and control*, ed. B.R.E Bloom, 417–435. Washington, DC: ASM Press.

Barros, M.E., Schor, N., and Boim, M.A. 2003. Effects of an aqueous extract from *Phyllanthus niruri* on calcium oxalate crystallization *in vitro*. *Urol. Res.* 30: 374–379.

Blumberg, B.S., Millman, I., Venkateswaran, P.S., and Thyagarajan, S.P. 1989. Hepatitis B virus and hepatocellular carcinoma—treatment of HBV carriers with *Phyllanthus amarus*. *Cancer Detect. Prev.* 14: 195–201.

Calixto, J.B., Santos, A.R., Cechinel, F.V., and Yunes, R.A. 1998. A review of the genus *Phyllanthus*: their chemistry, pharmacology, and therapeutic potential. *Med. Res. Rev.* 18: 225–258.

Campos, A.H., and Schor, N. 1999. *Phyllanthus niruri* inhibits calcium oxalate endocytosis by renal tubular cells: its role in urolithiasis. *Nephron.* 81: 393–397.

Chan, J., and Kauffman, S.H.E. 1994. Immune mechanisms of protection. In *Tuberculosis pathogenesis, protection and control*, ed. B.R.E Bloom, 415. Washington, DC: ASM Press.

Cicerello, E., Merlo, F., Gambaro, G., Maccatrozzo, L. et al. 1994. Effect of alkaline citrate therapy on clearance of residual renal stone fragments after extracorporeal shock wave lithotripsy in sterile calcium oxalate and infection nephrolithiasis patients. *J. Urol.* 151: 5–9.

Corrigan, E.M., Clancy, R.L., Dunkley, M.L., et al. 1998. Cellular immunity in recurrent vulvovaginal candidiasis. *Clin. Exp. Immunol.* 111: 574–578.

Dannenberg, A.M. 1993. Immunopathogenesis of pulmonary tuberculosis. *Hosp. Pract.* 15: 51–58.

Dirjomuljono, M., Kristyono, I., Tjandrawinata, R.R., and Nofiarny, L. 2008. Symptomatic treatment of acute tonsillo-pharyngitis patients with a combination of *Nigella sativa* and *Phyllanthus niruri* extract. *Int. J. Clin. Pharmacol. Ther.* 46: 295–306.

Dunkle, L.M., Arvin, A.M., Whitley, R.J., et al. 1991. Controlled trial of acyclovir for chickenpox evaluating time of initiation and duration of therapy and viral resistance. *N. Engl. J. Med.* 325: 1539–1544.

Egan, M.E., and Lipsky, M.S. 2000. Diagnosis of vaginitis. *Am. Fam. Physician* 62: 1095–1104.

Elbahnasy, A.M., Shalhav, A.L., Hoenig, D.M., et al. 1998. Lower caliceal stone clearance after shock wave lithotripsy or ureteroscopy: the impact of lower pole radiographic anatomy. *J. Urol.* 159: 676–682.

Fenton, M.J., and Vermeulen, M.W. 1996. Immunopathology of tuberculosis: roles of macrophages and monocytes. *Infect. Immun.* 64: 683–690.

Freitas, A.M., Schor, N., and Boim, M.A. 2002. The effect of *Phyllanthus niruri* on urinary inhibitors of calcium oxalate crystallization and other factors associated with renal stone formation. *BJU Int.* 89: 829–834.

Halim, H., and Saleh, K. 2005. The effectiveness of *Phyllanthus niruri* extract in the management of pulmonary tuberculosis. *Dexa Media* 18: 103–107.

Hanafi, M.S.M., and Hatem, M.E. 1991. Studies on the antimicrobial activity of *Nigella sativa* seed (black cumin). *J. Ethnopharmacol.* 34: 275–278.

Ihantola-Vormisto, A., Summanen, J., Kankaanranta, H., et al. 1997. Anti-inflammatory activity of extracts from leaves of *Phyllanthus emblica*. *Planta Med.* 63: 518–524.

Indira, B.K., Subba, R.K.V., and Subramanyam, M.V.G. 1973. Bonnisan a digestive corrective and anabolic tonic for newborns and infants. *Paediatr. Clin. India* 1: 39–43.

Jones, B.E., Oo, M.M., Taikwel, E.K., et al. 1997. CD4 cell counts in human immunodeficiency virus-negative patients with tuberculosis. *Clin. Infect. Dis.* 24: 988–991.

Kauffman, S.H.E. 1993. Immunity to intracellular bacteria. *Annu. Rev. Immunol.* 11: 129–163.

Khanna, A.K., Rizvi, F., and Chander, R. 2002. Lipid lowering activity of *Phyllanthus niruri* in hyperlipemic rats. *J. Ethnopharmacol.* 82: 19–22.
Krishnamurthy, M.S., Ferucci, P.G., Sankey, N., and Chandhoke. P.S. 2005. Is stone radiodensity a useful parameter for predicting outcome of extracorporeal shockwave lithotripsy for stones < or = 2 cm? *Int. Braz. J. Urol.* 31: 3–8; discussion 9.
Kumar, S., D., Kumar, R.R., Deshmukh, P.D., et al. 2008. Antidiabetic potential of *Phyllanthus reticulatus* in alloxan-induced diabetic mice. *Fitoterapia* 79: 21–23.
Lakshmipathi, G., and Venugopala, R.B. 1962. Clinical effects of a combination of indigenous herbs in otorhinolaryngologic infections. *J. Indian Med. Assoc.* 4: 174–176.
Lin, Y., Zhang, M., Hofman, F.M., et al. 1996. Absence of a prominent Th-2 cytokine response in human tuberculosis. *Infect. Immun.* 64: 1351–1356.
Ma'at, S. 1996. *Phyllanthus niruri* L. as an immunostimulator in mice. Diss., University of Airlangga, Surabaya.
Ma'at, S. 2001. Antimycotic and immunomodulator in mycosis: *Phyllanthus niruri* extract as a natural immunomodulator. Paper presented at the periodic scientific meeting of the Association of Human and Animal Mycological Medicine, Solo. Central Java, September 2011.
Madeb, R., Knopf, J., Golijanin, D., et al. 2007. Evidence for alpha receptors in the human ureter. In *Renal stone disease: first annual international urolithiasis research symposium, American Institute of Physics Conference Proceedings*, ed. A.P. Evan, J.E. Lingeman, and J.C. Williams Jr., 253–260. New York: Americal Institute of Physics (AIP).
Magliani, W., Conti, S., Cassone, A., et al. 2002a. New immunotherapeutic strategies to control vaginal candidiasis. *Trends Mol. Med.* 8: 121–126.
Magliani, W., Conti, S., Salati, A., et al. 2002b. New strategies for treatment of Candida vaginal infections. *Rev. Iberoam. Micol.* 19: 144–148.
Marya, S.K.S., and Gulati, S.K. 1978. Pilex (tablets and ointment): a clinical trial in patients with internal haemorrhoids. *Antiseptic* 1: 21–24.
Matin, S.F., Yost, A., and Streem, S.B. 2001. Extracorporeal shock wave lithotripsy: a comparative study of electrohydraulic and electromagnetic units. *J. Urol.* 166: 2053–2056.
Micali, S., Sighinolfi, M.C., Celia, A., et al. 2006. Can *Phyllanthus niruri* affect the efficacy of extracorporeal shock wave lithotripsy for renal stones? A randomized, prospective, long-term study. *J. Urol.* 176: 1020–1022.
Misger, M.S., Mir, M.A., Wani, N.A., and Rashid, P.A.1977. Pilex therapy in the treatment of haemorrhoids. *Ind. Med. Gaz. (XVI)* 9: 353–356.
Mishra, R.B. 1978. Studies on Tamalaki (*Phyllanthus niruri* Linn). Thesis, Dravyaguna IMS, BHU, Varanasi, India.
Misra, H.S. 1970. Indigenous drugs in the management of diarrhea. *Indian Pract.* 8: 549–551.
Moore, K.W., de Waal Malefyt, R., Coffman, R.L., and O'Garra, A. 2001. Interleukin-10 and the interleukin-10 receptor. *Annu. Rev. Immunol.* 19: 683–765.
Moshi, M.J., Lutale, J.J., Rimoy, G.H., et al. 2001. The effect of *Phyllanthus amarus* aqueous extract on blood glucose in non-insulin dependent diabetic patients. *Phytother. Res.* 15: 577–580.
Nadkarmi, K.M. 1993. *India materia medica*, Vol. 1, 947–948. Bombay: Popular Prakashan.
Naik, A.D., and Juvekar, A.R. 2003. Effects of alkaloidal extract of *Phyllanthus niruri* on HIV replication. *Indian J. Med. Sci.* 57: 387–393.
Nishiura, J.L., Campos, A.H., Boim, M.A., et al. 2004. *Phyllanthus niruri* normalizes elevated urinary calcium levels in calcium stone forming (CSF) patients. *Urol. Res.* 32: 362–366.
Notka, F., Meier, G.R., and Wagner, R. 2003. Inhibition of wild-type human immunodeficiency virus and reverse transcriptase inhibitor-resistant variants by *Phyllanthus amarus*. *Antiviral Res.* 58: 175–186.

Notka, F., Meier, G., and Wagner, R. 2004. Concerted inhibitory activities of *Phyllanthus amarus* on HIV replication *in vitro* and *ex vivo*. *Antiviral Res.* 64: 93–102.
Obek, C., Onal, B., Kantay, K., et al. 2001. The efficacy of extracorporeal shock wave lithotripsy for isolated lower pole calculi compared with isolated middle and upper caliceal calculi. *J. Urol.* 166: 2081–2084; discussion 2085.
Oxman, M.N., and Alani, R. 1993. Varicella and herpes zoster. In: *Dermatology in general medicine*, ed. T.B. Fitzpatrick, and A.Z. Eizen, 2534–2572. New York: McGraw-Hill.
Pace, K.T., Ghiculete, D., Harju, M., and Honey, R.J., University of Toronto Lithotripsy Associates. 2005. Shock wave lithotripsy at 60 or 120 shocks per minute: a randomized, double-blind trial. *J. Urol.* 174: 595–599.
Pareek, G., Armenakas, N.A., Panagopoulos, G., Bruno, J.J., and Fracchia, J.A. 2005. Extracorporeal shock wave lithotripsy success based on body mass index and Hounsfield units. *Urology* 65: 33–36.
Pearle, M.S., Lingeman, J.E., Leveillee, R., et al. 2005. Prospective, randomized trial comparing shock wave lithotripsy and ureteroscopy for lower pole caliceal calculi 1 cm or less. *J. Urol.* 173: 2005–2009.
Pettit, G.R., Cragg, G.M., Niven, M.L., and Nassimbeni, L.R. 1983. Structure of the principal antineoplastic glycosides of *Phyllanthus acuminatus* Vahl. *Can. J. Chem.* 61: 2630–2632.
Pettit, G.R., Schaufelberger, D.E., Nieman, R.A., Dufresne, C., and Saenz-Renauld, J.A. 1990. Antineoplastic agents. Isolation and structure of phyllanthostatin 6. *J. Nat. Prod.* 53: 1406–1413.
Pramayanti, I., Paraton, H., and Ma'at, S. 2005. Comparison of success rate of vaginal candidiasis treatment between ketoconazole and combination of ketoconazole-*Phyllanthus niruri* extract. *Dexa Media* 18: 97–102.
Preminger, G.M., Tiselius, H.G., Assimos, D.G., et al., EAU/AUA Nephrolithiasis Guideline Panel. 2007. Guideline for the management of ureteral calculi. *J. Urol.* 178: 2418–2434.
Radityawan, D. 2005. The immunomodulatory effect of *Phyllanthus niruri* L on serum IFN-γ level in pulmonary tuberculosis patients. *Dexa Media* 18: 94–96.
Rajeshkumar, N.V., Joy, K.L., Kuttan, G., et al. 2002. Antitumour and anticarcinogenic activity of *Phyllanthus amarus* extract. *J. Ethnopharmacol.* 81: 17–22.
Rajeshkumar, N.V., and Kuttan, R. 2000. *Phyllanthus amarus* extract administration increases the life span of rats with hepatocellular carcinoma. *J. Ethnopharmacol.* 73: 215–219.
Ramakrishnan, P.N., Murugesan, R., Palanichamy, S., and Murugesh, N. 1982. Oral hypoglycaemic effect of *Phyllanthus niruri* Linn. *Indian J. Pharm. Sci.* 44: 10–12.
Raphael, K.R., Sabu M.C., and Kuttan, R. 2002. Hypoglycemic effect of methanol extract of *Phyllanthus amarus* Schum & Thonn on alloxan induced diabetes mellitus in rats and its relation with antioxidant potential. *Indian J. Exp. Biol.* 40: 905–909.
Raveinal, R. 2003. The effect of natural immunomodulator (*Phyllanti* herb extract) administration on cellular immune response of patients with pulmonary tuberculosis. Thesis, University of Andalas, Padang, Indonesia.
Santos, A.R., De Campos, R.O., Miguel, O.G., et al. 2000. Antinociceptive properties of extracts of new species of plants of the genus *Phyllanthus* (*Euphorbiaceae*). *J. Ethnopharmacol.* 72: 229–238.
Santos, A.R., Filho, V.C., Yunes, R.A., and Calixto, J.B. 1995. Analysis of the mechanisms underlying the antinociceptive effect of the extracts of plants from the genus *Phyllanthus*. *Gen. Pharmacol.* 26: 1499–1506.
Sarisetyaningtyas, P.V., Hadinegoro, S.R., and Munasir, Z. 2006. Randomized controlled trial of *Phyllanthus niruri* Linn extract. *Paediatr. Indones.* 46: 77–81.

Schuler, T.D., Shahani, R., Honey, R.J., and Pace, K.T. 2009. Medical expulsive therapy as an adjunct to improve shockwave lithotripsy outcomes: a systematic review and meta-analysis. *J. Endourol.* 23: 387–393.
Sen, B., Dubey, S.D., Singh, V.P., and Tripathi, K. 2009. A study on shvasahara karma of Tamalaki (*Phyllanthus fraternus* Webster). *Ayu* 30: 42–46.
Sheir, K.Z., Madbouly, K., and Elsobky, E. 2003. Prospective randomized comparative study of the effectiveness and safety of electrohydraulic and electromagnetic extracorporeal shock wave lithotriptors. *J. Urol.* 170: 389–392.
Shimizu, M., Horie, S., Terashima, S., et al. 1989. Studies on aldose reductase inhibitors from natural products. II. Active components of a Paraguayan crude drug, "para-parai-mi," *Phyllanthus niruri*. *Chem. Pharm. Bull. (Tokyo)* 37: 2531–2532.
Smith, S., Jacobs, R.F., and Wilson, C.B. 1997. Immunobiology of childhood tuberculosis: a window on the ontogeny of cellular immunity. *J. Paediatr.* 131: 16–26.
Sobel, J.D. 1997. Vaginitis. *N. Engl. J. Med.* 337: 1896–1903.
Sorensen, C.M., and Chandhoke, P.S. 2002. Is lower pole caliceal anatomy predictive of extracorporeal shock wave lithotripsy success for primary lower pole kidney stones? *J. Urol.* 168: 2377–2382.
Soygur, T., Akbay, A., and Kupeli, S. 2002. Effect of potassium citrate therapy on stone recurrence and residual fragments after shockwave lithotripsy in lower caliceal calcium oxalate urolithiasis: a randomized controlled trial. *J. Endourol.* 16: 149–152.
Srividya, N., and Periwal, S. 1995. Diuretic, hypotensive and hypoglycaemic effect of *Phyllanthus amarus*. *Indian J. Exp. Biol.* 33: 861–864.
Sumino, Y., Mimata, H., Tasaki, Y., et al. 2002. Predictors of lower pole renal stone clearance after extracorporeal shock wave lithotripsy. *J. Urol.* 168: 1344–1347.
Sunardi, F., and Tjandrawinata, R.R. 2006. A study on antimicrobial activities of a herbal extract combination containing *Phyllanthus niruri* and *Nigella sativa*. *Medika* 32(12): 744–748.
Tangpukdee, N., Hiransaree, A., Krudsood S., et al. 2004. Phase-I clinical trial to evaluate the safety and tolerability of oral SN-1 extract formulation in healthy Thai volunteers. *J. Trop. Med. Parasitol.* 27: 71–78.
Thyagarajan, S.P., Jayaram, S., Valliammai, T., et al. 1990. *Phyllanthus amarus* and hepatitis B. *Lancet* 336: 949–950.
Thyagarajan, S.P., Subramanian, S., Thirunalasundari, T., Venkateswaran, P.S., and Blumberg, B.S. 1988. Preliminary study: effect of *Phyllanthus amarus* on chronic carriers of hepatitis B virus. *Lancet* 2: 764–766.
Tsao, T.C.Y., Chen C.H., Hong, J.H., et al. 2002. Shifts of T4/T8 lymphocytes from BAL fluid and peripheral blood by clinical grade in patients with pulmonary tuberculosis. *Chest* 122: 1285–1291.
Unander, D.W., Webster, G.L., and Blumberg, B.S. 1995. Usage and bioassays in *Phyllanthus* (*Euphorbiaceae*). IV. Clustering of antiviral uses and other effects. *J. Ethnopharmacol.* 45: 1–18.
Venkateswaran, P.S., Millman, I., and Blumberg, B.S. 1987. Effects of an extract from *Phyllanthus niruri* on hepatitis B and woodchuck hepatitis viruses: *in vivo* and *in vitro* studies. *Proc. Natl. Acad. Sci U. S. A.* 84: 274–278.
Vijayasarathy, V., and Sharma, L.K. 1981. Indigenous drug therapy for haemorrhoids. *Indian J. Med. Surg.* 21(1–2): 22–25.
Watford, W.T., Moriguchi, M., Morinobu, A., and O'Shea, J.J. 2003. The biology of IL-12: coordinating innate and adaptive immune responses. *Cytokine Growth Factor Rev.* 14: 361–368.
Witkin, S.S., Linhares, I., Giraldo, P., Jeremias, J., and Ledger, W.J. 2000. Individual immunity and susceptibility to female genital tract infection. *Am. J. Obstet. Gynecol.* 183: 252–256.

World Health Organization (WHO). 2003. *Treatment of tuberculosis: guidelines for national programmes* 3rd ed. WHO/CDS/TB/2003.313. Available at http://www.who.int/tb/publications/cds_tb_2003_313/en (accessed September 29, 2009).

Yang, C.M., Cheng, H.Y., Lin, T.C., Chiang, L.C., and Lin, C.C. 2005. Acetone, ethanol and methanol extracts of *Phyllanthus urinaria* inhibit HSV-2 infection *in vitro*. *Antiviral Res.* 67: 24–30.

20 Immunomodulatory Activity of Brahma Rasayana, an Herbal Preparation Containing *Phyllanthus emblica* as the Main Ingredient

Praveen K. Vayalil, Ramadasan Kuttan, and Girija Kuttan

CONTENTS

20.1 Immunomodulation by Plants 316
20.2 Immunomodulation in Ayurvedic Treatment 316
20.3 Brahma Rasayana 317
    20.3.1 Immunomodulatory Activity of BR 317
    20.3.2 Antioxidant Activity of BR 318
    20.3.3 Anticlastogenic Activity of BR 319
    20.3.4 Potential Uses of BR in Cancer Treatment 319
        20.3.4.1 Antitumor Action 319
        20.3.4.2 Inhibition of Metastasis 320
        20.3.4.3 Protective Effects of BR against Radiation- and Chemotherapy-Induced Damages 321
        20.3.4.4 Effect in Carcinogenesis 321
        20.3.4.5 Clinical Studies of BR 321
20.4 Conclusion 322
References 322

20.1 IMMUNOMODULATION BY PLANTS

Interactions among immune cells and those of immune cells with other tissue cells are highly complex and not yet completely understood. They are essential to maintain health, and when disturbed, disease may follow. Immunomodulating agents that are free from side effects and can be administered for a long duration, if possible throughout life, for continuous immune activation are highly desirable for the prevention of diseases, both infectious diseases and diseases of the aged, such as cancer. Uses of plants or herbal products to improve defenses against various diseases have a long history. There has been remarkable interest in the past few decades in plants that are capable of modulating the immune system in favor of the host. As a result, several laboratories around the world tested plant extracts, newly identified and existing natural products, for their immunomodulatory effects in both *in vitro* and *in vivo* experimental models, and there were some clinical trials to test their efficacy in humans. In the Indian system of medicine called *Ayurveda*, several plants or herbs, either alone or in combinations, have been used to treat a variety of human diseases. Experimental analysis of many of these herbs demonstrated potent immunomodulatory activity. Some of the well-known plants that are widely used in the clinics are *Withania sominifera, Asparagus recemosus, Phyllanthus emblica, Tinospora cordifolia, Allium sativum,* and others (Govindarajan et al., 2005).

20.2 IMMUNOMODULATION IN AYURVEDIC TREATMENT

The individual herbs used in Ayurveda that have strong immunomodulatory activity mainly fall into a category of the *Materia Medica* of Ayurveda called *Rasayanas*. Rasayana is a branch of Ayurveda. It literally means "the path of juice," which aims to nourish, restore, and balance the body functions. According to Ayurvedic scripts, Rasayana is generally used to rejuvenate the general health of the body to achieve its maximum potential. The restoration of the complete potential within the human body by maintaining or replenishing the "juiciness" is called Rasayana. One of the ways by which rejuvenation may be attained is through medicines. Rasayana medicines (medicinal rejuvenation) utilize a group of individual herbs or medicinal drug preparations that are customarily complex mixtures based mostly on herbal products (see Vayalil et al., 2002a, for a review of Rasayanas). According to classical Ayurvedic texts, Rasayana therapy arrests aging and enhances intelligence, memory, strength, youth, luster, sweetness of voice, and vigor. Rasayanas are aimed at nourishing *Sapta dhadu* or seven elements or structures of the human body (blood, lymph, bone marrow, bone, flesh, adipose tissue, and semen) and thus prevent degenerative changes and illness. Most of these Rasayanas may be consumed regularly as a food for maintaining balanced mental and physical health. They may also be used either alone or along with other modalities of treatment as an adjuvant.

As Rasayanas nourish blood, lymph, and bone marrow, the three major organ systems that regulate production, differentiation, and functioning of immune cells, it was speculated that Rasayanas may modulate the immune system favorably for patients. However, these speculations were not scientifically established until we initiated our research on Rasayanas in the early 1990s. In this review, we present various activities of a Rasayana preparation called *Brahma Rasayana* [BR] that signify its role as an immunomodulating herbal preparation and potential role as an adjuvant in cancer treatment.

20.3 BRAHMA RASAYANA

In the Ayurvedic scripts describing Rasayana, BR has been mentioned as the first and the principal Rasayana preparation. It is a herbal preparation that contains more than 40 herbs, and the main ingredients are *Phyllanthus emblica* and *Terminalia chebula,* which constitute about 20% and 6.7%, respectively, of the total ingredients present. In addition, to obtain a consistency similar to jam, clarified butter, jaggery (concentrated product of cane juice without separation of the molasses and crystals) and honey were also included. Details of the ingredients and preparation have been described elsewhere (Rekha et al., 2001a). It is generally used as a tonic and may be consumed by any age group. It is also prescribed for healthy individuals as a prophylactic medicine as well as to individuals with illness to boost their ability to fight a disease.

BR is nontoxic. In our clinical studies, BR was used at a dose of 50 g/day in three split doses for 1 month. In general, BR showed no significant toxicity or other side effects at this dosage level (Joseph et al., 1999). In animal studies, a dose of up to 500 mg/day for 1 month did not produce any toxic effects (Praveenkumar et al., 1994a).

20.3.1 Immunomodulatory Activity of BR

BR is an excellent immunomodulatory agent (Table 20.1). It is exceptional compared to other known immunomodulators as it activates immune function without altering the basic physiological parameters in experimental animals and in humans (Vayalil et al., 2002b). BR treatment in animals improved the total number of leukocytes, absolute number of polymorphonuclear (PMN) cells, and lymphocytes in the peripheral blood of mice (Praveenkumar et al., 1994a). This increase was within the normal physiological range. BR treatment did not alter other hematological parameters, such as hemoglobin levels or liver or kidney functions (data not published). Interestingly, the functions of various immune cells were activated significantly within 6 days of treatment with BR, such as lymphocyte blastogenesis, ADCC (antibody-dependent cellular cytotoxicity), ACC (antibody-dependent compliment-mediated cytotoxicity), natural killer (NK) cell activity, macrophage-mediated cytotoxicity, and so on (Kumar et al., 1999a, 1999b). These animal studies were further supported by our clinical studies in human volunteers (Joseph et al., 1999). Administration of BR for 10 days did not affect the total leukocyte or absolute counts of various white blood cells (WBCs) or other hematological parameters but significantly enhanced lymphocyte functions, such as lymphocyte blastogenesis in healthy volunteers with a marginal increase in serum cytokines, such as granulocyte-macrophage colony-stimulating factor (GM-CSF). In most cases, to prevent infectious disease or tumor development, a highly active functional immune system with a normal range of immune cells would be sufficient. Therefore, we proposed that BR may be useful as

TABLE 20.1
Immunomodulatory Activity of *Rasayanas*

WBC Counts	Organ Weights
Improved	Increased
• Total leukocyte counts	• Thymus
• Improved absolute lymphocyte counts	• Spleen
• Improved absolute PMNs	

Humoral Immunity	Cellular Immunity
Increased	Increased
• Circulating antibody titer	• Lymphocyte blastogenesis
• Plaque forming cells	• NK-cell activity
• ACC	• ADCC
	• Macrophage-mediated cytotoxicity

Bone Marrow	Cytokines
Enhanced	Increased
• Bone marrow cell proliferation	• IL-2
• CFU-s	• IFN-γ
• α-Naphthyl acetate esterase positive cells	• GM-CSF

a chemoprophylactic medicine for daily use as an inexpensive alternative to vaccines and preventive therapy for immunodeficiency disorders.

Another interesting observation was that treatment of animals with BR enhanced the proliferation of bone marrow cells in culture, and mitogens did not have any effect on the proliferation (Kumar et al., 1999a). There was also an increase in the number of esterase-positive cells, which are specific for cells of monocyte and macrophage lineage (Kumar et al., 1999b). It was demonstrated that BR enhanced bone marrow cellularity as well as increased the number of CFUs (colony-forming units) when bone marrow cells from these BR-treated animals were transferred to irradiation recipient mice (Rekha et al., 2001b). Thus, it may be inferred that Rasayanas have a positive influence on the regulatory systems to activate bone marrow stem cell proliferation and their self-renewal and differentiation.

20.3.2 Antioxidant Activity of BR

BR is a good scavenger of superoxide, hydroxyl, and nitric oxide radicals and inhibitors of lipid peroxide formation *in vitro* (Rekha et al., 2001a). Administration of BR for 5 days consecutively to animals has been demonstrated to inhibit nitric oxide radical production as well as PMA (phorbol myristate acetate)-induced superoxide generation in peritoneal macrophages (Rekha et al., 2001a). In addition, BR inhibited both serum and tissue lipid peroxide formation in animals treated with radiation or cyclophosphamide and serum lipid peroxides in healthy volunteers as well as patients undergoing chemotherapy or radiotherapy (Joseph et al., 1999).

BR imparts antioxidant effects not only by directly scavenging the free radicals or blocking the oxidation of macromolecules but also by enhancing the endogenous antioxidant enzyme systems. Oral administration of BR for 10 or 30 days enhanced the activity of liver antioxidant enzymes such as superoxide dismutase (SOD), catalase (CAT), glutathione-S-transferase (GST), and GPx (glutathione peroxidase) after cyclophosphamide (Rekha et al., 2001b) and radiation treatment (Rekha et al., 2001c). BR also enhanced liver and serum GSH (glutathione). Although Rasayanas are potent inhibitors of free radicals and lipid peroxide formation, their effect on protein oxidation is currently unknown and is an interesting area for further investigation as these are found to be one of the main causes of age-related disorders such as Alzheimer's disease.

20.3.3 Anticlastogenic Activity of BR

Diverse environmental factors, including toxic chemicals such as pesticides and ionizing radiation, constantly induce a wide range of molecular lesions in cells that can lead to a variety of cellular responses, such as inactivation of cells, chromosomal rearrangements and mutations, and subsequent cancer and hereditary diseases. Anticlastogenicity of compounds is generally measured by chromosome aberrations (CAs), sister chromatid exchanges (SCEs), and micronucleus (MN) formation. BR was found to protect mice exposed to ionizing radiation. BR treatment significantly decreased the radiation-induced chromosomal aberrations, such as chromosome breaks, chromatid breaks, and chromatid exchange, in the bone marrow cells. Similarly, BR treatment drastically reduced the number of polychromatic and normochromatic MN in the erythrocytes. These studies implicated the potential of BR as an anticlastogenic agent.

20.3.4 Potential Uses of BR in Cancer Treatment

A variety of biological effects of BR renders it a potential therapy for the prevention of cancer and as an adjuvant for the treatment of cancer. Some of the potential effects of BR against cancer are antitumor action, metastasis inhibition, protection against radiation- and chemotherapy-induced damages, and in carcinogenesis.

20.3.4.1 Antitumor Action

Like many of the immunomodulators, BR is not directly cytotoxic to tumor cells. BR did not induce cytotoxicity to tumor cells in standard *in vitro* cytotoxic assays. However, BR was found to have a significant inhibitory effect on ascites tumor development and solid tumor growth in mice during the treatment period (Praveenkumar et al., 1994a). One of the mechanisms of tumor reduction may be the enhanced immune functions within the host due to a tumor challenge, as BR augmented NK cell activity and ADCC in tumor-bearing animals very early compared to untreated tumor-bearing controls (Praveenkumar et al., 1994b). Cytokines may also play an important role in tumor reduction as BR enhanced the production of interleukin 2 (IL-2), interferon γ (IFN-γ), and GM-CSF (Rekha et al., 2001b). Not only are these cytotoxic to tumor cells but also they activate other immune cells, such as NK and T cells that mediate tumor cell cytotoxicity both *in vivo* in normal and tumor-bearing animals (Praveenkumar et al., 1994b). The other possible mechanism may be that the metabolites of the active components present in BR inhibit or interfere with tumor growth. These studies again emphasized the significance of BR in the prevention and progression of fatal diseases of the aged including cancer. However, further studies are required to disentangle the mechanism of tumor reduction by BR, such as induction of apoptosis, differentiation, and so on.

BR is clinically used and nontoxic and has several activities that are capable of inhibiting initiation, development, and progression of cancer; it is effective when given orally. Therefore, it is worthwhile to assess its cancer chemopreventive efficacy in high-risk groups and in the general population. Further, it is also worth exploring the use of Rasayanas in primary as well as secondary chemoprevention of cancer.

20.3.4.2 Inhibition of Metastasis

Metastasis is the major cause of death of individuals with cancer. Although several mechanisms were proposed to explain the metastatic process, the exact mechanism still remains elusive. Moreover, an effective treatment strategy against metastasis is still not available.

BR has been studied for antimetastatic activity using B16F-10 melanoma cells in C57BL/6 mice (Menon et al., 1997). Oral administration of BR significantly reduced lung tumor nodule formation. Similarly, the lung collagen hydroxyproline content and the serum sialic acid levels were also significantly low in BR-treated animals compared to the untreated controls. The life span of BR-treated animals was found to be significantly increased, suggesting that BR possesses antimetastatic activity against melanoma cells.

Another study evaluated BR for the inhibition of tumor development and prevention of metastasis *in vivo* using Copenhagen rats and a MAT-LyLu prostate cell model system (Gaddipati et al., 2004). BR treatment resulted in a decrease in tumor incidence, a delay in tumor occurrence, lower mean tumor volumes, and significant reduction in tumor weight and lung metastasis in comparison to untreated controls. They proposed that BR may be a potential source of lead chemopreventive compounds and might prove useful for the treatment of disorders such as human prostate cancer. Inhibition of metastasis was independent of the tumor type, which indicates that BR may be effective to bring down the tumor load irrespective of the tumor type.

The effects of BR on angiogenesis and regulation of molecular markers involved in angiogenesis using *in vivo* and *in vitro* models were also studied (Gaddipati et al., 2004). BR treatment showed a significant reduction in factor VIII expression, which indicates reduced angiogenesis. BR treatment also decreased proangiogenic factors like vascular endothelial growth factor (VEGF), matrix metalloproteinase 9 (MMP-9) and MMP-2. BR dose-dependently inhibited the proliferation, tube formation, cell migration, and attachment of HUVECs (human umbilical vascular endothelial cells) on Matrigel. This suggests that one of the mechanisms of inhibition of tumor growth and metastasis might be due to the inhibition of angiogenesis.

20.3.4.3 Protective Effects of BR against Radiation- and Chemotherapy-Induced Damages

Myelosuppression and free radical-mediated tissue damage have been found to be the major drawbacks in such cancer treatments as chemotherapy and radiotherapy. Therefore, the agents that are capable of protecting bone marrow or scavenging free radicals have been considered beneficial as an adjuvant therapy to the standard regimens. To explore the potential of BR as an adjuvant to cancer therapy, we tested whether BR could reduce the leukopenia and maintain bone marrow cellularity in animals exposed to cyclophosphamide and radiation (Rekha et al., 2000, 2001a, 2001b, 2001c; Praveenkumar et al., 1994a; Vayalil et al., 2002a). Our studies clearly demonstrated that oral administration of BR protected mice from leukopenia induced by cyclophosphamide (50 mg/kg) and radiation (400 rads). The bone marrow cellularity was also significantly high in BR-treated animals compared to that of controls. Moreover, BR protected mice from chemotherapy- and radiation-induced loss of body and organ weights. Similarly, we have shown that BR could augment the NK cell activity and ADCC in animals exposed to radiation compared with the untreated controls. The number of nodular colonies on the surface of the spleen on day 7 increased significantly in lethally irradiated recipients receiving bone marrow cells from animals treated with BR. Oral administration of BR also enhanced the serum level of IFN-γ, IL-2, and GM-CSF in normal and irradiated mice.

Administration of BR restored the tissue levels of SOD, CAT, and GSH after whole-body irradiation or chemotherapy. Similarly, elevated levels of serum and liver lipid peroxides after chemotherapy and radiation were also significantly reduced by treatment with BR (Rekha et al., 2000, 2001a, 2001b, 2001c; Praveenkumar et al., 1994a). These studies clearly demonstrated the potential of BR to be used as an adjuvant to current cancer therapies.

20.3.4.4 Effect in Carcinogenesis

Another important finding is that BR inhibited methylcholanthrene-induced sarcoma development in mice (Menon et al., 1996). Oral administration of 50 mg BR/mouse twice a week for 10 weeks inhibited methylcholanthrene-induced carcinogenesis by 80% and increased the survival of these animals drastically. Twelve of 15 animals of the BR-treated group survived the tumor, while all the control animals died within 120 days. These results indicate the potential use of this herbal drug preparation as a chemopreventive agent. However, further studies have to be performed to completely delineate the potential use of BR in cancer prevention and its mechanisms.

20.3.4.5 Clinical Studies of BR

Highly promising results *in vitro* and in animal studies prompted us to conduct a preliminary clinical study of the potential of BR in augmenting the immune function in healthy volunteers and the ability to protect the bone marrow from the toxic effects of chemotherapy and radiation.

Healthy human volunteers were given BR for 1 month at a dose of 50 g/day. The blood was analyzed for the improvement in the WBC counts and lymphocyte blastogenesis as a functional assay for the activation of the immune system both before the start of the treatment and every week thereafter for 1 month. The interesting observation is that treatment of human volunteers with BR did not increase the total number of leukocytes or differential WBC counts significantly. However, it enhanced the lymphocyte blastogenesis, suggesting increased functional activation of the immune system (Joseph et al., 1999).

A similar study was also conducted in cancer patients receiving chemotherapy and radiation (Joseph et al., 1999). Administration of BR accelerated the recovery of the hematopetic system, as seen by a rapid rise in total leukocytes. Both lymphocytes and neutrophils were significantly increased by Rasayana treatment. The nadir of WBCs was significantly higher in treated compared to untreated patients. The nadir of neutrophils was $2,830 \pm 964$ in treated patients and $1,791 \pm 922$ in untreated patients. The nadir of lymphocytes remained almost unchanged. Total number of consecutive days of leukopenia, neutropenia, and lymphopenia was also significantly reduced after the treatment. BR treatment also reduced serum lipid peroxidation, confirming its capacity of reducing oxidative stress induced by cancer treatment.

20.4 CONCLUSION

Emerging evidence shows the potential of BR to prevent cancer and be highly useful as an adjuvant to standard cancer treatment modalities. In addition, many of the activities of BR implicate its beneficial effects against degenerative changes occurring in the body and for maintaining health throughout the life span of the individual. Rasayanas may function by preventing damage to the cellular macromolecules, removing old and damaged or mutant cells, activating cell proliferation in the tissues, and maintaining homeostasis and self-renewal, detoxifying the harmful endogenous and exogenous toxic products, and enhancing the functions of each cellular component of the defense system. However, further studies of the mechanisms at the molecular level are essential, as is identification and characterization of active components required to fully appreciate the beneficial effects of BR.

REFERENCES

Gaddipati, J.P., Rajeshkumar, N.V., Thangapazham, R.L., et al. 2004. Protective effect of a polyherbal preparation, Brahma Rasayana against tumor growth and lung metastasis in rat prostate model system. *J. Exp. Ther. Oncol.* 4: 203–212.

Govindarajan, R., Vijayakumar, M., and Pushpangadan, P. 2005. Antioxidant approach to disease management and the role of "Rasayana" herbs of Ayurveda. *J. Ethnopharmacol.* 99: 165–178.

Joseph, C.D., Praveenkumar, V., Kuttan, G., and Kuttan, R. 1999. Myeloprotective effect of a non-toxic indigenous preparation Rasayana in cancer patients receiving chemotherapy and radiation therapy. A pilot study. *J. Exp. Clin. Cancer Res.* 18: 325–329.

Kumar, V.P., Kuttan, R., and Kuttan, G. 1999a. Effect of "Rasayanas," a herbal drug preparation on cell-mediated immune responses in tumour bearing mice. *Indian J. Exp. Biol.* 37: 23–26.

Kumar, V.P., Kuttan, R., and Kuttan, G. 1999b. Effect of "Rasayanas," a herbal drug preparation on immune responses and its significance in cancer treatment. *Indian J. Exp. Biol.* 37: 27–31.

Menon, L.G., Kuttan, R., and Kuttan, G. 1996. Inhibition of chemical induced carcinogenesis by Rasayana—an indigenous herbal preparation. *J. Exp. Clin. Cancer Res.* 15: 241–243.

Menon, L.G., Kuttan, R., and Kuttan, G. 1997. Effect of Rasayanas in the inhibition of lung metastasis induced by B16F-10 melanoma cells. *J. Exp. Clin. Cancer Res.* 16: 365–368.

Praveenkumar, V., Kuttan, R., and Kuttan, G. 1994a. Chemoprotective action of Rasayanas against cyclosphamide toxicity. *Tumori* 80: 306–308.

Praveenkumar, V., Kuttan, R., and Kuttan, G. 1994b. Effect of Rasayanas on normal and tumor-bearing animals. *J. Exp. Clin. Cancer Res.* 13: 67–70.

Rekha, P.S., Kuttan, G., and Kuttan, R. 2000. Effect of herbal preparation, Brahma Rasayana, in amelioration of radiation induced damage. *Indian J. Exp. Biol.* 38: 999–1002.

Rekha, P.S., Kuttan, G., and Kuttan, R. 2001a. Antioxidant activity of Brahma Rasayana. *Indian J. Exp. Biol.* 39: 447–452.

Rekha, P.S., Kuttan, G., and Kuttan, R. 2001b. Effect of Brahma Rasayana on antioxidant systems and cytokine levels in mice during cyclophosphamide administration. *J. Exp. Clin. Cancer Res.* 20: 219–223.

Rekha, P.S., Kuttan, G., and Kuttan, R. 2001c. Effect of Brahma Rasayana on antioxidant system after radiation. *Indian J. Exp. Biol.* 39: 1173–1175.

Thangapazham, R.L., Sharma, A., Gaddipati, J.P., Singh, A.K., and Maheshwari, R.K. 2006. Inhibition of tumor angiogenesis by Brahma Rasayana (BR). *J. Exp. Ther. Oncol.* 6: 13–21.

Vayalil, P.K., Kuttan, G., and Kuttan, R. 2002a. Protective effects of Rasayanas on cyclophosphamide- and radiation-induced damage. *J. Altern. Complement. Med.* 8: 787–796.

Vayalil, P.K., Kuttan, G., and Kuttan, R. 2002b. Rasayanas: evidence for the concept of prevention of diseases. *Am. J. Chinese Med.* 30: 155–171.

21 Triphala
An Ayurvedic Drug Formulation

Sandhya T. Das and K. P. Mishra

CONTENTS

21.1 Introduction 325
21.2 Beneficial Effects of Triphala 326
      21.2.1 Chondroprotectve and Antiarthritic Activity 326
      21.2.2 Antimicrobial and Anti-inflammatory Activity 326
      21.2.3 Antidiabetic Activity 326
      21.2.4 Antioxidant Activity 327
      21.2.5 Triphala in Cancer Therapy 327
References 329

21.1 INTRODUCTION

Triphala (TPL) is a herbal formulation consisting of the dried and powdered fruits of three plants—*Terminalia chebula, Emblica officinalis,* and *Terminalia bellerica*—in equal proportions (*Triphala* is a Sanskrit word that means three [*tri*] fruits [*phala*]). It is an important medicine of the "rasayana" group and is believed to promote health, immunity, and longevity (Jagetia et al., 2002). TPL has long been known in India to be an excellent laxative and to cure several digestive problems and other ailments of the intestine. It even seems to work as an effective antidiarrheal agent (Biradar et al., 2007). This formulation, rich in antioxidants, is a frequently used ayurvedic medicine to treat many diseases, such as anemia, jaundice, constipation, asthma, fever, and chronic ulcers. Most people practicing ayurvedic medicine consume TPL as a "health tonic." These properties of TPL that would have been otherwise just considered folklore are now being confirmed by scientists using the latest available technology. Furthermore, TPL is also under investigation for newer health effects.

21.2 BENEFICIAL EFFECTS OF TRIPHALA

21.2.1 CHONDROPROTECTVE AND ANTIARTHRITIC ACTIVITY

Formulations involving TPL, such as TPL guggulu and individual components of TPL (specifically *T. chebula*) have been shown to be chondroprotective, with significant inhibitory effects on hyaluronidase and collagenase, and hence could be used in developing medications for arthritis (Sumantran et al., 2007). More confirmation of its antiarthritic effects came from a study that found that administration reduced inflammation and alleviated several arthritic parameters in experimentally induced arthritis in mice (Rasool and Sabina, 2007).

21.2.2 ANTIMICROBIAL AND ANTI-INFLAMMATORY ACTIVITY

Easing inflammation seems to be a common mechanism by which TPL exerts its medicinal properties. Application of TPL ointment reduced inflammation and aided wound healing in a rat infection model. This study also emphasized the antimicrobial properties of TPL as it significantly reduced the bacterial load in the infected wounds in this model (Kumar et al., 2010). Other studies have also reiterated the antimicrobial properties of TPL. One study showed that TPL was capable of completely inhibiting bacterial growth in root canals of extracted human teeth (Prabhakar et al., 2010). Particularly interesting is a study that looked at the effects of aqueous and ethanol extracts of TPL and its individual components on bacterial isolates from HIV-infected patients. Some of the microbial agents tested were less susceptible to regular antibiotics but were inhibited by TPL as well as its individual components, especially *T. cehbula* (Srikumar et al., 2007). Others have shown that *T. belerica*, another constituent of TPL, effectively protected mice from experimental salmonellosis-induced death (Madani and Jain, 2008).

The attractiveness of this drug as a therapeutic agent for myriad disease conditions stems from the fact that it was found to be nongenotoxic in *in vitro* assays. Even *in vivo*, large doses of TPL were found to be nontoxic to mice given intraperitoneal injections of the drug. The intraperitoneal LD_{50} (dose lethal to 50% of a population) in this same study was found to be 280 mg/kg body weight, and the mice tolerated doses of 240 mg/kg body weight without reported adverse effects. The general health effects of TPL could also come from the fact it is capable of improving the body lipid profile by reducing cholesterol.

21.2.3 ANTIDIABETIC ACTIVITY

Hypercholesteremic rats pretreated with cholesterol showed a marked decrease in low, very low, and high-density lipoproteins as well as reduced free fatty acids (Saravanan et al., 2007). Since there is a profound association between hypercholesteremia and diabetes, it is only logical to wonder if TPL could help in the management of diabetes. In fact, that seems to be the case, with one study pointing out that oral administration of TPL extract to diabetic rats reduced their blood sugar levels significantly within 4 h (Sabu and Kuttan, 2002). This study further pointed out that the medicinal effects of TPL could well be because of its antioxidant property since free radicals have been implicated in several chronic diseases, including diabetes, heart disease, and cancer.

21.2.4 ANTIOXIDANT ACTIVITY

The discussion of TPL has to include how this drug exerts its various effects presumably by using its antioxidant properties. One study exploring the radical scavenging and antioxidant properties of TPL and its individual constituents *in vitro* revealed that *Embilica officinalis* was the component with the most antioxidant properties, followed by *Terminalia chebula* and *Terminalia Belerica*. All three components and TPL effectively inhibited γ-radiation-induced strand breaks in plasmid DNA (Naik et al., 2005). *In vivo*, *Terminalia chebula* effectively modulated the activity of several antioxidant enzymes in the liver and kidney of aged rats and significantly inhibited oxidative stress and age-induced damage (Mahesh et al., 2009).

Ayurvedic herbal formulations, including TPL, that have known antioxidant constituents have been known to have excellent free radical scavenging properties (Jagetia et al., 2004b). The ability of TPL to scavenge free radicals and act as an antioxidant could be the reason for its ability to protect mice from radiation-induced DNA damage and mortality, as has been reported by several studies (Jagetia et al., 2002, 2004a; Sandhya et al., 2006b).

21.2.5 TRIPHALA IN CANCER THERAPY

Studies have shown that intraperitoneal injection of TPL protected mice from whole-body γ-radiation-induced mortality. Intraperitoneal injection is not a convenient method of drug administration for practical purposes. More recent investigations have revealed that TPL protected mice significantly from whole-body γ-radiation when administered orally. This study further showed that TPL significantly prevented DNA damage induced by radiation using alkaline single-cell gel electrophoresis, which is a known highly sensitive technique to measure single-strand breaks. In addition, it was found that TPL modulated the activities of xanthine oxidase (XO) and superoxide dismutase (SOD), two important enzymes known to be affected by radiation exposure. In whole-body-irradiated mice, the activity of the free radical-producing enzyme XO was significantly increased, whereas that of the antioxidant enzyme SOD was greatly decreased. However, in whole-body-irradiated, TPL-fed mice, the alterations in these enzymes were reversed to near control, suggesting a role of TPL in prevention of radiation-induced changes.

TPL thus exerts its radioprotective abilities by modulating the activities of radiation-sensitive enzymes and protecting DNA from radiation-induced damage. These findings may prove useful in developing a TPL-based radioprotection regime. However, further research needs to be done to optimize the doses needed for application in clinical and other situations of radiation exposure.

Another medicinal property of TPL has been slowly emerging: the anticarcinogenic effects of TPL and its constituents. There have been some studies suggesting anticlastogenic and antitumor effects of *E. officinalis* (Nandi et al., 1997; Jose et al., 2001) and antiproliferative effect of *T. chebula* (Saleem et al., 2002).

Chemotherapeutic agents generally exhibit several side effects, and research is on a high-priority search to find anticancer drugs with minimal toxicity to normal cells. Due to growing research interest in the study of anticancer properties of several dietary phytochemicals, there is a greater realization of the role of natural compounds in diseases. Naturally occurring phytochemicals are widely used in the traditional Indian medicinal system of ayurveda for treatment of a variety of diseases. Some of the plant products have shown a remarkable antitumor property against various human cancers (Kawamori et al., 1999; Choi et al., 2001; Rege et al., 1999). However, detailed mechanisms of action of these phytochemicals at cellular and molecular levels are only now being elucidated (Yang et al., 2000; Choudhuri et al., 2005). However, there have been serious efforts at understanding the mechanisms of regulation of cell death and cell proliferation by phytochemicals that are central to developing these as drugs for more effective treatment protocols for cancer.

Experimental evidence is now available on TPL being effective in killing breast cancer cells *in vitro*, and it has been shown to significantly inhibit transplanted mouse tumors *in vivo* (Sandhya et al., 2006a). TPL has been shown to inhibit xenograft growth of human pancreatic tumor cells (Shi et al., 2008). Several groups are actively involved in understanding the scientific basis of the mode of action of TPL, and a detailed evaluation of the anticancer properties of TPL is under way in a variety of cancer types. TPL thus is emerging as a potential anticancer drug.

Investigations into the mechanisms of the effects of TPL on a breast cancer cell line (MCF 7) revealed that TPL was capable of killing tumor cells by apoptosis. However, TPL was found to be nontoxic to several normal cells of human origin, such as MCF-10 F, and cells of mouse origin, such as spleenocytes and hepatocytes, at similar concentrations. The observed cytotoxicity was found to correlate with increased apoptosis, generation of intracellular reactive oxygen species (ROS), and decrease in mitochondrial membrane potential, suggesting the involvement of an ROS-mediated apoptotic pathway in TPL-induced toxicity in tumor cells. However, TPL did not induce substantial ROS generation in the normal cells investigated. Thus, the differential effect of TPL on normal and tumor cells seems to be related to its ability to evoke a differential response in intracellular ROS generation *in vitro*.

Oral feeding of animals with a water extract of TPL resulted in a significant decrease in tumor growth in terms of tumor dimension. TUNEL (terminal deoxynucleotidyl transferase dUTP nickend labeling) assay results of excised tumor tissue indicated that induction of apoptosis was the most likely mechanism of TPL-induced tumor cell death *in vivo*.

In view of the indications that TPL acted as a prooxidant in tumor cells, investigations were carried out on the effect of TPL on tumor cells with different p53 gene status as it is known to be a redox-sensitive gene. Results from this study showed that breast cancer cell lines MCF 7 and T 47 D cells, differing in their p53 status, exhibited differential sensitivity to TPL. Since T 47 D (p53-mutated) cells were less sensitive to TPL compared to MCF 7 cells (p53 wild type), it was concluded that the p53 status of cancer cells formed an important factor in predicting the response of cancer cells to prooxidant drugs (Sandhya et al., 2006).

TPL has also been shown to be effective in inducing apoptosis in the human pancreatic cell line Capan-2 *in vitro* (Shi et al., 2008). Orally administered TPL inhibited the growth of Capan-2 xenografts in hairless mice. This study further showed that the growth inhibitory effects of TPL both *in vitro* and *in vivo* were mediated by the activation of ERK (extracellular signal regulated kinase) and p53. On the other hand, TPL failed to induce apoptosis or activate ERK or p53 in normal human pancreatic ductal epithelial cells.

There is a great need for developing drugs for tumor radiosensitization with minimal or no toxicity to normal cells and tissues. Since TPL was found to be nontoxic to normal cells and was found to protect mice from radiation-induced mortality, experiments were conducted to evaluate the combined effect of TPL and γ-irradiation on tumor cells. Data showed that postirradiation treatment with TPL remarkably sensitized human breast cancer cells MCF 7 to γ-radiation, whereas pretreatment of MCF 7 cells with TPL antagonized the cytotoxic effects of radiation *in vitro* (Das and Mishra, unpublished data). The combination treatment of MCF 7 cells with γ-radiation enhanced apoptosis compared to the individual treatments, probably involving an ROS-dependent mechanism. *In vivo*, combination treatment of TPL and γ-irradiation produced greater tumor growth delay in transplantable thymic lymphoma solid tumors compared to individual treatments. However, TPL and γ-radiation treatment of radio-resistant fibrosarcoma tumor showed no reduction in tumor growth. Results suggest that neither TPL nor radiation, either alone or in combination, was effective in inducing tumor growth delay. Thus, the ability of TPL to enhance radiation-induced toxicity in tumor cells and its ability to protect mice and normal cells from radiation may prove it is an ideal candidate with potential as a radiosensitizer for cancer radiotherapy in the clinic.

American Herbal Pharmacopoeia: Botanical Pharmacognosy – Microscopic Characterization of Botanical Medicines, edited by Roy Upton et al., records that this formulation has been in continuous use for 2300 years and provides a color photograph of the three whole fruits arranged together (p. xlvi). The book also describes the microscopical characters of the fruits and their seeds with drawings and photographs: *Phyllanthus embica* (pp. 535–538), *Terminalia bellerica* (pp. 646–649), and *T. chebula* (pp. 650–652).

REFERENCES

Biradar, Y.S., Singh, R., Sharma, K., et al. 2007. Evaluation of anti-diarrhoeal property and acute toxicity of TPL Mashi, an Ayurvedic formulation. *J. Herb. Pharmacother.* 7: 203–212.

Choi, J.A., Kim, J.Y., Lee, J.Y., et al. 2001. Induction of cell cycle arrest and apoptosis in human breast cancer cells by quercetin. *Int. J. Oncol.* 19: 837–844.

Choudhuri, T., Pal, S., Das, T., and Sa, G. 2005. Curcumin selectively induces apoptosis in deregulated cyclin D1-expressed cells at G2 phase of cell cycle in a p53-dependent manner. *J. Biol. Chem.* 280: 20059–20068.

Jagetia, G.C., Baliga, M.S., Malagi, K.J., and Kamath, S.M. 2002. The evaluation of the radioprotective effect of Triphala (an ayurvedic rejuvenating drug) in the mice exposed to gamma-radiation. *Phytomedicine* 9: 99–108.

Jagetia, G.C., Malagi, K.J., Baliga, M.S., Venkatesh, P., and Veruva, R.R. 2004a. Triphala, an ayurvedic rasayana drug, protects mice against radiation-induced lethality by free-radical scavenging. *J. Altern. Complement. Med.* 10: 971–978.

Jagetia, G.C., Rao, S.K., Baliga, M.S., and Babu, K. 2004b. The evaluation of nitric oxide scavenging activity of certain herbal formulations *in vitro*: a preliminary study. *Phytother. Res.* 18: 561–565.

Jose, J.K., Kuttan, G., and Kuttan, R. 2001. Antitumour activity of *Emblica officinalis*. *J. Ethnopharmacol.* 75: 65–69.
Kawamori, T., Lubet, R., Steele, V.E., et al. 1999. Chemopreventive effect of curcumin, a naturally occurring anti-inflammatory agent, during the promotion/progression stages of colon cancer. *Cancer Res.* 59: 597–601.
Kumar, M.S., Kirubanandan, S., Sripriya, R., and Sehgal, P.K. 2010. Triphala incorporated collagen sponge—a smart biomaterial for infected dermal wound healing. *J. Surg. Res.* 158: 162–170.
Madani, A., and Jain, S.K. 2008. Anti-salmonella activity of *Terminalia belerica*: *in vitro* and *in vivo* studies. *Indian J. Exp. Biol.* 46: 817–821.
Mahesh, R., Bhuvana, S., and Begum, V.M. 2009. Effect of *Terminalia chebula* aqueous extract on oxidative stress and antioxidant status in the liver and kidney of young and aged rats. *Cell Biochem. Funct.* 27: 358–363.
Naik, G.H., Priyadarsini, K.I., Bhagirathi, R.G., et al. 2005. *In vitro* antioxidant studies and free radical reactions of Triphala, an ayurvedic formulation and its constituents. *Phytother. Res.* 19: 582–586.
Nandi, P., Talukder, G., and Sharma, A. 1997. Dietary chemoprevention of clastogenic effects of 3,4-benzo(a)pyrene by *Emblica officinalis* Gaertn. fruit extract. *Br. J. Cancer* 76: 1279–1283.
Prabhakar, J., Senthilkumar, M., Priya, M.S., et al. 2010. Evaluation of antimicrobial efficacy of herbal alternatives (Triphala and green tea polyphenols), MTAD, and 5% sodium hypochlorite against *Enterococcus faecalis* biofilm formed on tooth substrate: an *in vitro* study. *J. Endod.* 36: 83–86.
Rasool, M., and Sabina, E.P. 2007. Antiinflammatory effect of the Indian ayurvedic herbal formulation Triphala on adjuvant-induced arthritis in mice. *Phytother. Res.* 21: 889–894.
Rege, N.N., Thatte, U.M., and Dahanukar, S.A. 1999. Adaptogenic properties of six rasayana herbs used in ayurvedic medicine. *Phytother. Res.* 13: 275–291.
Sabu, M.C., and Kuttan, R. 2002. Anti-diabetic activity of medicinal plants and its relationship with their antioxidant property. *J. Ethnopharmacol.* 81: 155–160.
Saleem, A., Husheem, M., Härkönen, P., and Pihlaja, K. 2002. Inhibition of cancer cell growth by crude extract and the phenolics of *Terminalia chebula* retz. fruit. *J. Ethnopharmacol.* 81: 327–336.
Sandhya, T., and Mishra, K.P. 2006. Cytotoxic response of breast cancer cell lines, MCF 7 and T 47 D to Triphala and its modification by antioxidants. *Cancer Lett.* 238: 304–313.
Sandhya, T., Lathika, K.M., Pandey, B.N., and Mishra, K.P. 2006a. Potential of traditional ayurvedic formulation, Triphala, as a novel anticancer drug. *Cancer Lett.* 231: 206–214.
Sandhya, T., Lathika, K.M., Pandey, B.N., et al. 2006b. Protection against radiation oxidative damage in mice by Triphala. *Mutat. Res.* 609: 17–25.
Saravanan, S., Srikumar, R., Manikandan, S., Jeya Parthasarathy, N., and Sheela Devi, R. 2007. Hypolipidemic effect of Triphala in experimentally induced hypercholesteremic rats. *Yakugaku Zasshi*.127: 385–388.
Shi, Y., Sahu, R.P., and Srivastava, S.K. 2008. Triphala inhibits both *in vitro* and *in vivo* xenograft growth of pancreatic tumor cells by inducing apoptosis. *BMC Cancer* 8: 294.
Srikumar, R., Parthasarathy, N.J., Shankar, E.M., et al. 2007. Evaluation of the growth inhibitory activities of Triphala against common bacterial isolates from HIV infected patients. *Phytother. Res.* 21: 476–480.
Sumantran, V.N., Kulkarni, A.A., Harsulkar, A., et al. 2007. Hyaluronidase and collagenase inhibitory activities of the herbal formulation Triphala guggulu. *J. Biosci.* 32: 755–761.
Upton, Roy et al. [Eds.]. 2011. *American herbal pharmacopoeia: Botanical pharmacognosy – microscopic characterization of botanical medicines*. Boca Raton, FL: CRC Press.
Yang, G.Y., Liao, J., Li, C., et al. 2000. Effect of black and green tea polyphenols on c-jun phosphorylation and $H(2)O(2)$ production in transformed and non-transformed human bronchial cell lines: possible mechanisms of cell growth inhibition and apoptosis induction. *Carcinogenesis* 21: 2035–2039.

22 Kalpaamruthaa
A Successful Drug against Various Ailments

P. Sachdanandam and P. Shanthi

CONTENTS

22.1 Introduction 332
22.2 Major Ingredients of KA 332
    22.2.1 *Semecarpus anacardium* 332
        22.2.1.1 Uses of *Semecarpus anacardium* 332
        22.2.1.2 Phytochemical Analysis of SA 333
        22.2.1.3 Toxicity Studies of the Drug SA 333
        22.2.1.4 Therapeutic Potential of SA 333
    22.2.2 *Emblica officinalis* 334
        22.2.2.1 Phytochemistry of EO 334
        22.2.2.2 Therapeutic Potential of EO 334
    22.2.3 Honey 334
        22.2.3.1 Phytochemistry of Honey 334
        22.2.3.2 Therapeutic Potential of Honey 334
    22.2.4 Flavonoids and Ascorbic Acid 334
    22.2.5 Conclusion 335
22.3 Kalpaamruthaa 335
    22.3.1 Phytochemical Analysis of the Drug KA 335
        22.3.1.1 HPTLC Analysis of Flavonoids 336
        22.3.1.2 HPTLC Analysis of Gallic Acid and Other Tannins 336
    22.3.2 Toxicity Studies of the Drug KA 336
    22.3.3 Therapeutic Potential of the Drug KA 340
        22.3.3.1 With Respect to Arthritis 340
        22.3.3.2 With Respect to Cancer 342
Acknowledgments 344
References 344

22.1 INTRODUCTION

Traditional Indian medicine has been used for pharmaceutical and dietary therapy for several millennia. A number of Indian medicinal herbs and many relevant prescriptions have been screened and used for treatment and preventing various types of ailments during long-term folk practice. Among these systems, choosing a correct combination is an important goal in any therapy. In this way, we have prepared a modified indigenous preparation named Kalpaamruthaa (KA). It is a modified indigenous preparation prepared in our laboratory and contains *Semecarpus anacardium* nut milk extract (SA) and dried *Phyllanthus emblica* (PE) fruit as major constituents. Before going to the details of KA, a summary description of these plants is given.

22.2 MAJOR INGREDIENTS OF KA

22.2.1 S*emecarpus anacardium*

Semecarpus anacardium Linn. (Anacardiaceae), commonly known as "marking nut," has high priority and applicability in the Siddha indigenous system of medicine against various ailments. It is well known for its medicinal value in Ayurveda, Siddha, and even homeopathy. Ayurveda describes *Semecarpus anacardium* as a potent drug against a variety of ailments, and it is popularly known as "Ardha Vaidhya" (Nadkarni and Nadkarni, 1976). The fruit is acrid, hot, sweetish, edible, aphrodisiac, and antihelminthic; it removes ascites, heals ulcers, strengthens the teeth, and is useful in insanity and asthma. Purified fruits are claimed to possess rejuvenating properties, increasing longevity, bringing a glow to the face, sweetness in tone, and improved vision (Gil et al., 1995).

22.2.1.1 Uses of *Semecarpus anacardium*

SA is used both internally and externally. The fruits, their oil, and the seeds have great medicinal value and are used to treat a wide range of diseases. Externally, the oil, mixed with coconut or sesame oil, is applied on wounds and sores to prevent pus formation. The topical application of its oil on swollen joints and traumatic wounds effectively controls pain. In glandular swellings and filariasis, the application of its oil facilitates draining the discharge of pus and fluids and eases the conditions.

Internally, SA is widely used in a vast range of diseases because of its multifarious properties. As it augments the agni (fire), it is extremely beneficial in diseases like hemorrhoids, colitis, diarrhea, dyspepsia, ascites, tumors, and worms, which are caused mainly due to weakened agni.

SA is the best rejuvenative (Rasayana) for skin ailments and vata disorders (vata is one of the three bodily humors) and as a preventive measure to increase body resistance. It augments the appetite, improves digestion, eliminates ama (inadequate digestion), and clears up srotasas—the microchannels of all the systems, hence facilitating the nourishment of all the tissues (dhatus).

22.2.1.2 Phytochemical Analysis of SA

Phytochemical studies of the Siddha preparation of SA from our laboratory, reported by Vijayalakshmi et al. (1996), was found to contain phenols, flavonoids, and carbohydrates. Thin-layer chromatographic (TLC), high-performance liquid chromatographic (HPLC), and high-performance thin-layer chromatographic (HPTLC) analysis of the nut and milk extract confirmed the presence of the compounds mentioned (Sahoo et al., 2008; Aravind et al., 2008; Shin et al., 1999; Nair et al., 2009; Mythilipriya et al., 2008a). Further analysis revealed the presence of iron, copper, sodium, calcium, and aluminum in traces. On the basis of chemical and spectral data, several biflavonoids have been characterized: semecarpuflavanone (Murthy, 1983); jeediflavanone (Murthy, 1985); and anacarduflavanone (Murthy, 1992), the first biflavanone to occur with a methylenedioxy group. Phytochemical examination (Gulati and Ohiman, 1984) revealed 3.68% total ash, 0.33% acid-insoluble ash, 11.27% alcohol-soluble extract, 11.84% water-soluble extract, and 12.71% moisture content in *Semecarpus anacardium* nuts.

22.2.1.3 Toxicity Studies of the Drug SA

In an acute toxicity study, the nut extract preparation does not show any detrimental, toxic external symptoms or mortality up to a dose of 2,000 mg/kg body weight in laboratory animals (Vijayalakshmi et al., 2000). The hematological picture does not show any significant adverse change, and histopathological studies carried out in vital organs did not reveal any significant pathological lesions, even when administered at a dose of 1,000 mg/kg body weight (Ghosh et al., 1981). The crude extracts were found to be very toxic. The LD_{50} (median lethal dose) of the chloroform extract of the *Semecarpus anacardium* nut was found to be 230 mg/kg body weight (Kesava Rao et al., 1979).

22.2.1.4 Therapeutic Potential of SA

Many pharmacological properties, such as antimicrobial, anti-inflammatory, anti-amebic (Patwardhan et al., 1988), antirheumatic (Vijayalakshmi et al., 1997), and anticancer (Sujatha and Sachdanandam, 2002) properties, have been attributed to the nuts of the *Semecarpus anacardium* plant. Various diseases, such as neurological disorders, cardiac troubles, enlargement of spleen, leprosy, and leukoderma (Kurup et al., 1979; Raghunath and Mitra, 1982; Sivarajan and Indira, 1994), have been effectively cured by *Semecarpus anacardium* nut extracts. The fruits and oil have been claimed to be highly efficacious in the treatment of neuritis (Charaka, 1941) and venereal disorders (Sushruta Samhita, 1938).

The milk extract of the nut used in our laboratory was found to be protective in adjuvant-induced arthritis (Ramprasath et al., 2005) and exhibited antitumor activity against various types of experimental tumors, such as breast cancer (Arathi and Sachdanandam, 2003; Premalatha and Sachdanandam, 1997); liver cancer (Premalatha and Sachdanandam, 1999); and leukemia (Sugapriya et al., 2008) and in breast cancer cell lines (Mathivadhani et al., 2007). The extract also possesses antioxidant, anti-inflammatory, hypoglycemic (Arulkumaran et al., 2007), antiarthritic, and anticancer activities (Nair et al., 2009). Selvam and Jachak (2004) reported anti-inflammatory activity and cyclo-oxygenase inhibitory properties.

22.2.2 EMBLICA OFFICINALIS

Emblica officinalis Gaertn (*Phyllanthus emblica* L.) (family Euphorbiaceae) (EO), popularly known as Amla, is a common household remedy used in the Indian indigenous system of medicine against several ailments.

22.2.2.1 Phytochemistry of EO

The fruits of EO are rich in tannins. The fruits have 28% of the total tannins distributed in the whole plant. The fruit contains two hydrolyzable tannins, emblicanin A and B, which have antioxidant properties; on hydrolysis one gives gallic acid, ellagic acid, and glucose, and the other gives ellagic acid and glucose. The fruit also contains phyllemblin 1 (Ghosal, 1996; *Wealth of Asia*, 1998; *Dictionary of Indian Medicinal Plants*, 1988).

22.2.2.2 Therapeutic Potential of EO

The EO fruits have been reported to possess expectorant, purgative, spasmolytic, antibacterial, hypoglycemic (Jamwal et al., 1959; Jayashri and Jolly, 1993), hepatoprotective (Achilya et al., 2004; Jose and Kuttan, 2000), and hypolipidemic (Thakur and Mandal, 1984) activity. The aqueous extract has been reported to have antipyretic, laxative, and tonic properties and showed antibacterial activity (Vinayagamoothy, 1982).

22.2.3 HONEY

Honey, a most assimilable carbohydrate compound, is a singularly acceptable, practical, and most effective remedy to generate heat, create and replace energy, and to form certain tissues (http://www.benefits-of-honey.com/health-benefits-of-honey.html).

22.2.3.1 Phytochemistry of Honey

Honey is a supersaturated solution of sugars; its main contributors are fructose (38%) and glucose (31%). It also has minor constituents, such as flavonoids and phenolic acids like hydroxycinnamates (caffeic, p-coumaric, and ferulic acids) (Cherchi et al., 1994); certain enzymes (glucose oxidase, catalase); ascorbic acid (White, 1975); organic acids (Cherchi et al., 1994); amino acids and proteins (White and Rudyj, 1978). Flavonoids—pinobanksin, pinocembrin, quercetin, chrysin, galangin, luteolin, and kaempferol were reported in honey (Gheldof and Engeseth, 2002; National Honey Board, 2002).

22.2.3.2 Therapeutic Potential of Honey

The medicinal properties of honey have been exploited for a variety of medicinal and nutritional purposes (Nagai et al., 2001). Therefore, it is usually mixed with medicinal preparations (Khare, 2004). Antiviral, antifungal, antibacterial, and antiseptic properties have also been found in honey (Ensminger and Ensminger, 1986).

22.2.4 FLAVONOIDS AND ASCORBIC ACID

Interaction between flavonoids and ascorbic acid has been well documented. Ascorbate is reported to have flavonoid-protective and flavonoid-enhancing activities

(Sorata et al., 1988; Chen et al., 2004). In turn, vitamin present in suboptimal concentration is stabilized by flavonoids (Satoh et al., 1999). Flavonoids and vitamins coexist as active substances in some fruits and vegetables. By virtue of the strong antioxidant characteristics, they portray a wide spectrum of biological activities (Cai et al., 2004).

22.2.5 CONCLUSION

Thus, these facts indicate that combining *Semecarpus anacardium* with the water-soluble antioxidant vitamin C could be a useful strategy to improve the activity of the drug. To promote intellect, prevent senility, and impart longevity, honey was also added to this preparation to improve the efficacy of each ingredient to make the combination effective against a variety of ailments.

22.3 KALPAAMRUTHAA

22.3.1 PHYTOCHEMICAL ANALYSIS OF THE DRUG KA

Phytochemical studies carried out in our laboratory revealed the presence of flavonoids, sterols, triterpenoids, carbohydrates, and proteins in KA. The same study was carried out in SA as well as EO, which were used as raw materials in the formulation.

Also, drug analysis of KA showed the presence of other components, such as minerals (Fe, Ca, Mg, P, Na, K and Zn); vitamins (ascorbic acid, thiamine, riboflavin, niacin, pyridoxine, cyanocobalamine, and folic acid); and fiber (Table 22.1).

TABLE 22.1
Preliminary Photochemical Analysis of SA, EO and KA

Tests	SA	EO	KA
Carbohydrates			
Molisch's test	+	+	+
Proteins and amino acids			
Millon's and Biuret tests	+	+	+
Alkaloids			
Mayer's, Wager's, and Dragendroff's tests	−	+	+
Steroids/triterpenoids			
Liberman-Burchard's test	+	+	+
Polyphenols (flavonoids and tannins)			
Ferric chloride and lead acetate	+	+	+
Saponins			
Foam test	−	+	+

22.3.1.1 HPTLC Analysis of Flavonoids

HPTLC analysis of KA, SA, and EO was carried out to prove the presence of flavonoids in the raw materials as well as in the formulation. Scanning of SA and *Semecarpus anacardium* nut at 254 nm revealed the presence of 10 well-resolved peaks with Rf values of 0.09, 0.12, 0.26, 0.40, 0.50, 0.56, 0.60, 0.69, 0.83, and 0.92, which almost matched in KA, with a lesser difference in the Rf values. Also, scanning *Semecarpus anacardium* nut and SA at 550 nm revealed the presence of 10 and 12 well-resolved peaks, respectively. Scanning of EO also revealed the peaks at 254 nm in the Rf values 0.22, 0.37, 0.47, 0.59, 0.69, and 0.84 and four peaks at 550 nm in the Rf values 0.22, 0.37, 0.47, 0.71 (Figure 22.1a). All the ferric chloride reagent-positive phenolic constituents were present in the KA formulation. These findings prove the presence of flavonoids in the raw materials as well as in the formulation. Also, the analysis standardizes the KA formulation for its phytoconstituents by comparing with its constituents SA and EO (Figure 22.1b).

22.3.1.2 HPTLC Analysis of Gallic Acid and Other Tannins

HPTLC analysis to test the presence of tannins and gallic acid in EO, SA, and KA was done. The profile of EO and KA revealed the presence of gallic acid in comparison with the standard gallic acid, and the Rf value was 0.55 at both ultraviolet (UV) 254- and 550-nm detection; SA did not show the presence of gallic acid. Apart from gallic acid, two more tannins were observed at Rf values 0.35 and 0.74, which were observed at both UV 254 and 550 nm. Thus, by HPTLC analysis, the presence of gallic acid and other tannins in EO and the presence of phenolic compounds in SA and *Semecarpus anacardium* nut were standardized in the KA formulation (Figure 22.1c).

22.3.2 Toxicity Studies of the Drug KA

KA was evaluated for its behavioral and toxicological effects and its consequence on biochemical and histological variations. Acute and subacute toxicity studies were done on Wistar albino rats. During an acute toxicity study (72 h), there were no adverse effects found in the general behavior and mortality at any given dose level (50–2,000 mg/kg body weight). In a subacute toxicity study (30 days), KA (50, 100, 250, and 500 mg/kg body weight) did not impart any gross changes in hematological and biochemical parameters with the exception of a transient rise in hemoglobin, leukocyte count, free fatty acid, and plasma and urine creatinine and a significant decrease in blood glucose, total cholesterol, triglycerides, and phospholipid levels. The changes observed were significant only at the highest dosage of 500 mg/kg body weight. Further, histopathological examination of vital organs showed normal architecture, suggesting no morphological disturbances; hence, KA was considered to be safe and nontoxic (Mythilipriya et al., 2007b).

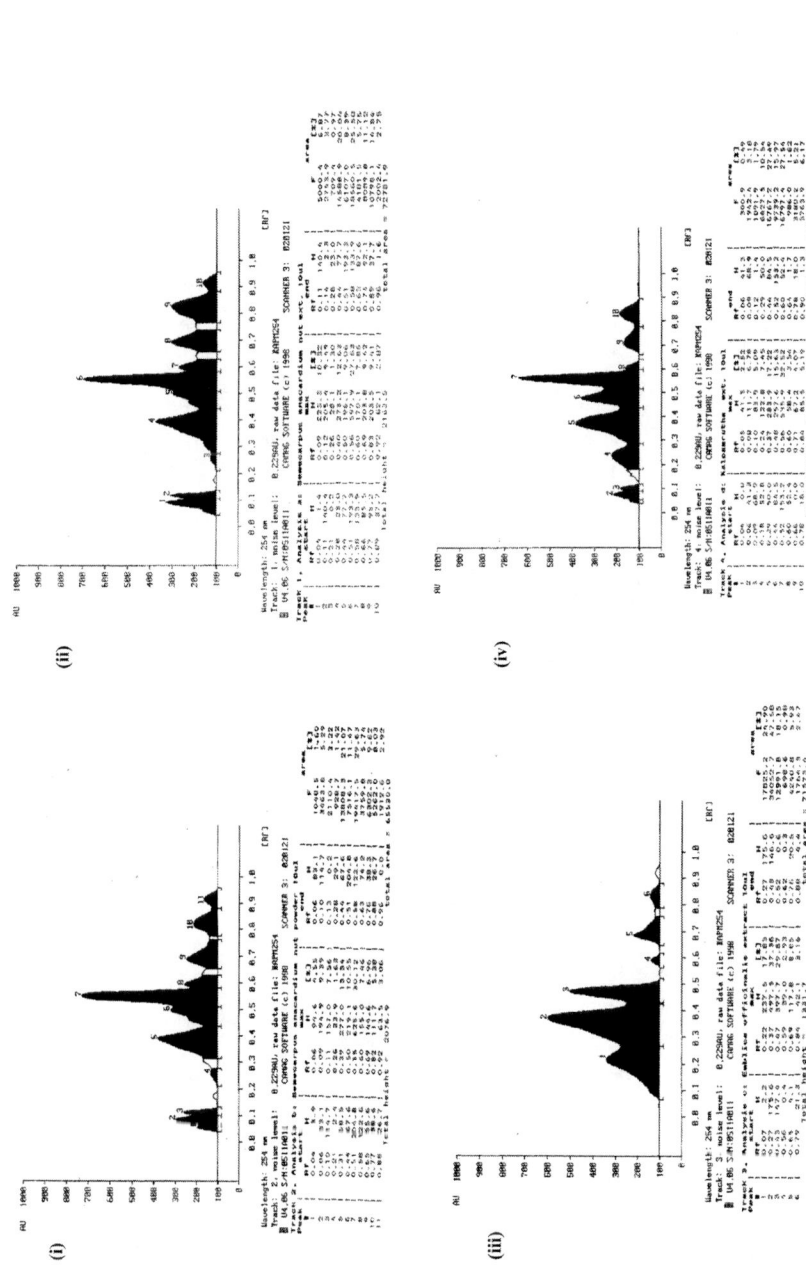

FIGURE 22.1 (a) HPTLC analysis of flavonoids at 254 nm: (i) *Semecarpus anacardium* nut; (ii) *Semecarpus anacardium*; (iii) *Emblica officinalis*; and (iv) Kalpamruthaa. (b) HPTLC analysis of flavonoids at 550 nm: (i) *Semecarpus anacardium* nut; (ii) *Semecarpus anacardium*; (iii) *Emblica officinalis*; and (iv) Kalpamruthaa. (c) HPTLC analysis of gallic acid at 254 nm: (i) Standard gallic acid; (ii) *Semecarpus anacardium*; (iii) *Emblica officinalis*; and (iv) Kalpamruthaa. *(continued)*

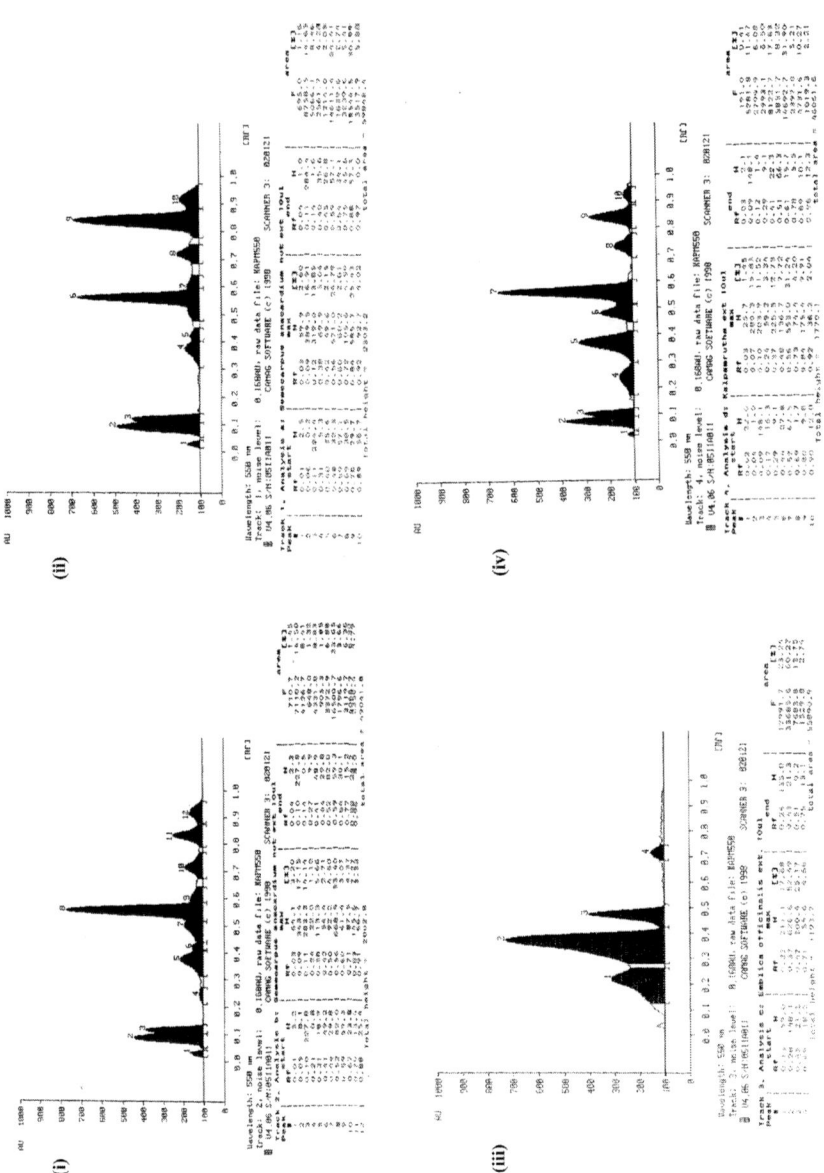

FIGURE 22.1 (continued)

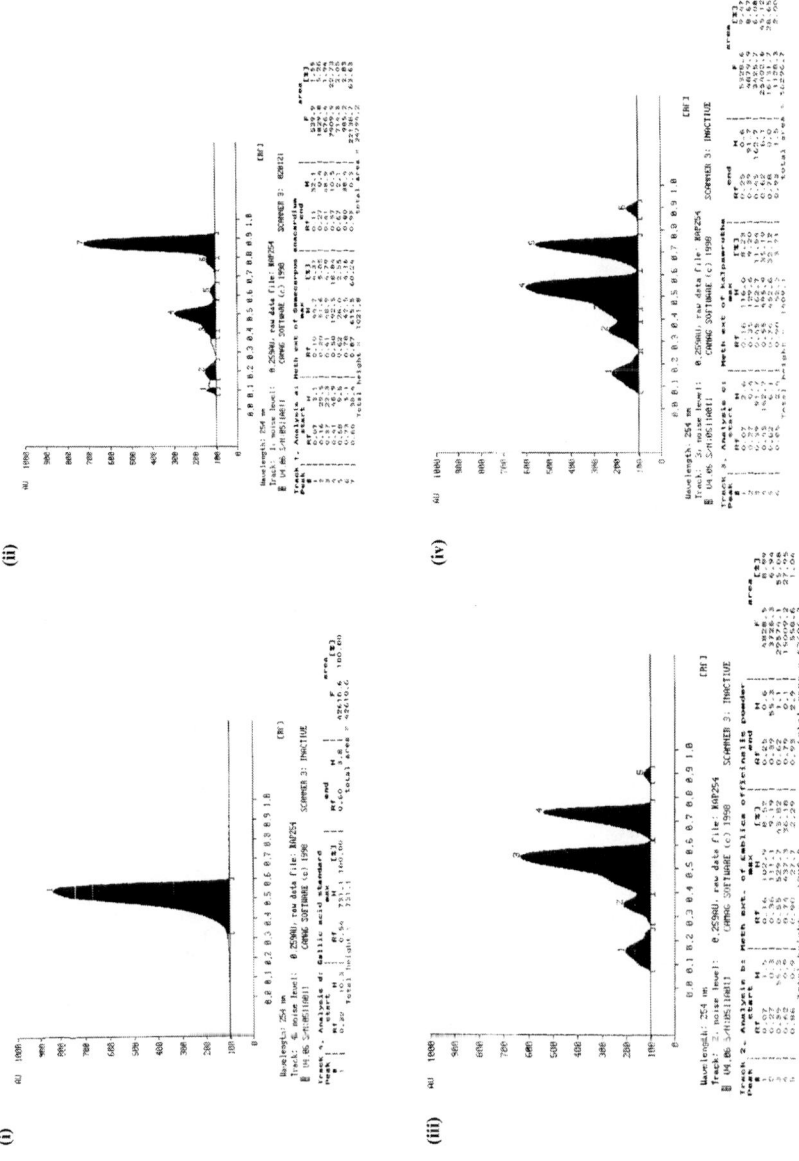

FIGURE 22.1 (continued)

22.3.3 Therapeutic Potential of the Drug KA

22.3.3.1 With Respect to Arthritis

Rheumatoid arthritis (RA) is a chronic inflammatory autoimmune disease characterized by a series of pathological processes of the joints, such as leukocyte infiltration, pannus formation, and extensive destruction of the articular cartilage and bone (Yeom et al., 2006). RA affects about 1% of the adult population worldwide (Gabriel, 2001). The advantage of AIA (adjuvant-induced arthritis) over another experimental arthritis model consists in the exactly defined stages of the development of arthritis elicited by injection of killed mycobacteria in paraffin oil into rats (Billingham, 1983). After this initiation of AIA, it is possible to distinguish between the acute stage from day 0 to day 7 and the chronic evolution with recurrent inflammatory bouts, resulting in periarticular, articular, and bone lesions (Pearson, 1956). So, this model was used to test the antiarthritic potential of the drug KA despite the use of nonsteroidal anti-inflammatory drugs (NSAIDs) by about 1.5% of the population of the world. NSAIDs are often associated with severe side effects, the most common being gastrointestinal bleeding and peptic ulcers (Corley et al., 2003).

The analgesic, antipyretic, and ulcerogenic effects of the drug KA and SA were evaluated, and it was found that KA exhibited better analgesic and antipyretic activities than sole SA therapy. No significant ulceration was detected on either SA or KA treatment. A standard anti-inflammatory drug, diclofenac, was also used for comparison. The enhanced antiulcerogenic activity and greater antineoceptive activity exhibited by KA were achieved via inhibition of prostaglandin synthesis. The protection afforded by KA is probably due to the presence of a range of phytochemicals, including flavonoids, tannins, ascorbic acid, hydroxycinnamates, and others. It can be concluded that traditional medicine formulae such as KA are able to relieve inflammation and pain and correct overall symptoms associated with RA favorably without producing the toxic side effects associated with long-term administration when compared to the standard reference drug diclofenac sodium (Mythilipriya et al., 2007a).

RA is a prevalent and debilitating disease that affects the joints; in turn, blood-derived cells infiltrate the affected joints, and on activation generate reactive oxygen species (ROS) and reactive nitrogen species (RNS), resulting in oxidative stress. The drug KA was evaluated for its synergistic antioxidant potential in AIA in rats. Levels and activities of ROS, RNS, myeloperoxidase, lipid peroxide, and enzymic and nonenzymic antioxidants were determined in control and experimental groups of rats. KA showed a more profound antioxidant potential than sole SA treatment. KA enhanced the antioxidant status in AIA more than treatment with only KA, proving to be an important therapeutic modality in the management of RA and thereby instituting the role of oxidative stress in the clinical manifestations of the disease RA. The profound antioxidant efficacy of KA might be due to the synergistic effect of polyphenols such as flavonoids, and tannins and other compounds such as vitamin C and hydroxycinnamates present in KA (Mythilipriya et al., 2007c).

Several studies have indicated that lysosomes at least partly mediate acute and chronic inflammation in joints and in the synovial fluid of those with RA (Safina et al., 1992). One of the characteristic features of AIA in rats is the correlation between the development of inflammatory processes and the release of lysosomal enzymes in the extracellular compartments (Weissmann, 1972). The inflammatory processes of AIA also cause alterations in the metabolism of connective tissue macromolecular components, thereby bringing about a change in the levels of glycoproteins. So, alterations in the activity levels of lysosomal enzymes and changes in the levels of glycoprotein components were determined in control and experimental groups of rats. The activities of lysosomal enzymes and amino transferases and levels of plasma protein-bound carbohydrate components of glycoproteins were elevated in arthritic rats when compared to normal rats. After administration of KA, the activities of lysosomal enzymes and aminotransferases and protein-bound carbohydrate component levels were significantly normalized. The effect might be speculated due to the combined interactions of phytochemicals such as flavanones, tannins, gallic acid, and vitamin C and other components present in KA that could be attributed to the effectiveness of the drug in lysosome membrane stabilization, lysosomal enzyme inhibition, or decreased infiltration of leukocytes. This was further confirmed by radiological studies (Mythilipriya et al., 2008b).

The cytokines play a central role in the inflammatory articular process, including the synovial proliferation and cartilage destruction in RA, and understanding the role of these cytokines in turn exploits them as therapeutic targets in RA. So, the role of KA in reducing the pathological lesions caused by the proinflammatory cytokines in AIA in rats was studied. The protein expressions of tumor necrosis factor α (TNF-α) and interleukin 1β (IL-1β), the levels of acute-phase proteins, immunoglobulins, and the radiological, histopathological, and electron microscopic changes in control and experimental animals were analyzed. Both SA and KA significantly regulated the inflammation in arthritic joints by reducing extracellular matrix (ECM) degradation and cartilage and bone destruction via downregulating the levels of TNF-α and IL-1β and the levels of acute-phase proteins, with appreciable increase in the levels of immunoglobulins in arthritic rats. Of both the drugs, KA exhibited a more profound effect than SA-only treatment, and the enhanced effect of KA might be attributed to the combined effect of the flavonoids, tannins, vitamin C, and other phytoconstituents present in the drug (Mythilipriya et al., 2009). Antiangiogenic potential of the drug KA was also determined through expression of vascular endothelial growth factor (VEGF) and matrix metalloproteinase (MMP) 1, 9, and 13 at the transcriptional and translational levels.

22.3.3.1.1 Conclusion
Based on these studies, it can be suggested that the anti-inflammatory, antiarthritic, and antiangiogenic mechanisms of KA might be related to their *in vitro* membrane stabilization, inhibitory action of proteinases, and protein denaturation and via *in vivo* antioxidant, inhibitory effect on prostaglandin synthesis, downregulation of proinflammatory cytokines (IL-1β and TNF-α), ECM-degrading enzymes (MMPs and lysosomal enzymes), and angiogenic factors (hypoxia-inducible factor 1α [HIF-1α], VEGF, fibroblast growth factor 2 [FGF-2], and platelet-derived growth factor [PDGF]) and upregulation of TIMPs (tissue inhibitor metalloproteinases) in the AIA model. This enhanced effect of KA might be attributed to the cumulative effects of polyphenols such as flavonoids, tannins, and its constituents

gallic and ellagic acids and other compounds such as vitamin C present in KA. Consequently, from this observation one can propose conveniently that KA can be a novel formulation containing a combination of herbal drugs holding the potential to become the therapy of choice in the future due to the effective combinatorial effect achieved by the phytochemicals present in KA.

22.3.3.2 With Respect to Cancer

Cancer accounts for a high morbidity and high mortality rate throughout the world. Cancer of the breast is common in women in developed countries, and more than 40% of all breast cancer cases are found in developing countries (Ray et al., 2000). The biochemical evaluation of glycoprotein components, marker enzymes, and lysosomal enzymes of liver, kidney, and blood of control and experimental rats was carried out to study the therapeutic efficacy of KA on 7,12-dimethylbenz(a)anthracene- (DMBA) induced mammary carcinoma-bearing rats. On administration of KA, the levels of these enzymes and the changes in the body weights and volume were significantly normalized at a dosage of 300 mg/kg body weight.

Lipids, lipoproteins, and lipid-metabolizing enzymes have been associated with the risk of breast cancer. So, a study examined the variations in lipids, lipoproteins, and lipid-metabolizing enzymes in cancerous animals and the effect of KA on the lipid metabolism. KA was more effective in reversing the altered lipid profile and the lipid-metabolizing enzymes. This increased effect is due to the synergistic activities of various components of KA, such as EO and honey, other than SA. The pharmacological property of the herbal preparation may be due to the hypolipidemic property and the cytoprotective effect of the drug on the deteriorated cell membrane, which is a crucial condition in the cancerous state. This property of the herbal preparation may be due to the presence of flavonoids, which have been reported to reduce the levels of serum triglycerides, and due to the presence of ascorbic acid, which is known to protect the deranged lipid and lipoprotein metabolism against oxidative damage (Veena et al., 2006).

A study was also undertaken to examine whether EO rich in vitamin C content acting synergistically in combination with SA can enhance the antioxidant and anticancer effect of SA in mammary carcinoma-bearing rats. The DMBA-treated rats showed a decline in the activities of mitochondrial enzymes as well as mitochondrial antioxidant enzymes and an increase in lipid peroxide levels. KA treatment significantly reversed the upsurge in the lipid peroxide levels and the altered antioxidant status. The protective effect of KA was most probably due to the presence of phenolic compounds. Hydrogen-donating hydroxyl groups present in the aromatic ring of phenolic compounds might have effectively scavenged the ROS, thereby decreasing the level of lipid peroxides. Immunomodulatory activities on humoral and cellular immunity were also studied by hemagglutination (HA) titer, delayed-type hypersensitivity (DTH), and phagocytic index. KA enhanced the HA titer, phagocytic index, and DTH, which indicates that KA triggers both humoral and cell-mediated responses to a great extent (Arulkumaran et al., 2007). Messenger RNA (mRNA) and protein expressions of MMPs 2 and 9 and TIMP 2 were performed to assess the malignant progression, and morphological changes were analyzed by transmission electron microscopic (TEM) studies. The extent of apoptosis was investigated using Bcl 2, Bax, Cyt C, and p53 expression; mitochondrial membrane potential; and DNA fragmentation in MCF cell line. The effect of KA on regulation of MMPs (matrix metalloproteases) 2 and 9 established the anti-invasive and antiangiogenic properties of the drug at the phase of ECM (extra cellular matrix) degradation in rats with mammary carcinoma rats. KA, SA, and PE exhibited a cytotoxic effect on MCF-7 and MDA-MB-231 breast cancer cell lines, which was confirmed by MTT (3-(4,5-dimethyl thiazol-2-yD-2,5-diphenyl tetrazolium bromide), Trypan blue, and LDH (lactate dehydrogenase) leakage assays and by DNA fragmentation, showing significant increase in the percentage of inhibition of cells, positive staining of cells with Trypan blue, increased percentage of LDH leakage, and increased DNA fragmentation in drug-treated cell lines. KA, SA, and PE induced cell death, as evidenced by accumulation of cells at the sub-G0/G1 phase of the cell cycle by PI-labeling flow cytometric analysis in the MCF-7 breast cancer cell line. This in turn exploits the cytotoxic and apoptotic effects of the drug by suppressing the proliferation of cancer cells.

It might be concluded that, based on these *in vitro* and *in vivo* studies, KA has been found to exhibit enhanced membrane-stabilizing, anti-invasive, and apoptotic properties. It can be emphasized that KA possesses profound anticarcinogenic activity, which might be attributed to the synergistic effects of the polyphenols such as flavonoids, tannins, gallic acid, ellagic acid, vitamin C, pyrogallol, hydroxycinnamate, and other components present in KA.

Hepatocellular carcinoma (HCC) is the most common primary epithelial malignancy and is one of the most common malignancies in the world (Washington, 1999). AFB-1 (acid-fast bacillus 1) is a potent hepatotoxic and hepatocarcinogenic mycotoxin that has been postulated to play a role in the etiology of human liver cancer. Hence, an investigation was carried out to assess the effect of KA on AFB-1-mediated HCC in rats by acting as a strong antineoplastic agent and exhibiting its apoptotic potential through PI3 kinase-mediated AKT dephosphorylation in p53-dependent or -independent manner. *In vivo* studies were carried out to establish the efficacy of the drug as a potent antioxidant, immunomodulatory agent (Umarani et al., 2008). *In vitro* studies were carried out with two different HCC cell lines (HepG2 and Hep3B) with different p53 status. The drug showed enhanced apoptotic activity as studied by DAPI staining, Tunnel assay, and flow cytometric analysis of subdiploid cell population. Cell cycle analysis was also carried out by immunoblotting of cell cycle regulatory proteins, cyclins, and cyclin-dependent kinase levels.

22.3.3.2.1 Conclusion

Accordingly, it can be emphasized that KA possesses profound anticarcinogenic activity in both *in vivo* and *in vitro* HCC models. The spectrum of these properties, such as antiangiogenic, anti-invasive, membrane-stabilizing, antineoplastic, mixed-function oxidase- (MFO) modifying, and apoptotic properties might be attributed to the presence of several phenolic compounds, like flavonoids, tannins, ellagic acid, gallic acid, and vitamin C and other components present in the drug, which act synergistically with and counteract the deleterious effect, if any, by any internal components and impart antineoplastic activity.

ACKNOWLEDGMENTS

We authors acknowledge the help of Ms. H. Haseena Banu and Ms. S. Kaladevi with a deep sense of gratitude.

REFERENCES

Achilya, G.S., Wadodkar, S.G., and Dorle, A.K. 2004. Evaluation of hepatoprotective effect of Amalkadi Ghrita against carbon tetrachloride-induced hepatic damage in rats. *J. Ethnopharmacol.* 90: 229–232.

Arathi, G., and Sachdanandam, P. 2003. Therapeutic effect of *Semecarpus anacardium* Linn. nut milk extract on carbohydrate metabolizing and mitochondrial TCA cycle and respiratory chain enzymes in mammary carcinoma rats. *J. Pharm. Pharmacol.* 55: 1283–1290.

Aravind, S.G., Arimboor, R., Rangan, M., Madhavan, S.N., and Arumughan, C. 2008. Semi-preparative HPLC preparation and HPTLC quantification of tetrahydroamentoflavone as marker in *Semecarpus anacardium* and its polyherbal formulations. *J. Pharm. Biomed. Anal.* 48: 808–813.

Arulkumaran, S., Ramprasath, V.R., Shanthi, P., and Sachdanandam, P. 2007. Free radical quenching and immunomodulatory effect of a modified Siddha preparation, Kalpaamruthaa. *J. Health. Sci.* 53: 170–176.

Billingham, M.E. 1983. Models of arthritis and the surge for anti-arthritic drugs. *Pharmacol. Ther.* 2: 389–428.

Cai, Y., Luo, Q., Sun, M., and Corke, H. 2004. Antioxidant activity and phenolic compounds of 112 traditional Chinese medicinal plants associated with anti-cancer. *Life Sci.* 74: 2157–2184.

Charaka, S. 1941. In *Vimnasthana*, 3rd ed. Bombay: Nirnaya Sagar Press 7.

Chattopadhyaya, M.K., and Khare, R.L. 1969. Isolation of anacardic acid from *Semecarpus anacardium* Linn. and study of its anti-helminthic activity. *Ind. J. Pharm.* 31: 104–105.

Chen, C.Y., Milbury, P.E., Kwak, H.K., et al. 2004. Avenanthramides and phenolic acids from oats are bioavailable and act synergistically with vitamin C to enhance hamster and human LDL resistance to oxidation. *J. Nutr.* 134: 1459–1466.

Cherchi, A., Spanedda, L., Tuberoso, C., and Cabras P. 1994. Solid phase extraction and high performance liquid chromatographic determination of organic acids in Honey. *J. Chromatogr. A* 669: 59–64.

Corley, D.A., Kerlikowske, K., Verma, R., and Buffler, P. 2003. Protective association of aspirin/NSAIDS and esophageal cancer: a systematic review and meta analysis. *Gastroenterology* 124: 47–56.

Dictionary of Indian Medicinal Plants. 1988. CIMAP, Lucknow.

Ensminger, A.H., and Esminger, M.K.J. 1986. *Food for health: a nutrition encyclopedia.* Clovis, CA: Pegus Press.

Gabriel, S.E. 2001. The epidemiology of rheumatoid arthritis. *Rheum. Dis. Clin. North Am.* 27: 269–281.

Gheldof, N., and Engeseth N.J. 2002. Antioxidant capacities of honey from various floral sources based on the determination of oxygen radical absorbance capacity and inhibition of *in vitro* lipoprotein oxidation in human serum samples. *J. Agri. Food Chem.* 50: 3050–3055.

Ghosal, S. 1996. Active constituents of *Emblica* officinals: Part 1. The chemistry and antioxidaive effects of two new hydrolysable tannins, emblicanin A and B. *Indian J. Chem.* 35B: 941–948.

Ghosh, D., Thejomoorthy, P., Shetty, B.M.W., and Veluchamy. G. 1981. Certain pharmacological studies risk SKx (a coded anticancer Siddha preparation) with special reference to its toxicity. *J. Res. Ayur. Siddha* 2: 150–159.
Gil, R.R., Lin, L.Z., Cordell, A., et al. 1995. Anacardoside from the seeds of *Semecarpus anacardium*. *Phytochemistry* 39: 405–407.
Gulati, G.S., and Dhiman, A.K. 1984. 'Bhallataka'-a clinical and pharmacolognost study. *J. Sci. Res. Pl. Med.* 5: 35–39.
Jamwal, K.S., Sharma, I.P., and Chopra, C.L. 1959. Pharmacological investigation on the fruits of Emblica officinalis. *J. Sci. Ind. Res.* 18c: 180–181.
Jayashri, S., and Jolly C.I. 1993. Phytochemical antibacterial and pharmacological investigations on *Momordica chirantia* and *Emblica officinalis*. *Ind. J. Pharm. Sci.* 1: 6–13.
Jose, J., and Kuttan, R. 2000. Hepatoprotective activity of *Emblica officinalis* and chyavanaprash. *J. Ethnopharmacol.* 72: 135–140.
Kesava Rao, K.V., Gothoskar, S.V., Chitnis, M.P., and Ranadive, K.J. 1979. Toxicological study of *Semecarpus anacardium* nut extract. *Indian J. Physiol. Pharmacol.* 23: 115–120.
Khare, C.P. 2004. *Encyclopedia of Indian medicinal plants*, 200–203. Berlin: Springer-Verlag.
Kurup, P.N.V., Ramdas, V.N.K., and Joshi, P. 1979. *Handbook of medicinal plants*, 32. New Delhi, India: Oxford and IBM Publishing.
Mathivadhani, P., Shanthi, P., and Sachdanandam, P. 2007. Effect of *Semecarpus anacardium* nut extract on ECM and proteases in mammary carcinoma rats. *Vasc. Pharmacol.* 46: 419–426.
Murthy, S.S.N. 1983. A bioflavonoid from *Semecarpus anacardium*. *Phytochemistry* 22: 1518–1520.
Murthy, S.S.N. 1985. Jeediflavanone—a bioflavonoid from *Semecarpus anacardium*. *Phytochemistry* 24: 1065–1069.
Murthy, S.S.N. 1992. New biflavonoid from *Semecarpus anacardium* Linn. *Clin. Acta Turcica* 20: 33–37.
Mythilipriya, R., Shanthi, P., and Sachdanandam, P. 2007a. Analgesic, antipyretic and ulcerogenic properties of an indigenous formulation—Kalpaamruthaa. *Phytother. Res.* 21: 574–578.
Mythilipriya, R., Shanthi, P., and Sachdanandam, P. 2007b. Oral acute and subacute toxicity studies with Kalpaamruthaa, a modified indigenous preparation on rats. *J. Health Sci.* 53: 351–358.
Mythilipriya, R., Shanthi, P., and Sachdanandam P. 2007c. Restorative and synergistic efficacy of Kalpaamruthaa, a modified Siddha preparation, on an altered antioxidant status in adjuvant induced arthritic rat model. *Chem. Biol. Interact.* 168: 193–202.
Mythilipriya, R., Shanthi, P., and Sachdanandam, P. 2008a. Synergistic effect of Kalpaamruthaa on antiarthritic and antiinflammatory properties—its mechanism of action. *Inflammation* 31: 391–398.
Mythilipriya, R., Shanthi, P., and Sachdanandam, P. 2008b. Therapeutic effect of Kalpaamruthaa, a herbal preparation on adjuvant induced arthritis in Wistar rats. *Inflammopharmacology* 16: 21–35.
Mythilipriya, R., Shanthi, P., and Sachdanandam, P. 2009. Ameliorating effect of Kalpaamruthaa, a Siddha preparation in adjuvant induced arthritis in rats with reference to changes in proinflammatory cytokines and acute phase proteins. *Chem. Biol. Interact.* 179: 335–343.
Nadkarni, K.M., and Nadkarni, A.K. 1976. *Indian materia medica*, 322. Bombay: Popular Prakashan.
Nagai, T., Sakai, M., Inoue, R., Inoue, H., and Suzuki, N. 2001. Antioxidative activities of some commercial honeys, royal jelly and propolis. *Food Chem.* 75: 237–240.

Nair, P.K., Melnick, S.J., Wnuk, S.F., et al. 2009. Isolation and characterization of an anticancer catechol compound from *Semecarpus anacardium*. *J. Ethnopharmacol.* 122: 450–456.
National Honey Board. 2002. International honey markets (www.nhb.org/intl/).
Patwardhan, B., Ghoo, R.B., and David, S.B. 1988. A new anerobic inhibitor of herbal origin. *Ind. J. Pharm. Sci.* 50: 130–132.
Pearson, C.M. 1956. Development of arthritis, periarthritis and periostitis in rats given adjuvance. *Proc. Soc. Exp. Biol. Med.* 91: 95.
Premalatha, B., and Sachdanandam, P. 1997. Modification of *Semecarpus anacardium* Linn. nut milk extract on rat serum alpha-fetoprotein level in aflatoxin B1 mediated hepatic tumourigenesis. *Fitoterapia* 70: 279–283.
Premalatha, B., and Sachdanandam, P. 1999. *Semecarpus anacardium* L. nut extract administration induces the *in vivo* antioxidant defence system in aflatoxin B1 mediated hepatocellular carcinoma. *J. Ethnopharmacol.* 66: 131–139.
Raghunath, S., and Mitra, R. 1982. *Pharamacognosy of indigenous drugs*, 185. New Delhi, India: Oxford and IBH Publishing.
Ramprasath, V.R., Shanthi, P., and Sachdanandam, P. 2005. Evaluation of antioxidant effect of *Semecarpus anacardium* Linn. nut extract on the components of immune system in adjuvant arthritis. *Vasc. Pharmacol.* 42: 179–186.
Ray, A., Ratnakar, N., Murthy, N.S., and Sharma, B.K. 2000. Adrenocorticotropic hormone and growth factor receptors in breast cancer. *Indian J. Exp. Biol.* 38: 663–668.
Safina, A.F., Korolenko, T.A., Mynkina, G.I., Dushkin, M.I., and Krasnoselskaya, G.A. 1992. Liver and serum lysosomal enzymes activity during zymosan-induced inflammation in mice. *Agents Actions Suppl.* 38: 370–375.
Sahoo, A.K., Narayanan, N., Sahana, S., Rajan, S.S., and Mukherjee, P.K. 2008. In vitro antioxidant potential of *Semecarpus anacardium* L. *Pharmacology Online* 3: 327–335.
Satoh, K., Ida, Y., Ishihara, M., and Sakagami, H. 1999. Interaction between sodium ascorbate and polyphenols. *Anticancer Res.* 19: 4177–4186.
Selvam, C., and Jachak, S.M. 2004. A cyclooxygenase (COX) inhibitory biflavonoid from the seeds of *Semecarpus anacardium*. *J. Ethnopharmacol.* 95: 209–212.
Shin, Y.G., Cordell, G.A., Dong, Y., et al. 1999. Rapid identification of cytotoxic alkenyl catechols in *Semecarpus anacardium* using bioassay-linked high performance liquid chromatography-electrospray/mass spectrometric analysis. *Phytochem. Anal.* 10: 208–212.
Sivarajan, V.V., and Indira, B. 1994. *Ayurvedic drugs and their plant sources*, 184. New Delhi, India: Oxford and IBH Publishing.
Sorata, Y., Takahama, U., and Kimura, M. 1988. Co-operation of quercetin with ascorbate in the protection of photosensitized lysis of human erythrocytes in the presence of hematoporphyrin. *Photochem. Photobiol.* 48: 195–199.
Sugapriya, D., Shanthi, P., and Sachdanandam, P. 2008. Restoration of energy metabolism in leukemic mice treated by a Siddha drug—*Semecarpus anacardium* Linn. nut milk extract. *Chem. Biol. Interact.* 9: 43–58.
Sujatha, V., and Sachdanandam, P. 2002. Recuperative effect of *Semecarpus anacardium* Linn. nut milk extract on carbohydrate metabolizing enzymes in experimental mammary carcinoma-bearing rats. *Phytother. Res.* 16: S14–S18.
Sushrutha Samhita. 1938. *Sutrasthana*, 45: 122. Bombay: Niranaya Sagar Press.
Thakur, C.P., and Mandal, K. 1984. Effect of *Emblica officinalis* on cholesterol-induced atherosclerosis in rabbits. *Ind. J. Med. Res.* 79: 142–146.
Umarani, M., Shanthi, P., and Sachdanandam, P. 2008. Protective effect of Kalpaamruthaa in combating the oxidative stress posed by aflatoxin B 1 induced hepatocellular carcinoma with special reference to flavonoid structure -activity relationship. *Liver. Inter.* 28: 200–213.

Veena, K., Shanthi, P., and Sachdanandam, P. 2006. Anticancer effects of Kalpaamruthaa on memory carcinoma in rats with reference to glycoprotein components, lysosomal and marker enzymes. *Biol. Pharm. Bull.* 29: 565–569.

Vijayalakshmi, T., Muthulakshmi, V., and Sachdanandam, P. 1996. Effect of milk extract of *Semecarpus anacardium* nut on adjuvant arthritis-A dose dependent study in albino rats. *Gen. Pharmacol.* 27: 1223–1226.

Vijayalakshmi, T., Muthulakshmi, V., and Sachdanandam, P. 1997. Salubrious effect of *Semecarpus anacardium* against lipid peroxidative changes in adjuvant arthritis in rats. *Mol. Cell. Biochem.* 175: 65–69.

Vijayalakshmi, T., Muthulakshmi, V., and Sachdanandam, P. 2000. Toxicity studies on biochemical parameters carried out in rats with Serankottai nei, a Siddha drug milk extract of *Semecarpus anacardium* nut. *J. Ethnopharmacol.* 69: 9–15.

Vinayagamoothy, T. 1982. Antibacterial activity of some medicinal plants of Sri Lanka. *Ceylon J. Sci. Biol. Sci.* 11: 50–55.

Washington, K. 1999. Pathology of primary and secondary liver tumours. In *Malignant liver tumours: current and emerging therapies*, ed. P.A. Clavien, S. Breitenstein, J. Belghiti, et al., 3–26. Malden, MA: Blackwell Science.

Wealth of Asia. 1998. CD-ROM, NISCOM, New Delhi.

Weissmann, G. 1972. Lysosomal mechanisms of tissue injury in arthritis. *N. Engl. J. Med.* 286: 141–147.

White, J.W., and Rudyj, O.N. 1978. The protein content of honey. *J. Apic. Res.* 17: 234–238.

White, J.W., Jr. 1975. Physical characteristics of honey. In *Honey: a comprehensive survey*, ed. E. Crane, 57–160. London; Heinemann.

Yeom, M.J., Lee, H.C., Kim, G.H. et al. 2006. Anti-arthritic effects of *Ephedra sinica* STAPF herb-acupuncture: inhibition of lipopolysaccharide-induced inflammation and adjuvant induced polyarthritis. *J. Pharmacol. Sci.* 100: 41–50.

Index

A

Aamalakee, see *Phyllanthus emblica*
Aamalaki, see *Phyllanthus emblica*
abscesses, 66
accession numbers, 82, 103–104
acid-insoluble ash parameters
  Phyllanthus emblica, 59
  Phyllanthus fraternus, 45
adduct formation inhibition
  Phyllanthus amarus, 179
  Phyllanthus emblica, 188
adjunct therapies, *x*
adulteration, 64
agarose gel electrophoresis, 77
aging disorders, 58, see also Antiaging effects
Agrobacterium tumefaciens, 62
Ajjhada, see *Phyllanthus fraternus*
alcohol-soluble extractive parameters
  Phyllanthus emblica, 59
  Phyllanthus fraternus, 45
aldose reductase inhibition, 246–247
alexiteric properties, 58
alkaloids
  Phyllanthus amarus, 107
  Phyllanthus genus, 120
Amalaka, see *Phyllanthus emblica*
Amalakam, see *Phyllanthus emblica*
Ambal, see *Phyllanthus emblica*
Ambla, see *Phyllanthus emblica*
Amioki, see *Phyllanthus emblica*
Amla, see *Phyllanthus emblica*
Amlaki, see *Phyllanthus emblica*
analgesic properties
  Phyllanthus sellowianus, 67
  Phyllanthus stipulatus, 67
anemia, 58
angiogenesis, 198, 199
Anisonema, 23
Anogeissus latifolia, 32
antiaging effects
  cardiovascular disease, 272–273
  dementia, 272
  diabetes, 272
  free radicals, 269–271
  fundamentals, 267–269, 273–274
antiangiogenic effects, see Proapoptotic and antiangiogenic effects, *Phyllanthus urinaira*
antiarthritic activity, 326, see also Arthritis
anticancer activity
  adduct formation inhibition, 188
  apoptosis induction, 190
  carcinogen-metabolizing enzymes, 187–188
  cell growth and multiplication, effect on, 189
  chemical carcinogenesis, inhibition, 183–184, 186
  clastogenicity inhibition, 189
  DNA adduct formation inhibition, 188
  fundamentals, 183
  mechanism of action, 187–189
  medicinal preparations, 190
  mutagenesis inhibition, 188
  Phyllanthus urinaria, 197–200
  radioprotective effect, 187
  transplanted tumors, 186
  tumor cell proliferation, 186–187
anticancer studies
  adduct formation inhibition, 179
  apoptosis induction, 179
  cdc25 phosphatase activity, 180
  cellular macromolecules inhibition, 179
  chemically induced carcinogenesis, 172–174
  clastogenicity inhibition, 179
  fundamentals, 171
  mechanism of action, 178–180
  mutagenicity inhibition, 179
  phase I enzymes inhibition, 178
  phase II enzymes, 178
  transcription factors, 179–180
  transplanted tumors, 174
  virally induced cancers, 174–176
anticlastogenic activity, 318–319
antidiabetic effects and activities
  clinical trials, 299, 302–303
  diabetes, 238–244
  Phyllanthus amarus, 64
  Phyllanthus sellowianus, 67
  Phyllanthus stipulatus, 67
  Triphala, 326
antigenotoxic effects, see Chemoprotective, genotoxic, and antigenotoxic effects
anti-hepatitis B virus infections, 208–209, see also Hepatitis B infections
antihepatitis properties
  clinical trials, anti-hepatitis B virus infections, 208–209
  fundamentals, 214
  hepatitis C, 212
  HIV, 212–213
  lead optimization challenge *vs.* clinical efficacy, 209–212

mechanism, anti-hepatitis B virus activity, 208
Phyllanthus amarus, 207–212
Phyllanthus niruri, 206–212
safety studies, 207–208
antihepatoxic mechanism, 165–166
anti-inflammatory activity and properties
  Entox, 154
  formulations, 154
  fundamentals, 149–150, 155
  Phyllanthus amarus, 64, 150–151
  Phyllanthus corcovadensis, 153–154
  Phyllanthus debilis, 152
  Phyllanthus emblica, 151–152
  Phyllanthus polyphyllus, 152–153
  Phyllanthus reticulatus, 153
  Phyllanthus singampattiyana, 154
  Phyllanthus species, 166
  Phyllanthus tenellus, 154
  Phyllanthus urinaria, 153
  Septilin, 154–155
  Triphala, 326
antimicrobial activity, 326
antioxidant activity and properties, 58, 318–319, 327
antioxidative stress, 165
antipyretic properties
  Phyllanthus amarus, 64
  Phyllanthus emblica, 58
antiseptic properties
  Phyllanthus amarus, 64
  Phyllanthus sellowianus, 67
  Phyllanthus stipulatus, 67
antitumor action, 319–320
antiviral activities and properties
  chemical constituents, 222
  classification, 220–222
  compounds, 224
  description, 220–222
  enterovirus, 230
  flavonoids, 223
  fundamentals, 219–220, 224–225, 230
  hepatitis B virus, 225–226
  herpesvirus, 226–229
  metabolites, 224
  Phyllanthus amarus, 64
  phytochemcial sieving, 223
  properties, 220–222
  tannins, 223
Aonla, *see Phyllanthus emblica*
APG II System classification
  Phyllanthus amarus, 60–61
  Phyllanthus emblica, 49
  taxonomy, *Phyllanthus* genus, 4
apoptosis induction
  Phyllanthus amarus anticancer studies, 179
  Phyllanthus emblica anticancer activity, 190

Phyllanthus urinaria, 195–196
Aranelli, *see Phyllanthus acidus*
arthritis, 340–342, *see also* Antiarthritic activity
arylnapthalenes, 126
aryltetralins, 125–126
ascorbic acid
  Kalpaamruthaa, 334–335
  Phyllanthus Emblica, xiii
assays
  Phyllanthus amarus, 42–43
  Phyllanthus emblica, 39–40
Astanga Hrdaya, xiii
asthma
  Phyllanthus emblica, 58
  Phyllanthus reticulatus, 66
astringent properties, 64
Ayurvedic treatment, 316–317, *see also specific treatment*
Azobacter chroococcum, 52, 54
Azospirillum brasilens, 52, 54

B

bacterial infections, *xiii*
Bahuphala, *see Phyllanthus fraternus*
barley, 93–94
Basic Local Alignment Search Tool (BLAST), 79
Bazarmani, *see Phyllanthus maderaspatensis*
benzene derivatives, 127
Bhiuavate, *see Phyllanthus gardnerianus; Phyllanthus virgatus*
Bhoi Amali, *see Phyllanthus fraternus*
Bhonyaabbali, *see Phyllanthus debilis*
Bhonyaanvali, *see Phyllanthus amarus*
Bhonyamali, *see Phyllanthus fraternus*
Bhony amari, *see Phyllanthus fraternus*
Bhoomi amalaki, *see Phyllanthus amarus*
Bhoomyaamalakee, *see Phyllanthus amarus*
Bhooyimabi, *see Phyllanthus urinaria*
Bhuaamla, *see Phyllanthus amarus*
Bhuaamlaki, *see Phyllanthus amarus*
Bhu amla, *see Phyllanthus fraternus*
Bhuamlaki, *see Phyllanthus maderaspatensis*
Bhudhatri, *see Phyllanthus fraternus*
Bhui Amala, *see Phyllanthus fraternus*
Bhui amla, *see Phyllanthus amarus*
Bhuiamla, *see Phyllanthus amarus*
Bhuiavala, *see Phyllanthus amarus*
Bhuiavla, *see Phyllanthus amarus*
Bhuiawali, *see Phyllanthus fraternus*
Bhuin Amla, *see Phyllanthus fraternus*
Bhuinamla, *see Phyllanthus amarus*
Bhuinanvalah, *see Phyllanthus debilis*
Bhumamla, *see Phyllanthus fraternus*
Bhumi amalaki, *see Phyllanthus fraternus*
Bhumyamalaki, *see Phyllanthus fraternus*
Bhumyavli, *see Phyllanthus urinaria*

Index

Bhupatri, *see Phyllanthus debilis*
Bhupushpi, *see Phyllanthus debilis*
bile conditions, 107
biliousness, 58
biochemical mechanisms, diabetes, 236–237
bitter properties, 64
black-berried featherfoil, *see Phyllanthus reticulatus*
black catnip, *see Phyllanthus amarus*
black-honey shrub, *see Phyllanthus reticulatus*
BLAST (Basic Local Alignment Search Tool), 79
bleeding gums, 66
blood pressure, *see* Hypertension
blood uric acid level
    Phyllanthus sellowianus, 66
    Phyllanthus stipulatus, 66
botanical classification
    Phyllanthus amarus, 60–61
    Phyllanthus emblica, 49
Brahma Rasayana
    anticlastogenic activity, 318–319
    antioxidant activity, 318–319
    antitumor action, 319–320
    Ayurvedic treatment, 316–317
    cancer treatment, potential uses, 319–322
    carcinogenesis, 321
    chemotherapy damage protection, 321
    clinical studies, 321–322
    fundamentals, 317, 322
    immunomodulation, 316–317
    immunomodulatory activity, 317–318
    metastasis inhibition, 320
    radiation damage protection, 321
branches, tree terminology, 75
branch length, tree terminology, 75
bronchial asthma, 304–305
bronchitis, 58
burns, 66

C

calyx, 4
CAM assay, 198
cancer and cancer therapy
    Brahma Rasayana, 319–322
    Kalpaamruthaa, 342–343
    Phyllanthus urinaria, 194–196
    Triphala, 327–329
Candida albicans, 66, *see also* Vaginal candidiasis
cane peas senna, *see Phyllanthus amarus*
capillary electrophoresis, 145
carcinogenesis, 321
carcinogen-metabolizing enzymes, 187–188
cardiovascular disease, 272–273
carminative properties, 58

carry me seed, *see Phyllanthus acuminates*; *Phyllanthus amarus*
cataracts, delay, 248–251
cdc25 phosphatase activity, 180
cell growth and multiplication, 189
cellular macromolecules inhibition, 179
Cenchrus ciliaris, 54
Chamberbitter, *see Phyllanthus urinaria*
Chancoa Piedra, *see Phyllanthus niruri*
Chanco piedra, *see Phyllanthus amarus*
charcoal, sources of, 58
Chattu, *see Phyllanthus emblica*
Chayvanprasha, 268
chemical analysis, 200–201
chemical carcinogenesis, inhibition, 183–184, 186
chemical composition and constituents
    Phyllanthus amarus, 64
    Phyllanthus emblica, 57–58
    Phyllanthus orbicularis, 222
chemically induced carcinogenesis, 172–174
chemoprotective, genotoxic, and antigenotoxic effects
    background, 256–257
    fundamentals, 255–256, 261–262
    mechanism of action, 259–261
    Phyllanthus amarus, 257–258
    Phyllanthus emblica, 257
    Phyllanthus spp., 258–259
    radiation-induced damage, 258–259
chemotaxonomic significance, 132
chemotherapy damage protection, 321
chest diseases, 58
chew sticks, 66
China, nutritive value, 58
Chirukizhukanelli, *see Phyllanthus urinaria*
Chittusiri, *see Phyllanthus indo-fischeri*
chondroprotective activity, 326
chromatographic system, *see also* Thin-layer chromatography (TLC)
    Phyllanthus amarus, 42–43
    Phyllanthus emblica, 40
chronic dysentery, 58
Chrysopogon fulvus, 54
Chukanna-kizha-nelli, *see Phyllanthus urinaria*
chyawanprash, 57
Cicca Linnaeus, 23
cinnamic acid derivatives, 120, 122
cladistics, 72
cladograms, 92–94
classification, 220–222
clastogenicity inhibition
    Phyllanthus amarus, 179
    Phyllanthus emblica, 189
climate, 62
clinical studies and trials
    antidiabetes, 299, 302–303

Brahma Rasayana, 321–322
bronchial asthma, 304–305
combination, herbs and medication, 303–305
fundamentals, 289–290, 308
hemorrhoid cases, 305
immunomodulatory effects, 290–303
pulmonary tuberculosis, 290–291, 294–295
safety, 306–308
tolerability, 306–308
tonsillopharyngitis, 303–304
urolithiasis, 297–299
vaginal candidiasis, 295–296
varicella zoster infection, 296–297
Clustal W, 80
cold, 107
combination, herbs and medication, 303–305
complications, diabetes, *see* Diabetes and diabetic complications
compounds, *Phyllanthus orbicularis,* 224
conjunctivitis
   Phyllanthus emblica, 58
   Phyllanthus reticulatus, 66
conservation, *x,* 111–112, *see also* Genetic diversity and conservation
construction
   Phyllanthus emblica, 58
   Phyllanthus reticulatus, 66
cooling properties, 58
cosmeceutical products
   Phyllanthus emblica, 57
   Phyllanthus spp., 48
costs, *see* Economics
cough, 58
coumarins, 120, 122
Creole senna, *see Phyllanthus amarus*
Cronquist System
   Phyllanthus amarus, 60
   Phyllanthus emblica, 49
cropping system, 54
cultivars
   Phyllanthus amarus, 62
   Phyllanthus emblica, 51–52
cultivation
   fundamentals, 65, 67
   Phyllanthus acidus, 65
   Phyllanthus amarus, 65
   Phyllanthus debilis, 65
   Phyllanthus fraternus, 65
   Phyllanthus indofischeri Bennet, 66
   Phyllanthus maderaspatensis, 65
   Phyllanthus minicus, 65
   Phyllanthus niruri, 65
   Phyllanthus piscatorum, 66
   Phyllanthus reticulatus, 66
   Phyllanthus sellowianus, 66–67
   Phyllanthus stipulatus, 66
   Phyllanthus urinaria, 65

cultivation, economics, and marketing
   adulteration, 64
   APG II System, 49, 60–61
   botanical classification, 49, 60–61
   chemical composition, 57–58, 64
   China, nutritive value, 58
   climate, 62
   Cronquist System, 49, 60
   cropping system, 54
   cultivars, 51–52, 62
   diseases, 54–55, 63
   distribution, 49, 59–60
   economics, 56, 64
   fertilizers and manures, 54, 63
   fruit development, 55–56
   fundamentals, 48–49, 59
   genetic diversity and conservation, 51, 61
   grading, 57
   habit, 50–51, 61
   harvesting, 55–56, 63–64
   irrigation, 53–54, 63
   marketing, 56, 64
   nutritive value, fruit, 58
   origin, 49, 59–60
   packing, 57
   pests, 54–55, 63
   Phyllanthus emblica, 49–59
   planting, 53
   products from, 57
   propagation, 52–53, 62
   pruning, 53
   quality parameter limits, 59
   safety issues, 64
   soil, 62
   soil and climate, 52
   storage, 57
   transplanting, 62
   uses, 58, 64
   weed control, 63
   weeding, 54
   yield, 55–56, 63–64

D

dementia, 272
dendogram, 81
deobstruent properties, 64
descriptions, *see also* General characters, *Phyllanthus* species
   Phyllanthus acidus, 25
   Phyllanthus amarus, 26–27, 41
   Phyllanthus debilis, 27
   Phyllanthus emblica, 28, 38
   Phyllanthus fraternus, 28–29
   Phyllanthus gardnerianus, 29
   Phyllanthus indo-fischeri, 30
   Phyllanthus maderaspatensis, 30

Index

Phyllanthus niruri, 31
Phyllanthus orbicularis, 220–222
Phyllanthus pinnatus, 31–32
Phyllanthus polyphyllus, 32
Phyllanthus reticulatus, 32–33
Phyllanthus rheedii Wight, 33
Phyllanthus rotundifolius, 33
Phyllanthus urinaria, 34
Phyllanthus virgatus G. Forst., 34–35
Deye do, *see Phyllanthus amarus*
diabetes and diabetic complications
    aldose reductase inhibition, 246–247
    antidiabetic effects and activities, 238–244
    biochemical mechanisms, 236–237
    cataracts, delay, 248–251
    complications, 245–251
    Emblica officinalis, 238, 240–241
    fundamentals, 235–236
    herbal drugs, 237–238
    human studies, 245
    hypoglycemic effects, 238, 240–244
    medicinal plants, 237–238
    molecular mechanisms, 236–237
    pathophysiology, 236–237
    pharmacological interventions and limitations, 237
    Phyllanthus amarus, 64, 107, 242
    Phyllanthus debilis, 243–244
    Phyllanthus emblica, 58, 238, 240–241
    Phyllanthus fraternus, 243
    Phyllanthus niruri, 241
    Phyllanthus reticulatus, 66, 242
    Phyllanthus rheedii, 244
    Phyllanthus sellowianus, 244
    Phyllanthus simplex, 243
    Phyllanthus sp., 272
    sorbital accumulation, 248
    type 2 models, experimental studies, 244
diarrhea
    Phyllanthus amarus, 64
    Phyllanthus emblica, 58
    Phyllanthus reticulatus, 66
Diarun plus, 240
Diasperus Kuntze, 23
Diasulin, 240
dibenzylbutyrolactones, 124
Dichanthium annulatum, 54
diepoxylignans, 125
Dihar, 240
distance scale, tree terminology, 75
distribution
    economically important hot spots, 107–108
    fundamentals, 104
    geographic hot spots, 104–107
    Phyllanthus acidus, 25
    Phyllanthus amarus, 27, 59–60, 107
    Phyllanthus debilis, 27–28

Phyllanthus emblica, 28, 49, 107
Phyllanthus fraternus, 29
Phyllanthus gardnerianus, 29
Phyllanthus indo-fischeri, 30
Phyllanthus maderaspatensis, 31
Phyllanthus niruri, 31
Phyllanthus pinnatus, 32
Phyllanthus polyphyllus, 32
Phyllanthus reticulatus, 33
Phyllanthus rheedii Wight, 33
Phyllanthus rotundifolius, 34
Phyllanthus urinaria, 34
Phyllanthus virgatus G. Forst., 34–35
phytochemistry, *Phyllanthus* genus, 132
diterpenes, 130
diuretic properties
    Phyllanthus amarus, 64
    Phyllanthus emblica, 58
    Phyllanthus sellowianus, 67
    Phyllanthus stipulatus, 67
DNA adduct formation inhibition, 188
DNA bar coding, 110–111
DNA level variation, 72–73
doshas, balancing, *xiii*
dropsy, 64
dyes
    Phyllanthus emblica, 57–58
    Phyllanthus reticulatus, 66
    Phyllanthus spp., 48
dysentery
    Phyllanthus amarus, 64
    Phyllanthus emblica, 58
    Phyllanthus reticulatus, 66
dysmenorrhea, 66
dyspepsia
    Phyllanthus amarus, 64
    Phyllanthus emblica, 58

E

ear infections, 66
economically important hot spots
    fundamentals, 107
    Phyllanthus amarus, 107
    Phyllanthus emblica, 107
economics and marketing
    Phyllanthus amarus, 64
    Phyllanthus emblica, 56
edema, 64
ellagic acid, 201
Elrageig, *see Phyllanthus amarus*
Emblic, *see Phyllanthus emblica*
Emblica officinalis, see also *Phyllanthus emblica*
    antiaging effects, 269, 271
    cardiovascular disease, 273
    dementia, 271
    diabetes, 238, 240–241

general characters, 5
Phyllanthus genus, 23
Phyllanthus indo-fischeri, 30
phytochemistry, 334
therapeutic potential, 334
Emblic myrobalan, *see Phyllanthus emblica*
enterovirus, 230
Entox, 154
Epistylium Swartz, 23
epoxy lignans, 125
Escherichia coli, 206

F

fabric dyes, 57
face cream/pack, 57
FASTA program, 79–80
febrifugal properties, 64
fertilizers and manures
    Phyllanthus amarus, 63
    Phyllanthus emblica, 54
fever
    Phyllanthus amarus, xiii
    Phyllanthus reticulatus, 66
firewood, 66
flavanones, 124
flavones, 124
flavonoids
    Kalpaamruthaa, 334–336
    Phyllanthus amarus, 107
    Phyllanthus genus, 122–124
    Phyllanthus orbicularis, 223
flavonols, 122–123
flowering and fruiting season, *see also* Fruits
    Phyllanthus acidus, 25
    Phyllanthus amarus, 27
    Phyllanthus debilis, 27
    Phyllanthus emblica, 28
    Phyllanthus gardnerianus, 29
    Phyllanthus indo-fischeri, 30
    Phyllanthus maderaspatensis, 30
    Phyllanthus pinnatus, 32
    Phyllanthus polyphyllus, 32
    Phyllanthus reticulatus, 33
    Phyllanthus rheedii Wight, 33
    Phyllanthus rotundifolius, 33
    Phyllanthus urinaria, 34
    Phyllanthus virgatus G. Forst., 34–35
flowers
    Phyllanthus genus, 4
    Phyllanthus acidus, 6
    Phyllanthus amarus, 5
    Phyllanthus carpentariae, 20
    Phyllanthus debilis, 13
    Phyllanthus emblica, 5
    Phyllanthus fraternus Webster, 15
    Phyllanthus fuernrohirii, 20

Phyllanthus gardnerianus, 12
Phyllanthus gasstroemii, 19
Phyllanthus gongyloides, 18
Phyllanthus gradyi, 17
Phyllanthus grandisepalus, 20
Phyllanthus gunnii, 19
Phyllanthus hebecarpus, 20
Phyllanthus hirtellus, 20
Phyllanthus indicus, 19
Phyllanthus indofischeri Bennet., 16
Phyllanthus lacunarius, 19–20
Phyllanthus lawii, 10
Phyllanthus longipedicellatus, 17
Phyllanthus macraei, 12
Phyllanthus maderspatensis, 11
Phyllanthus missionis, 13
Phyllanthus niruri, 6
Phyllanthus pinnatus, 14
Phyllanthus polyphyllus, 9
Phyllanthus reticulatus, 9
Phyllanthus rheedii, 10
Phyllanthus rotundifolius, 12
Phyllanthus salesiae, 18
Phyllanthus scabrifolius, 16
Phyllanthus similis, 21
Phyllanthus subcrenulatus, 21
Phyllanthus tenellus, 10
Phyllanthus urinaria, 8
flu, 107
fodder, 48, 58
food
    Phyllanthus acidus, 65
    Phyllanthus emblica, 57
    Phyllanthus spp., 48
foreign matter parameters
    Phyllanthus emblica, 59
    Phyllanthus fraternus, 45
formulations, anti-inflammatory properties, 154
free radicals, 165, 269–271
fruits, *see also* Flowering and fruiting season
    Phyllanthus genus, 4
    Phyllanthus acidus, 6
    Phyllanthus amarus, 5
    Phyllanthus carpentariae, 20
    Phyllanthus debilis, 13
    Phyllanthus emblica, 5, 55–56
    Phyllanthus fuernrohirii, 20
    Phyllanthus gardnerianus, 12
    Phyllanthus gasstroemii, 19
    Phyllanthus gongyloides, 18
    Phyllanthus gradyi, 17
    Phyllanthus grandisepalus, 20
    Phyllanthus gunnii, 19
    Phyllanthus hebecarpus, 20
    Phyllanthus hirtellus, 21
    Phyllanthus indicus, 19
    Phyllanthus lacunarius, 20

Index

Phyllanthus lawii, 10
Phyllanthus longipedicellatus, 17
Phyllanthus macraei, 12
Phyllanthus maderspatensis, 11
Phyllanthus narayanswamii, 11
Phyllanthus niruri, 8
Phyllanthus polyphyllus, 9
Phyllanthus reticulatus, 9
Phyllanthus rheedii, 10
Phyllanthus rotundifolius, 13
Phyllanthus salesiae, 18
Phyllanthus scabrifolius, 16
Phyllanthus similis, 21
Phyllanthus spp., 48
Phyllanthus subcrenulatus, 21
Phyllanthus tenellus, 10
Phyllanthus urinaria, 8
Phyllanthus virgatus, 11
fuel
   Phyllanthus emblica, 58
   Phyllanthus spp., 48
fungal infections, 66
furniture, 58

G

gale-wind grass, see Phyllanthus amarus
gall bladder stones, xiii, 64
gallic acid, 336
gas chromatography and mass spectrometry (GC-MS), 145
gastric disorders, xiii
GC-MS, see Gas chromatography and mass spectrometry
gel documentation, 78
GenBank accession numbers, 82–83, 103–104
general characters, Phyllanthus species, see also Descriptions
   fundamentals, 4
   Phyllanthus abnormis, 15
   Phyllanthus acidus, 5–6
   Phyllanthus acuminates, 14
   Phyllanthus amarus, 4–5
   Phyllanthus beillei Hutch., 16
   Phyllanthus caesiifolius, 14
   Phyllanthus capillaris Schum. & Thonn., 15
   Phyllanthus caroliniensis, 14
   Phyllanthus carpentariae, 20
   Phyllanthus debilis, 13
   Phyllanthus emblica, 5
   Phyllanthus fraternus Webster, 14–15
   Phyllanthus fuernrohirii, 20
   Phyllanthus gardnerianus, 12
   Phyllanthus gasstroemii, 19
   Phyllanthus gentryi, 14
   Phyllanthus gongyloides, 18–19
   Phyllanthus gradyi, 16–17
   Phyllanthus grandisepalus, 20
   Phyllanthus gunnii, 19
   Phyllanthus hebecarpus, 20
   Phyllanthus hirtellus, 20–21
   Phyllanthus indicus, 19
   Phyllanthus indofischeri Bennet., 16
   Phyllanthus kozhikodianus, 9
   Phyllanthus lacunarius, 19–20
   Phyllanthus lawii, 10
   Phyllanthus longipedicellatus, 17
   Phyllanthus macraei, 12
   Phyllanthus maderspatensis, 10–11
   Phyllanthus mirabilis, 14
   Phyllanthus missionis, 13
   Phyllanthus muellerianus (Kuntze) Excell, 14
   Phyllanthus muellerianus (O Ktze) Exell, 15
   Phyllanthus myrtifolius, 9
   Phyllanthus narayanswamii, 11
   Phyllanthus niruri, 6, 8
   Phyllanthus odontadenius, 15
   Phyllanthus pinnatus, 13–14
   Phyllanthus polyphyllus, 8–9
   Phyllanthus pulcher Wall. Ex Müll. Arg., 14
   Phyllanthus reticulatus, 9
   Phyllanthus rheedii, 9–10
   Phyllanthus rotundifolius, 12–13
   Phyllanthus salesiae, 18
   Phyllanthus scabrifolius, 16
   Phyllanthus similis, 21
   Phyllanthus speciosus Jacq, 13
   Phyllanthus subcrenulatus, 21
   Phyllanthus sublanatus Schum. & Thonn., 15
   Phyllanthus tenellus, 10
   Phyllanthus urinaria, 8
   Phyllanthus virgatus, 11
genetic diversity and conservation
   Phyllanthus amarus, 61
   Phyllanthus emblica, 51
genetic resources, Southern India
   accession numbers, 103–104
   conservation, 111–112
   distribution, 104–107
   DNA bar coding, 110–111
   economically important hot spots, 107–108
   fundamentals, 98, 102, 104
   geographic hot spots, 104–107
   harvesting impact, Phyllanthus amarus, 108–110
   pharmacological activities, 99–102
   Phyllanthus amarus, 107
   Phyllanthus emblica, 107
   sequenced regions, 103–104
   species adulteration, 110–111
   taxonomic incongruities, 110–111
   ultilization implications, 111–112

genitourinary system, 64
genotoxic effects, *see* Chemoprotective, genotoxic, and antigenotoxic effects
geographic hot spots, 104–107
Glomus fasciculatum, 54
gonorrhea
   Phyllanthus emblica, 58
   Phyllanthus reticulatus, 66
grading, 57
Graine en bas fievre, *see Phyllanthus amarus*
Gulf leaf-flower, *see Phyllanthus fraternus*

H

habit
   Phyllanthus genus, 4
   Phyllanthus amarus, 61
   Phyllanthus emblica, 50–51
hair oils and dyes, 57
Harfarauri, *see Phyllanthus acidus*
Hariphal, *see Phyllanthus acidus*
harvesting
   Phyllanthus amarus, 63–64, 108–110
   Phyllanthus emblica, 55–56
Hazarmani, *see Phyllanthus urinaria*
headache, 66
head diseases, 58
heart disease, 58
Helicobacter pylori, 153
hemorrhage, 58
hemorrhoid cases, 305
hepatitis B infections, *see also* Anti-hepatitis B virus infections
   Phyllanthus amarus, 208–209
   Phyllanthus orbicularis, 225–226
hepatitis C infections, 212
hepatitis infections, 107
hepatoprotective properties
   anti-inflammatory activity, 166
   antioxidative stress, 165
   fundamentals, 157–158, 166
   mechanism, antihepatoxic activities, 165–166
   Phyllanthaceae, 158–163
   Phyllanthus amarus, 64
   Phyllanthus amarus Schum. and Thonn., 159–161
   Phyllanthus debilis Ex ., 161
   Phyllanthus emblica, 161
   Phyllanthus kozhikodianus Sivarajan & Manilal, 162
   Phyllanthus maderaspatensis, 162–163
   Phyllanthus niruri, 163
   Phyllanthus reticulatus, 164
   Phyllanthus urinaria, 164
herbal drugs, 237–238
herbs, synoptic key, 24
herpesvirus, 226–229

high performance liquid chromatography (HPLC), 140–145
homology, 81
honey, 334
HPLC, *see* High performance liquid chromatography
human studies, 245
hurricane weed, *see Phyllanthus amarus*
hypertension, 107
hyphenated techniques, *Phyllanthus* genus
   capillary electrophoresis, 145
   fundamentals, 139–140, 145
   gas chromatography and mass spectrometry, 145
   high performance liquid chromatography, 140–145
hypoglycemic effects, 238, 240–244
Hyponidd, 240–241

I

identification manual
   herbs, synoptic key, 24
   Phyllanthus acidus, 25
   Phyllanthus amarus, 25–27
   Phyllanthus debilis, 27–28
   Phyllanthus emblica, 28
   Phyllanthus fraternus, 28–29
   Phyllanthus gardnerianus, 29
   Phyllanthus indo-fischeri, 29–30
   Phyllanthus maderaspatensis, 30–31
   Phyllanthus niruri, 31
   Phyllanthus pinnatus, 31–32
   Phyllanthus polyphyllus, 32
   Phyllanthus reticulatus, 32–33
   Phyllanthus rheedii Wight, 33
   Phyllanthus rotundifolius, 33–34
   Phyllanthus urinaria, 34
   Phyllanthus virgatus G. Forst., 34–35
   scientific classification, 23
   shrubs, synoptic key, 25
   synoptic key, 24–25
   trees, synoptic key, 25
identifications
   Phyllanthus amarus, 41–42
   Phyllanthus emblica, 39
   Phyllanthus fraternus, 44–45
immunomodulation, 316–317
immunomodulatory activity, 317–318
immunomodulatory effects
   antidiabetes, 299, 302–303
   combination, herbs and medication, 303–305
   pulmonary tuberculosis, 290–291, 294–295
   urolithiasis, 297–299
   vaginal candidiasis, 295–296
   varicella zoster infection, 296–297

Index

Indian Gooseberry, *see Emblica officinalis; Phyllanthus emblica*
industrial products
   Phyllanthus emblica, 57
   Phyllanthus spp., 48
infantile diarrhea, 66
inflammation, 58
ink
   Phyllanthus emblica, 57
   Phyllanthus reticulatus, 66
integrated medicine, *x*
internal transcribed spacer (ITS) region, 73–74, 93–94
in vivo anticancer effect, 197–200
irrigation
   Phyllanthus amarus, 63
   Phyllanthus emblica, 53–54
ITS, *see* Internal transcribed spacer (ITS) region

J

Jamaican gooseberry tree, *see Phyllanthus acuminates*
Jangli amla, *see Phyllanthus amarus*
Jar amla, *see Phyllanthus amarus; Phyllanthus fraternus*
jaundice
   Phyllanthus amarus, 64, 107
   Phyllanthus emblica, 58
   Phyllanthus spp., 205–206

K

Kaddu nelli, *see Phyllanthus virgatus* G. Forst.
Kalpaamruthaa
   arthritis, 340–342
   cancer, 342–343
   Emblica officinalis, 334
   flavonoids, 334–336, 336
   fundamentals, 332
   gallic acid, 336, 336
   HPTLC analysis, 336
   phytochemical analysis, 333, 335–336
   phytochemistry, 334
   Semecarpus anacardium, 332–334
   tannins, 336, 336
   therapeutic potential, 333–334, 340–343
   toxicity studies, 333, 336
Kattukilanelli, *see Phyllanthus reticulatus*
Kattunelli, *see Phyllanthus polyphyllus*
Keela nelli, *see Phyllanthus amarus*
Keelanelli, *see Phyllanthus fraternus*
Keezhanelli, *see Phyllanthus fraternus*
Keezharnelli, *see Phyllanthus amarus*
Kempu nelanelli, *see Phyllanthus urinaria*
kidney stones
   Phyllanthus amarus, xiii, 64

Phyllanthus sellowianus, 66
Phyllanthus stipulatus, 66
Kilanelli, *see Phyllanthus amarus; Phyllanthus debilis*
Kilarnelli, *see Phyllanthus polyphyllus*
Kilkkayanelli, *see Phyllanthus amarus*
Kirganelia Jussieu, 23
Kirunelli, *see Phyllanthus amarus; Phyllanthus fraternus*
Kizanelli, *see Phyllanthus fraternus*
Kizhararnelli, *see Phyllanthus amarus*
Kizhkkayinelli, *see Phyllanthus amarus*
Kizhukai nelli, *see Phyllanthus fraternus*
Kondapachaari, *see Phyllanthus polyphyllus*
Krishna-kamboji, *see Phyllanthus reticulatus*
Krishna neli, *see Phyllanthus polyphyllus*

L

lactones, 126–127
Lavaliphala, *see Phyllanthus acidus*
laxative properties
   Phyllanthus emblica, 58
   Phyllanthus sellowianus, 67
   Phyllanthus stipulatus, 67
leaves
   Phyllanthus genus, 4
   Phyllanthus acidus, 6
   Phyllanthus amarus, 5
   Phyllanthus carpentariae, 20
   Phyllanthus debilis, 13
   Phyllanthus emblica, 5
   Phyllanthus fraternus, 44, 45
   Phyllanthus fraternus Webster, 15
   Phyllanthus fuernrohirii, 20
   Phyllanthus gardnerianus, 12
   Phyllanthus gasstroemii, 19
   Phyllanthus gongyloides, 18
   Phyllanthus gradyi, 16
   Phyllanthus grandisepalus, 20
   Phyllanthus gunnii, 19
   Phyllanthus hebecarpus, 20
   Phyllanthus hirtellus, 20
   Phyllanthus indicus, 19
   Phyllanthus indofischeri Bennet., 16
   Phyllanthus lacunarius, 19
   Phyllanthus lawii, 10
   Phyllanthus longipedicellatus, 17
   Phyllanthus macraei, 12
   Phyllanthus maderspatensis, 11
   Phyllanthus missionis, 13
   Phyllanthus narayanswamii, 11
   Phyllanthus niruri, 6
   Phyllanthus pinnatus, 13
   Phyllanthus polyphyllus, 9
   Phyllanthus reticulatus, 9

Phyllanthus rheedii, 10
Phyllanthus rotundifolius, 12
Phyllanthus salesiae, 18
Phyllanthus scabrifolius, 16
Phyllanthus similis, 21
Phyllanthus subcrenulatus, 21
Phyllanthus tenellus, 10
Phyllanthus urinaria, 8
Phyllanthus virgatus, 11
lignans
 Phyllanthus amarus, 107
 phytochemistry, *Phyllanthus* genus, 124
liver cancer, 107
liver-related disorders, *xiii*

M

macroscopic identifications
 Phyllanthus amarus, 41
 Phyllanthus emblica, 39
 Phyllanthus fraternus, 44–45
madaras nelli, *see Phyllanthus maderaspatensis*
Madras leaf flower, *see Phyllanthus maderaspatensis*
Mahidhatrika, *see Phyllanthus fraternus*
malaria, *xiii*
Manikanni, *see Phyllanthus polyphyllus*
manure, *see* Fertilizers and manures
 Phyllanthus emblica, 58
Mapatan, *see Phyllanthus amarus*
marketing, 64
Mascarene Island leaf-flower, *see Phyllanthus tenellus*
materials and methods, 76–80
matrix metalloproteinase 2 activity, 199–200
mechanism of action
 anticancer activity, 178–180, 187–189
 anti-hepatitis B virus activity, 208
 antihepatoxic activities, 165–166
 chemoprotective, genotoxic, and antigenotoxic effects, 259–261
 proapoptotic effect, 196–197
medicinal plants, 237–238
medicinal preparations, 190
mental problems, 66
metabolic disorders, 58
metabolic memory, 245
metabolites, 224
metastasis inhibition, 320
microscopic identifications
 Phyllanthus amarus, 42
 Phyllanthus emblica, 39
 Phyllanthus fraternus, 44
miscellaneous compounds, 132
mobile phase
 Phyllanthus amarus, 42
 Phyllanthus emblica, 39

molecular mechanisms, 236–237
monoterpenes, 130
Mousetail plant, *see Phyllanthus myrtifolius*
mouth inflammation, 58
mouthwash, 65
MSA, *see* Multiple-sequence alignment (MSA)
mulch, 58
multiple-sequence alignment (MSA), 80, 83, 84–92
mutagenesis inhibition
 Phyllanthus amarus anticancer studies, 179
 Phyllanthus emblica anticancer activity, 188
myalgia, 58
Mycobacterium tuberculosis, 153

N

Nallapurgudu, *see Phyllanthus reticulatus*
Natural System of classification, 3–4
nausea, 58
Nela nelli, *see Phyllanthus amarus*
Nelanelli, *see Phyllanthus amarus; Phyllanthus fraternus*
Nelausiri, *see Phyllanthus amarus; Phyllanthus maderaspatensis*
Nela usirika, *see Phyllanthus fraternus*
Nela virika, *see Phyllanthus amarus*
Nelavusuri, *see Phyllanthus amarus*
Nelli, *see Phyllanthus emblica*
Nelli kayi, *see Phyllanthus emblica*
Nellikka, *see Phyllanthus emblica*
Nellka, *see Phyllanthus emblica*
neo-Darwinism, 72
neolignans, 126
nickel hyperaccumulator, 221
Nigella sativa
 safety studies, 307
 tonsillopharyngitis, 303–304
Niruri, *see Phyllanthus virgatus* G. Forst.
niruriside, 213
nodes, tree terminology, 75
nontimber forest products, 49
nutraceutical products
 Phyllanthus emblica, 57
 Phyllanthus spp., 48
nutritive value, fruit, 58

O

obesity, *xiii*
Office of Alternative Medicine, *x*
ophthalmia, 64
origin
 Phyllanthus amarus, 59–60
 Phyllanthus emblica, 49
Otaheite gooseberry, *see Phyllanthus acidus*

Index

ovary
    Phyllanthus genus, 4
    Phyllanthus rotundifolius, 13

P

Pachaari, *see Phyllanthus pinnatus*
packing, 57
pain
    Phyllanthus amarus, xiii, 107
    Phyllanthus reticulatus, 66
Panjhuli, *see Phyllanthus reticulatus*
Pansheuli, *see Phyllanthus reticulatus*
paralysis, 66
pathophysiology, 236–237
PCR, *see* Polymerase chain reaction (PCR)
peptic ulcer, 58
perient properties, 58
Perunelli, *see Phyllanthus indo-fischeri*
pests
    Phyllanthus amarus, 63
    Phyllanthus emblica, 54–55
petals, 4
pharmaceutical products
    Phyllanthus emblica, 57
    Phyllanthus spp., 48
pharmacological activities, 99–102
pharmacological interventions and limitations, 237
pharmacopoeial status
    fundamentals, 38, 45
    Phyllanthus amarus, 40–43
    Phyllanthus emblica, 38–40
    Phyllanthus fraternus, 43–45
phase I enzymes inhibition, 178
phase II enzymes, 178
phenolics, 127
Phyllanthus genus, 4
Phyllanthus genus
    characters, 23–24
    descriptions, 25–35
Phyllanthus genus
    fundamentals, 3, 23
    general characteristics, 4
    general characters, 4
    herbs, synoptic key, 24
    identification manual, xiii, 23–35
    Natural System of classification, 3–4
    number of species, xiii
    Phyllanthus acidus, 25
    Phyllanthus amarus, 25–27
    Phyllanthus debilis, 27–28
    Phyllanthus emblica, 28
    Phyllanthus fraternus Webster, 28–29
    Phyllanthus gardnerianus, 29
    Phyllanthus indo-fischeri, 29–30
    Phyllanthus maderaspatensis, 30–31

    Phyllanthus niruri, 31
    Phyllanthus pinnatus, 31–32
    Phyllanthus polyphyllus, 32
    Phyllanthus reticulatus, 32–33
    Phyllanthus rheedii Wight, 33
    Phyllanthus rotundifolius, 33–34
    Phyllanthus urinaria, 34
    Phyllanthus virgatus G. Forst., 34–35
    scientific classification, 23
    shrubs, synoptic key, 25
    synoptic key, 24–25
    taxonomy, 3–21
    trees, synoptic key, 25
Phyllanthus genus, hyphenated techniques
    capillary electrophoresis, 145
    fundamentals, 139–140, 145
    gas chromatography and mass spectrometry, 145
    high performance liquid chromatography, 140–145
Phyllanthus genus, phytochemistry
    alkaloids, 120
    arylnapthalenes, 126
    aryltetralins, 125–126
    benzene derivatives, 127
    chemotaxonomic significance, 132
    cinnamic acid derivatives, 120, 122
    coumarins, 120, 122
    dibenzylbutanes, 124
    dibenzylbutyrolactones, 124
    diepoxylignans, 125
    distribution, 132
    diterpenes, 130
    epoxy lignans, 125
    flavanones, 124
    flavones, 124
    flavonoids, 122–124
    flavonols, 122–123
    fundamentals, 120, 132
    lactones, 126–127
    lignans, 124
    miscellaneous compounds, 132
    monoterpenes, 130
    neolignans, 126
    phenolics, 127
    sesquiterpenes, 130
    simple lactones, 126–127
    simple phenolics, 127
    steroidal compounds, 128
    tannins, 128
    terpenoids, 128, 130, 132
    triterpenes, 130
Phyllanthus species
    antiaging effects, 267–274
    antihepatitis properties, 212–213
    anti-inflammatory activity, 149–155

chemoprotective, genotoxic, and
    antigenotoxic effects, 255–262
clinical trials, 289–308
diabetes and diabetic complications, 235–251
general characters, 4–21
genetic resources, 97–112
hepatitis B, hepatitis C, and HIV infections,
    205–214
hyphenated techniques, 139–145
toxicity studies, 279–285
trade issues, 110–111
Phyllanthus abnormis, 15
Phyllanthus acidus
    antiaging effects, 268
    antiviral activities, 220
    cultivation, 65
    gas chromatography and mass spectrometry,
        145
    GenBank accession numbers, 83
    general characteristics, 5–6
    hyphenated capillary electrophoresis, 145
    identification manual, 25
    monoterpenes, 130
Phyllanthus acuminatus
    antiaging effects, 268
    distribution and chemotaxonomic
        significance, 132
    GenBank accession numbers, 83
    general characteristics, 14
    sesquiterpenes, 130
    traditional uses, 194
Phyllanthus ajmerianus, 105
Phyllanthus amarus, 209–212, *see also*
    Phyllanthus niruri auct. Non.
    adduct formation inhibition, 179
    adulteration, 64
    antiaging effects, 268, 270–271
    antidiabetic effects, 238, 302
    antihepatitis properties, 206–207
    anti-inflammatory properties, 64, 150–151,
        166
    antioxidant stress and hepatoprotection, 165
    antiviral activities, 224
    APG II System, 60–61
    botanical classification, 60–61
    chemical composition, 64
    chemoprotective, genotoxic, and
        antigenotoxic effects, 256–258
    climate, 62
    clinical trials, anti-hepatitis B virus
        infections, 208–209
    Cronquist System, 60
    cultivars, 62
    cultivation, 65
    dementia, 271
    diabetes, 242
    dibenzylbutanes, 124

diseases, 63
distribution, 59–60
economically important hot spots, 107
economics, 64
fertilizers and manures, 63
fundamentals, *xiii,* 59
general characteristics, 4–5
genetic diversity and conservation, 61
habit, 61
harvesting, 63–64, 108–110
hepatitis B virus, 226
hepatitis C, 212
hepatoprotective properties, 64, 159–161
HIV/AIDS, 213
human studies, 245
identification manual, 25–27
irrigation, 63
jaundice, 205–206
lead optimization challenge *vs.* clinical
    efficacy, 209–212
marketing, 64
mechanism, anti-hepatitis B virus activity,
    208
origin, 59–60
pests, 63
pharmacopoeial status, 40–43
Phyllanthus fraternus Webster, 29
Phyllanthus niruri, 31, 163
Phyllanthus urinaria, 34
propagation, 62
safety issues, 64
safety studies, 207–208, 307
soil, 62
steroidal compounds, 128
toxicity studies, 280, 281, 281–283
trade issues, 111
traditional uses, 194
transplanted tumors, 186
transplanting, 62
type 2 diabetic models, 244
uses, 64
weed control, 63
yield, 63–64
Phyllanthus amarus, anticancer studies
    adduct formation inhibition, 179
    apoptosis induction, 179
    cdc25 phosphatase activity, 180
    cellular macromolecules inhibition, 179
    chemically induced carcinogenesis, 172–174
    clastogenicity inhibition, 179
    fundamentals, 171
    mechanism of action, 178–180
    mutagenicity inhibition, 179
    phase I enzymes inhibition, 178
    phase II enzymes, 178
    transcription factors, 179–180
    transplanted tumors, 174

Index

virally induced cancers, 174–176
Phyllanthus anisobulos, 126
Phyllanthus arenarius
  distribution and chemotaxonomic
    significance, 132
  monoterpenes, 130
Phyllanthus armarus
  alkaloids, 120
  cardiovascular disease, 273
  distribution and chemotaxonomic
    significance, 132
  gas chromatography and mass spectrometry,
    145
  hots spots and conservation, 98, 105
  hyphenated HPLC, 141–142
  transcription factor, 179
Phyllanthus asperulatus, 29
Phyllanthus asperulatus Hutch., 31
Phyllanthus beillei Hutch., 16
Phyllanthus caesiifolius, 14
Phyllanthus capillaris Schum. & Thonn., 15
Phyllanthus caroliniensis
  antiaging effects, 268, 271
  general characteristics, 14
  steroidal compounds, 128
Phyllanthus carpentariae, 20, *see also*
  Phyllanthus grandisepalus;
  Phyllanthus hebecarpus
Phyllanthus chamacristoides, 220, 224
Phyllanthus clarkei
  GenBank accession numbers, 83
  phylogenetic analysis, 82–83, 92–93
Phyllanthus corcovadensis, 273
  anti-inflammatory properties, 153–154
  steroidal compounds, 128
Phyllanthus debilis
  antiaging effects, 270–271
  antidiabetic effects, 243
  anti-inflammatory properties, 152
  cultivation, 65
  diabetes, 243–244
  distribution and chemotaxonomic
    significance, 132
  GenBank accession numbers, 83
  general characteristics, 13
  hots spots and conservation, 98
  hyphenated HPLC, 142
  identification manual, 27–28
  jaundice, 205–206
  trade issues, 111
  transcription factor, 179
Phyllanthus discoides
  alkaloids, 120
  dibenzylbutanes, 124
Phyllanthus discolor
  antiviral activities, 220
  GenBank accession numbers, 83

Phyllanthus embergeri, 142
Phyllanthus emblica, see also Brahma Rasayana;
  Emblica officinalis
  antiaging effects, 268
  antidiabetic effects, 238
  anti-inflammatory properties, 151 152, 166
  APG II System, 49
  botanical classification, 49
  chemical composition, 57–58
  chemical constituents, 222
  chemoprotective, genotoxic, and
    antigenotoxic effects, 256–257
  China, nutritive value, 58
  coumarins and cinnamic acid derivatives,
    122
  Cronquist System, 49
  cropping system, 54
  cultivars, 51–52
  diabetes, 238, 240–241
  diseases, 54–55
  distribution, 49
  distribution and chemotaxonomic
    significance, 132
  diterpenes, 130
  economically important hot spots, 107
  economics and marketing, 56
  fertilizers and manures, 54
  flavones, 124
  fruit development, 55–56
  fundamentals, *xiii,* 49
  GenBank accession numbers, 83
  general characteristics, 5
  genetic diversity and conservation, 51
  grading, 57
  habit, 50–51
  harvesting, 55–56
  hepatitis B infection, 208
  hepatoprotective properties, 161
  HIV/AIDS, 213
  hots spots and conservation, 98, 106
  hyphenated HPLC, 140–141
  identification manual, 28
  irrigation, 53–54
  lactones, 126
  monoterpenes, 130
  nutritive value, fruit, 58
  origin, 49
  packing, 57
  pests, 54–55
  pharmacopoeial status, 38–40
  Phyllanthus amarus, 107
  Phyllanthus indo-fischeri, 30
  Phyllanthus niruri, 163
  Phyllanthus polyphyllus, 32
  planting, 53
  products from, 57
  propagation, 52–53

pruning, 53
quality parameter limits, 59
sesquiterpenes, 130
soil and climate, 52
steroidal compounds, 128
storage, 57
tannins, 128
toxicity studies, 280, 281, 283–284
uses, 58
utilization and conservation, 111, 112
weeding, 54
yield, 55–56
Phyllanthus emblica, anticancer activity
  adduct formation inhibition, 188
  apoptosis induction, 190
  carcinogen-metabolizing enzymes, 187–188
  cell growth and multiplication, effect on, 189
  chemical carcinogenesis, inhibition, 183–184, 186
  clastogenicity inhibition, 189
  DNA adduct formation inhibition, 188
  fundamentals, 183
  mechanism of action, 187–189
  medicinal preparations, 190
  mutagenesis inhibition, 188
  radioprotective effect, 187
  transplanted tumors, 186
  tumor cell proliferation, 186–187
Phyllanthus epiphyllanthus, 220
Phyllanthus flexuosus
  diterpenes, 130
  lactones, 127
  steroidal compounds, 128
Phyllanthus fluitans, 268
Phyllanthus fraternus
  antiaging effects, 271
  antidiabetic effects, 243
  bronchial asthma, 304–305
  cultivation, 65
  diabetes, 243
  GenBank accession numbers, 83
  general characteristics, 14–15
  hots spots and conservation, 98
  identification manual, 28–29
  jaundice, 205–206
  pharmacopoeial status, 43–45
  Phyllanthus amarus, 160, 161
  Phyllanthus niruri, 31
  Phyllanthus urinaria, 34, 164
  phylogenetic analysis, 82–83, 92–93
  toxicity studies, 285
  trade issues, 111
Phyllanthus fuernrohirii, 20
Phyllanthus gardnerianus
  general characteristics, 12
  identification manual, 29

Phyllanthus gasstroemii, see also *Phyllanthus gunnii*
  antihepatitis properties, 207
  general characteristics, 19
Phyllanthus gentryi, 14
Phyllanthus gongyloides, 18–19
Phyllanthus gradyi, 16–17
Phyllanthus grandisepalus, 20, see also *Phyllanthus carpentariae*; *Phyllanthus hebecarpus*
Phyllanthus graveolens, 83
Phyllanthus gunnii, see also *Phyllanthus gasstroemii*
  antihepatitis properties, 207
  general characteristics, 19
Phyllanthus hebecarpus, 20, see also *Phyllanthus carpentariae*; *Phyllanthus grandisepalus*
Phyllanthus hirtellus
  general characteristics, 20–21
Phyllanthus indicus, 19
Phyllanthus indofischeri
  cultivation, 66
  general characteristics, 16
  hots spots and conservation, 98, 106
  utilization and conservation, 111
Phyllanthus kozhikodianus, see also *Phyllanthus rheedii*
  general characteristics, 9
  hots spots and conservation, 98, 105
  toxicity studies, 285
  trade issues, 111
Phyllanthus kozhikodianus Sivarajan & Manilal, 162
Phyllanthus lacunarius, 19–20
Phyllanthus lawii, 10
Phyllanthus longipedicellatus, 17
Phyllanthus macraei, 12
Phyllanthus madagascariensis, 83
Phyllanthus maderaspatensis
  antiaging effects, 270
  chemoprotective, genotoxic, and antigenotoxic effects, 257
  cultivation, 65
  general characteristics, 10–11
  hepatoprotective properties, 162–163
  hots spots and conservation, 98
  identification manual, 30–31
  steroidal compounds, 128
  toxicity studies, 285
  trade issues, 111
Phyllanthus minicus, 65
Phyllanthus mirabilis, 14, 268
Phyllanthus missionis, 13
Phyllanthus muellerianus
  general characteristics, 14–15
  steroidal compounds, 128

Index

Phyllanthus multiflorus, 142
Phyllanthus myrtifolius
  arylnapthalenes, 126
  general characteristics, 9
  hyphenated HPLC, 142
  lignans, 124
Phyllanthus narayanswamii, 11
Phyllanthus niruri
  antiaging effects, 268, 270–271
  antidiabetic effects, 238
  antihepatitis properties, 206–212
  anti-inflammatory properties, 166
  antiviral activities, 224
  aryltetralins, 126
  cardiovascular disease, 273
  chemically induced carcinogenesis, 174
  cultivation, 65
  diabetes, 241
  dibenzylbutanes, 124
  dibenzylbutyrolactones, 124
  distribution and chemotaxonomic significance, 132
  epoxy and diepoxylignans, 125
  Escherichia coli, 206
  favanones, 124
  flavones, 124
  GenBank accession numbers, 83
  general characteristics, 6, 8
  hepatitis B infection, 208–209
  hepatoprotective properties, 163
  hyphenated HPLC, 141–142
  identification manual, 31
  immunomodulatory activity, 290
  jaundice, 205–206
  lactones, 127
  lignans, 124
  neolignans, 126
  Phyllanthus amarus, 27
  Phyllanthus fraternus Webster, 29
  phylogenetic analysis, 82–83, 92–93
  pulmonary tuberculosis, 291, 294
  safety studies, 207, 306–307
  steroidal compounds, 128
  tonsillopharyngitis, 303–304
  toxicity studies, 280, 281, 281–283
  urolithiasis, 297–299
  vaginal candidiasis, 295–296
  varicella zoster infection, 296–297
Phyllanthus niruri auct. Non., *see also Phyllanthus amarus*
Phyllanthus nummulariifolius, 83
Phyllanthus odontadenius, 15
Phyllanthus orbicularis
  antiaging effects, 270
  antiviral activities, 220
  chemical constituents, 222
  chemoprotective, genotoxic, and antigenotoxic effects, 257
  classification, 220–222
  compounds, 224
  description, 220–222
  enterovirus, 230
  flavonoids, 223
  fundamentals, 219–220, 224–225, 230
  hepatitis B virus, 225–226
  herpesvirus, 226–229
  metabolites, 224
  phytochemcial sieving, 223
  properties, 220–222
  tannins, 223
Phyllanthus oxyphyllus
  diterpenes, 130
  epoxy and diepoxylignans, 125
  GenBank accession numbers, 83
  steroidal compounds, 128
Phyllanthus pinnatus, see also Phyllanthus wightianus Muell; *Reidia floribunda* Wight
  general characteristics, 13–14
  identification manual, 31–32
Phyllanthus piscatorum, 66
Phyllanthus polyphyllus
  antiaging effects, 271
  anti-inflammatory properties, 152–153
  GenBank accession numbers, 83
  general characteristics, 8–9
  identification manual, 32
  phase II enzymes, 178
  Phyllanthus indo-fischeri, 30
Phyllanthus pulcher, 14
Phyllanthus reticulatus
  antiaging effects, 271
  antidiabetic effects, 242
  anti-inflammatory properties, 153, 166
  coumarins and cinnamic acid derivatives, 122
  cultivation, 66
  diabetes, 242
  distribution and chemotaxonomic significance, 132
  epoxy and diepoxylignans, 125
  flavones, 124
  GenBank accession numbers, 83
  general characteristics, 9
  hepatoprotective properties, 164
  hyphenated HPLC, 144
  identification manual, 32–33
  steroidal compounds, 128
  toxicity studies, 284
  triterpenes, 130
Phyllanthus rheedii, see also Phyllanthus kozhikodianus
  antiaging effects, 270

antidiabetic effects, 244
anti-inflammatory properties, 166
cardiovascular disease, 273
diabetes, 244
GenBank accession numbers, 83
general characteristics, 9–10
hots spots and conservation, 105
identification manual, 33
toxicity studies, 285
Phyllanthus rotundifolius
general characteristics, 12–13
identification manual, 33–34
Phyllanthus salesiae, 18
Phyllanthus salviaefolius
gas chromatography and mass spectrometry, 145
monoterpenes, 130
Phyllanthus scabrifolius
general characteristics, 16
hots spots and conservation, 105
Phyllanthus sellowianus
antidiabetic effects, 244
coumarins and cinnamic acid derivatives, 122
cultivation, 66–67
diabetes, 244
favanones, 124
flavones, 124
steroidal compounds, 128
Phyllanthus similis
antihepatitis properties, 207
general characteristics, 21
Phyllanthus simplex, see also Phyllanthus virgatus
antidiabetic effects, 243
diabetes, 243
lactones, 127
Phyllanthus singampattiyana
antiaging effects, 271
anti-inflammatory properties, 154
steroidal compounds, 128
Phyllanthus speciosus Jacq, 13
Phyllanthus stipulatus, 66
Phyllanthus subcrenulatus, 21
Phyllanthus sublanatus, 15
Phyllanthus tenellus
antiaging effects, 271
antihepatitis properties, 207
anti-inflammatory properties, 154
chemoprotective, genotoxic, and antigenotoxic effects, 257
GenBank accession numbers, 82
general characteristics, 10
hyphenated HPLC, 142
phylogenetic analysis, 81–82, 92–93
toxicity studies, 285
Phyllanthus thymoides, 207

Phyllanthus urinaria
antiaging effects, 268, 270–271
anti-inflammatory properties, 153
antioxidant profile, 261
antioxidant stress and hepatoprotection, 165
antiviral activities, 224
aryltetralins, 125–126
chemical constituents, 222
chemically induced carcinogenesis, 174
coumarins and cinnamic acid derivatives, 122
cultivation, 65
dibenzylbutanes, 124
dibenzylbutyrolactones, 124
distribution and chemotaxonomic significance, 132
epoxy and diepoxylignans, 125
general characteristics, 8
hepatitis B infection, 209
hepatitis B virus, 226
hepatoprotective properties, 164
herpesvirus, 226
hots spots and conservation, 98
hyphenated HPLC, 142, 144
identification manual, 34
lactones, 127
lignans, 124
monoterpenes, 130
phase I enzyme inhibition, 178
Phyllanthus fraternus Webster, 29
steroidal compounds, 128
toxicity studies, 280, 281, 283
trade issues, 111
traditional uses, 194
transcription factor, 179
Phyllanthus urinaria, proapoptotic and antiangiogenic effects
angiogenesis, 198
angiogenesis, tumor and CAM assay, 198
apoptosis induction, 195–196
chemical analysis, 200–201
ellagic acid, 201
fundamentals, 201–202
matrix metalloproteinase 2 activity, 199–200
mechanism, proapoptotic effect, 196–197
traditional uses, 194
tumor and CAM assay, 198
vascular endothelial cells, 198
viability of cancer cells, 194–196
in vivo anticancer effect, 197–200
Phyllanthus ussurensis, 271
Phyllanthus verminatus, 130
Phyllanthus virgatus, see also Phyllanthus simplex
antiaging effects, 270
dibenzylbutanes, 124

Index

dibenzylbutyrolactones, 124
flavonols, 123
general characteristics, 11
identification manual, 34–35
neolignans, 126
Phyllanthus watsonii, 128
Phyllanthus wightianus, see also Reidia floribunda
Phyllanthus xpallidu, 220
phylogenetic analysis, 72
    agarose gel electrophoresis, 77
    analysis, 81
    barley, 93–94
    BLAST, 79
    cladistics, 72
    cladograms, 92–94
    Clustal W, 80
    DNA level variation, 72–73
    FASTA, 79–80
    fundamentals, 71–72
    gel documentation, 78
    internal transcribed spacer region, 73–74, 93–94
    materials and methods, 76–80
    multiple-sequence alignment, 80, 83, 84–92
    Phyllanthus clarkei, 82–83, 92–93
    Phyllanthus fraternus, 82–83, 92–93
    Phyllanthus niruri, 82–83, 92–93
    Phyllanthus tenellus, 81–82, 92–93
    phylogenetic systematics, 72
    phylogenetic trees, 74–76
    phylograms, 92
    polymerase chain reaction, 78–79
    procedure, 77–78
    results, 81–83, 92–94
    Saccharomyces cerevisiae, 93
    sister groups/taxa, 93
    taxonomy, 72
    tree terminology, 75
phylogenetic trees, 74–76
phylograms, 92
phytochemcial sieving, 223
phytochemical analysis
    Kalpaamruthaa, 335–336
    Semecarpus anacardium, 333
phytochemistry
    alkaloids, 120
    arylnapthalenes, 126
    aryltetralins, 125–126
    benzene derivatives, 127
    chemotaxonomic significance, 132
    cinnamic acid derivatives, 120, 122
    coumarins, 120, 122
    dibenzylbutanes, 124
    dibenzylbutyrolactones, 124
    diepoxylignans, 125
    distribution, 132
    diterpenes, 130
    Emblica officinalis, 334
    epoxy lignans, 125
    flavanones, 124
    flavones, 124
    flavonoids, 122–124
    flavonols, 122–123
    fundamentals, 120, 132
    honey, 334
    lactones, 126–127
    lignans, 124
    miscellaneous compounds, 132
    monoterpenes, 130
    neolignans, 126
    phenolics, 127
    sesquiterpenes, 130
    simple lactones, 126–127
    simple phenolics, 127
    steroidal compounds, 128
    tannins, 128
    terpenoids, 128, 130, 132
    triterpenes, 130
polymerase chain reaction (PCR)
    DNA-based technique, 73
    phylogenetic analysis, 78–79
pomace, 57
powder identifications, 45
powders, 57
Practical Anti-Tumor Herb Medicine, 194
proapoptotic and antiangiogenic effects, *Phyllanthus urinaira*
    angiogenesis, 198
    angiogenesis, tumor and CAM assay, 198
    apoptosis induction, 195–196
    chemical analysis, 200–201
    ellagic acid, 201
    fundamentals, 201–202
    matrix metalloproteinase 2 activity, 199–200
    mechanism, proapoptotic effect, 196–197
    traditional uses, 194
    tumor and CAM assay, 198
    vascular endothelial cells, 198
    viability of cancer cells, 194–196
    in vivo anticancer effect, 197–200
procedures
    phylogenetic analysis, 77–78
products, 57
propagation
    Phyllanthus amarus, 62
    Phyllanthus emblica, 52–53
protection, plant habitats, x
pruning, 53
Pseudomonas aeruginosa, 152
Pterocarpus santalinus f., 32
pulmonary tuberculosis, 290–291, 294–295

Q

quality control, raw plant material, *ix*
quality parameter limits, 59

R

Raacha usiri, *see Phyllanthus acidus*
radiation damage, 258–259, 321
radioprotective effect, 187
Rajanyamalakadi, 245
Ranavali, *see Phyllanthus maderaspatensis*
RECK, 199–200
reference collections, *ix*
reference solutions
    Phyllanthus amarus, 42
    Phyllanthus emblica, 39
refrigerant properties, 58
Reidia floribunda, *see also Phyllanthus*
    pinnatus; Phyllanthus wightianus
reproductive organs diseases, 58
roof binders, 66
root, tree terminology, 75
root identifications, 44

S

Saccharomyces cerevisiae
    phylogenetic analysis, 93
    topoisomerase II activity inhibition, 180
Sadahazuramani, *see Phyllanthus amarus*
safety studies and issues, 64, 207–208
Salmonella typhimurium
    chemoprotective, genotoxic, and
        antigenotoxic effects, 256
    Phyllanthus amarus, 160, 257
    Phyllanthus emblica, 257
scaled branches, tree terminology, 75
scientific classification, 23
Scrubby spurge, *see Phyllanthus gasstroemii*)
seeds
    Phyllanthus genus, 4
    Phyllanthus amarus, 5
    Phyllanthus carpentariae, 20
    Phyllanthus debilis, 13
    Phyllanthus emblica, 5
    Phyllanthus fraternus, 15
    Phyllanthus fuernrohirii, 20
    Phyllanthus gardnerianus, 12
    Phyllanthus gongyloides, 19
    Phyllanthus gradyi, 17
    Phyllanthus grandisepalus, 20
    Phyllanthus hebecarpus, 20
    Phyllanthus hirtellus, 21
    Phyllanthus indicus, 19
    Phyllanthus lacunarius, 20
    Phyllanthus lawii, 10

Phyllanthus longipedicellatus, 17
Phyllanthus macraei, 12
Phyllanthus maderspatensis, 11
Phyllanthus missionis, 13
Phyllanthus narayanswamii, 11
Phyllanthus niruri, 8
Phyllanthus polyphyllus, 9
Phyllanthus reticulatus, 9
Phyllanthus rheedii, 10
Phyllanthus rotundifolius, 13
Phyllanthus salesiae, 18
Phyllanthus scabrifolius, 16
Phyllanthus subcrenulatus, 21
Phyllanthus urinaria, 8
Phyllanthus virgatus, 11
Semecarpus anacardium
    phytochemical analysis, 333
    therapeutic potential, 333
    toxicity studies, 333
Septilin, 154–155
sequenced regions, 103–104
sesquiterpenes, 130
shampoo, 57
Shivappukinelli, *see Phyllanthus urinaria*
Shka-nin-du, *see Phyllanthus amarus*
shrubs, synoptic key, 25
similarity, 81
sister groups/taxa, 93
skin afflictions
    Phyllanthus amarus, 64
    Phyllanthus emblica, 58
snakebites, 66
soil and climate, 52, 62
sorbital accumulation, 248
sores, 66
sore throat
    Phyllanthus acidus, 65
    Phyllanthus reticulatus, 66
spasms, 66
species, defined, 72
species adulteration, 110–111
stamens
    Phyllanthus genus, 4
    Phyllanthus rotundifolius, 13
Star gooseberry, *see Phyllanthus acidus*
stem identifications, 44–45
steroidal compounds, 128
stomachic properties, 64
Stonebreaker, *see Phyllanthus niruri*
stone breaker, *xiii*, *see also Phyllanthus*
    amarus
storage, 57
Stylosanthes hamata, 54
Sudhe, *see Phyllanthus emblica*
supplementary medicine, *x*
supportive therapies, *x*
synoptic key, 24–25

DATE DUE

wild barley, 93–94
wounds, 66

X

Xylophylla, 23

Y

Ya-tai-bai, *see Phyllanthus amarus*
yield
    Phyllanthus amarus, 63–64
    Phyllanthus emblica, 55–56

Index

T

Taamalakee, see *Phyllanthus amarus*
Tahitian gooseberry tree, see *Phyllanthus acidus*
Tamlaki, see *Phyllanthus urinaria*
tanning, 58
tannins
    Kalpaamruthaa, 336
    Phyllanthus orbicularis, 223
    phytochemistry, *Phyllanthus* genus, 128
taxonomic incongruities, 110–111
taxonomy
    APG II system classification, 4
    fundamentals, 3
    general characters, 4
    Natural System of classification, 3–4
    phylogenetic analysis, 72
terpenoids, 128, 130, 132
test solutions
    Phyllanthus amarus, 42
    Phyllanthus emblica, 39
therapeutic potential
    Emblica officinalis, 334
    honey, 334
    Kalpaamruthaa, 340–343
    Semecarpus anacardium, 333
thin-layer chromatography (TLC), *see also* Chromatographic system
    Phyllanthus amarus, 42
    Phyllanthus emblica, 39
timber, 48
TLC, see Thin-layer chromatography (TLC)
tobacco substitute, 66
tonsillopharyngitis, 303–304
tooth powder, 57
topology, tree terminology, 75
total ash parameters
    Phyllanthus emblica, 59
    Phyllanthus fraternus, 45
toxicity studies
    fundamentals, 279–280, 285
    Kalpaamruthaa, 336
    Phyllanthus amarus, 281–283
    Phyllanthus emblica, 283–284
    Phyllanthus fraternus, 285
    Phyllanthus kozhikodianus, 285
    Phyllanthus maderaspatensis, 285
    Phyllanthus niruri, 281–283
    Phyllanthus reticulatus, 284
    Phyllanthus rheedii, 285
    Phyllanthus tenellus, 285
    Phyllanthus urinaria, 283
    Semecarpus anacardium, 333
transcription factors, 179–180
transplanted tumors
    Phyllanthus amarus anticancer studies, 174
    Phyllanthus emblica anticancer activity, 186
transplanting, 62
trees, synoptic key, 25
tree terminology, 75
Trichoderma viride, 52
Triphala
    antiarthritic activity, 326
    antidiabetic activity, 326
    anti-inflammatory activity, 326
    antimicrobial activity, 326
    antioxidant activity, 327
    cancer therapy, 327–329
    chondroprotective activity, 326
    fundamentals, 325
triterpenes, 130
Tropical leaf flower, see *Phyllanthus* pulcher
tuberculosis, 107
tumor cell proliferation, 186–187
tumors, 198

U

Uchchi usiri, see *Phyllanthus virgatus* G. Forst.
Uchiusiri, see *Phyllanthus gardnerianus*
ulcerative stomatitis, 58
ulcers, 64
ultilization implications, 111–112
unscaled branches, tree terminology, 75
urinary conditions, 107
urolithiasis, 297–299
Usiri, see *Phyllanthus emblica*
Usirikaya, see *Phyllanthus emblica*

V

vaginal candidiasis, 295–296, *see also Candida albicans*
vaginitis, *xiii*
varicella zoster infection, 296–297
vascular endothelial cells, 198
viability of cancer cells, 194–196
Viernes santo, see *Phyllanthus amarus*
virally induced cancers, 174–176
Virohelp, 210
Vitamin C, 58

W

water clarification, 58
water-conducting pipes, 58
water-soluble extractive parameters
    Phyllanthus emblica, 59
    Phyllanthus fraternus, 45
weed control, 63
weeding, 54